The Arts of Ornamental Geometry

A Persian Compendium on Similar and Complementary Interlocking Figures

Studies and Sources in Islamic Art and Architecture

Supplements to Muqarnas

Sponsored by the Aga Khan Program for Islamic Architecture at Harvard University and the Massachusetts Institute of Technology, Cambridge, Massachusetts

VOLUME XIII

www.brill.com/muqs

The Arts of Ornamental Geometry

A Persian Compendium on Similar and Complementary Interlocking Figures

Fī tadākhul al-ashkāl al-mutashābiha aw al-mutawāfiqa
(Bibliothèque nationale de France, Ms. Persan 169, fols. 180r–199r)

A Volume Commemorating Alpay Özdural

Edited by

Gülru Necipoğlu

With Contributions by

Jan P. Hogendijk, Elaheh Kheirandish, Gülru Necipoğlu, Alpay Özdural, and Wheeler M. Thackston

BRILL

LEIDEN | BOSTON
2017

This paperback was originally published in hardback in 2017.

The Library of Congress has cataloged the earlier hardback edition as follows:

Names: Özdural, Alpay, honouree. | Necipoğlu, Gülru, editor. | Thackston, W. M. (Wheeler McIntosh), 1944- translator, editor. | Necipoğlu, Gülru. Ornamental geometries. | Hogendijk, J. P. Mathematical classification of the contents of an anonymous Persian compendium on decorative patterns. | Kheirandish, Elaheh. Early tradition in practical geometry. | Özdural, Alpay. Interlocks of similar or complementary figures. Selections.
Title: The arts of ornamental geometry : a Persian compendium on similar and complementary interlocking figures = Fī tadākhul al-ashkāl al-mutashābiha aw al-mutawāfiqa (Bibliothèque Nationale de France, Ms. Persan 169, fols. 180r-199r) : a volume commemorating Alpay Özdural / edited by Gülru Necipoğlu ; with contributions by Jan P. Hogendijk, Elaheh Kheirandish, Gülru Necipoğlu, Alpay Özdural, and Wheeler M. Thackston.
Other titles: Fī tadākhul al-ashkāl al-mutashābiha aw al-mutawāfiqa. | Fī tadākhul al-ashkāl al-mutashābiha aw al-mutawāfiqa. English.
Description: Leiden ; Boston : Brill, 2017. | Series: Muqarnas, supplements ; volume XIII | Includes bibliographical references and index.
Identifiers: LCCN 2015048613 (print) | LCCN 2015049379 (ebook) | ISBN 9789004315204 (E-book) | ISBN 9789004301962 (hardback : alk. paper)
Subjects: LCSH: Generation of geometric forms--Early works to 1800. | Geometry in art--Early works to 1800. | Repetitive patterns (Decorative arts)--Early works to 1800. | Shapes--Design--Art--Early works to 1800. | Manuscripts, Persian--France--Paris--Facsimiles. | Fī tadākhul al-ashkāl al-mutashābiha aw al-mutawāfiqa. | Bibliothèque nationale de France.
Classification: LCC QA464 (ebook) | LCC QA464 .A78513 2017 (print) | DDC 745.40955--dc23
LC record available at http://lccn.loc.gov/2015048613

A copy of the full, unedited original book manuscript *Interlocks of Similar or Complementary Figures: Collaboration of Mathematicians and Artisans in the Islamic World* by Alpay Özdural is freely available online at https://doi.org/10.6084/m9.figshare.5255416.

Typeface for the Latin, Greek, and Cyrillic scripts: "Brill". See and download: brill.com/brill-typeface.

ISSN 0921-0326
ISBN 978-90-04-67971-9 (paperback, 2023)
ISBN 978-90-04-30196-2 (hardback)
ISBN 978-90-04-31520-4 (e-book)

This book is printed on acid-free paper.

CONTENTS

BRILL

GÜLRU NECİPOĞLU

IN MEMORY OF ALPAY ÖZDURAL AND HIS UNREALIZED BOOK PROJECT

This volume came into being in response to the unfortunate circumstances surrounding the submission, in 2003, of a book manuscript for the *Supplements to Muqarnas* series by Alpay Özdural, who died suddenly just a few weeks later. During the last decade of his life, Professor Özdural published a series of articles that formed the foundation stones of his monumental book manuscript, titled "Interlocks of Similar or Complementary Figures: Collaboration of Mathematicians and Artisans in the Islamic World,"[1] a project he had discussed over the years with Margaret Ševčenko, the managing editor for *Muqarnas* at that time. His detailed study focuses on a celebrated anonymous Persian primary source in the Bibliothèque nationale de France, namely, fols. 180r–199r of Ms. Persan 169. This work, hereafter referred to as the *Anonymous Compendium*, first caught the attention of two historians of Islamic art and architecture in the early 1970s. The pioneering studies of Mitkhat Sagadatdinovich Bulatov and Wasma'a Chorbachi[2] on this unique document inspired a series of publications that only briefly analyzed its contents.[3] A Russian translation was published in 1978 (reprinted in 1988), and a modern Persian edition appeared in Iran in 1990–91,[4] but an English translation accompanied by a facsimile and transcription of the original medieval Persian text had not been attempted.[5] Özdural's endeavor in this direction was therefore especially valuable.

However, efforts to publish his book manuscript posthumously were impeded because revisions suggested by anonymous reviewers specializing in the history of mathematics could not be realized due to the author's untimely death. Moreover, Özdural's references and citations were in many cases either incomplete or improperly cited, which proved to be an insurmountable obstacle.[6] As editor of *Studies and Sources in Islamic Art and Architecture: Supplements to Muqarnas*, I searched in vain for a historian of mathematical sciences who might be willing to edit and revise the manuscript, even as a co-author. The present collaborative volume is therefore a compromise. While bringing to light selected sections from Özdural's manuscript, it is complemented by three related chapters dedicated to his memory by an interdisciplinary and international research team comprising two distinguished historians of science and myself. Each of these chapters contains individual interpretations of the Persian *Anonymous Compendium* that constitutes the focal point of our collective study. I would like to thank Jan P. Hogendijk and Elaheh Kheirandish for agreeing to contribute not only essays but also invaluable suggestions on other aspects of this timely publication. The English translation of the primary source published in the present volume is based on the one originally prepared by Özdural, who collaborated with the translator Zaka Siddiqi in interpreting the Persian text, but includes modifications made by the authors of the three chapters.[7] I am particularly grateful to Wheeler M. Thackston, who graciously edited and revised the final English translation, checking it against the original Persian text. Thackston also prepared the transcribed edition of the Persian text, which Özdural did not include in his book manuscript. We would also like to express our gratitude to the staff of the Reproductions Department at the Bibliothèque nationale de France, for granting us permission to publish the facsimile of Ms. Persan 169, folios 180r–199r, and to Ms. Sara Yontan, Conservator of the Turkish Collections of the Bibliothèque nationale de France, for the assistance and advice she so kindly provided over the years it took to complete this volume.

 | DOI 10.1163/9789004315204_002

CONTRIBUTIONS TO THE PRESENT VOLUME

The undated *Anonymous Compendium* is illustrated with geometrical constructions that are accompanied by texts in Persian explaining how to draw those complex figures. The heading in red ink on the first page has reasonably been assumed to be its title and this is indeed the way the work has traditionally been referred to in the secondary literature. Consisting of two vertical lines in Arabic written from bottom to top in the right margin, the heading reads: *Fī tadākhul al-ashkāl al-mutashābiha aw al-mutawāfiqa* (see the reduced-scale facsimile, fol. 180r [1]). Özdural proposed translating this as "On Interlocks of Similar or Complementary Figures," a modified version of which we have adopted here: "On Similar and Complementary Interlocking Figures."[8] Although in describing this primary source the late author observed that "we cannot even call it a treatise since it lacks a predetermined structure, an argument, or at least a logical sequence," he nonetheless referred to it in his manuscript as the "anonymous treatise."[9] Precisely because of its character as a loosely grouped collection of geometrical constructions, we have chosen to refer to it in this publication as the "*Anonymous Compendium*."

Each of the chapters written for the present volume by members of the aforementioned research team—Elaheh Kheirandish, Jan P. Hogendijk, and myself—sheds light on certain aspects of the *Anonymous Compendium*, without aiming to analyze it comprehensively. These essays serve as a segue to selected parts of Özdural's book manuscript. My own chapter, titled "Ornamental Geometries: A Persian Compendium at the Intersection of the Visual Arts and Mathematical Sciences," expands upon preliminary observations made on this primary source in my 1995 book *The Topkapı Scroll: Geometry and Ornament in Islamic Architecture*.[10] Taking into consideration questions debated in subsequent studies, I reassess the historical contexts of the *Anonymous Compendium* and examine its contents in light of new information. After analyzing its relationship to relevant works on surveying and practical geometry that are bound together with it in the same codex, this chapter turns to its wider implications for convergences between the applied mathematical sciences and artisanal-architectural practice in the realm of geometric ornament. It argues that intersections and mediations between the practice-oriented constructive geometry of practitioners and the theoretical geometry of geometer-astronomers cannot simply be reduced to a one-way transmission of knowledge.

The next chapter, by Elaheh Kheirandish, titled "An Early Tradition in Practical Geometry: The Telling Lines of Unique Arabic and Persian Sources," evaluates the early Islamic tradition of practical geometry by focusing on old and new primary sources, with a particular emphasis on the legacy of the renowned Iranian mathematician-astronomer Abu'l-Wafa' al-Buzjani (ca. 940–98). She considers the "contexts" and previously overlooked "revelations" of these sources, critically assessing the assumptions made in modern interpretations of the relationship between mathematicians, artisans, and surveyors. Kheirandish argues for the close relationship of practical geometry and surveying, while also re-examining issues of dating and provenance on the basis of recently discovered manuscripts and new arguments.

The final chapter, by Jan P. Hogendijk, focuses on the mathematical aspects of the *Anonymous Compendium*. Titled "A Mathematical Classification of the Contents of an Anonymous Persian Compendium on Decorative Patterns," his essay classifies the material compiled in this unparalleled document, so as to familiarize the readership of the present volume with its problems. Hogendijk analyzes only selected examples of geometrical constructions from each category of his classification, cross-referencing earlier studies on these particular exemplars and leaving the mathematical analysis of the rest to future research projects. His focused study aims to reach an understanding of the relationship between the unusual contents of the *Anonymous Compendium* and those of standard scholarly mathematical treatises in the medieval Islamic tradition.

A PRÉCIS OF ALPAY ÖZDURAL'S UNPUBLISHED BOOK MANUSCRIPT

Özdural's monographic study comprises three chapters: the first, titled "Preliminaries" (pages 1–31), provides a general introduction; the second, "Analyses" (pages 32–256), consists of a lengthy mathematical analysis of all the geometrical constructions contained in the *Anonymous Compendium*; and "Synthesis," the third chapter (pages 257–73), summarizes the author's main arguments concerning what he calls the collaboration between mathematicians and artisans in the medieval Islamic world. Özdural's three chapters are accompanied by his acknowledgements and bibliography, as well as an extensive appendix focusing directly on the *Anonymous Compendium* itself. In the present volume, the reader will find the introductory chapter, "Preliminaries," which offers a succinct overview of earlier publications on the subject and Özdural's own arguments, along with his bibliography and the contents of his entire appendix, comprising his drawings, the Persian facsimile, and the English translation made by the translator Zaka Siddiqi under Özdural's supervision (which was edited and revised by Wheeler M. Thackston). We have omitted the interdependent chapters titled "Analyses" and "Synthesis," since the readers' reports recommended revisions that could not be implemented without the author's personal intervention. However, for those who wish to obtain a copy of the full, unedited original book manuscript of Özdural, where he discusses the mathematical properties of all geometrical constructions in the *Anonymous Compendium* as well as the step-by-step method for drawing each one, we have made his work available online at https://doi.org/10.6084/m9.figshare.5255416.[11]

In the section of the present volume titled "Translation, Transcription, and Drawings," the English translation and Persian transcription of the *Anonymous Compendium* have been combined with Özdural's redrawings of the sixty-one figures in this primary source, so that readers would not have to turn back and forth to compare the interrelated texts and images. In some instances, Özdural extrapolated larger ornamental panels based on the redrawings; he also occasionally included drawings of comparative materials. These supplementary drawings follow each of the sixty-one redrawn figures and the associated translated and transcribed texts. This means that the English translation cannot be read as a continuous document, unless one skips the intervening drawings. Our decision to juxtapose the translation and transcription with the drawings may be justified by the fragmented character of the *Anonymous Compendium* itself, where the explanatory texts that accompany individual geometrical constructions lack a linear narrative.

Since Özdural's drawings lacked captions and were seamlessly integrated into the chapter titled "Analyses," in this volume they have been furnished with descriptive labels and brief commentaries, summarized from his much longer unpublished work. His redrawings of the geometrical constructions are particularly valuable because they incorporate the uninked incised lines that generated the constructions drawn in black and/or red ink on folios 180r–199r of Ms. Persan 169. The incised lines, which provide precious information on the construction process, are invisible in the photographed facsimile since they were marked on the paper surface with the sharp points of a divider-compass or stylus. Özdural scrupulously recorded those uninked incised marks as dotted lines in his redrawings. The present volume concludes with a reduced-scale facsimile reproduction of the *Anonymous Compendium*, presented in its original format, in which the pages are turned from left to right. The reader will thus find the first page of the facsimile [folio 180r] on page 374. (Throughout the volume references to the facsimile will be made by listing the folio and construction number (e.g., fol. 180r [1]). Özdural's redrawings of the sixty-one constructions and the supplementary drawings he made related to them are cited as plates: for example, the first construction, on folio 180r, is referred to as pl. 1.1, while the supplementary drawings related to that construction are referred to as pls. 1.2, 1.3, 1.4, 1.5, 1.6, and 1.7.)

Özdural began to examine the *Anonymous Compendium* in some earlier articles, in which he discussed selected geometrical constructions by providing his own redrawings and developing them into larger ornamental patterns, supplemented with comparative visual materials and translations of the Persian texts. This is precisely the approach he followed in the "Analyses" chapter of his book manuscript for all sixty-one geometrical drawings of the *Anonymous Compendium*. Indeed, he announced the plans for his future book project in his 1996 *Muqarnas* article, titled "On Interlocking Similar or Corresponding Figures and Ornamental Patterns of Cubic Equations":

> In a forthcoming publication I plan to translate the complete document and study its contents using information gathered from other works on mathematics, particularly *Geometric Constructions* [by Abu'l-Wafa' al-Buzjani] and other drawn material on ornamental geometry, particularly design scrolls, for comparative purposes. This article is a preliminary assessment of the mathematical content of *Interlocking Figures*, highlighting certain points of interest, particularly the use of cubic equations in the ornamental arts.[12]

In his book manuscript, Özdural dated this primary source (now referred to as *Interlocks of Similar or Complementary Figures*, abbreviated as *Interlocks of Figures*) to circa 1300 and hypothetically ascribed it to Ilkhanid Iran, probably Tabriz, on the basis of circumstantial evidence to which I shall return.[13] The unpublished sections of his book manuscript are somewhat speculative when he makes debatable inferences about the "collaboration between mathematicians and artisans" in a series of meetings that, in his view, must have led to the compilation of this document. The "Synthesis" chapter encapsulates his principal thesis concerning the *Anonymous Compendium*, which lacks the kind of systematic ordering more typical of a formal treatise:

> The main argument of the present study is that this work is in fact the documentation of a series of meetings between mathematicians and artisans and thus evidences the intimate link between theory and praxis in the architectural arts. At the beginning of the study this hypothesis is assumed to be true, and the constructions are analyzed accordingly. Now I reverse the order and demonstrate the truth of the hypothesis while assessing the *Interlocks of Figures*.
>
> All 61 constructions are not of the same character. More to the point, they can be grouped under two distinguishable categories. Some of them concern basic principles or skills of geometry, whereas others are related directly to ornamental patterns and the tools used for executing them. The first group of constructions gives the impression of being offered by mathematicians who were giving instructions on general issues in relation to specific topics. The topics, however, seem to be selected without any plan or organization. This haphazardness suggests that those constructions were merely *ad-hoc* answers to the varied needs of a group of persons in a gathering. We may describe such gatherings as "teaching sessions." We also observe that constructions regarding the ornamental patterns are grouped somewhat loosely around certain topics, which seem to have been chosen again not according to a preconceived plan but depending on the composition of the gatherings and the arbitrary inclinations of the participating persons. These sorts of gatherings, at which the ornamental patterns and pertinent tools were discussed, can be classified then as "working sessions."[14]

In other words, Özdural differentiates "teaching sessions," whose main objective was to prove a theorem or show a technique, from "working sessions," which were primarily dedicated to creating geometrical constructions intended for ornamental patterns. He notices a shift from the "teaching mode to the working mode" on fol. 185r, but admits the absence of a clear-cut organizing principle in this work, which sometimes combines the two modes.[15] His brief "Synthesis" ends with a conclusion reiterating the character of this primary source as an anonymous work without a single author:

> All the specific observations and interpretations discussed above suggest that *Interlocks of Figures* was the fruit of the collaboration of mathematicians and artisans at a series of meetings. It was apparently the task of a professional scribe to record the discussions that took place during those informal meetings. He was not usually involved in these discussions. The composition of those gatherings might have changed from session to session. Sometimes the participants were brilliant mathematicians and accomplished artisans; and in some cases, they were combined ideally in the same person. These meetings served as the venue for artisans to seek advice from mathematicians concerning fresh ornamental patterns or when they encountered a problem related to the application of geometry. Mathematicians, on the other hand, regarded the ornamental arts as a rewarding and enjoyable field in which to apply their expertise.[16]

This conclusion is largely based on hypotheses presented in Özdural's "Preliminaries" (published as chapter 4 in the present volume), which further develops assertions he makes in his articles. It is necessary to state at the outset that, in actuality, the non-uniform character of the *Anonymous Compendium* makes it difficult to determine with certainty the specific purpose(s) and author(s) of each geometrical construction, despite Özdural's understandable effort to do so in his "Analyses," where he attempts to construct a narrative sequence of "sessions." It is not even certain that this work can be interpreted as a record of meetings showing the "collaboration" between mathematicians and artisans, which were written down in successive sessions by a scribe, an arguable conclusion that the chapters of this volume will reconsider from different angles.

Less speculative are Özdural's mathematical analyses of each geometrical construction compiled in the *Anonymous Compendium*, which he groups under the following categories, although other classifications are also possible:[17]

(I) "Transforming polygons and star polygons by the cut-and-paste method"
(II) "Constructing the regular pentagon"
(III) "Finding the center of a circle"
(IV) "Defining polygons by way of angular modules"
(V) "Twin isosceles quadrangles facing opposite directions"
(VI) "Four rotating isosceles quadrangles in a square"
(VII) "Integration of different types of star polygons"
(VIII) "Miscellaneous patterns"
(IX) "*Muqarnas* plans"

In "Preliminaries," Özdural distinguishes the *Anonymous Compendium* from the work immediately preceding it in the Paris Codex (Ms. Persan 169), namely, a Persian translation of Abu'l-Wafa' al-Buzjani's well-organized and systematic treatise on the practical applications of geometry, titled *Tarjuma-i Kitāb-i Aʿmāl-i handasiyya* (Translation of the Book on Geometrical Constructions [henceforth the *Geometrical Constructions*][18]). Proposing that Abu'l-Wafa''s treatise was familiar to those responsible for the geometrical constructions compiled about three centuries later in the *Anonymous Compendium*, Özdural believed that it may have inspired Ilkhanid mathematicians and skilled artisans to work together in designing ornamental patterns: "This collaboration probably led to the creation of our document."[19] He argues that the *Anonymous Compendium*, seemingly recorded by an anonymous professional scribe in the Ilkhanid period, was recopied in mid-fifteenth-century Timurid Iran. This implies that he differentiates the time of its composition, at the turn of the fourteenth century (i.e., ca. 1300), from its subsequent transcription and compilation in the Paris Codex (Ms. Persan 169).

Özdural agreed with Edgar Blochet's suggestion that the entire codex may have been copied in the hand of Abu Ishaq ibn ʿAbdallah Kubanani Yazdi (fl. 1442–43 to post-1483), who added a postface to the Persian translation of Abu'l-Wafa''s practical geometry treatise included in the same codex.[20] He accepts the identification in previous publications of Kubanani Yazdi with the Timurid-era mathematician-astronomer known as Shaykhzada Hadim Burhani, who was active in 1442–43 at Sari, and in 1459 at Kirman and Hormuz.[21] However, Özdural makes a new proposal: that this individual was one of three contemporaneous Timurid scholars involved in that particular Persian translation of Abu'l-Wafa''s *Geometrical Constructions*. The other two figures were the master geometer (*muhandis*) Ustad Shams al-Din Abu Bakr Shah ibn Hajji Taj-al-Din and his late father, the great master (*ustād*) and scholar Shaykh Najm al-Din Mahmud Shah. It was the son who commissioned Kubanani Yazdi to complete his father's translation, parts of which had been lost.[22]

In his "Acknowledgements," Özdural thanked Eva Maria Subtelny, "who was of great help in translating" this difficult postface "concerning the Persian translation of Abu'l-Wafa''s work." Subtelny's judicious rereading of this complicated passage, which had been misinterpreted in previous studies, prompted Özdural to propose a Timurid date for the completed Persian text of Abu'l-Wafa''s practical geometry treatise: "We can infer…that he [Abu Ishaq Kubanani Yazdi], Shams al-Din Abu Bakr, and Najm al-Din Mahmud (the three mathematicians involved with that particular Persian translation of [Abu'l-Wafa''s] *Geometrical Constructions*) lived in the mid-fifteenth century." Misinterpreting the same passage, Bulatov dated both the translation of Abu'l-Wafa''s treatise, and the *Anonymous Compendium*, long assumed to be an appendix to the translation, to the eleventh century. Another questionable assumption, repeated in several publications including Blochet's catalogue entry, has been the presumed relationship of Kubanani Yazdi to the late Najm al-Din Mahmud, thought by many to be his sibling rather than his dear colleague, due to the way the author of the postface deferentially referred to him as "dearest brother (*barādar*)."[23] However, if he and Najm al-Din Mahmud were indeed siblings, then it seems rather odd that the postface does not refer to the presumed brother and nephew of Kubanani Yazdi in more intimate terms.[24]

Özdural proposed that this Timurid-era mathematician-astronomer not only completed the lost sections of the Persian translation of Abu'l-Wafa''s treatise but also copied that work and the *Anonymous Compendium* that followed it. As mentioned above, he furthermore conjectures that Kubanani Yazdi probably copied all the works assembled in the Paris Codex (as formerly proposed by Blochet), which is written in a single hand.[25] While this conjecture may not be unreasonable, it has yet to be verified by comparing the handwriting of the Paris Codex with other known manuscripts attributed to Kubanani Yazdi.[26] Özdural admits that although the former cataloguers of the codex, Edgar Blochet and Francis Richard, agreed on attributing the *nastaʿlīq* script of its treatises to the same hand, they differed in dating the volume to the seventeenth and sixteenth centuries, respectively. Özdural prefers Priscilla Soucek's expert re-dating of the calligraphy of the Paris Codex, on stylistic grounds, to the mid-fifteenth century.[27]

Whether or not the codex was recopied again later in the sixteenth or seventeenth century, which seems rather unlikely, I find it worth pointing out that datable items assembled in it do not go beyond the fifteenth century. The plausibly Timurid codex came to the Bibliothèque nationale de France from the collection of the French traveler Jean de Thévenot (d. 1667), who acquired it in Istanbul. The *ex libris* on fol. 2r bears the name of its otherwise unknown former Ottoman owner, "*Sinān Çāvūş, ṣāhib al-faḳr*." As noted in Turkish on folio 1r, the volume contains 201 folios (*ʿaded-i varāḳ iki yüz bir tamām*). A table of contents (*fihrist*) on folio 1v, which lists only epistles at the beginning of the codex, is presumed to have been written in his hand.[28]

Özdural acknowledged the possibility that a few textless geometrical constructions appearing at the end of the *Anonymous Compendium* might have been added in the Timurid period, as previously suggested by Lisa Golombek and Donald Wilber, who nevertheless accepted Bulatov's eleventh-century dating for this work.[29] He believed that the *Anonymous Compendium* actually ends on fol. 196v, at the bottom of which are written the twenty-eight letters of the Arabic alphabet with their accompanying numerical values (the *abjad* system of numbering, commonly used for labeling points on geometrical constructions). Özdural suggested that the following constructions on fol. 197r were probably added in the mid-fifteenth century by Kubanani Yazdi and that he may also have been their author. Özdural attributed the textless final construction on fol. 199r, which differs in drafting technique and, unlike the others, covers a whole page, to a later phase: according to him, it possibly came into the possession of Kubanani Yazdi, who inserted it at the time the volume was being bound.[30]

Finally, a key contribution made in Özdural's book manuscript is the hypothetical identification that he proposes for Abu Bakr al-Khalil al-Tajir, whose name appears twice in the *Anonymous Compendium* (on fols. 187r and 189r), as well as in two other treatises of the Paris Codex, as I had noted in my 1995 book. He suggests that this otherwise unidentified individual may have been the father of ʿAli ibn al-Khalil al-Tajir, who was the copyist of an Arabic manuscript on surveying dated 728 (1327–28), now kept at the Astan-i Quds-i Razavi Library in Mashhad. If so, Özdural infers, Abu Bakr al-Khalil al-Tajir probably flourished sometime "in the late thirteenth and early fourteenth centuries," thereby suggesting a date circa 1300 for the *Anonymous Compendium*. This tempting yet conjectural identification, too, requires reexamination. It is assessed in the chapters of the present volume, along with Özdural's proposed redating and reattribution of this primary source and his deliberations concerning the "collaboration" between mathematicians and artisans.[31]

In conclusion, all three chapters of this volume, much like the subsequent selections from Özdural's unpublished book manuscript, scrutinize the modalities of the engagement of medieval mathematicians and artisans with practical geometry, in light of the *Anonymous Compendium*. After more than a decade of deferment in the realization of this long-anticipated interdisciplinary publication project, it is a pleasure to bring the present volume to completion. I hope that this study may inspire further research on the *Anonymous Compendium* and on the unresolved implications concerning the intersection, or lack thereof, between mathematical knowledge and artisanal expertise. Our study is also timely because the exponentially growing interest in geometric patterning in contemporary architectural practice has recently led many architects and designers to study Islamic forms of surface decoration as an inspiration for tessellation and parametrically derived forms.[32] Far from being intended as an exhaustive study of this fascinating document at the interface between diverse fields, our collaborative volume aims to make it more accessible, thereby highlighting its multifaceted relevance to historians of art and science, as well as to contemporary mathematicians, physicists, artists, and architects.

Gülru Necipoğlu, Editor
Aga Khan Professor of Islamic Art
Director of the Aga Khan Program for Islamic Architecture,
Department of History of Art and Architecture,
Harvard University

NOTES

Author's note: For references following note 1, which fully cites Alpay Özdural's related publications, see the bibliography accompanying my chapter on page 74.

1. Alpay Özdural was an associate professor teaching design and architectural history in the Faculty of Architecture, Eastern Mediterranean University, Gazi Magusa, North Cyprus. Among his research interests were architectural photogrammetry, the history of mathematics, and historical metrology. He wrote several articles relevant to the subject of the present volume: "Omar Khayyam, Mathematicians, and *Conversazioni* with Artisans," *Journal of the Society of Architectural Historians* 54, no. 1 (Mar. 1995): 54–71; "On Interlocking Similar or Corresponding Figures and Ornamental Patterns of Cubic Equations," *Muqarnas* 13 (1996): 191–211; "A Mathematical Sonata for Architecture: Omar Khayyam and the Friday Mosque of Isfahan," *Technology and Culture* 39, no. 4 (1998): 699–715; "Mathematics and Arts: Connections between Theory and Practice in the Medieval Islamic World," *Historia Mathematica* 27, no. 2 (2000): 171–201; and "The Use of Cubic Equations in Islamic Art and Architecture," in *Nexus IV: Architecture and Mathematics*, ed. José Francisco Rodrigues and Kim Williams (Fucecchio, Italy: Kim Williams Books, 2002), 165–79.
2. In 1971, Wasma'a K. Chorbachi independently discovered the same primary source in the course of her doctoral research. See Bulatov 1988 (first ed. 1978); Chorbachi 1989a; Chorbachi 1989b; Chorbachi and Loeb 1992.
3. Subsequent publications are cited in Alpay Özdural, "Preliminaries," the first chapter of his unpublished book, which has been lightly edited and reproduced as chapter 4 of the present volume, and in his bibliography, also published in this volume, as well as in the individual bibliographies of the three accompanying chapters by Jan P. Hogendijk, Elaheh Kheirandish, and myself.
4. A. B. Vildanova's partial translation into Russian, which skips some confusing sections of the *Anonymous Compendium*, is published as an appendix in Bulatov 1988, 315–40. The modernized Persian text edited by ʿAli Riza Jaẕbi is included in Abū'l-Wafāʾ al-Būzjānī 1990–91 (1369). On the problematic aspects of the latter edition, including its modernized language, its inaccurate title, and the misattribution of the entire Persian translation of Abu'l-Wafa''s practical geometry treatise singly to Abū Isḥāq al-Kūbanānī, who is also assumed in the same publication to be the author of the appended *Anonymous Compendium*, see Kheirandish's chapter in this volume ("An Early Tradition in Practical Geometry: The Telling Lines of Unique Arabic and Persian Sources").
5. In his acknowledgements Özdural mentions the support of Oleg Grabar, then a member of the Institute for Advanced Studies in Princeton, who encouraged him to study "the involvement of mathematicians in Islamic architecture and architectural arts." As a visiting member of the Institute, Özdural carried out fruitful research and pursued "highly stimulating" academic discussions as he was preparing the book manuscript.
6. Some of the problems noted by the anonymous reviewers—which, given the circumstances, Özdural could not respond to or address—may have been due to the fact that the book manuscript was not yet in its final form when he submitted it for consideration in the *Supplements to Muqarnas* series. He was criticized for inconsistencies in his transliteration of Persian and Arabic words, and for the omission of proper citations for the primary and secondary sources. Özdural also added letters to his redrawings to label points not labeled in the original Persian manuscript, but did not sufficiently explain in his commentary what he had done. Throughout his mathematical analyses, he frequently mixed his English translation of the *Anonymous Compendium* with his own commentary without distinguishing between the two, and there were also errors of interpretation in the mathematical commentary itself. Had Özdural lived, he most likely would have agreed to implement the suggested revisions and add the missing references, so as to remedy the insufficient acknowledgement of previous scholarship, whose findings he often loosely paraphrased without proper citation.
7. Özdural was not trained in Persian or Arabic, languages for which he sought the help of translators. In his "Acknowledgements," Özdural expressed his gratitude to "Zaka Siddiqi (from the University of Petroleum and Minerals, Dhahran, Saudi Arabia), who provided the literary translations of the accompanying texts of the geometrical constructions"

in the anonymous Persian compendium. He also thanked Jan P. Hogendijk, "who kindly offered his unpublished study" on this source and later corrected Özdural's translations, and Taner Avcı, "for his help in translating the Arabic passages from Abu'l-Wafa'"s *On the Geometric Constructions Necessary for the Artisan.*" (This was Özdural's translation of *Kitāb Fī mā yaḥtāju ilayhi al-ṣāniʿ min aʿmāl al-handasa*; see also n. 18 below, and Kheirandish's chapter, esp. n. 1.)

8. The heading was translated in Chorbachi 1989b as "On Interlocking Similar and Congruent Figures," which I adopted in my book on the *Topkapı Scroll*: see Necipoğlu 1995. Özdural initially translated the heading as "On Interlocking Similar or Corresponding Figures." See Özdural 1996, in which he cites Chorbachi's article and my book. He proposed yet another modification in the book manuscript he submitted in 2003: "On Interlocks of Similar or Complementary Figures." In this volume we have translated the title *Fī tadākhul al-ashkāl al-mutashābiha aw al-mutawāfiqa* as "On Similar and Complementary Interlocking Figures"—using "and" rather than the literal "or" for the Arabic "*aw*"—because in English the use of "or" could suggest that the *Anonymous Compendium* was about either similar or complementary interlocking figures when it is about both.
9. See Özdural, "Preliminaries," the first chapter of his unpublished book, reproduced as chapter 4 in the present volume.
10. I discussed the *Anonymous Compendium* in two chapters, titled "Theory and Praxis: Uses of Practical Geometry," and "Manuals of Practical Geometry and the Scroll Tradition," in Necipoğlu 1995, 123, 133, 146–50 (figs. 108–14), 168–73. See also my earlier remarks in Necipoğlu 1992.
11. Interested readers can find Alpay Özdural's unpublished book manuscript, titled "Interlocks of Similar or Complementary Figures: Collaboration of Mathematicians and Artisans in the Islamic World," available online at https://doi.org/10.6084/m9.figshare.5255416.
12. See Özdural 1996, 203n7. In a later article, he says: "Since I am preparing an edition and English translation of the whole *Interlocks of Figures*, the Persian texts of the quoted passages are not provided here" (Özdural 2000, 173n7). However, the book manuscript he eventually submitted lacked an edition of the Persian text. He also remarked that "the provenance of *Interlocks of Figures* and the milieu in which it was written will be discussed below, and in my English translation with commentary, to appear in due course" (Özdural 2000, 172n5).
13. See Özdural, "Preliminaries" (chapter 4 in the present volume). Özdural first dated this primary source to ca. 1206 in Özdural 1995. After the publication of my book that year, in which I drew attention to uninked construction lines by reconstructing two examples (Necipoğlu 1995, 147, fig. 109a–b; and 149, fig. 113a–b), he visited Paris, and changed his mind after examining the *Anonymous Compendium* and other works compiled in Ms. Persan 169. As he explains in his 2003 book manuscript (236n177), he had initially attributed a T-square depicted and described in the *Anonymous Compendium* to Ibn Razzaz al-Jazari (d. ca. 1206), on the basis of which he had dated the work to early thirteenth-century Diyarbakır, but then revised this view upon studying the Paris manuscript closely. In a more recent article (Özdural 2000), he dated the *Anonymous Compendium* to ca. 1300 and referred the reader to his forthcoming book manuscript.
14. Özdural, "Synthesis" (third chapter of his unpublished book), 257. See n. 11 above.
15. Özdural, "Analyses" (second chapter of his unpublished book), 87–88. See n. 11 above.
16. Özdural, "Synthesis" (third chapter of his unpublished book), 272–73.
17. For an alternative classification, see Jan P. Hogendijk's chapter in this volume ("A Mathematical Classification of the Contents of an Anonymous Persian Compendium on Decorative Patterns").
18. Abu'l-Wafa'"s treatise (both the original Arabic text and the Persian translation) will henceforth be referred to throughout this volume as the *Geometrical Constructions.*
19. See Özdural, "Preliminaries," in this volume.
20. The postface and an English translation are provided in Kheirandish's chapter.
21. Previous publications referring to Abu Ishaq Kubanani Yazdi's contribution to the Paris Codex include Blochet 1912, no. 772, 41–47. Blochet hypothetically attributed the copying of Abu'l-Wafa'"s practical geometry treatise and the entire Paris Codex to Abu Ishaq Kubanani Yazdi, which he believed to have been recopied in the seventeenth century (46–47). Richard 1989, 186–87, attributed only the translation of Abu'l-Wafa'"s treatise to Abu Ishaq Kubanani Yazdi, dating the Paris Codex to ca. the sixteenth century. On Abu Ishaq Kubanani Yazdi, see also Munzavī 1969 (1348), 1:143, 152, 318, 368, cited in Richard 1989,186–87, and in Necipoğlu 1995,167–69, 180n115. Kheirandish's chapter includes a more up-to-date discussion of the primary sources and secondary literature on Abu Ishaq Kubanani Yazdi.
22. See Özdural, "Preliminaries," in this volume. In her chapter, Kheirandish introduces new information and unpublished primary sources on Abu Ishaq Kubanani Yazdi, discussing his role as the "co-translator" of the Persian rendering of Abu'l-Wafa'"s practical geometry treatise that is included in the Paris Codex, and establishing the new dating of post-1483 for his activity as a scholar from the colophon of a manuscript by Kubanani.
23. See Özdural, "Preliminaries," in this volume (n. 22), where he comments on the erroneous translation of the postface in Woepcke 1855, 359; and in Bulatov 1988, 89. The latter presumed that Abu Ishaq Kubanani Yazdi was the brother of the deceased Najm al-Din Mahmud. Blochet 1912, 46, likewise identified these two individuals as brothers.
24. See Kheirandish's chapter, where she observes that in his letters Abu Ishaq Kubanani also referred to him and

another figure as "my brother"; referring to associates in this way was common practice at the time. However, she cautiously adds that determining whether Kubanani was actually the first translator's brother remains an open question.

25. See Özdural, "Preliminaries," in this volume.
26. In her chapter Kheirandish highlights the need for a comparison of scripts and identifies relevant manuscripts of Abu Ishaq Kubanani Yazdi, some of which have only recently been brought to light, that are in need of this kind of comparative analysis.
27. See Özdural, "Preliminaries," in this volume. See also Blochet 1912, 41–47; and Richard 1989, 183–87. According to Blochet, although the Paris Codex was probably recopied in the seventeenth century, its linguistic and orthographic archaisms suggest a pre-thirteenth-century date for its original version, which may have been copied in the Timurid period by Abu Ishaq Kubanani Yazdi. Özdural mentions that Soucek's re-dating of the manuscript to the mid-fifteenth century is based on an oral communication.
28. I have identified the eight items listed in the partial table of contents as those with which the codex begins. The eighth item refers collectively to three consecutive treatises on surveying (*miṣāḥa*), which are listed separately by the cataloguers of the codex (nos. 9–11): Blochet 1912, 41–44; Richard 1989, 185.
29. See Golombek and Wilber 1988, 1:159. I too accepted this possibility but dated the *Anonymous Compendium* between the eleventh to thirteenth centuries in Necipoğlu 1995, 168–69.
30. See Özdural, "Preliminaries," in this volume.
31. See Özdural, "Preliminaries" (in this volume), and Özdural 2000, 172n4, where he thanks his informants in Iran for this reference: "According to the information kindly provided by Mohammed Bagheri and Mehran Akhbarifar (both of Tehran), Abū Bakr al-Khalīl is cited as the father of the copyist of a manuscript in Mashhad." He draws the conclusion that "since the manuscript was copied in 1327–1328, Abū Bakr al-Khalīl must have been active around 1300." For the manuscript in Mashhad, see Kheirandish's chapter. Both Kheirandish and I question the assumed father-and-son relationship in our respective chapters. Kheirandish has contacted Özdural's informants, who confirmed having referred to a father-son relationship in their communications with him, based on a "conjecture"; hence, this information remains unsubstantiated. Özdural's attribution of the *Anonymous Compendium* to Tabriz is also conjectural, based on the importance of that city as the Ilkhanid capital where major construction projects were undertaken: see Özdural, "Preliminaries," in this volume.
32. On the current fascination with geometric ornament, tessellation, and parametricism, accelerated by new computational techniques of modeling and production, see Hashim Sarkis, "Inscription: On the Surface of Exchange between Writing, Ornament, and Tectonic in Contemporary Architecture," in *Histories of Ornament: From Global to Local*, edited by Gülru Necipoğlu and Alina Payne (Princeton, N.J.: Princeton University Press, 2016), 34–43.

GÜLRU NECİPOĞLU

ORNAMENTAL GEOMETRIES: A PERSIAN COMPENDIUM AT THE INTERSECTION OF THE VISUAL ARTS AND MATHEMATICAL SCIENCES

An anonymous primary source in Persian that abruptly begins with the Arabic heading *Fī tadākhul al-ashkāl al-mutashābiha aw al-mutawāfiqa* (On Similar and Complementary Interlocking Figures) has fascinated historians of Islamic art and mathematics ever since its discovery in the 1970s by Mitkhat S. Bulatov and Wasma'a K. Chorbachi (fig. 1).[1] These two art historians demonstrated its significance as a rare medieval document (hereafter referred to as the *Anonymous Compendium*) that points to intersections between mathematical and artisanal knowledge, particularly in the field of geometric ornament.[2] Thereafter, it was widely acknowledged for its intriguing, though still debated, implications concerning the design procedures used in creating various geometric patterns and the relationship, or lack thereof, between the visual arts and applied mathematical sciences.[3]

The present chapter reconsiders this unique source, which was briefly discussed in my 1995 book *The Topkapı Scroll: Geometry and Ornament in Islamic Architecture.*[4] That book analyzed its ornamental patterns in comparison with those compiled in later design scrolls, the earliest extant example being the *Topkapı Scroll*, attributable to western or central Iran in the Timurid-Turkmen period (late fifteenth to early sixteenth century).[5] It is therefore pertinent to scrutinize the *Anonymous Compendium* more closely in light of the new information and debates that have emerged in the twenty years since my book was published. Starting with a reexamination of the historical contexts in which this primary source came into being and geometric ornament rose to prominence, I discuss its relationship to relevant works on surveying and practical geometry that are included in the Paris Codex (Bibliothèque nationale de France,

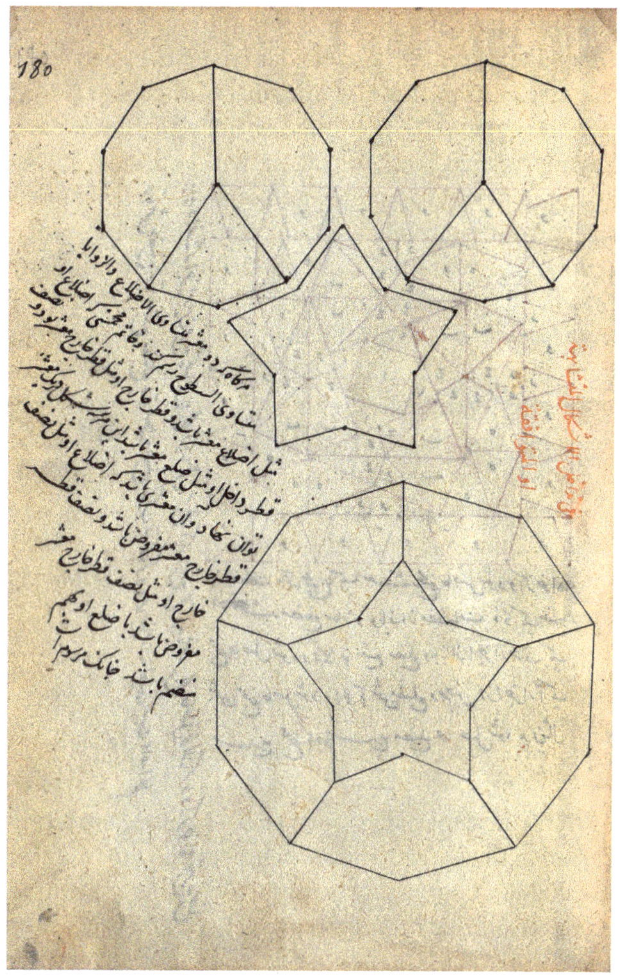

Fig. 1. A "dividing and assembling" construction in two stages, showing the transformation of a pentagonal star and two congruent decagons into a larger decagon containing a central pentagonal star. From the *Anonymous Compendium*. Paris, Bibliothèque nationale de France, Ms. Persan 169, fol. 180r [1]. (Photo: Bibliothèque nationale de France)

 | DOI 10.1163/9789004315204_003

Ms. Persan 169). Following an analysis of the contents of the *Anonymous Compendium*, I conclude with an interpretation of its wider implications for mediations and convergences between the realms of the applied mathematical sciences and artisanal-architectural practice.

The primary sources considered in this chapter explicitly or implicitly allude to intersections between those realms. With a few exceptions, however, historians of Islamic art and science have tended to resist entertaining the possibility that geometer-astronomers, artisan-designers, and master builder-architects could meaningfully communicate with or learn from one another through their engagement in interconnected epistemic activities. That tendency stems in part from the assumption that the domains of theoretical and practical geometry remained entirely separate, each having its own methods, along with a preference to attribute nearly complete autonomy to artisanal practice.[6] Another related reason is a prevalent stance against interpretation, compounded by the commonly presumed illiteracy of medieval Islamic practitioners of the visual arts as lowly handworkers, segregated from contemporaneous cultural-scientific trends.[7] Taking into consideration documented instances of boundary crossing, this chapter highlights the agency of technologies and individuals who mediated the realms of theory and practice. Even if those individuals may have been exceptional, the cultures of learning and circles of knowledge that rippled around them were often far-reaching.

CLUES CONCERNING THE *ANONYMOUS COMPENDIUM*'S HISTORICAL CONTEXTS

The unicum copy of the *Anonymous Compendium* is preserved in the Paris Codex, which, according to its latest cataloguer, comprises twenty-two Persian and three Arabic treatises on mathematical sciences.[8] Most of these works lack a preface, a title, and an author, and the majority are abridged extracts or translations of longer works in Arabic or Persian, datable from the late tenth to the mid-fifteenth century. The treatises address topics such as astronomy, with a work on the astrolabe and a poem on mansions of the moon by the geometer-astronomer Nasir al-Din al-Tusi (d. 1274), who directed the Ilkhanid observatory at Maragha in northwestern Iran. Related works include diagrams representing the qibla direction from principal cities in Iran and from Isfahan in particular, and a work containing a diagram on the qibla direction from Yazd, located southeast of Isfahan. There are also arithmetic problems; sine and cosine tables; extracts from a treatise on pairs of numbers; problems based on Book II of Euclid's *Elements*; treatises on surveying (*misāḥa*, the science of measurement) and practical geometry; a diagram showing the derivation of cabbalistic letters from five secant circles; astrological tables; and a treatise on finger-reckoning arithmetic by the Timurid scholar Sharaf al-Din ʿAli Yazdi "al-Muʿammāʾī" (d. ca. 1454).

The Paris Codex also contains an abridged Persian translation of a treatise in Arabic on practical geometry (ca. 991–92 to 998) by Abu'l-Wafa' al-Buzjani (d. 998), who wrote works on arithmetic, geometry, trigonometry, and astronomy in Buyid Baghdad.[9] Titled *Tarjuma-i Kitāb-i Abī al-Wafāʾ Muḥammad b. Muḥammmad al-Būzjānī dar Aʿmāl-i handasiyya* (Translation of the Book of Abu'l-Wafāʾ al-Buzjani on Geometrical Constructions [henceforth *Geometrical Constructions*][10]), this treatise is followed by our anonymous Persian compendium. The postface of Abu'l-Wafa''s *Geometrical Constructions* was written by the Iranian mathematician-astronomer Abu Ishaq b. ʿAbdallah Kubanani Yazdi (fl. 1442–43 to post-1483), who states that he "completed" the partial translation of this work, which had been started by a beloved late colleague, upon the request of the latter's son. Kubanani Yazdi's works and letters, which were compiled by a disciple, show that this bilingual scholar wrote scientific and literary works in Persian and Arabic.[11] The latest datable treatises of the Paris Codex were authored around the same time in the Timurid-Turkmen period, by the scholars Mahmud Haravi (ca. 1436) and the aforementioned Sharaf al-Din ʿAli Yazdi. Interestingly, the latter's *nisba* (component of a name indicating relation or origin) is connected with the city of Yazd like that of Kubanani Yazdi.

Since all the treatises assembled in the Paris Codex are written in the same hand, it has been suggested that they may have been copied by or for Kubanani Yazdi.[12] The contents of this collection (*majmūʿa*) point to the special interest of its owner, perhaps Kubanani himself, in the practical application of mathematics to a wide variety of fields. I find it significant in this regard that he has recently been identified as the probable owner of another codex, comprising a collection of mathematical-astronomical treatises copied not by him, but by several different scribes. A verse in Persian appears as an annotation on one of the folios in this codex. According to a statement in Arabic (written by someone else), that verse had been copied by Kubanani: "Script (*khaṭṭ*) of the master of great mathematicians (*ustāẕ akābir al-riyāżīyīn*), who revived the traces of the sciences of past philosophers, His Excellency Abu Ishaq b. ʿAbdallah al-Kubanani."[13]

The *Anonymous Compendium* was believed to have been an appendix to the abridged Persian translation of Abu'l-Wafa''s practical geometry treatise in the Paris Codex and both works were ascribed by Bulatov to the early eleventh century.[14] Disputing this presumed connection, Chorbachi noted the greater complexity of the geometrical constructions in the *Anonymous Compendium,* which she dated to Iraq circa 1200 (late twelfth to early thirteenth century). This hypothetical date was further extended to circa 1300 (late thirteenth to early fourteenth century) by Alpay Özdural, who attributed the *Anonymous Compendium* to Mongol-Ilkhanid Iran.[15] Whether it was put together as an informal record of gatherings attended by geometers and artisans, or as a teaching manual, as some scholars have speculated, must remain an open question.[16] Nor can we be certain of its date and place of production. By whom and for what purpose it was initially compiled is not disclosed in this primary source, unlike Abu'l-Wafa''s *Geometrical Constructions*, where, as we shall see, the author implies that the intended audiences of his treatise were artisans, surveyors, and other interested parties, whether or not this expectation was fulfilled.

In my 1995 book I drew attention to an otherwise unknown individual, whose name appears not only in the *Anonymous Compendium* but also in two other Persian treatises of the Paris Codex, which likewise bear Arabic headings in red ink, with no further information as to the author. It seemed to me particularly important at that time to determine the identity of this person. The *Anonymous Compendium* refers to him as "Abu Bakr al-Khalil al-Tajir" and "Abu Bakr al-Khalil" (fols. 187r [25], and 189r [31]). One of the two other works that mentions him begins with the heading *Fī Qismat al-muthallathāt kullihā bi'l-awtār* (On the Division of Triangles by Chords). A compilation of practical geometry problems related to surveying, it focuses on the measurement (*misāḥa*) and division of triangles, quadrangles, and other polygons (fig. 2a). The work ends with a section on numbers and numerical tables, one of which is attributed in an Arabic caption to Abu Bakr, whose name is cited differently: "This problem was derived by Abu Bakr b. al-Khalil, the pride of geometers, by intellectual power" (*Ikhrāj hādhihi al-mas'ala li-mufakhkhar al-muhandisīn Abī Bakr b. al-Khalīl bi'l-quwwa al-fikriyya*) (fol. 118v [fig. 2b]). The title of the second work, which suggests that it may have been a compilation extracted from a longer treatise, is named after him as *Lil-aʿmāl Abī Bakr al-Khalīl al-Tājir al-Raṣadī* (Of the Constructions by Abu Bakr al-Khalil al-Tajir al-Rasadi) (fig. 2c). I previously proposed that Abu Bakr, whose *nisba* (*al-raṣadī*) suggests his specialization in observational astronomy, must have been a practice-oriented geometer-astronomer familiar to some degree with both theoretical and practical geometry.[17] A closer look at his works in the two treatises of the Paris Codex supports this inference. After briefly summarizing the contents of these works, which deserve more detailed analysis in the future, I will consider new information on Abu Bakr that was brought to light by Özdural.

Two anonymous treatises and further traces of Abu Bakr's identity

"On the Division of Triangles" starts with problems explicating the division of triangles in ratio/proportion (*nisbat*) and moves on to the division of quadrangles into triangles. It also explores the relationship between a hexagon and a triangle, and between a heptagon and a triangle inscribed within a circle (fig. 2a). Labeled section (*faṣl*), problem (*mas'ala*), and example (*miṯāl*), much like its counterparts in the *Anonymous Compendium,* the geometrical constructions of this work also include analyses of proportional relationships between polygons with equal areas. Examples include the area of a square to that of a regular octagon, and the area

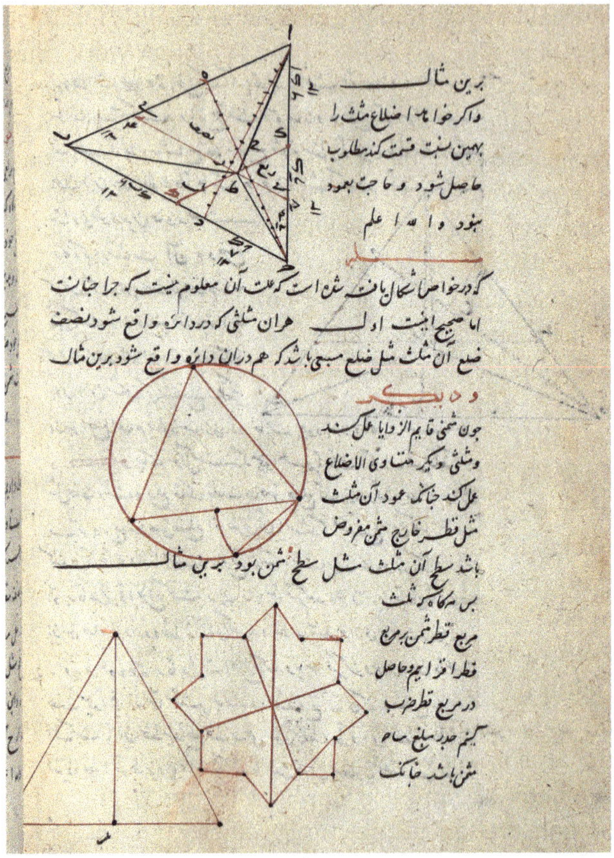

Fig. 2a. Page with geometrical constructions; from top to bottom: division of a triangle; a circle inscribed with a triangle half of whose side equals the side of a heptagon; an equilateral triangle and a right-angled octagonal star with equal areas. From "On the Division of Triangles." Paris, Bibliothèque nationale de France, Ms. Persan 169, fol. 114v. (Photo: Bibliothèque nationale de France)

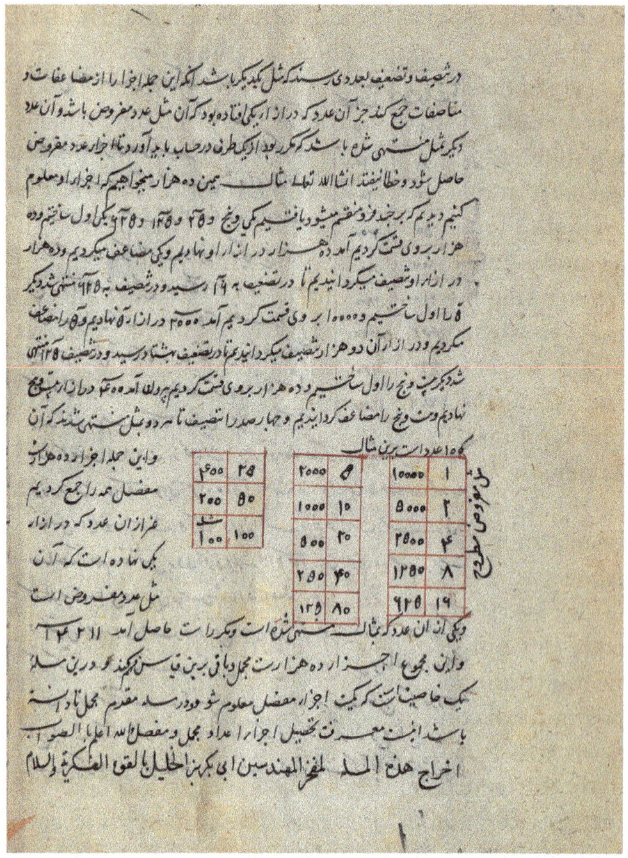

Fig. 2b. Page with numerical tables attributed in the last line to "Abu Bakr b. al-Khalil." From "On the Division of Triangles." Paris, Bibliothèque nationale de France, Ms. Persan 169, fol. 118v. (Photo: Bibliothèque nationale de France)

of an equilateral triangle to that of a right-angled octagonal star (fig. 2a). The latter example is significant because star polygons are typically omitted from surveying and practical geometry treatises, including that of Abu'l-Wafa' (with the exception of pentagonal stars, which are introduced in his last chapter, on divisions of the sphere). As we shall see, several constructions in the *Anonymous Compendium* also address the topic of polygons and star polygons with equal areas. "On the Division of Triangles" culminates in a section on numbers and numerical tables, some of which are attributed to Nicomachus of Gerasa and Pythagoras (fol. 116r).

Since the previously mentioned numerical table is ascribed to Abu Bakr b. al-Khalil, he may have been the author of some of the other constructions compiled in this source, given their similarity to those of the second treatise, which is specifically named after him. "Constructions by Abu Bakr" demonstrates his ability to combine arithmetical (numerical) and geometrical methods, neither of them provided with proofs (fig. 2b). While both methods coexisted in the field of surveying (*miṣāha*, measurement), practitioners of the visual arts and architecture generally relied on the latter, a constructive geometry that subordinated the computation of numbers to the manipulation of geometrical shapes.[18] This work

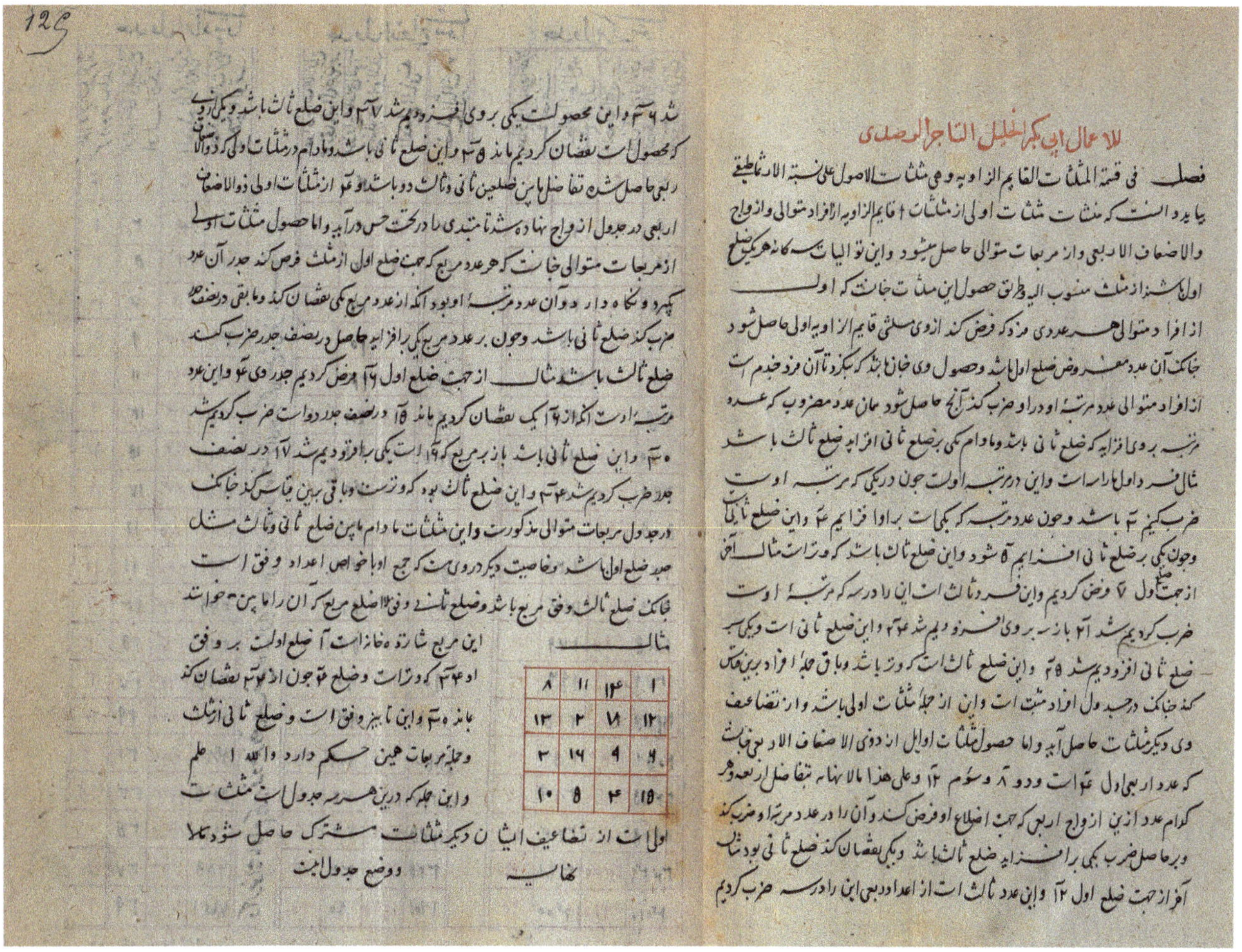

Fig. 2c. Title page and facing page with a magic square. From "Constructions by Abu Bakr." Paris, Bibliothèque nationale de France, Ms. Persan 169, fols. 124v–125r. (Photo: Bibliothèque nationale de France)

cites Book X of Euclid's *Elements* (on irrational magnitudes), and also makes a cross-reference to Ptolemy's astronomical "tables" (*jadval*), in keeping with Abu Bakr's expertise in observational astronomy. These citations show his familiarity, at least to a certain degree, with theoretical geometry treatises. The title page and the page facing it begin with the division of right-angled triangles, pertaining to "the method of arithmetical ratios/proportions" (*al-uṣūl ʿalā nisbat al-arithmāṭīqī*), whose properties relate to "harmonious numbers" (*aʿdād-i wafq*), as seen in an accompanying numerical table. The table is a square with sixteen sub-squares or "compartments" (*khāna,* lit. cell), constituting a magic square of order four (fig. 2c), a topic on which Abu'l-Wafa' wrote one of the earliest treatises.[19] The next two pages present diverse numerical tables for the most important dimensions of right-angled triangles, extracted by Abu Bakr [b.] al-Khalil al-Tajir in the "arithmetical method" (*uṣūl al-arithmāṭīqī*), for use in "constructions of surveying" (*aʿmāl al-miṣāḥa*) (fols. 125v–126r). The following section explains how harmonious numbers are a property of squares, produced by means of right-angled triangles.

Another section, on the measurement of arcs and the division of circles, is once again accompanied by numerical tables. It differentiates the method of "astronomers/astrologers" (*aṣḥāb-i nujūm, munajjimīn*), based on the division of circles into degrees, minutes, and

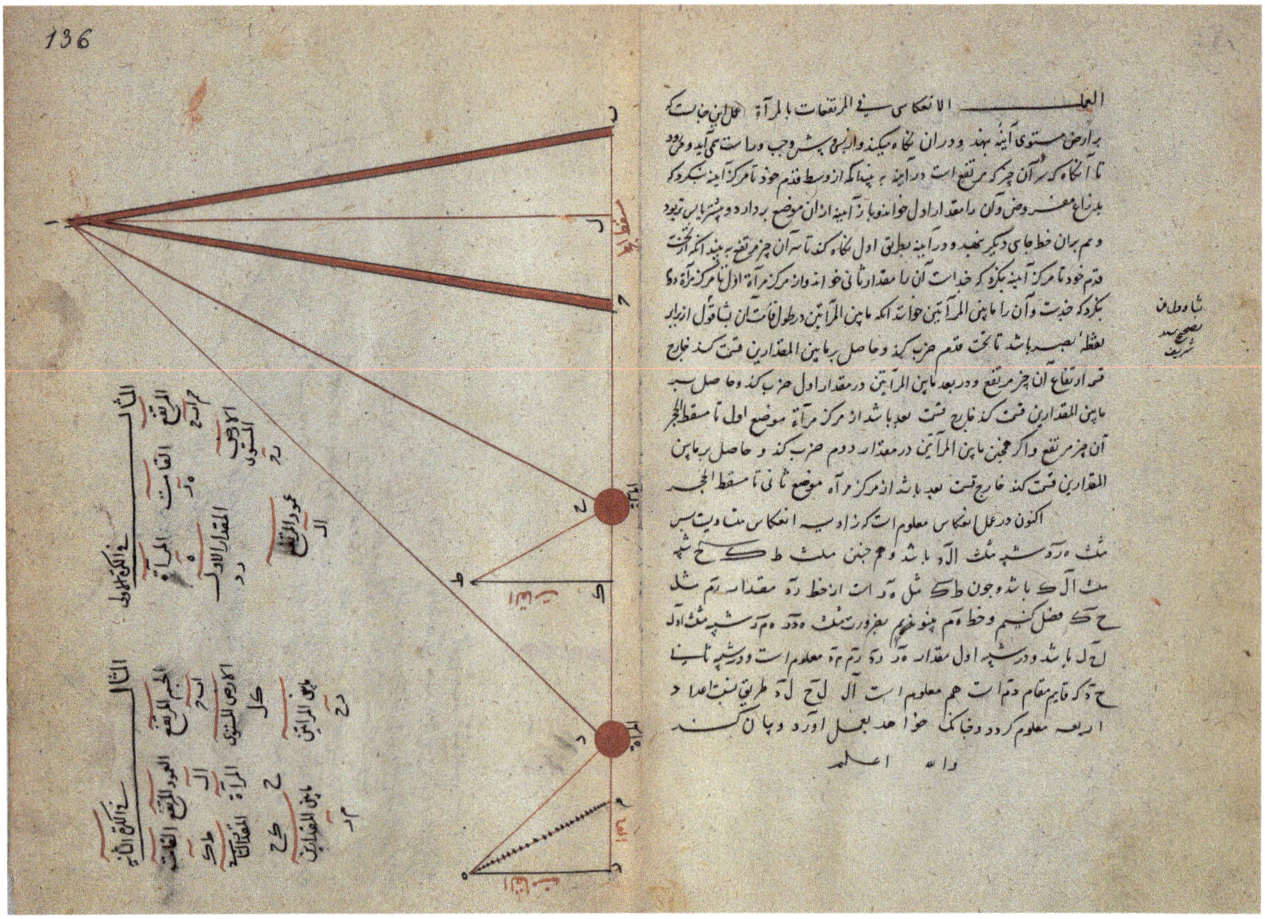

Fig. 3. Approximating the measurement of an inaccessible height by means of reflections on two mirrors. From "Constructions by Abu Bakr." Paris, Bibliothèque nationale de France, Ms. Persan 169, fols. 135v–136r. (Photo: Bibliothèque nationale de France)

seconds, from that of "surveyors" (*massāḥ*), referred to as the "method of simplification and approximation" (*ṭarīq-i tashīl va taqrīb*) (fol. 129r). Abu Bakr firmly asserts his preference for triangulation methods in surveying: "know that the best kind of surveying is that in which one divides surfaces/areas into triangles." This statement is followed by tables with measurements of triangles, and a geometrical construction showing the measurement of a quadrangle by means of triangulation (fols. 131r–132r). The subsequent section deals with approximate construction methods used in surveying (*aʿmāl-i miṣāḥa ast bi-taqrīb*). It focuses on measuring the sides of diverse polygons inscribed in a circle by means of a central triangle whose apex coincides with that circle's midpoint, as is illustrated in two diagrams of a heptagon and an octagon. Accompanying lists give approximate (*taqrīb*) measurements of the ratios/proportions (*nisbat*) of the sides and diameters for all regular polygons inscribed in a circle, from the triangle to the decagon (fols. 132v–133r).

"Constructions by Abu Bakr" culminates with diagrams on applied surveying problems related to the construction of pulleys; the measurement of inaccessible heights by means of reflections on two mirrors (fig. 3); the measurement of the height of "a wall, tree, minaret, mountain, or something else," illustrated with the case of a mountain (fig. 4); and the measurement of the depth of a well (fig. 5). The last three diagrams make use of triangulation techniques, based on proportionally related "similar" (*shabīh*) triangles. One of the texts

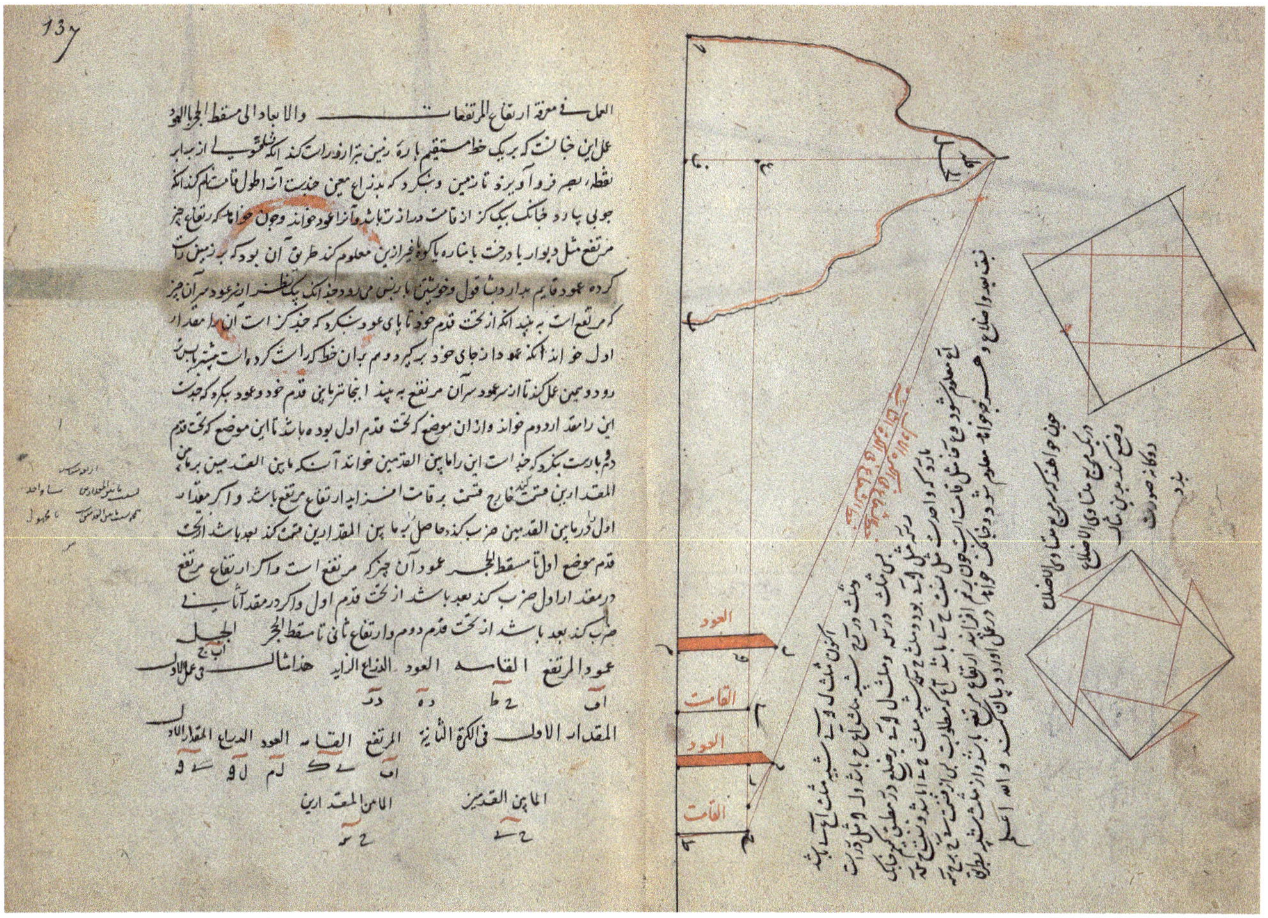

Fig. 4. Approximating the measurement of the height of a mountain, accompanied by two alternative solutions for composing a square from three squares. From "Constructions by Abu Bakr." Paris, Bibliothèque nationale de France, Ms. Persan 169, fols. 136v–137r. (Photo: Bibliothèque nationale de France)

refers to this common mode of surveying as the "method of the ratio/proportion of square numbers" (*ṭarīq-i nisbat-i aʿdād-i arbaʿa*), that is, numbers for which square roots exist (fig. 3). The other two surveying operations also involve the "method of ratio/proportion" (*ṭarīq-i nisbat*), based on similar triangles and triangulation. Their textual commentaries, aligned with the diagonal and/or vertical outlines of accompanying composite figures, recall the multi-directionality of texts in the *Anonymous Compendium* (figs. 4 and 5). The treatise ends abruptly, with the measurements and drawings of scalene triangles, in keeping with Abu Bakr's preference for triangulation methods in surveying. A marginal gloss accompanying the diagram showing the use of two mirrors indicates that the word "plummet" was corrected by Sayyid Sharif (*shāqūl, min taṣḥīḥ-i Sayyid Sharīf*) (fig. 3). This name can almost certainly be identified as the celebrated Timurid philosopher-theologian and mathematician Sayyid al-Sharif al-Jurjani (ca. 1339–1413).[20] If so, the original text by Abu Bakr must date earlier than al-Jurjani's gloss.

Admitting that the date and provenance of the *Anonymous Compendium* "remain uncertain," I tentatively ascribed it in my book to sometime between the eleventh and thirteenth centuries, with the proviso that it could be more precisely dated if Abu Bakr's identity were ascertained.[21] Following up my insight, Özdural discovered a clue on the basis of which he hypothetically identified "Abu Bakr al-Khalil al-Tajir" as the father of "ʿAli b. al-Khalil al-Tajir," who, in 728 (1327–28), copied

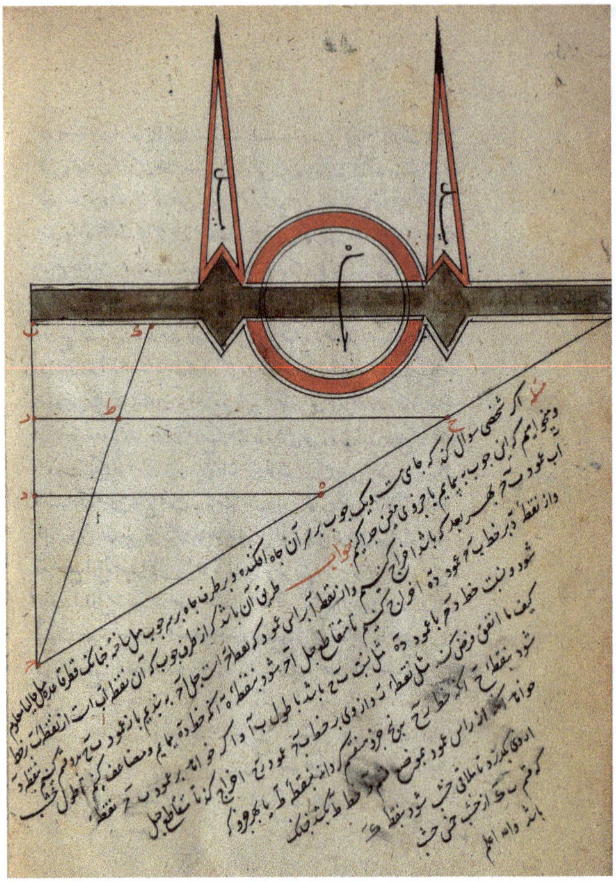

Fig. 5. Approximating the measurement of the depth of a well. From "Constructions by Abu Bakr." Paris, Bibliothèque nationale de France, Ms. Persan 169, fol. 137v. (Photo: Bibliothèque nationale de France)

an Arabic manuscript on surveying now kept at the Astan-i Quds-i Razavi Library in Mashhad. This tantalizing clue encouraged him to propose that Abu Bakr probably flourished around the late thirteenth to early fourteenth centuries, thereby suggesting a date circa 1300 for the *Anonymous Compendium.*[22]

The same codex in Mashhad includes the Persian translation of another Arabic treatise on surveying that lacks a colophon but was transcribed in the same hand, by ʿAli b. al-Khalil al-Tajir. These works are discussed in the present volume by Elaheh Kheirandish, who questions the father-and-son relationship between Abu Bakr and ʿAli that was inferred by Özdural. We learn from her that this inference had only been proffered to Özdural as "conjecture," by his Iranian informants, without concrete documentation.[23] Nonetheless, it is indeed striking that both men were involved in the applications of practical geometry to the field of surveying. Moreover, as noted above and further elaborated below, extracts from Abu Bakr's works that are preserved in the Paris Codex resonate in several ways with the contents of the *Anonymous Compendium.* Coupled with the conspicuous similarity of names, the circumstantial evidence is compelling enough, in my view, to posit a familial relationship. Besides, it was common in medieval Iran for mathematical training to run within families. To give an example, Abu'l-Wafa', who was born in Buzjan near Nishapur in northeastern Iran, initially studied numbers and arithmetic with his paternal and maternal uncles before moving to Baghdad.[24]

Moreover, the different spelling of Abu Bakr's name, as "Abu Bakr b. al-Khalil" in the work titled "On the Division of Triangles," finds a hitherto unnoticed parallel in a collection of mathematical works in Saint Petersburg, where his name is similarly cited as "Abu Bakr b. al-Khalil al-Tajir." He was the copyist of this Arabic codex, which features a fragment of a treatise on surveying illustrated with geometrical constructions and authored by the famous Iraqi geometer-astronomer and physician Ibn al-Haytham (d. ca. 1040). In his discussion of the treatise, Rushdi Rashid states that "we know nothing about this copyist, except that he had a somewhat mediocre grasp of orthography. The number of his mistakes in fact suggests that Abu Bakr did not belong to the learned classes."[25]

Judging by the works of Abu Bakr compiled in the Paris Codex, I suggest that he was one of those in-between scholar-practitioners who mediated theory and practice with ingenuity. Although these works betray his less than thorough knowledge of Arabic grammar and orthography, he nevertheless had a fine command of Persian. It is not unreasonable to hypothesize that the correct version of his name was Abu Bakr b. al-Khalil al-Tajir al-Rasadi, given that it was not uncommon for the "ibn" to be left out during copying. If so, it is likely he was the brother of ʿAli b. al-Khalil al-Tajir, rather than his father, and that they were both active towards the end and possibly beyond the reign of the Ilkhanids (r. 1256–1335), whose territories encompassed Iran and

Iraq. These bilingual individuals, then, appear to have flourished in the 1300s, each copying—and in the case of Abu Bakr also authoring—Arabic and Persian works on surveying and practical geometry that were illustrated with geometrical constructions, although we do not know where.[26]

Abu Bakr and the connection to Ibn al-Haytham

Ibn al-Haytham's fragmentary surveying treatise in the aforementioned Arabic Saint Petersburg Codex, which was copied by Abu Bakr, is titled *Fī uṣūl al-misāḥa* (On the Principles of Measurement).[27] This illustrious Iraqi scientist flourished in Basra and Baghdad, the capital of the Abbasid caliphate, which was then under the tutelage of the Buyid dynasty, spending his later years in the Fatimid capital, Cairo.[28] Another work transcribed by Abu Bakr in that codex, titled *al-Fawā'id al-bahā'iyya fī al-qawā'id al-ḥisābiyya* (Notable Uses of Arithmetic Rules), is a treatise on geometry, algebra, and arithmetic that includes a chapter on surveying methods.[29] It was written by the mathematician Muhammad Ibn al-Khawwam al-Baghdadi (ca. 1245–1324), who worked in Baghdad as well as Isfahan and was a pupil of the first director of the Maragha observatory, Nasir al-Din al-Tusi (d. 1274).[30] That treatise is followed by Ibn al-Haytham's piece, and another work titled *al-Ḥisāb* (Arithmetic), by the mathematician-engineer Abu Bakr b. Muhammad b. al-Husayn al-Karaji (ca. 953–ca. 1029), who prospered in Buyid Baghdad.

Interestingly, "Ibn-i Haytham" (i.e., Ibn al-Haytham) is mentioned in one of the commentaries in the *Anonymous Compendium*, which refers to his now-lost treatise on a right-angled triangle with particular proportions that he constructed "by means of conic sections, a hyperbola and parabola." It is tempting to conjecture that the author of this commentary may have been Abu Bakr, who copied Ibn al-Haytham's treatise on surveying. Textual commentaries in the *Anonymous Compendium* refer not only to Ibn-i Haytham and Abu Bakr [b.] al-Khalil al-Tajir, but also to Euclid's *Elements*, indicating that its author(s) knew at least the subjects covered in some treatises on theoretical geometry. Ibn al-Haytham used intersecting conic sections in three theoretical works—two treatises on constructing a regular heptagon, and another piece titled "Completion of the Conics." He also wrote short essays on the properties of the right-angled triangle with respect to its height, as well as on the construction of conic sections by mechanical means.[31] This major scientist composed at least four works on surveying methods, revealing his readiness to address the needs of practitioners. One of these is an essay titled "On Knowing the Height of Upright Objects, on the Altitude of Mountains and the Height of Clouds." Another version, titled "On the Determination of the Height of Mountains," lacks proofs and was intended for practitioners.[32]

Ibn al-Haytham's treatise "On the Principles of Measurement," which Abu Bakr copied, was meant "to provide surveyors with a rigorous but elementary geometrical manual of their art." Everything in it is reduced to linear measurements (rather than curves), with polygons divided into triangles and polyhedrons into pyramids. This testifies to the prevalence of triangulation methods in surveying, as noted in Abu Bakr's own treatise on the same subject. In Ibn al-Haytham's "Principles of Measurement," geometric magnitudes consisting of lengths, areas, and volumes are measured from given rectilinear dimensions or angles. As we shall see, the *Anonymous Compendium* likewise employs both rectilinear and angular measurements in the construction of proportionally related interlocking figures. Such modular measurements in Ibn al-Haytham's surveying treatise are referred to in terms of arbitrary units (such as the cubit), expressed as rational or irrational numbers, through which the author introduces a "numerical concept of the ratios of magnitudes," with some degree of approximation.[33] Accompanied by scientific proofs, his manual ends in an excerpt summarizing "practical procedures" that, in his own words, was intended for use by surveyors. Since they "do not make use of any proofs when making their measurements," he hoped that "its content may be accessible and easy for someone who wishes to acquire instruction in the art of measurement and to make use of the practical procedures."[34] Ibn al-Haytham's statement echoes a similar remark concerning artisanal practices in Abu'l-Wafa''s late tenth-century *Geometrical Constructions*, to which I shall turn later. The *Anonymous Compendium*, too, differentiates, though not consistently, between geometrically accurate and approximate solutions, thereby linking the worlds of the scholar-practitioner and artisan-designer.

The genre of "mixed" mathematical sciences, combining knowledge (*ʿilm*) and action (*ʿamal*), has only recently attracted the attention of historians of Islamic science.[35] Already theorized in the early classification of sciences by the philosopher-scientist Abu Nasr al-Farabi (d. 950–51), who flourished in Abbasid Baghdad and passed away in Damascus, the tradition of "mixed" sciences provides a fruitful framework within which to contextualize the *Anonymous Compendium*. Al-Farabi classified seven fields of mathematics (arithmetic, geometry, optics, astronomy, music, weights, and mechanics), each comprising theoretical (*al-naẓarī*) and practical (*al-ʿamalī*) branches; among these practical geometry was used by carpenters, metal workers, masons, and surveyors. Abu al-Hasan al-ʿAmiri (d. 992), who enjoyed the patronage of the Buyid vizier Ibn al-ʿAmid (d. 971), states that geometry is easier to understand than arithmetic, as it is concerned with "sensual prototypes," and is more far-reaching in its practical applications, "for without its aid the arithmetician cannot extract irrational roots, nor the surveyor determine the shapes of sites, nor human reason calculate the extent of the seas and the height of mountains." He adds that geometry is also useful to "all gifted architects, carpenters, sculptors, and goldsmiths, and used for the manufacture of astronomical instruments."[36] Before proceeding further, it will be helpful to consider Abu Bakr's personal involvement in the authorship of some of the geometrical constructions in the *Anonymous Compendium*.

Constructions by Abu Bakr in the Anonymous Compendium

Two geometrical constructions in the *Anonymous Compendium* are attributed by accompanying texts to Abu Bakr. The first concerns the method for drawing a regular pentagon with a fixed compass opening equal to the chord (fol. 187r [25]). This pentagon construction, ascribed by its Arabic caption to "Abu Bakr al-Khalil al-Tajir," is based on a sophisticated approximation, in accordance with his competence in the approximative methods employed in surveying. Because Abu Bakr's pentagon is inscribed with a pentagonal star, it is particularly relevant for composing interlocking star-and-polygon patterns. This construction is accompanied by several others on drawing regular pentagons with a fixed compass opening equal to a given altitude, or side, or the radius of a circumscribing circle (fols. 186r–187v).

One of those pentagons contains an annotation explaining that the method for its construction was "deduced, not transmitted," a comment implying that the constructions compiled in the *Anonymous Compendium* were the result of a mix of methodological continuity and innovation (fol. 186bis(r) [21]). The simpler, exact methods for drawing a regular pentagon with a fixed compass opening that are explicated in Abu'l-Wafaʾ's *Geometrical Constructions* are not repeated here, an omission suggesting a prior knowledge of those methods. Only one of his pentagon constructions, slightly modified, appears in a tiny marginal drawing of the *Anonymous Compendium*.[37] It was apparently added later, close to the edge of a page, with a caption written in a different hand than that of the mid-fifteenth-century copyist (fol. 186r [20], pls. 20.1–4). The script of this document is otherwise generally uniform, with the exception of another caption in the same hand that accompanies a pentagonal star inscribed within a semicircle, which is likewise drawn close to the edge of a page (fol. 186bis(v) [24]). Rendered in black ink, both figures and their captions were seemingly inserted by the same person, as a visual commentary or gloss, next to two of the original pentagon constructions.

Unlike Abu'l-Wafaʾ's *Geometrical Constructions*, a substantial part of the *Anonymous Compendium* deals with ornamental patterns. The repeat-units for such patterns combine diverse polygons, the stars associated with them (often named after the corresponding polygons), and the fragments of these figures, all of which fit seamlessly together. The textual instructions for the generally square and rectangular repeat-units describe geometrical procedures for inscribing proportionally related similar and complementary figures, which in turn interlock with one another. In comparison with the seemingly generic audience of artisans, surveyors, and other interested parties targeted by Abu'l-Wafaʾ's practical geometry treatise, the *Anonymous Compendium* addresses the concerns of a more specialized audience with relatively advanced skills, probably scholar-practitioners, artisan-designers, and master builder-architects. The genre of ornamental star-and-polygon patterns scrutinized in this source had been evolving since the eleventh century, having now reached a stage that involved the cross-examination and rationalization of their ratios/proportions, along with a growing concern for precision.

Such a process of cross-examination is implied by the textual commentary of the other geometrical construction that names Abu Bakr (fig. 6):

> Masters (*ustādān*) examine the proportion (*nisbat*) of this repeat-unit (*ʿaqd*). Abu Bakr [b.] al-Khalil examined it in a few ways. The constructions were made. One of these [constructions] is the one that has been explained (fol. 189r [31]).

The drawing corresponding to this text is on the previous page, where it is stated that:

> Some practitioners of geometry (*aṣḥāb-i handasa*) held that [if] we want the distance between two parallel squares to be equal to the side of the square in the middle, then it is necessary that magnitude SN be equal to [R]J. The proportion (*nisbat*) of it [the repeat-unit] is such that, since the square of OF is assumed, they draw this repeat-unit (*ʿaqd*) in another way.... If one of these two squares is known, the other is found by means of the proportion [of the repeat-unit] (fol. 188v [30]).

The "masters" (*ustādān*), including Abu Bakr, who examined the proportions of this geometrical construction may or may not have been the same individuals as the "practitioners of geometry" (*aṣḥāb-i handasa*) mentioned on the previous page. The latter could have belonged to another time and space, when they formulated that particular construction method. As the text is ambiguous, the "masters" and "practitioners of geometry" may also have been one and the same.[38] Whatever the actual circumstances, it is telling that not all of the constructions made by Abu Bakr were recorded. The texts cited above hint at investigational processes that led to the modification of preexisting designs and the formulation of alternative construction methods, only some of which finally made it to the *Anonymous Compendium*.

In this particular example, a single geometric diagram has been provided to describe two different approaches to a given problem, one rectilinear, the other angular (i.e., curvilinear). The square repeat-unit with four-fold rotational symmetry consists of two large parallel squares creating a border around a third rotated central square, whose sides extend to form four kite-shaped quadrangles (fig. 6). Özdural has suggested that the purpose of delineating both constructions on the same figure was to compare the differing proportions of two variants of the same pattern (pls. 30 and 31). Jan Hogendijk explains that the first version (fol. 188v [30]) is based on a given rectilinear unit repeated in many line segments and distances, which is derived from the sides of a right-angled triangle (1:2:√5), while the second version is generated by a given angle called a "line [number] four" that is repeated throughout (fol. 189r [31]).[39]

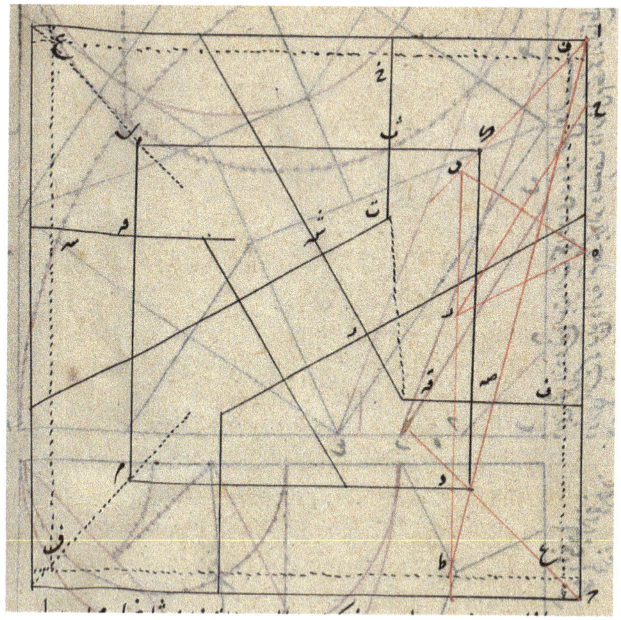

Fig. 6. A geometrical construction examined by Abu Bakr, composed of parallel concentric squares forming a border around a central rotated "four-legged" square surrounded by rotating kites and non-convex hexagons. From the *Anonymous Compendium*. Paris, Bibliothèque nationale de France, Ms. Persan 169, fol. 188v [30]. (Photo: Bibliothèque nationale de France)

The instructions in the *Anonymous Compendium* address practitioners in their own specialized technical language; allusions to drawing a "line [number] four" (*khaṭṭ-i chahār*), "line [number] five" (*khaṭṭ-i panj*), "line [number] six" (*khaṭṭ-i 6*), "line [number] seven" (*khaṭṭ-i haft*), and "line [number] eight" (*khaṭṭ-i 8*) constitute a shorthand for designating various angle measurements, their numbers corresponding to angles formed by the radii of uninked circular construction lines.[40] The text of the same construction also refers to "four [number] 6 kites" (*chahār turunj 6*), each of which is further subdivided into a smaller kite and a chevron (a non-convex hexagon) (fol. 189r). The quadrangles called "[number] 6 kites" combine two congruent right-angled triangles (30°-60°-90°), drawn by means of a "set square [number] 6" (*gūnyā-6*). We shall see that elsewhere in the *Anonymous Compendium* a template is provided for such a set square (fol. 184v [14]).[41] It is likely that the repeat-unit analyzed by Abu Bakr was based on a preexisting pattern. Early variants of this geometrical construction appear at the mausoleum of the Rum (Anatolian) Seljuk sultan ʿIzz al-Din Kaykavus in Sivas (1217), and in a madrasa at Zuzan in eastern Iran (1218) that was built by a Khwarazmshahid governor on the eve of the Mongol invasions (fig. 6). Both variants feature an additional hexagonal star motif, unlike a later Timurid version at the Gawhar Shad Mosque in Mashhad (1418), which comes closer to the pattern in the *Anonymous Compendium* (pl. 30.3b).[42]

Simpler versions of this square pattern, without the outer border, contain a smaller rotated square at the center, whose extended sides touch the framing square to create four kites or four right-angled triangles along each side. It is known today among some traditional Iranian builders as *chahār turunj* (lit. "four kites"), or *chahār langa* (lit. "four-legged"), in reference to the four extended sides or "legs" of the rotated central square.[43] An early example with four rotating kites, mentioned by Chorbachi, appears in one of the twin Seljuk tomb towers at Kharraqan in Iran dated 486 (1093): it forms the invisible underlying scheme that structures repeat patterns in cut brick. The same scheme is used as a decorative motif in its own right in the woodwork door of the Imam Ibrahim Mosque at Mosul in northern Iraq, which is dated 498 (1104) (pl. 2.6).[44] Subsequent examples decorate the woodwork minbars and screens of several late thirteenth- to early fourteenth-century mosques in Anatolia, produced after the 1243 subjugation of the Rum Seljuks by the Mongols.[45]

This versatile pattern, which comprises a central rotated four-legged square, is featured in many square repeat-units in the *Anonymous Compendium* (fols. 180v [2], 182v [10], 188r [28], 188v [30], 189v [33], 191r [36], 192v [41], 194v [48]). One of these has an outer border formed by two parallel squares, but lacks a rotated central square (fol. 192v [41]). The accompanying text explains that it is based on a given unit of measurement, from which one derives its "small and large proportions" (*nisbat-i kuchak va buzurg,*). Not all of these geometrical constructions seem to have been intended as repeat-units for decorative patterns. As a group, they may be interpreted as explorations of the properties of the rotated "four-legged" square as a dynamic proportioning device, which could play a role in generating ornamental panels or in setting out ground plans and elevations. In some of these constructions, the "four-legged" square forms an underlying generative scheme, indicated in dotted lines, that structures a decorative pattern (see fig. 17), or as a design motif surrounded by four right-angled triangles that are proportionally subdivided into smaller units, including kites, rhombuses, and triangles (see fig. 12).

The *Anonymous Compendium* also features a series of constructions composed of kites with a two-fold rotational symmetry that are framed in rectangular repeat-units. One of these is based on a "difficult" (*mushkil*) right-angled triangle that "falls outside Euclid's *Elements* and concerns the science of conics (*ʿilm-i makhrūṭāt*)." It therefore required the use of a "moving ruler" (fol. 185v [17, 18]). (As the *Elements* deals only with polygons drawn by ruler-and-compass, conic sections and polygons with seven, nine, eleven, thirteen and fourteen sides fall outside its scope). The special right-angled triangle on which the proportion of that repeat-unit is based has been "established on the backside of this folio" (fol. 185r [16]). The rectangular frame of the repeat-unit is composed of two such triangles, whose shared hypotenuse divides the pattern into two congruent parts. Cutting off a triangle from the tip of each of those triangles creates two quadrangular kites facing in opposite directions, which are further subdivided into three kites. Replicating the repeat-unit yields a composition combining irregular hexagonal stars, hexagons, and kites (pl. 16.3).[46] This pattern and others like it have indirect relevance for ornamental patterns, in that they investigate proportional relationships between triangles, quadrangles, kites, and their derivatives.

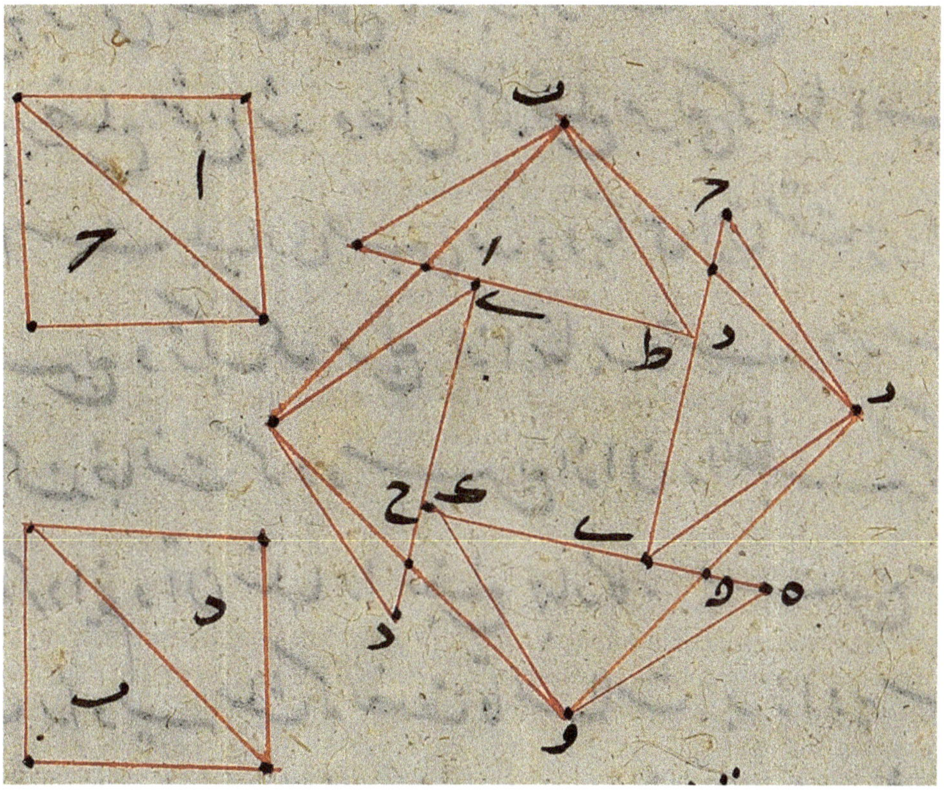

Fig. 7. Abu'l-Wafa''s solution for composing a square from three squares, consisting of four triangles rotating around a central rotated "four-legged" square. From the *Geometrical Constructions*. Paris, Bibliothèque nationale de France, Ms. Persan 169, fol. 171v. (Photo: Bibliothèque nationale de France)

Remarkably, two variations on a square containing a rotated "four-legged" square, surrounded by four right-angled triangles, are drawn side by side in the margin of a page in the surveying treatise titled "Constructions by Abu Bakr." Initially these drawings seem to bear no obvious connection to the text and illustration on that page, which concern how to measure the height of a mountain by approximation (fig. 4). Labeled "Two alternative solutions for the problem of composing a square from three squares," the drawings address the problem posed by artisans to geometers at a gathering attended by Abu'l-Wafa'. As we shall see in the next section, the solution to that problem involved a similar geometrical diagram based on "square numbers," that is, numbers having square roots (fig. 7). Since Abu'l-Wafa''s chapter on the division and composition of squares addressed questions most frequently posed by artisans and surveyors to geometers, this diagram must have been particularly relevant for practitioners in those fields. Applied problems discussed by the author in the same chapter include the cutting and assembling of bricks, and the division of estates among heirs.[47]

Özdural plausibly identified one of the two diagrams in "Constructions by Abu Bakr" as the solution of Abu'l-Wafa' and attributed the other one to Abu Bakr himself (fig. 4, pl. 2.4). Noting the latter diagram's similarity to a geometrical construction in the *Anonymous Compendium*, he suggested that that one too might have been drawn by Abu Bakr (fol. 182v [9], pls. 9.1 and 9.2a–b). Although Özdural surmised that the paired diagrams in "Constructions by Abu Bakr" were arbitrarily inserted by the author into one of the margins of his surveying treatise, they are actually fully integrated with that page, rather than having been added as an afterthought.

Moreover, the surveying problem accompanying both diagrams is related to them.[48] This problem, on the measurement of a mountain by approximation, involves a triangulation technique similar to the one using two mirrors; the aforementioned text on the latter technique refers to that procedure as the "method of the ratio/proportion of square numbers" (*ṭarīq-i nisbat-i aʿdād-i arbaʿa*) (fig. 3). Thus, the paired diagrams that are also based on "square numbers" were not an irrelevant insertion to this page (fig. 4). The prominence of the rotated "four-legged" square in the *Anonymous Compendium*, then, accords well with Abu Bakr's expertise in surveying methods dominated by triangulation and the proportional division of polygons, particularly triangles and quadrangles. He may well have contributed to some of the patterns with rotated "four-legged" squares that constitute a special subgroup in this source.

Whatever the unknown circumstances and contexts of its production, the *Anonymous Compendium* likely drew upon geometrical constructions formulated in now-lost earlier Arabic and Persian treatises.[49] As a matter of fact, freely copying and reworking older problems was a typical characteristic of medieval practical geometry manuals circulating in the Latin West and in the Islamic lands.[50] Nor do ornamental patterns in extant buildings or artifacts that find counterparts in this document allow us to confirm its date and provenance. After all, those patterns could have been drawn later, contemporaneously with, or even before their appearance on particular monuments, whether directly copied from them or from previous drawings. Briefly put, the contextual circumstances in which the *Anonymous Compendium* was compiled must remain open questions until more concrete additional evidence is garnered in the future. That said, even if it included patterns from earlier monuments and extracts from previously written sources, it appears that the unicum copy in the Paris Codex was compiled at a discrete point in time and in a particular place, rather than having been formed additively over several centuries. This is strongly suggested by the relative consistency of its language, its technical terminology, and the types of issues addressed in its textual commentaries, which persistently focus on ratios/proportions (*nisbat*) and close approximations of geometrically correct constructions.

To my mind, a provisional date in the 1300s seems more than likely for the initial compilation of the *Anonymous Compendium*, given its connection with Abu Bakr and extracts from other works of his included in the same codex. This first compilation, in turn, was in all probability copied again, in a revised version, around the mid-fifteenth century by or for Kubanani Yazdi. It is hoped that, with future research, additional clues concerning the historical contexts of this fascinating source may come to light, particularly vis-à-vis the familial relationship and scholarly profiles of Abu Bakr b. al-Khalil al-Tajir al-Rasadi and ʿAli b. al-Khalil al-Tajir, as well as the specific locales in which they flourished. For the time being, however, a close reading of the *Anonymous Compendium* in relation to Abu'l-Wafa''s *Geometrical Constructions* can furnish fruitful insights.

ABU'L-WAFA''S TREATISE AND THE PROLIFERATION OF GEOMETRIC ORNAMENT

On a page in the *Anonymous Compendium* featuring two constructions, the text for the lower geometrical diagram explains that, "the ratio of the length and width of this repeat-unit is determined by geometrical method (*bi-ṭarīq-i handasa*) from the octagon" (fol. 187v [27], pls. 27.1–2). The construction at the top of that page illustrates a "simple proportion" (*nisbat-i sahl*), which reflects the working method of "some artisans" (*baʿżī az-ṣunnāʿ*) that is "extremely close" to the required ratio/proportion (fol. 187v [26], pl. 26.1). The juxtaposition of a geometrically exact construction with an approximation discovered by artisans indicates that the author(s) of this source were cognizant of alternative ways of achieving a particular construction. A similar awareness characterizes the celebrated passage in the tenth chapter of Abu'l-Wafa''s *Geometrical Constructions*, in which the author states that he had participated in one of the gatherings (*majālis*) attended by some artisans (*al-ṣunnāʿ*) and geometers (*al-muhandisīn*) where the differing geometrical construction methods preferred by each group became manifest. When the artisans inquired about the construction of a square from three squares, a skilled geometer easily responded. However, none of the artisans were satisfied with that geometrically correct construction, based on proofs, because the artisan wants to divide those squares into pieces from which a single square can be assembled.[51] In other words, artisans favored the method of "dividing and assembling," also common among surveyors, which has been compared to the "cut-and-paste" technique of modern computer technology.[52]

Abu'l-Wafa' stated that the aim of his treatise was to expose the correct rules of practical geometry because some artisans made errors in dividing and assembling geometrical figures. While geometers steeped in theory lacked practical experience in finding the approximate geometrical constructions needed for the work of "artisans and surveyors" (*al-ṣunnāʿ wa al-māsiḥ*), the latter were merely concerned with practical aspects and had no knowledge of or interest in the linear constructions of geometers demonstrated by proofs, which sometimes led them to make mistakes. Likewise, when artisans and surveyors asked skillful geometers about how to divide or assemble geometrical figures, the latter were not always successful in finding practical solutions.[53]

As mentioned earlier, Abu'l-Wafa' composed the *Geometrical Constructions* circa 991–92 to 998. It was commissioned by and dedicated to the Buyid amir Baha' al-Dawla (r. 988–1012) during the tenure of the Abbasid caliph al-Qadir (r. 991–1031) in Baghdad. Prior to this Abu'l-Wafa' had written a treatise on applied arithmetic, which he dedicated to Baha' al-Dawla's father, ʿAdud al-Dawla (r. 949–83). Referred to by his protégé, the man of letters and philosopher Abu Hayyan al-Tawhidi, as "Abu'l-Wafa' *al-muhandis*," this geometer-astronomer was firmly ensconced in the political and intellectual-scientific circles of Buyid Baghdad.[54] His treatise on arithmetic aimed to rectify errors committed by secretaries (*al-kuttāb*), administrators (*al-ʿummāl*, sing. *al-ʿāmil*), surveyors (*al-massāḥ*), and others. Some of the applied problems in it concern calculations related to surveying, such as those necessary for the measurement of domes; the treatise also explains how to prepare cost estimates for buildings and dams, and how to compute wages for mortaring, bricklaying, and filling a well.[55] The emphasis on numerical calculation differs from the visual primacy of the diagrammatic drawings in Abu'l-Wafa''s *Geometrical Constructions*, where no such administrative problems are addressed.

Both treatises exemplify the abiding dedication of the author and his royal patrons to fostering the practical applications of arithmetic and geometry. These works were commissioned from Abu'l-Wafa' at the height of the Buyids' royal and elite patronage of the mathematical sciences, when they reigned over a territory roughly corresponding to modern Iraq and western Iran.[56] It has been shown that interest in geometry and algebra (combining computational and geometrical problems) peaked in the late tenth century, which saw the "emergence of a new mathematical community, whose members [were] linked by exchanges of letters, and face to face conversations in the numerous salons (*majālis*), or even the courts of Kings, Viziers, and around patrons." Geometry was one of the topics discussed in royal and vizierial assemblies in an era characterized by a multiplicity of courtly and scientific centers, among which Baghdad reigned supreme.[57]

Tenth- and eleventh-century geometers worked competitively on conic sections, using parabolas and hyperbolas to solve challenging theoretical problems, including the trisection of the triangle, as well as the construction of the regular heptagon and nonagon. In so doing they went beyond Euclidean ruler-and-compass methods in solving these problems.[58] Simpler, application-oriented versions of such theoretical problems found their way into not only Abu'l-Wafa''s practical geometry treatise but also the *Anonymous Compendium*. Yet unlike the latter, the *Geometrical Constructions* does not deal with the ornamental patterns that were just being developed at that time in, around, and beyond Baghdad, when the capital of the Abbasid caliphate remained under the governance of the Buyid dynasty (ca. 945–1055). An early example of this distinctive mode of geometric ornament can be found in the star-and-polygon patterns with interlaced strapwork, combining octagonal and six-pointed stellate figures, that appear in the frontispiece of a Koran manuscript produced in Baghdad in 391 (1000–1001) and signed by the renowned calligrapher Ibn al-Bawwab.[59] Having flourished during the Buyid period, this calligrapher probably died in 1022, before the emergence in 1055 of the Turkic dynasty of the Great Seljuks (r. 1037–1194) as the new guardians of the Abbasid caliphs (fig. 8).

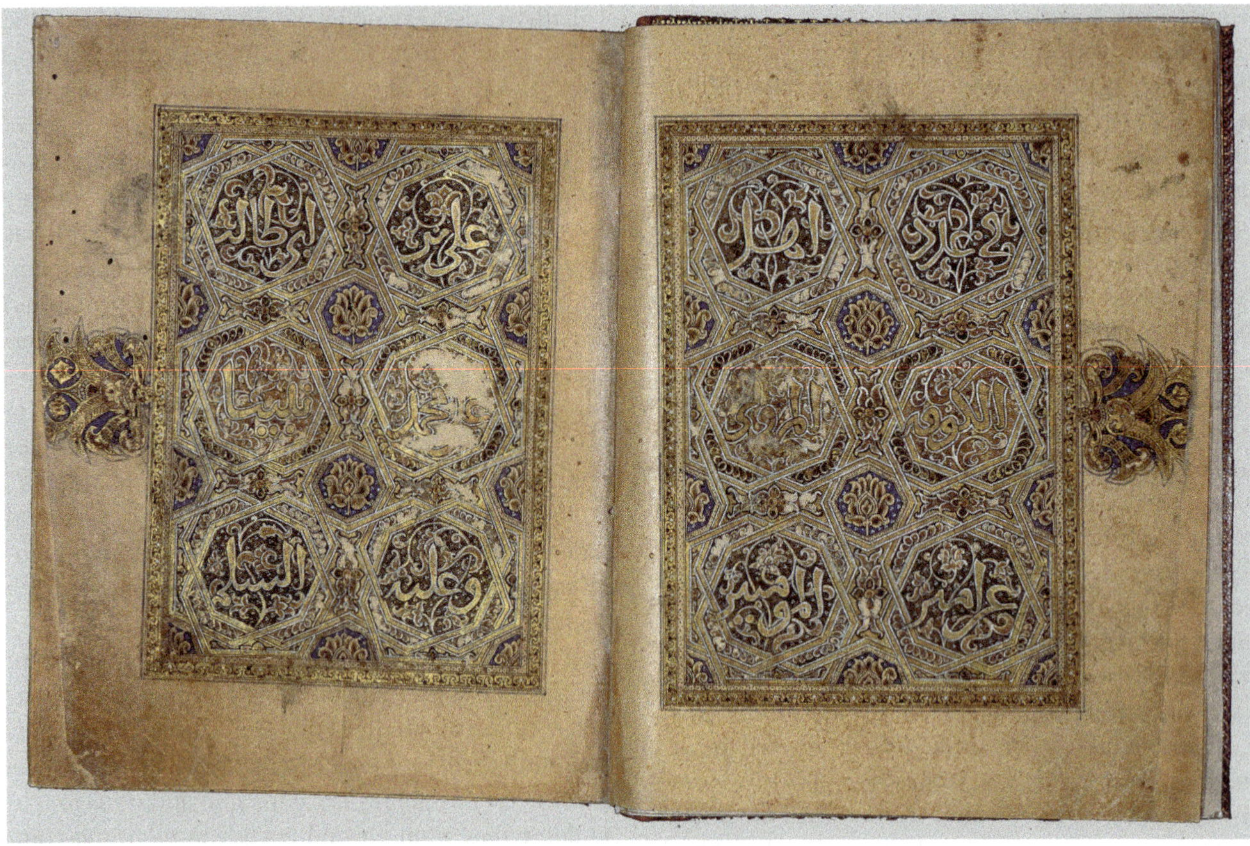

Fig. 8. Double-page frontispiece with strapwork star-and-polygon panels. Koran manuscript, signed by Ibn al-Bawwab in Baghdad, 391 (1000–1001). Dublin, Chester Beatty Library, Ms. Is 1431, fols. 7v–8r. (Photo: © The Trustees of the Chester Beatty Library, Dublin)

While Baghdad continued to be the seat of the caliphate, the Seljuk sultans based their capital in Isfahan. It was particularly in territories ruled by the Great Seljuks and the late Abbasid caliphs that ornamental geometry rose to unprecedented prominence between the late eleventh and early thirteenth centuries. Two-dimensional ornaments were dominated by interlocking star-and-polygon patterns, often combined with ribbon-like strapwork that resembles weaving in its "over-under" relationship. These were generally accompanied by their three-dimensional counterparts, particularly *muqarnas* and stellate vaults. Geometric patterns with or without interlaced strapwork came to be used alone or in combination with calligraphy and other motifs, non-figural (generally vegetal) or figural. For example, a painted and gilded enamel (*mīnā'ī*) ceramic bowl from late twelfth- to early thirteenth-century Seljuk Iran depicts a couple wearing robes with aniconic and figural versions of star-and-polygon strapwork designs (fig. 9). Such patterns proliferated in multiple media, including pottery, textiles, manuscript illumination, wall painting, woodwork, metalwork, stonework, carved stucco, cut brickwork, and eventually polychrome mosaic tilework, which emerged in the early thirteenth-century on the eve of the Mongol invasions.

Interestingly, a pattern in the *Anonymous Compendium* has been identified by Hogendijk as the rectangular repeat-unit of a star-and-polygon design that decorates a tympanum arch in the North Dome (1088–89) of the Seljuk-era Friday Mosque in Isfahan (fig. 10a). Comprising regular heptagons (which were still rare at such an early date), this pattern is complemented by others that cover the surfaces of eight shallow tympanum arches just under the transition zone of the dome,

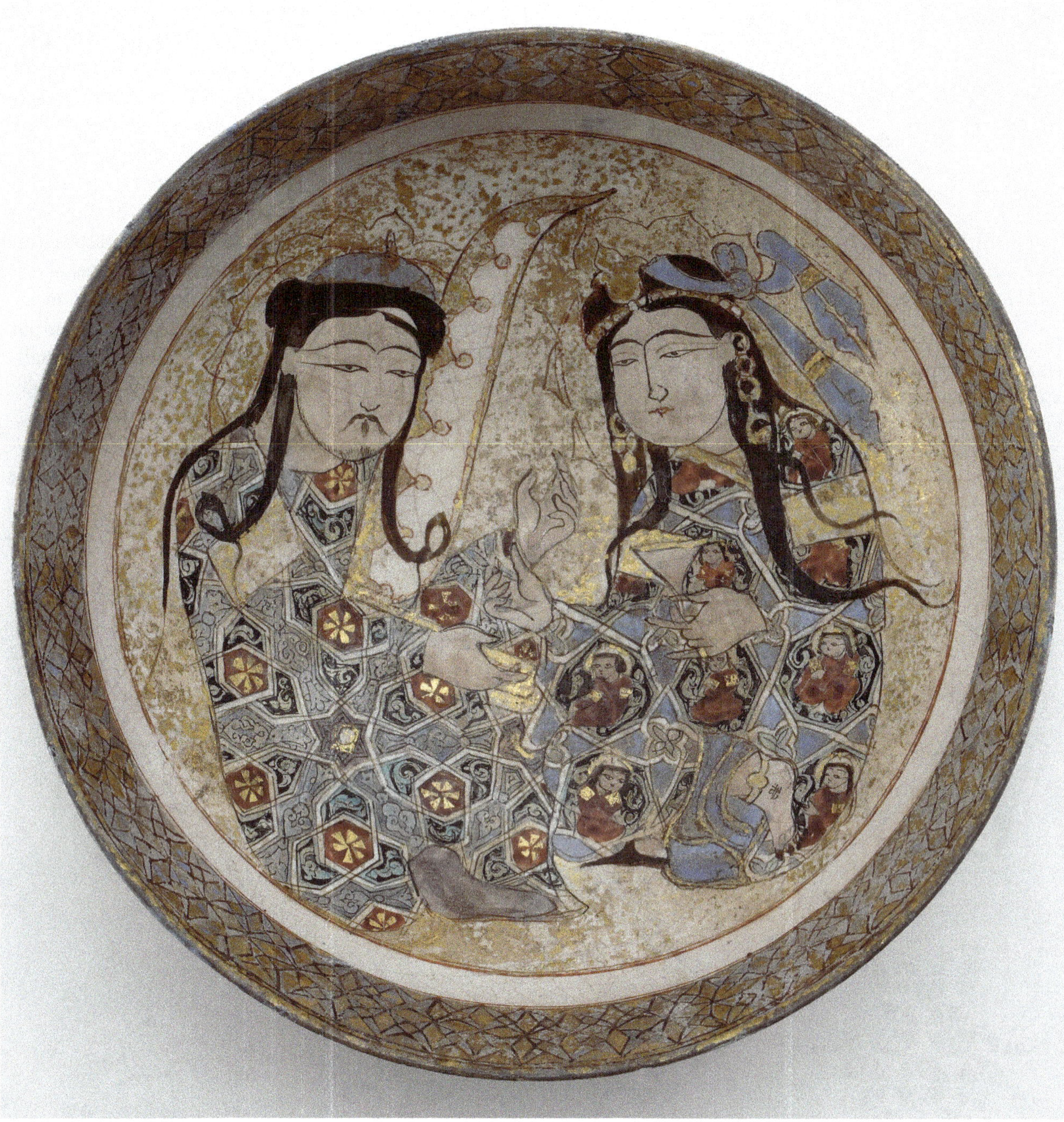

Fig. 9. *Mina'i* ceramic bowl depicting a couple wearing robes with aniconic and figural versions of strapwork star-and-polygon patterns. Seljuk Iran, late twelfth to early thirteenth century. Freer Gallery of Art, Smithsonian Institution, Washington, D.C.: Purchase, F.1938.12. (Photo: Freer Gallery of Art)

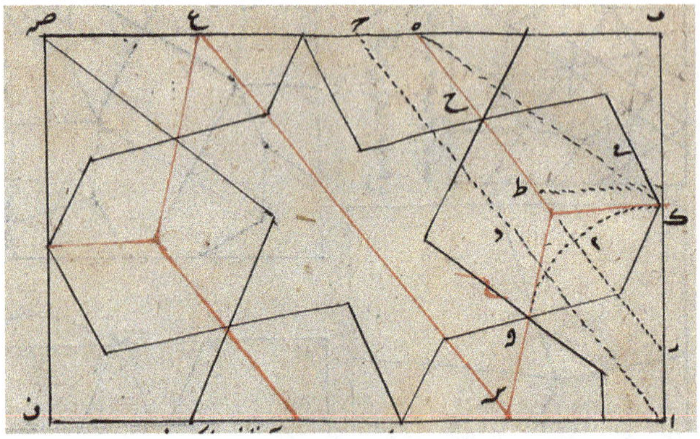

Fig. 10a. Rectangular repeat-unit of a geometric construction with regular heptagons, irregular hexagons, and two different kinds of six-pointed stellate forms. From the *Anonymous Compendium*. Paris, Bibliothèque nationale de France, Ms. Persan 169, fol. 192r [37]. (Photo: Bibliothèque nationale de France)

which rests on *muqarnas* squinches. Hogendijk has shown that replicating the mirror reflection of that repeat-unit yields a composition with interlocking regular heptagons, irregular hexagons, and two kinds of six-pointed stellate forms (see figs. 8–10, pp. 152–153).[60] The corresponding design on the tympanum arch is executed in cut and molded brick, accompanied by carved terracotta and painted plaster inlays, which allowed a greater precision in the motifs. Comparable patterns on the remaining tympanum arches fuse sequences of different polygons and star polygons, with or without interwoven strapwork. Their interlocking individual motifs are inscribed with geometric patterns on a smaller scale, turning these complex patterns into unusually early examples of two-layered compositions (fig. 10b).

One of the arched panels is a variant of the heptagonal pattern analyzed by Hogendijk, this time combining

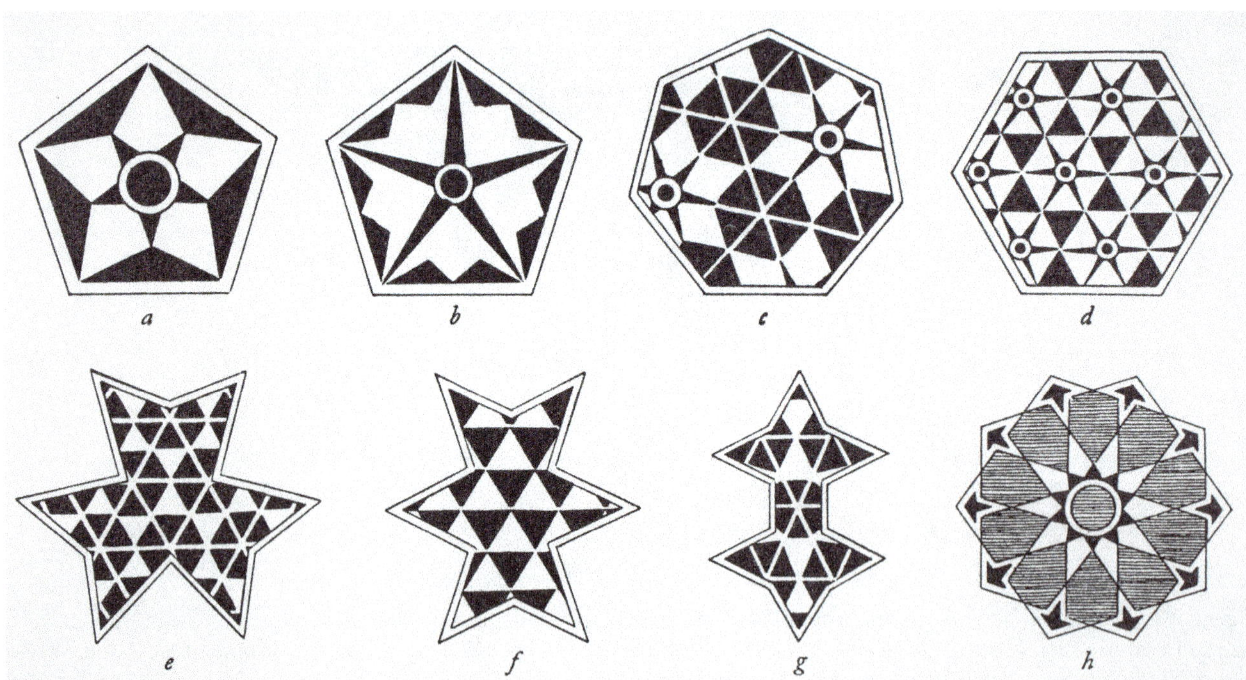

Fig. 10b. Drawing of some carved terracotta and painted plaster inlays used in the ornamental patterns of various tympanum arches. North Dome (1088–89) of the Friday Mosque in Isfahan. (Drawing: Eric Schroeder, in Pope and Ackerman, *A Survey of Persian Art*, vol. 2, p. 1290, fig. 468[a–h])

regular heptagons, irregular hexagons, and five- and six-pointed stellate forms with interwoven strapwork (fig. 11a).[61] Another fascinating tympanum pattern without the strapwork interweave displays five- and ten-fold symmetries, composed as it is of decagonal stars forming rosettes, two kinds of pentagons inlaid with pentagonal and decagonal stars, kites, and filler units (fig. 11b and c).[62] These arched tympana, surmounted by *muqarnas* squinches, demonstrate the inventiveness of two- and three-dimensional geometric patterning under the Great Seljuks during the last quarter of the eleventh century. Given the close links between their capital, Isfahan, and the seat of the Abbasid caliphate, Baghdad, it is not unreasonable to assume that monuments and decorated artifacts in the latter city boasted similar geometric ornaments. The earliest preserved examples of such innovative designs in Baghdad, which was sacked by the Mongols in 1258, appear in brick buildings datable from the late twelfth to the early thirteenth centuries (see figs. 20a and b).

The few specimens of geometric strapwork designs from Fatimid Syria and Egypt, mostly dating from the second half of the twelfth century and limited to portable objects in woodwork (particularly minbars, mihrabs, and cenotaphs) hardly challenge the fact that the densest geographical and temporal concentration of this ornamental idiom overlapped with the dominions of the Abbasid caliphate and its vassals during the era generally referred to as the "Seljuk period," roughly spanning the eleventh to mid-thirteenth centuries.[63] It

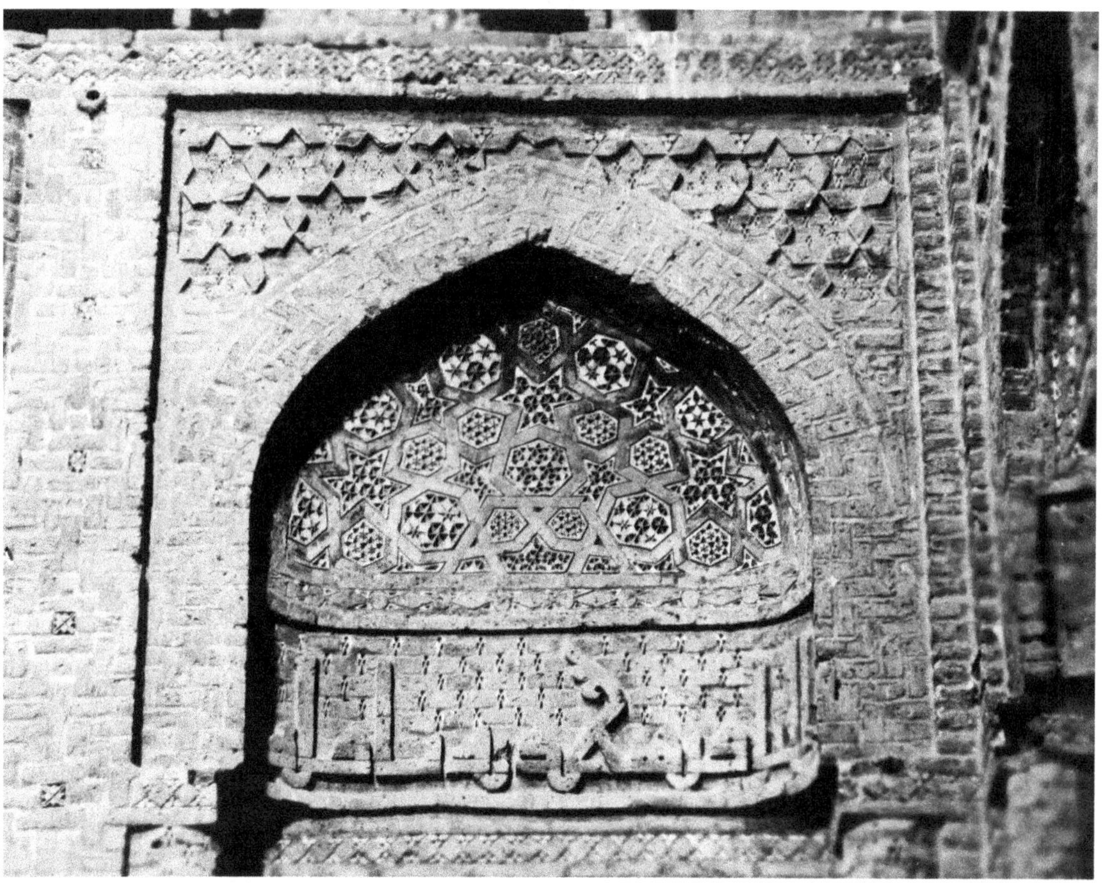

Fig. 11a. Arched tympanum with strapwork pattern combining regular heptagons, irregular hexagons, and five- and six-pointed stellate forms. North Dome (1088–89) of the Friday Mosque in Isfahan. Smithsonian Institution, Freer Gallery of Art and Arthur M. Sackler Gallery Archives, File 24, 115, A.U. Pope, FSA A.6 04.24.115. (Photo: Freer Gallery of Art and Arthur M. Sackler Gallery Archives)

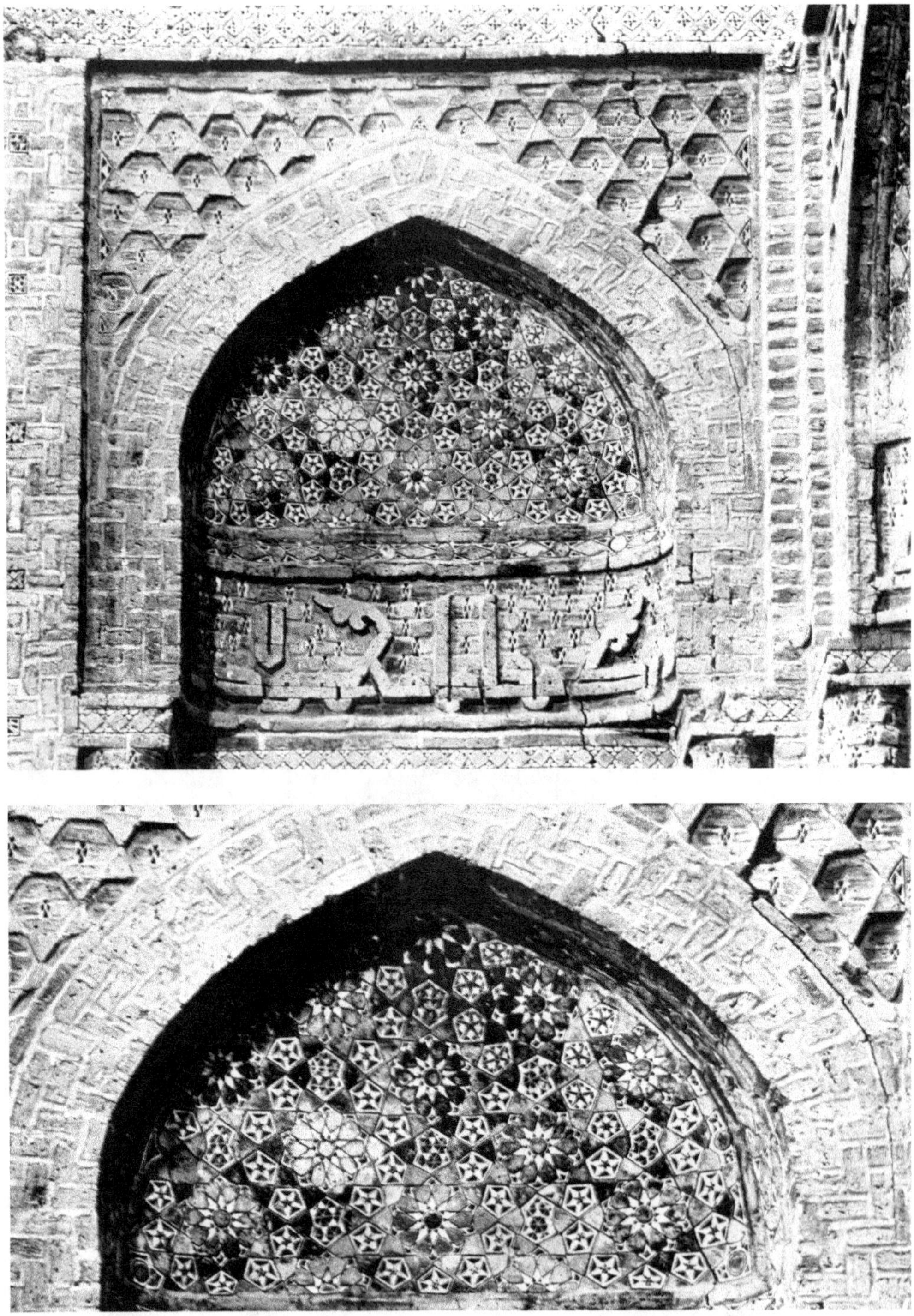

b

c

Fig. 11b–c. Tympanum pattern combining decagonal stars, two kinds of pentagons overlaid with pentagonal and decagonal star motifs, and filler units. North Dome (1088–89) of the Friday Mosque in Isfahan. Photo taken by Myron B. Smith, dated 1935. Smithsonian Institution, Freer Gallery of Art and Arthur M. Sackler Gallery Archives, File MBH 846. (Photo: Freer Gallery of Art and Arthur M. Sackler Gallery Archives)

was then that two- and three-dimensional geometric ornament, including the *muqarnas*, achieved iconic status through monumentalization and canonization in buildings whose fabric conspicuously made "geometry manifest," both externally and internally.[64] Geometric patterning was most widely adopted in the architectural monuments of dynastic polities affiliated with the late Abbasid caliphate in Baghdad, especially the Great Seljuks (r. 1037–1194) and their successors in Anatolia, the Jazira (northern Iraq), Syria, and Egypt: namely, the Rum Seljuks (r. 1077–1307), Artuqids (r. 1102–1408), Zangids (r. 1127–1251), Ayyubids (r. 1169–ca. 1250/60), and Mamluks (1250–1517). Other Abbasid vassals further east and west were likewise among the earliest dynasties to embrace geometric ornament, including the Ghaznavids (r. 977–1186) and Ghurids (r. ca. 1000–1215) in Khurasan, Afghanistan, and Northern India, as well as the Almoravids (r. ca. 1030s–1147) in North Africa and Spain.[65]

While this trend coincided with the so-called Sunni revival, the early associations of star-and-polygon patterns cannot be conflated with clear-cut religious orientations, given the multi-sectarian constituents of the polities within which it was adopted. Even the capital of the Abbasid caliphate and its environs abounded in shrines commemorating ʿAlid imams and shaykhs, venerated by Shiʿi and Sunni devotees alike.[66] Eventually shedding whatever its original associations may have been, this mode of ornament was integrated into various traditions of local knowledge as it spread further through specific "actor-networks" combining human and non-human agents: patrons, artisans, architects, portable objects, practical geometry manuals, technical drawings, and technologies in multiple media, including woodwork.[67]

Variants of Abu'l-Wafa''s practical geometry treatise: Its contents and audiences

In this connection, it is noteworthy that one of the alternative titles found in extant manuscripts of Abu'l-Wafa''s *Geometrical Constructions* explicitly refers to woodwork (joinery). The title of the earliest surviving Arabic copy, in Cairo, dated circa 600 (1203), is *Kitāb al-nijāra fī ʿamal al-masṭara wa al-birkār wa al-kūnyā* (The Book of Joinery on Construction with Ruler, Compass, and Set Square).[68] Two undated Persian translations, hypothetically assumed to have been transcribed in the eleventh-twelfth and the twelfth-thirteenth centuries, are both titled "This Is the Book of Master (*ustād*) Abu'l-Wafa' Muhammad b. Muhammad al-Buzjani, On What Artisans and Artificers (*ṣunnāʿ va muḥtarifa*) Need from Geometrical Constructions (*aʿmāl-i handasa*), and They Call This the Book of Joinery (*Kitāb-i Nijārat*)."[69] Artisans are even more explicitly mentioned in the title of a fifteenth-century Arabic copy of this practical geometry treatise. Made for the library of the Timurid ruler Ulugh Beg (d. 1449), renowned as a mathematician-astronomer and founder of an observatory in Samarqand, it is called *Kitāb Fī mā yaḥtāju ilayhi al-ṣāniʿ min aʿmāl al-handasa* (Book on the Geometrical Constructions Necessary for the Artisan).[70]

The longer Arabic preface of this luxury manuscript, which was dedicated to Ulugh Beg around 1447–49, is only partly preserved in the other copies. In it Abu'l-Wafa' explains that he completed this work upon the order of his master, the Buyid amir Baha' al-Dawla, who "requested that I substantiate (*ithbāt*) the notions (*al-maʿānī*) that were discussed in the presence of His Excellency, concerning the geometrical constructions (*al-aʿmāl al-handasiyya*) most commonly employed by artisans (*al-ṣunnāʿ*), leaving out all the demonstrations and proofs (*mujarradan min al-ʿilal wa al-barāhīn*), in order to make these easy for artisans (*al-ṣunnāʿ*) to use and to familiarize them with their method (*ṭarīqatahu*)."[71] The discussion may well refer to one of the gatherings (*majālis*) attended by the author himself, where artisans/surveyors and geometers participated. In my view, it is possible that this gathering took place within the precincts of the Buyid palace (*Dār al-Mamlaka*) in Baghdad, in the presence of Abu'l-Wafa''s royal patron, to whom the treatise was dedicated. Such an interpretation is indirectly supported by the tenth chapter, where the author describes an assembly at which he demonstrated the errors of artisans with proofs that his treatise substantiates. Whatever the location of that gathering, the preface testifies to the ruler's role in chanelling scientific knowledge to practical applications, as he commissioned the *Geometrical Constructions* not merely for his own edification, but primarily for the use of practitioners.[72]

Although the alternative titles of this treatise point to its uses in joinery, its contents address a much wider array of topics, such as land surveying, the equal division among inheritors of estates through which a road passes, construction, brick-cutting, and burning mirrors refracting light. Perhaps the meaning of "carpentry" (*al-nijāra*) in those titles encompassed a broader semantic range, as captured in a passage by the Maghribi polymath Ibn Khaldun (d. 1406), whose use of the term evoked a tradition that assigned the epithet of "carpenter" (*al-najjār*) to famous geometers of the Alexandrian school, a tradition that possibly reflected then-current Islamic artisanal practices:

> In view of its origin, carpentry needs a good deal of geometry of all kinds. It requires either a general or specialized knowledge of proportion and measurement, in order to bring the forms [of things] from potentiality into actuality in the proper manner, and for the knowledge of proportions one must have recourse to the geometrician. Therefore the leading Greek geometricians were all master carpenters. Euclid, the author of the *Book of the Principles* on geometry, was a carpenter and was known as such. The same was the case with Apollonius, the author of the book on *Conic Sections*, and Menelaus, and others.[73]

There is not much information about the reception of Abu'l-Wafa''s practical geometry treatise, but its five surviving Arabic copies and three versions in Persian, seemingly resulting from at least two different translations, attest to its circulation.[74] The humbleness of some of these manuscripts suggests that they had an audience beyond elite circles. The eight known copies are complemented by two Arabic commentaries and one in Persian. The first Arabic commentary was by Kamal al-Din Ibn Yunus Mawsili (ca. 1156–1242), an internationally renowned Iraqi jurist and professor of mathematics based in Mosul, a city that had considerable scholarly exchanges with Baghdad, where he himself studied. Titled *Sharḥ al-Aʿmāl al-handasiyya li-Abī al-Wafāʾ* (Commentary on the *Geometrical Constructions* by Abu'l-Wafa'), it is dedicated to the founder of the Ayyubid dynasty in Syria-Egypt, Salah al-Din Yusuf b. Ayyub (Saladin, r. 1174–93). The dedicatee's family had served the Zangid rulers of Mosul, who in turn were governors (*atabeg*) of the Great Seljuks.[75] The commentator, Ibn Yunus, states that he had encountered no other work like Abu'l-Wafa''s *Geometrical Constructions*, thus affirming its rarity. The fifteenth-century deluxe copy belonging to Ulugh Beg incorporates another Arabic commentary in its margins, which provides the missing proofs for its geometrical constructions, as does that of Ibn Yunus, and both learned commentators frequently cite Euclid's *Elements*. The much later Persian commentary by Muhammad Baqir Yazdi, dated 1031 (1621), also adds proofs and contextualizes problems within the tradition of Euclid's *Elements*, in response to the request of an unnamed companion.[76]

Ibn Yunus's commentary differs in three respects from Abu'l-Wafa''s practice-oriented treatise, whose intended purpose is thereby distorted. First, he provides demonstrations by proofs unlike Abu'l-Wafa', who deliberately omits them, heeding his Buyid patron's wish to make the treatise useful for practitioners. Secondly, Ibn Yunus employs conic sections to solve problems such as trisecting an angle or constructing a regular heptagon, and criticizes his predecessor for not having given exact solutions based on conics. Thirdly, Ibn Yunus refers to the applied uses of geometry in woodwork, with which he was apparently familiar, but entirely ignores the practical needs of artisans.[77]

Abu'l-Wafa''s *Geometrical Constructions* begins by introducing three basic drafting tools—the ruler, compass, and set square—along with some preliminary procedures of constructive geometry. It then moves on to the construction of regular polygons, often with a fixed compass opening to reduce errors, showing polygons inscribed in and circumscribed by a circle. These operations are followed by more difficult ones, such as inscribing polygons in one another, dividing triangles and quadrangles, assembling a square from smaller squares, and dividing a square into smaller squares. The work continued with two chapters (now lost), one on dividing irregular quadrangles, the other on tangent circles, ending with a discussion of the division of spheres by polygons (polyhedrons). The treatise displays the author's familiarity with theoretical geometry and practical applications, as well as his ingenuity in teaching construction techniques in a simplified manner. The pedagogical intention behind it is clearly declared, though without specifying or even limiting its intended audiences: "We shall present these methods so that the correct ones may be distinguished from the false ones, and someone who looks into this subject will not make a mistake by accepting a false method, God willing."[78]

It may, of course, be argued that Abu'l-Wafa''s (and his royal patron's) well-intentioned objective of accommodating the needs of artisans and surveyors, whose errors he severely criticized, does not necessarily mean that practitioners actually made use of his *Geometrical Constructions*.[79] This argument was advanced by George Saliba in a critical review of my book on the *Topkapı Scroll*:

> [O]ne cannot fail to note that the overall concern of Buzjani was to demonstrate the error of artisans, and he makes every attempt to set them right. In that sense such books could be read as instructions for artisans. But that claim has to be documented by evidence that artisans did indeed consult those books and that they followed the instructions in them. Similarities of drawings or discussions of subject matter that has a similarity to the drawings found in anonymous manuscripts and scrolls is not enough to "contextualize" the relationship between the theorists and artisans. And no quantity of pictures from manuscripts in juxtaposition to pictures from scrolls or the like will make that point any clearer.[80]

The transmission of knowledge rarely leaves a paper trail. It is therefore difficult to document and contextualize the relationship between theory and practice beyond concrete correspondences in design methods. Arabic and Persian written sources generally provide scarce information on artisanal-architectural working methods, which were mostly transmitted orally and visually, by means of geometric drawings on paper or templates in other materials. Categorically dismissing the visual evidence, as Saliba does, therefore misses the point. Technical drawings can sometimes shed more light than textual commentaries on exchanges between theoretical geometry and the ingenious working methods of practitioners. As I have shown in my book, the similarities between recurring types of practical design methodologies that are visually recorded in Abu'l-Wafa''s *Geometrical Constructions*, in the *Anonymous Compendium*, and in subsequently compiled textless scrolls document a genealogical trajectory. These successive primary sources constitute a rare paper trail, exposing the prevalence of a tradition of constructive geometry that certainly differs from "academic geometry."[81] This is the geometry of practitioners, in which diagrams are just as informative as, if not more so than, texts. The power of diagrammatic drawings to visually convey information is its crux, with or without the crutch of accompanying texts.

While we do not know as much as we would like about the dynamics of the relationship between theorists and practitioners, Abu'l-Wafa''s *Geometrical Constructions* does mention "gatherings" (*majālis*) he attended, in one of which the two groups interacted and had to reckon with one another's different methods. Such interactions, also insinuated in the brief texts of the *Anonymous Compendium,* surely left their imprint on the thoughts and actions of both groups, neither of which could have remained completely insusceptible to "influence." However, some scholars prefer to dismiss the enterprise of Abu'l-Wafa''s treatise as a mere "academic" exercise.[82] Why then, one might ask, did he go to the trouble of writing a whole manual on practical geometry methods, instead of a presumably more prestigious theoretical geometry treatise accompanied by learned proofs, which would have garnered him more accolades? The treatise clearly responds to objections made by practitioners to purely theoretical solutions and attempts to convey to at least clever and receptive artisans/surveyors the difference between exact and approximate solutions in a simplified manner. Abu'l-Wafa' accomplishes this by using the demonstrations of visually graspable diagrams as a form of proof and a cognitive resource.

The author took it for granted that artisans and surveyors derived the geometrical notions on which they based their work from geometers, without being concerned about proofs justifying correctness; they consequently sometimes committed errors.[83] At the gathering he described, geometers are shown responding to problems posed by artisans, who already possessed well-established constructive geometry methods of their own, with which he was apparently quite familiar.[84] This implies a reciprocal relationship, in which the agency and preferred design methods of practitioners did not necessarily play a subordinate role to that of the geometers from whom they sought advice. It was the ultimate kinship and convergence of partly shared methodologies of geometrical construction that brought artisans/surveyors and geometers together in the first place, thus initiating a relationship between art and science. Their mutual interactions, in turn, went beyond a one-way transmission of knowledge. Abu'l-Wafa' *al-muhandis* writes about several artisanal construction methods that bear witness to his talent in bridging theory with practice. He theorizes with visual proofs the correctness or inaccuracy of practical solutions discovered by artisans, with whom he communicates by means of their own intuitional constructive logic. Through this translation effort, his treatise synthesizes theoretical and practical geometry in a way that incorporates traces of artisanal invention, thereby opening a space in between for future mediations and translations.

For instance, the *Geometrical Constructions* includes descriptions of tools used in surveying and building crafts, such as the plummet (*shāqūl*) and set square (Arabic *kūnyā*, Persian *gūnyā*), which were needed for correcting the corners of buildings. The author's discussion of how to verify the straightness of a ruler and the precision of a set square provides both geometrically correct and approximate artisanal methods.[85] Abu'l-Wafa''s chapter on regular polygons inscribed within a circle specifically alludes to skilled and clever artisans who use multiples of thirty for approximating the sides of polygons. He explains that artisans prefer dividing the circumference of a circle with a fixed compass opening to approximate by trial and error, with much difficulty, the arcs of as many line segments as desirable, unlike geometers, who favor constructions based on the central angles of polygons inscribed in a circle. Both methods, based on lines and angles respectively, are employed in the geometrical constructions of the *Anonymous Compendium*. Given the purpose of his treatise, as commanded by his royal patron, Abu'l-Wafa' expresses an obligation to provide correct methods for the "artisanal technique" (*ṭarīq ṣināʿī*) of dividing circles into arcs, the ideal being an accurate as well as a practical solution. It has been noted that the treatise includes non-Euclidean methods for the construction of a regular pentagon, the trisection of an angle and an arc, and close approximations for drawing a regular heptagon and a nonagon in a circle.[86]

For problems dealt with in subsequent chapters that artisans and surveyors found particularly difficult, Abu'l-Wafa' tends to give several alternative construction methods. These involve inscribing polygons within one another, dividing triangles, and dividing quadrangles so that their parts have a given proportional relationship to one another. In the tenth chapter, on the "division of quadrangles (i.e., squares) and their assemblage" (*qismat al-murabbaʿāt wa ta'līfahā*), he addresses especially demanding geometrical constructions that required some knowledge of even and odd numbers, as well as square roots (fig. 7). Abu'l-Wafa' states that he has explained those constructions in a sufficiently suitable way for someone who possesses a little intelligence and subtlety in the field of geometry. Most challenging was the division of a square into a certain number of squares, and constructing a square equal to a given number of squares, the number of which is the sum of two unequal "square numbers" (square roots). One such problem involves constructing a square brick from ten bricks (*khisht*), the solution for which is based on the number ten comprising two square numbers, nine and one.[87]

Studies have shown that in this chapter Abu'l-Wafa' was innovative in his application of number theory to geometry in order to establish basic "scientific principles" (*qānūn ʿilmī*) intended to make their application easier than cumbersome trial-and-error methods.[88] He explains that there are two kinds of numbers: square numbers (i.e., 4 or 36), for which square roots exist (here, 2 and 6, respectively), and non-square numbers. The latter consist of two categories: first, numbers that are the sum of two square numbers (such as 13, combining 9 and 4; or 41, combining 16 and 25); and second, prime numbers not constituted by two square numbers (such as 7 or 11).[89] The latter are more difficult cases, for which Abu'l-Wafa' promises to provide very simple construction methods.

Correspondences between the *Geometrical Constructions* and Abu Bakr's constructions in the Paris Codex suggest that he may have been familiar with his predecessor's treatise (figs. 4, 6, and 7). Whether this was the case or not, the constructions of both authors inventively fuse geometry and numbers in surprisingly uncomplicated ways. We have, for instance, encountered square numbers in the surveying treatise comprising Abu Bakr's constructions as well, where the division of quadrangles and their relationship with "similar" (*shabīh*, i.e., *mutashābiha*) right-angled triangles plays an important role, along with complementary or "harmonious numbers" (*aʿdād-i wafq*; Arabic sing.*ʿadad al-mutawāfiqa*) (fig. 3).[90] Remarkably, the Paris Codex includes the extract of a treatise in Persian, without title or author, on pairs of numbers that are composite or prime with respect to each other, or in which one number evenly divides into the other (*tadākhul*).[91] The terms *tadākhul*, *mutashābiha*, and *mutawāfiqa* strikingly resonate with the Arabic heading of the *Anonymous Compendium*: *Fī tadākhul al-ashkāl al-mutashābiha aw al-mutawāfiqa* (On Similar and Complementary Interlocking Figures). As implied by its heading, this document illustrates the division and assembling of figures, inscribed in and/or interlocking with one another, whose harmonious proportions are based on geometrically expressed numerical ratios.

THE CONSTRUCTIONS AND COMMENTARIES OF THE *ANONYMOUS COMPENDIUM*

Overview of contents and preliminary construction problems

Lacking theoretical proofs, the geometrical constructions of the *Anonymous Compendium* are accompanied instead by brief textual commentaries that explain how to draw a figure or composite figures using drafting tools and basic procedures with which prior familiarity is assumed. The only exceptions are advanced professional devices, such as set squares and the T-square, the construction methods for which are verbally described and visually illustrated. Special attention is given to constructing difficult polygons (the pentagon, decagon, heptagon, and nonagon) and their stars, approximations of conic sections, and ratios/proportions. As Chorbachi has perceptively observed in her step-by-step reconstruction of the drawing procedures of ten selected patterns from the *Anonymous Compendium*, their properties were guided by particular ratios/proportions:

> Nearly each of these examples of geometric design exemplifies a different and specific construction procedure to mark the unique or critical points required to establish the geometric subdivisions defining the geometric figures that interlock to tessellate space…All the major and minor proportions are then derived by using the module of measurement set or specified at the starting point of the design process and which ensures that the proportions relate geometrically to one another. This leads to the strong proportional characteristics of the small geometric interlocking shapes that in turn make Islamic geometric design a modular art.[92]

In fact, the textual commentaries emphasize ratios and proportionality more than the related concepts of symmetry and tessellation, which have preoccupied most modern studies on Islamic geometric ornament. Unlike theoretical geometry treatises, the *Anonymous Compendium* cites only one theorem without proof, focusing instead on operations to be performed in order to reach a desired goal. The drawing techniques are based on the logic of iteration, a sequence of steps generated by a given common ratio/proportion. This involves a dynamic process in which each figure emerges from an unfolding of forms based on geometrical procedures that knit them together. Differing from standard geometry manuals, the *Anonymous Compendium* visually articulates the beginning of the construction process by means of red or dotted construction lines that are densely concentrated at one section of a repeat-unit (usually a corner or side, but a few times in the middle). In some patterns the important points are boldly marked, an unusual convention also seen in several drawings of the two treatises in the Paris Codex associated with Abu Bakr and in the Persian translation of Abu'l Wafa''s *Geometrical Constructions* (fig. 2a).

The first page of the *Anonymous Compendium*, which lacks a preamble, features a "dividing and assembling" construction that explores the proportional relationship between two convex decagons and a pentagonal star (a non-convex decagon) (fig. 1). The textual commentary on it recalls that of a previously unnoted problem in the treatise "On the Division of Triangles," concerning the proportional relationship (rapport) of an equilateral triangle and octagonal star with equal areas (fig. 2a):

> When a right-angled octagon[al star] (*mus̱amman-i qā'imu'z-zavāyā'*) is constructed and another equilateral triangle is constructed, such that the altitude of that triangle equals the outer diameter (*quṭr-i khārij*) of the assumed octagon[al star], the area of that triangle equals the area of the octagon[al star], as in this example (*mis̱āl*).

The text of that treatise goes on to explain the measurement (*misāḥa*) of these two figures and the relative proportions of their diagonals. Its reference to the "outer" and "inner" diameters of the star (i.e., of the concentric circles defining the outer and inner vertices of the star) echoes the commentary that accompanies the *Anonymous Compendium*'s first construction. Here too the star is named after the polygon corresponding to it and thus referred to as a "pentagonal star" (*khātim-i mukhammas*, lit. pentagonal seal). However, the primary concern of the latter construction is to demonstrate visually, by means of a graphic representation in two steps, the way in which similar and complementary figures with equal areas are assembled so as to interlock with one another (fig. 1).

The top of that page in the *Anonymous Compendium* depicts two identical regular decagons and a pentagonal star, all sharing equal sides and angles, thereby ensuring the complementarity of both their linear dimensions and areas. The objective here is to divide the two congruent decagons so as to assemble their parts in a larger decagon that contains the pentagonal star at its center. First, each of the identical decagons is divided by an inverted Y-shape (composed of three radii) into three pieces consisting of a kite and two double kites. The second step is to assemble the large decagon at the bottom of the page by fitting together pieces of the two decagons along the edges and vertices of the central pentagonal star (pls. 1.1–5). This construction involves figures based on angles of the pentagon and decagon, whose similarities to Penrose tilings (discovered in the 1970s) were noted by Chorbachi and Loeb. However, as they observed, the textual commentary "does not explicitly deal with the matter of non-periodicity." While the relative proportions of the radii and sides of the polygons are provided, "the step by step construction instructions are not given; neither is the theoretical significance of the given proportions."[93]

The construction procedure and unarticulated ornamental potential of these interlocking figures were presumably familiar to the author(s) as part of an oral and visual tradition of constructive geometry (pls. 1.6–7). The concise technical language of the *Anonymous Compendium* only occasionally hints at the aesthetic criteria that must surely have played a part in the creation and cognition of ornamental patterns. A notable example is the second construction, which was judged to be "extremely elegant" (*bi-ghāyat laṭīf ast*) (fol. 180v [2]). This too is an exercise in "dividing and assembling," here involving a large square inscribed with two parallel squares and a fourth, rotated square with extended sides at the center that contains yet another rotated square. The geometrical construction with nested squares is based on an operation of "quadrature," or rotated squares, whose dynamic geometry played a conspicuous role in Gothic architectural practice. Studies have shown that the rotated square and equilateral triangle, as well as their derivatives such as the octagon, governed the proportions of many extant architectural drawings of Gothic towers and façades, "even when the structure in question has no obvious triangular, square, or octagonal form."[94]

The second construction of the *Anonymous Compendium* is divided by triangulation methods into similar and complementary figures, mostly triangles, kites, rhombuses, and trapezia. From the pieces of the large outer square "one tetragonal [square] star, one octagonal star, one nonagonal star, and one decagonal star can be obtained." Once again, the stars are named after their polygons. Alternatively, "if one wants," two squares, two octagons, two nonagons, and two decagons can also be composed (pls. 2.1–7). The geometric exercise demonstrates the proportional interrelatedness of the polygons, the stars associated with them, and the fragments obtained by subdividing the polygons and star polygons; this is what makes these figures fit together in an interlocking pattern. It also introduces the "four-legged" rotated square, surrounded by rotating kites, at its very center. As we have already seen, that scheme plays a significant role in the subsequent constructions of the *Anonymous Compendium*—both as a dynamic proportioning device with four-fold rotational symmetry, and as an ornamental motif (see figs. 12 and 17). Those constructions include the one whose proportions were examined by Abu Bakr. Similarly composed of parallel squares containing a central, rotated, "four-legged" square, that design had decorative potential (fig. 6). Hence, the "dividing and assembling" constructions of the *Anonymous Compendium* cannot be easily dismissed as unrelated to Islamic ornamental patterns. This group of constructions is related to procedures such as dividing figures in proportion and assembling interlocking shapes, which govern the geometrical principles of decorative star-and-polygon patterns.[95]

Several "dividing and assembling" constructions concentrated at the very beginning of this source explore the rapport of polygons with equivalent areas, such as the equilateral triangle and hexagonal star, as well as the square and rectangle. These constructions are mixed with others, demonstrating the relationship between the circle and its arcs, and the division of triangles and a hexagon by triangulation methods (fols. 181r–183r). Textless versions of "dividing and assembling" problems appear later, on two pages. The first pair, depicting squares and octagons of equal area, was seemingly inserted as an afterthought in the spaces left empty on each side of the main construction involving a pentagon (fol. 186(bis)r). The second set, on the penultimate page of the document, was possibly added by the fifteenth-century copyist (fol. 197r).

The next section, bearing the heading "preliminary" (*muqaddima*), begins on fol. 183v [12] with the only theorem cited in the *Anonymous Compendium*: "For every triangle inscribed in a circle, the ratio (*nisbat*) of the angles of that triangle to each other is equal to the ratio of [the opposite] arcs of the chords of the angles." This section examines the proportional relationship between the angles of triangles inscribed in a circle to the arcs of chords subtended by those angles (fols. 183v–190r). Some of the problems focus on constructing right-angled triangles and kites with special ratios/proportions, and on construction methods for the pentagon and pentagonal star.

Problems examining the construction of ornamental patterns

Constituting more than half of the sixty-one constructions in the *Anonymous Compendium*, the problems considered thus far culminate in ornamental patterns that emerge only when their abbreviated repeat-units are replicated by symmetry operations (e.g., mirror-reflection, glide-reflection, translation, or rotation) (fols. 190v–196v). The instructions that accompany them address the problem of how to assemble figures (in modern parlance, tessellations or tiles) so that they interlock in a single pattern. The term *ʿaqd* (lit. knot), translatable as pattern or repeat-unit, more often appears in the instructions for ornamental designs, or preliminary procedures related to their construction. These mostly square and rectangular repeat-units comprise diverse polygons, which, together with their stars and fragments, fit together without any gaps and overlaps, like the pieces of a jigsaw puzzle. While in most cases the length-to-width ratio of a repeat-unit is given, in some constructions only one side is assumed and the other sides of the "compartment of the repeat-unit" (*khāna*, or *khāna-i ʿaqd*) are geometrically determined in the construction process (fols. 190r [34], and 195v [54]).

Toward the end of the *Anonymous Compendium* there are fewer commentaries. Some drawings have brief texts that only provide the required ratios/proportions, while others have none, presumably because the relevant methods have already been covered. One such example is the square repeat-unit of an orthogonal ornamental pattern with four-fold symmetry whose textual commentary does not fully explicate the final pattern generated by it, probably because that procedure was obvious (see fig. 15a [top]). The repeat-unit is composed of a central cross-axis with eight arms and four squares, one at each corner, that are further subdivided, their outer corners coinciding with the centers of four quarter-octagonal stars. According to the accompanying text, when the required proportion is achieved, all the horizontal, vertical, and diagonal arms of the latticework have the same width. The final pattern is obscured due to the addition of red construction lines, which visually demonstrate the underlying proportions, but must be erased when the repeat-unit is replicated. The fully drawn out pattern is composed of two types of interlaced octagons, regular and elongated, combined with regular octagonal stars. Judging by the textual commentary, which refers to the "most correct proportions" (*aṣaḥḥ-i nisbathā*) of this repeat-unit, Chorbachi proposed that a previously established pattern had probably "caused problems, raising the issue of the correctness of the proportions." Indeed, early examples of this design appear in mid-eleventh- to early twelfth-century brick monuments in Khurasan, and subsequently in the stone mihrabs and portals of late twelfth- to thirteenth-century monuments in Anatolia.[96]

This also seems to have been the rationale for some other repeat-units, such as two variants combining decagonal and dodecagonal stars. One of these is an incomplete pattern, with emendations to an unsatisfactory composition indicated in dotted lines (see fig. 14; pls. 54.1–2). As Özdural has shown, the other pattern provides a perfect solution for that problem by substituting the imperfect central pentagonal stars with arrow-shaped motifs (see figs. 15a and b; pls. 56.1–2). He hypothesizes that the last three pages of the *Anonymous Compendium* were added by the fifteenth-century copyist, presumably Kubanani Yazdi. The final construction is a full-page repeat-unit for an ornamental pattern that combines rotating swastikas with pentagonal and decagonal stars. It features five- and ten-fold rotational symmetries, which were explored on the first page of the *Anonymous Compendium* (see fig. 21a, pl. 61.2). The swastika pattern strongly resembles a fifteenth-century mosaic tile panel at the Friday Mosque in Yazd and may have been added in the Timurid-Turkmen period, as has been proposed in several studies. Yet versions of the same pattern, combining swastikas with octagonal stars, appear in Seljuk Anatolia at the Sırçalı Medrese (1243) in Konya, and at the Gök Medrese (1271) in Sivas, just around the time of the Mongol-Ilkhanid invasion. Another example, with octagonal stars, was used much earlier in Seljuk Iran, at one of the twin tomb towers in Kharraqan (1067–68).[97]

The last construction in the *Anonymous Compendium*, then, is not necessarily out of character with the other ornamental patterns included in it. Despite their somewhat haphazard sequence, the contents of this source obey a relatively coherent organizational logic. The application of constructive geometry methods to the repeat-units of ornamental patterns, primarily in the second half of the *Anonymous Compendium*, is often directed at solving difficult problems rather than at creating facile versions of a by-then common design repertoire. As such, this primary source comes closer to a technical manual than to a catalogue of decorative compositions, as in a design scroll or pattern book. Ornamental geometry is presented in it as a technology to be mastered, an art of construction with its own mathematical rules and procedures. Geometry thus functions here as a constructional design tool with which to draw and transform figures, while at the same time maintaining their inbuilt proportionality.

Repeat-units, ratios/proportions, and drafting tools

As we have seen, many problems in the *Anonymous Compendium* concern preliminary constructions and proportional relationships, which elucidate underlying principles and processes rather than the finished product. One of the texts, for instance, offers the option of inserting an additional tetragonal star (*khātim-i chahār*) into an already described pattern (fol. 190v [35]). Combining a heptagonal star with hexagonal and octagonal stars, its repeat-unit does not yield an attractive ornamental composition because it was not meant to be a fully worked out design, but rather an exercise exploring methods for connecting stars with different geometries (pls. 35.1–3).[98] Similar to the pentagonal star with which the *Anonymous Compendium* begins, its tetragonal counterpart and other stars constitute knot-centers or nodal-points (*khātim*, lit. seal) in patterns with multiple rotational symmetries. The stars bind together composite figures that tessellate the plane. This might partly explain the search for close approximations of exact solutions, which could allow greater flexibility in fusing different types of polygons and star polygons.

The texts accompanying star-and-polygon compositions often explain the "ratio/proportion of this repeat-unit" (*nisbat-i īn ʿaqd*), or "the procedure for drawing this repeat-unit" (*ṭarīq-i kashīdan-i īn ʿaqd*).[99] Many problems concerning the construction of ornamental patterns are thus modeled as repeat-units, made up of vertices and nodes, and the lines (edges) that connect them. In some texts, "fitting" composite figures into the quadrangular frames of repeat-units emerges as a major concern. For example, "If line KT perpendicular [to HV] is made equal to side ZH, then it will fit [the repeat-unit]" (fol. 195r [49]); or "If magnitude M′C′ is made equal to half of GK′, then a reliable (*maʾmūn*) repeat-unit comes out. If it is not a *maʾmūn* repeat-unit, magnitude M′C′ is less than half of GK′" (fol. 193v [44]). In another example, the addressee is instructed to check whether "point E" falls on the midpoint of the assumed square frame of the repeat-unit (see fig. 17): "If it is on the midpoint of the square, then this is the required proportion, but if it is not on the midpoint of the square," then additional operations need to be performed "in order to find the center of the heptagon, which is required, on the side of the square" (fol. 194v [48]). In yet another case, the addressee is urged to check whether a certain dimension is "greater or lesser" than the required one, and if so, to modify it by "addition" or "subtraction" (fol. 184r [13]). Elsewhere it is stated that if a particular intersection "does not happen, the construction is erroneous" (fol. 186bis(v) [23]).

I pointed out in my book on the *Topkapı Scroll* that three constructions in the *Anonymous Compendium* "composed of squares and rhombuses do recall orthogonal *muqarnas* vault projections, such as the one represented on the thirteenth-century Takht-i Sulayman tablet," discovered at the site of an Ilkhanid palace (ca. 1270). I interpreted those textless constructions as proto-*muqarnas* patterns that testify to the intimate link between tessellations in planar and spatial geometry (see fig. 23 [middle and bottom]):

> These two-dimensional patterns suggest that the *muqarnas* must have evolved from a similar kind of geometry, involving the proportional division of space into closely packed compartments without leaving any gaps.... The regular division of space into similar and congruent three-dimensional cells, or space tilings, was a natural extension of experiments with plane tilings. Just as the simplest orthogonal *muqarnas* compositions exhibit an affinity to two-dimensional surface patterns generated by rectilinear grids ..., the far more complex radial *muqarnas* projections of the Topkapı and Tashkent scrolls can be read as spatial interpretations of two-dimensional patterns with interlocking stars and polygons. Extending the idea of filling a plane with congruent stars and polygons into three dimensions must have been a natural outcome of exercises in practical geometry and mensuration manuals, which proceed from constructions in the plane to constructions in space....This affinity explains why *muqarnas* projections (three-dimensional space-filling patterns) are always juxtaposed with geometric interlaces (two-dimensional surface-filling patterns) in Islamic scrolls that brought experiments initiated by early medieval designers to their logical conclusion with congruent tilings, both planar and spatial.[100] (See fig. 24a–b.)

Özdural went one step further, reasonably proposing that those three patterns in the *Anonymous Compendium* are, in fact, designs for *muqarnas* vaults (fol. 184v [15], 194r [46, 47]). Based on their similarity to the plan of the Ilkhanid *muqarnas* quarter vault incised on a plaster slab discovered at the Takht-i Sulayman Palace, mentioned in the quotation above, and to extant examples in monuments from the same period, he interpreted the numerals (1 to 3, and 1 to 4) written on these geometrical constructions as a numbering system indicating the corbelling order of *muqarnas* tiers (pls. 15, 46, and 47). Özdural identified these patterns on paper as examples of an archaic, non-radial type of early *muqarnas* and concluded that they are the oldest surviving examples (ca. 1300) after the Takht-i Sulayman plate (ca. 1270). To those drawings one can add an even earlier comparable *muqarnas* vault pattern that was incised on a stone slab on the façade of an Armenian monastery twenty to twenty-five years before the Takht-i Sulayman example.[101]

The inclusion of *muqarnas* projections in the *Anonymous Compendium* brings its contents closer to the orbit of architectural practice. Although the term *ʿaqd* is not consistently used throughout this source, it does appear in the text that accompanies the previously mentioned design decorating a tympanum arch in the North Dome in Isfahan: we learn from it the "procedure for constructing the ratio/proportion of this repeat-unit" (*ṭarīq-i ʿamal-i nisbat-i īn ʿaqd*) (fig. 10a). Interestingly, the text of that construction starts on the previous page, which depicts a T-square (*gūnyā misṭar*) used for rare proportions and angles that could not be drawn with a ruler and compass (fols. 191v–192r [36 and 37]). I therefore find it plausible to propose that the heptagonal North Dome pattern, based on "three-sevenths of a right angle (angle BAG)," was drawn in this repeat-unit by means of a T-square (rather than a set square, as postulated by Özdural: see pl. 37.1). This is all the more likely because the repeat-unit lacks incised construction lines and the text omits any mention of a "line [number] seven," corresponding to the angle measurement of a heptagon, which is referred to in the commentary of another heptagonal pattern discussed below (fol. 194v [48], see fig. 17).

Hogendijk's mathematical analysis of the North Dome pattern shows that each of its two types of six-pointed stellate forms is framed by a hexagon drawn in red ink. Dotted lines in black ink visually articulate the steps to be taken in the drawing process, which starts at the upper right corner of the repeat-unit. Replicating its mirror reflection yields an underlying generative grid of contiguous hexagons outlined in red ink (fig. 10a, pls. 37.1–3). The repeat-unit itself is bisected into two congruent trapezia by an oblique line in red ink, constituting an axis of dynamic symmetry. Each trapezium, in turn, is divided into three quadrangles, which define the sides of the generative hexagonal grid. Lines joining the midpoints of the hexagons created the final pattern seen at the North Dome, from which the construction lines were omitted. The textual commentary focuses on alternative methods for finding the midpoints of the two regular half-heptagons that are lined up along the upper and lower edges of the repeat-unit and coincide with the endpoints of the oblique symmetry axis drawn in red ink. According to Hogendijk, since the brief written instructions accompanying the pattern do not sufficiently explicate its design and application method, the *Anonymous Compendium* must have been part of an oral tradition of instruction.[102] While this is certainly likely, I would argue that experienced artisan-designers and master builder-architects would have visually comprehended the logic of incompletely annotated patterns such as this one and of textless patterns as well. That is why the first part of the commentary, before moving on to alternative procedures, ends with the expression "[the rest of] the construction would be easy," as do the instructions for another pattern (fol. 182r [8]). The texts accompanying some repeat-units, on the other hand, simply say "complete the construction," without further explanation.

With the exception of square patterns in which rotated "four-legged" squares drawn in ink structure the composition, the rectangular repeat-unit of the heptagonal North Dome pattern is the only example of its kind in the *Anonymous Compendium* to include a hexagonal generative grid outlined in ink. Another interesting exception is also a heptagonal pattern, whose square repeat-unit features a rotated "four-legged" square with four rotating irregular kites drawn in a dotted black ink line. The latter structures the final decorative pattern composed of four composite polygons (double pentagons) rotating around a central swastika motif (see fig. 17). In this case, the centers of the four heptagonal half-stars lined up along each edge of the square repeat-unit coincide with the dotted lines of the "four-legged" square, and their angles are referred to as "lines [number] seven." In the same repeat-unit, the heptagonal stars are framed by half-hexagons drawn in red ink, thereby combining the dotted "four-legged" square scheme with a hexagonal generative grid. Other construction lines in red ink, which pass through the vertices of four interlocking double-pentagons at the center, define a rotated square subdivided into four squares. Because one vertex (point) of each of the four heptagonal stars protrudes beyond the square frame, those vertices overlap in the final composition created by replicating the repeat-unit (pl. 48.3). However, stars such as these with overlapping points were not uncommon, as can be seen in the repeat-unit of another decorative pattern in the *Topkapı Scroll* (see fig. 19a).

To summarize, ornamental repeat-units in the *Anonymous Compendium* have only uninked incised construction lines, with the exception of these two heptagonal patterns, which feature inked polygonal grids. As we have seen, this source sometimes offers several possible solutions to a given problem. One such example refers to two kinds of procedures for establishing the desired proportion of a pattern, "rectilinear" (*khaṭṭī*) and "circular" (*dawrī*), based on the triangle and circle, respectively (fol. 188r [28]). Both proportioning strategies were described in Abu'l-Wafa''s *Geometrical Constructions*, which recommended that artisans draw polygons by the simple method of inscribing them in a circle subdivided into equal arcs corresponding to their central angles. For the second method, involving the proportional division of straight lines, he offered techniques similar to those elucidated in Euclid's *On Divisions of Figures*, as well as in *On Cutting Lines in Ratio* and *Cutting Surfaces in Ratio*, both by Apollonius of Perga. These works had been translated into Arabic and were already available in tenth-century Baghdad. It is interesting to note that the Paris Codex includes an Arabic excerpt of theorems without proofs from Book II of Euclid's *Elements*; these concern the division of a straight line into proportional segments, whose relationships will hold between rectangles and squares formed from that line and its segments.[103]

The circular method, relying on arcs corresponding to angle measurements, is based on the theorem quoted above in the *Anonymous Compendium*, which is accompanied by a diagram depicting a triangle whose angle ratios (1:2:4) are related to the heptagon (fol. 183v [12]). The theorem is followed on the next page by a related construction, illustrating a triangle whose angle ratios (1:2:6) add up to nine and relate to the nonagon (fol. 184r [13]). The simple procedure described involves dividing the circumference of the circle into nine parts (arcs), then obtaining the desired triangle by joining "the chords of the arcs of one part, 2 parts, and 6 parts." This procedure is complemented by a more complicated one that is required if one wants to construct the same triangle on a given line; although unstated, this might be required for fitting it between two established points in a repeat-unit. An application of the same theorem to the dodecagon is also provided—that is, a triangle (3:4:5) whose angles add up to twelve. It is obtained by dividing the circumference of a circle into twelve parts and joining "the chords of 3, 4, and 5 parts."

The same page is accompanied by a tabulated numerical table, which gives all triangles the sum of whose angles are in nine parts (fol. 184r [13]). The table starts with the triangle (1:2:6) used in the construction itself, and is accompanied by six others for constructing nonagons.[104] It recalls numerical tables for the measurements of triangles and other polygons that are included in the surveying treatise titled "Constructions by Abu Bakr." However, numbers corresponding to angle measurements are rarely marked on the geometrical constructions of the *Anonymous Compendium*, the only exceptions being the present construction (fol. 183r [13]) accompanying the tabulated table, and two others, whose commentaries refer to specific numerical ratios related to angles (fol. 184v [14], 190r [34]).

One of those commentaries explains how to draw special right-angled triangles within a semicircle using a fixed compass opening (fol. 184v [14], pl. 14.1). The purpose of this construction has been interpreted as providing templates for drafting tools, which are referred to as "set square [number] 5" (*gūnyā-5*) and "set square [number] 6" (*gūnyā-6*). Those set squares approximated angle measurements in terms of modules equaling six degrees. "Set square [number] 5," corresponding to 36°-54°-90° (degrees in ratio 6:9:15), facilitated the construction of regular pentagons and decagons; "set square [number] 6," which corresponded to 30°-60°-90° (degrees in ratio 5:10:15), simplified the procedure for drawing regular hexagons and dodecagons. The text explains that a third triangle in the same diagram becomes "[number] 15," that is, a set square of 24°-66°-90° (degrees in ratio 4:11:15), which could be used for drawing regular fifteen-sided polygons. While this geometrical construction may have provided templates for producing set squares, it also elucidates the interrelated angle ratios of different types of polygons by placing the equivalent hypotenuses of two intersecting right-angled triangles along a semicircle's diameter. That relational aspect becomes apparent in the text, according to which the diagram's objective is to construct "*gūnyā-5*" by means of the compass opening of the radius of "*gūnyā-6*." As Hogendijk observes, the same construction provides an unstated approximation of the side of a regular pentagon inscribed in the semicircle.[105] It is not a coincidence, then, that this diagram is followed by a series of related constructions, beginning with the properties of special triangles (fol. 185r–v] and continuing with methods for drawing regular pentagons (fols. 186r–187r).

Although some commentaries considered above refer to proportions derived from conic sections, in actuality, the intention is to closely approximate those proportions by using special drafting tools. To give an example, the text mentioning a treatise by "Ibn-i Haytham" on a special right-angled triangle, which he drew by means of two intersecting conic sections, explicates the proportion of a repeat-unit based on the same triangle but constructed in a simpler method, providing a practical shortcut (fig. 12, fol. 191r [36]):

> However, the objective can be achieved here with the aid of this T-square (*misṭara gūnyā*). As we said in the preceding "preliminary," the objective of our repeat-unit (*ʿaqd*) is four right-angled "pinecone figures" (*shakl-i ṣanawbarī*) surrounding a right-angled equilateral and equiangular quadrilateral.

The four figures, also referred to as "pinecone-like quadrilaterals" (*murabbaʿ-i ṣanawbarī*) later in the same text, are special kites that surround a rotated central "four-legged" square, lined up along the edges of the repeat-unit. The rotating kites are obtained by cutting off a triangle from the tip of each of the four right-angled triangles defined by the extended sides of the central "four-legged" square. Each kite is further subdivided into three quadrangles. The T-square, composed of a set square (*gūnyā*) and ruler (*misṭar*), is shown superimposed on that particular repeat-unit. This drafting tool for the mechanical drawing of conic sections is excluded from Abu'l-Wafa''s practical geometry treatise, which lists only the ruler, compass, and set square.

The following page separately depicts the T-square (fol. 191v [36]), which was already introduced in an earlier construction with no explanation (fol. 190r [34]). The text accompanying the depiction describes how to construct that tool, with which many "difficult repeat-units" (*ʿuqūd-i mushkil*) based on conic sections could be drawn (the term *ʿuqūd* here is the plural of *ʿaqd*). It immediately becomes clear that this was the inference of the scribe, who seems not to have been familiar with the tool: "This is presently the deduction of the scribe (*kātib*); whether or not it is so is not known." If this is the voice of the *Anonymous Compendium*'s original compiler around the 1300s, then the geometrical construction may have been copied from an earlier source depicting an unfamiliar tool. Alternatively, if it is the voice of the fifteenth-century copyist, this would mean that the tool was apparently no longer familiar in the Timurid-Turkmen period. The commentary goes on with a description of how to construct "the ruler in the same way as the alidade of an astrolabe" and erect in the middle of it a perpendicular ruler at a right angle, similar to the vertex of the alidade of the "boat astrolabe" (*usṭurlāb-i zawraqī*) (pls. 36.1′, 36.3). The text ends with a reassuring remark, apparently in a voice different from and earlier than the one above (i.e., prior to the 1300s or 1400s), for it now confirms that if the tool is correctly constructed, "many extraordinary, difficult proportions" (*nisbathā-yi gharīb-i mushkil*) can be obtained with it.[106]

The text of the earlier rectangular pattern superimposed with the same T-square similarly explains that "the proportion of this repeat-unit" (*nisbat-i īn ʿaqd*) is based on kites (*ṭurunj*) drawn using that drafting tool (fol. 190r [34]). An annotation announces the discovery of an alternative method of approximation, using angles:

> However we have found a construction by approximation (*taqrīb*). When we divide the right angle into 9 equal parts, four parts of it will constitute angle AGB, and 5 parts will constitute angle BGD. This is a very close approximation (*ghāyat taqrīb*).

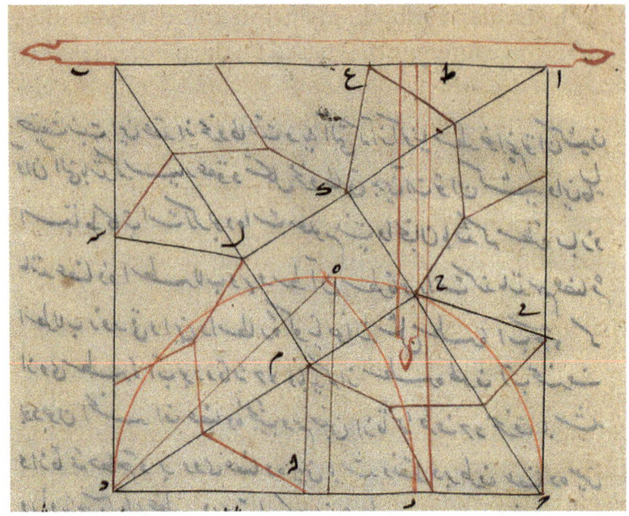

Fig. 12. T-square superimposed on a square repeat-unit with a central rotated "four-legged" square, surrounded by four subdivided right-angled triangles. From the *Anonymous Compendium*. Paris, Bibliothèque nationale de France, Ms. Persan 169, fol. 191r [36]. (Photo: Bibliothèque nationale de France)

The repeat-unit of the same pattern appears earlier in the *Anonymous Compendium*, where two other methods for its construction are explained on a single page, to which I briefly referred above (fol. 187v [26 and 27]). The text accompanying the drawing at the bottom of that page (no. 27 [pl. 27]) states that the ratio/proportion of the length to width of this repeat-unit (*ʿaqd*) has been determined using the "geometrical method" (*bi-ṭarīq-i handasa*) from the octagon. The text written next to the drawing at the top of the page (no. 26 [pl. 26]) explains how to construct a right-angled triangle that is labeled "easy proportion" (*nisbat-i sahl*). An annotation written upside down in between both texts explicates this practical way of approximating, reflecting the then current working method of artisans, based on rectilinear proportions instead of angular ones derived from the circle: "Some artisans (*baʿżī az-ṣunnāʿ*) draw this repeat-unit (*ʿaqd*) such that its length is seven parts and the width six parts. This is extremely close [to the required ratio] (*ʿaẓīm nazdīk ast*)." The approximation entails dividing

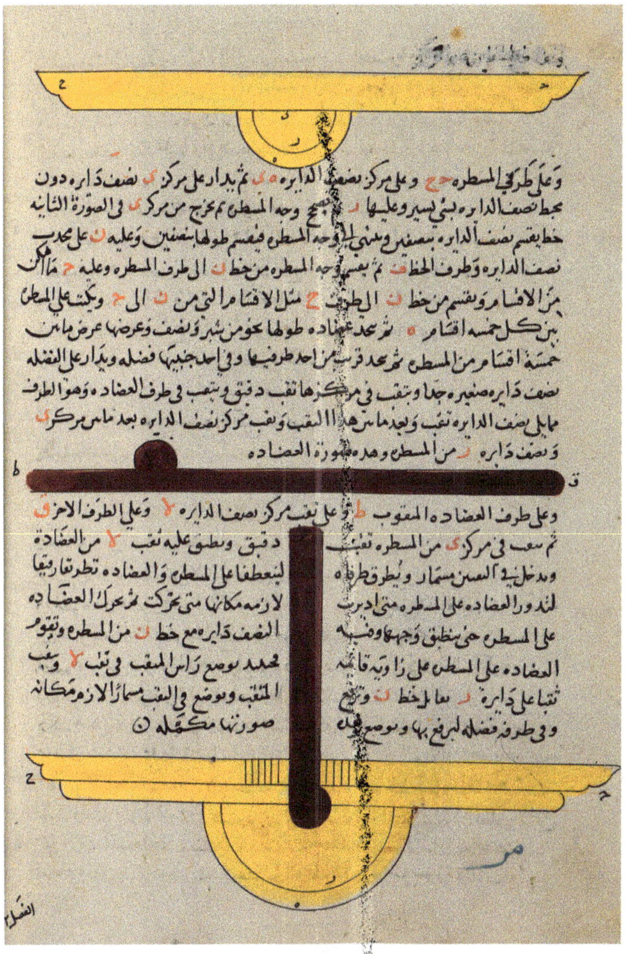

Fig. 13a. Drawing of a T-square with a movable alidade and semicircular projecting sextant. Ibn al-Razzaz al-Jazari, "A Compendium of Science and Useful Practice in the Mechanical Arts." University of Oxford, Bodleian Library, Ms. Greaves 27, fol. 108v. (Photo: The Bodleian Library)

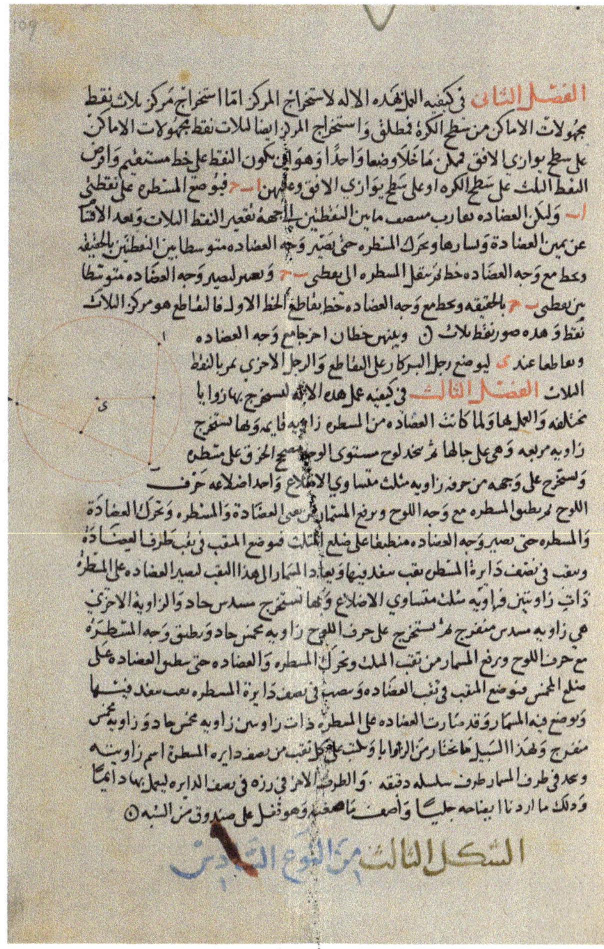

Fig. 13b. Diagram showing how to determine the center of three points marked on a circle by using a T-square. Ibn al-Razzaz al-Jazari, "A Compendium of Science and Useful Practice in the Mechanical Arts." University of Oxford, Bodleian Library, Ms. Greaves 27, fol. 109r. (Photo: The Bodleian Library)

the lines corresponding to the length and width of the repeat-unit into seven and six equal parts, respectively, i.e., a construction that corresponds to the numerical ratio 7:6.

A drafting tool comparable to the T-square described above, but this time featuring a movable alidade with a semicircular projecting sextant at its center, is described and illustrated in a treatise by the mechanical engineer Badiʿ al-Zaman Abu al-ʿIzz Ismaʿil b. al-Razzaz al-Jazari (fl. ca. 1200) titled *Al-Jāmiʿ bayna al-ʿilm wa'l-ʿamal al-nāfiʿ fī ṣināʿat al-ḥiyal* (A Compendium of Science and Useful Practice in the Mechanical Arts). Al-Jazari's famous tract was written in 1206 and dedicated to the Artuqid ruler of Diyarbakır (the Artuqids were a Seljuk successor state based in eastern Anatolia and northern Syria-Iraq). Shifting the position of that T-square's alidade enabled the setting out of angles corresponding proportionally to two segments of the sextant's semicircular arc, which was pierced with holes showing different angles (fig. 13a). Another diagram of al-Jazari (fig. 13b) elucidates how to determine the center of three points on the surface of a sphere or a circle by using that drafting tool, provided that they are not in a straight line. This explains why a construction illustrating the

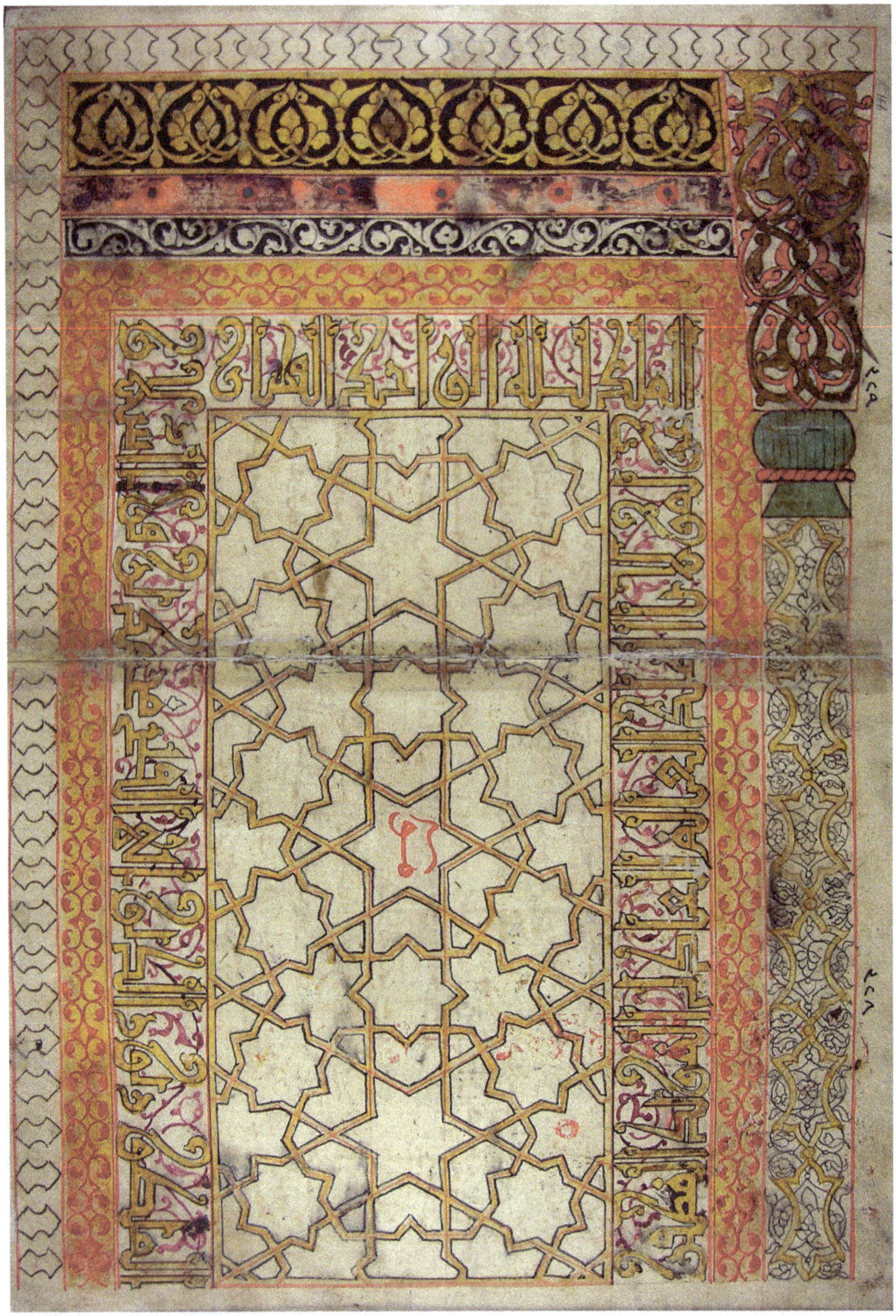

Fig. 13c. Template for a cast bronze door decorated with interlocking hexagonal and octagonal stars and filler units. Ibn al-Razzaz al-Jazari, "A Compendium of Science and Useful Practice in the Mechanical Arts." Istanbul, Topkapı Palace Library, Ms. A. 3472, fols. 165v–166r. (Photo: Topkapı Palace Library)

same problem is included under the North Dome pattern in the *Anonymous Compendium*, the repeat-unit of which was most probably drawn by using the T-square depicted on the previous page (fol. 192r [38]), as I have proposed above. Al-Jazari adds that with the aid of that tool "other angles in [general] use, acute and obtuse, may also be determined." Moreover, using the same tool, he designed a door with interlocking star-and-polygon patterns for the Artuqid royal palace in Diyarbakır, the cast bronze sheets of which were fixed by decorative nails over a wooden base (fig. 13c).[107] He boasted that this was truly a masterpiece, whose composition perfectly fit the doorframe, with no "half- or quarter-stars nor any incomplete pieces, except for two half-stars" [at the centers of the upper and lower edges].[108]

Al-Jazari was a geometer-engineer and master craftsman fully conversant with the branches of his trade, including metalwork. The primacy of visual illustration (*mithāl*, example or model) over textual description is captured in a revealing passage of his treatise, which was commissioned by his Artuqid royal patron, who asked him to record the mechanical devices he had made. In that passage, the bronze door is meticulously described as having a latticework (*shabaka*) pattern of hexagonal and octagonal stars (*khātim*) joined by filler units. Its star-and-polygon pattern resembled in type those that appeared on carpenter's work for tongue-and-groove joinery (*kārzawān*).[109] Drawing attention to the shared language of star-based ornamental patterns in metalwork and woodwork, this statement brings to mind the previously discussed title of a nearly contemporary Arabic copy (dated 1209) of Abu'l-Wafa''s *Geometrical Constructions* in Cairo, namely *Kitāb al-Nijāra* (Book of Joinery).

In this fascinating passage, which indicates the significance of mechanical drawings, al-Jazari warns the reader that his abbreviated step-by-step textual description in Arabic is less informative than the accompanying set of diagrammatic visual representations (*mithāl*). This is often the case in the illustrations of the Persian *Anonymous Compendium* as well, which are sometimes labeled *miṡāl*. Hence, al-Jazari's technical drawings were not mere embellishments but rather visual clarifications of the text. He explains that in making the drawings of the door and its individual components, he did not aim at completeness but instead wished to present a clear picture of the parts in relation to the whole. He thus communicated the visual information necessary for experts already familiar with the theory and practice of the mechanical arts. Less accomplished artisans, however, could still copy his drawings as templates, just as his treatise may have contributed to the appreciation of clever geometrical expertise by al-Jazari's royal patron and his circle of courtiers.

The invention of the T-square as a shortcut for obtaining complex proportions may account for the increasing sophistication and precision of geometric ornament around the thirteenth century, when earlier irregularities in star-and-polygon patterns, often comprising figures with unequal sides and angles, were considerably reduced. Concern for greater exactitude involved not only the internal proportions of interlocking motifs, now dominated by regular polygons and stars, but also the correspondence of the length-to-width proportions of their repeat-units with the edges of given surfaces. This important aesthetic criterion is affirmed when al-Jazari proudly points to the absence of half- and quarter- stars, or any other incomplete pieces, in the door that he designed and produced. The *Anonymous Compendium* may have partly addressed the same aesthetic criterion through its point-joining instructions, which explain how to construct reliable repeat-units for ornamental patterns whose "small and large proportions" closely approximated geometrically correct ones. These dimensionless repeat-units transformed patterns into replicable templates. Besides establishing a suitable drawing technique for the investigative exploration and representation of patterns, the precision of repeat-units would also help to reduce disjunctions and distortions amplified by their application on large-scale surfaces, and to adjust their motifs to the borders of given supports, including the curvature of arch spandrels.[110]

ORNAMENTAL GEOMETRY IN THE *ANONYMOUS COMPENDIUM* AND THE *TOPKAPI SCROLL*

I previously interpreted the term *ʿaqd* as a synonym for *girih* (lit. knot), which more appropriately characterizes ornamental patterns compiled in the *Topkapı Scroll* and its successors. Sometimes featuring kufic inscription panels and architectural plans drawn on squared or triangular grids, such design scrolls were prepared by master builder-architects, unlike the *Anonymous Compendium*, whose author(s) included practice-oriented geometer-astronomers such as Abu Bakr.[111] The term *girih* is widely used in Persian sources from at least the mid-sixteenth century onwards, contrary to the assumption that it did not appear until the nineteenth century. The semantic range of *ʿaqd* closely corresponds to that of the terms *girih* or *girih-bandī* (knotted, joined, bound, fastened together). Star-and-polygon patterns are identified in sixteenth- and seventeenth-century Persian sources on the visual arts as being in the *girih* mode; other variants include *girih-bandī*, *girih-bandī-yi Rūmī*, *band-i Rūmī*, and *girih-i Rūmī*.[112] Some of these terms associate the *girih* mode with Anatolia (*Rūm*), where geometric ornament became widespread under the Rum Seljuk sultanate, between the early twelfth and mid-thirteenth centuries. Following the Mongol conquest of Anatolia in 1243, when the Rum Seljuks became vassals of the Ilkhanid dynasty, this type of geometric ornament continued to flourish in that region and far beyond.[113] The *girih* mode underwent sophisticated elaborations during the post-Mongol era, from the mid-thirteenth through the early sixteenth centuries, when regionally specific ornamental idioms became normative in the Islamic East and West alike. Geometric patterns would subsequently be marginalized by floral ornaments in the Eastern Islamic lands, despite their continuous afterlife well into the present.[114]

It has been argued that the *girih* mode was not essentially different from early Islamic geometric ornament during the Umayyad period (661–750), largely based on Late Antique prototypes. From this perspective, the only later Islamic additions were a new preference for straight lines instead of curves, and two kinds of innovation: "the increase in their complexity," and "their utilization of radial symmetries that are not sextal or octal" (i.e., hexagonal and octagonal). Indeed, recent studies have demonstrated the relatively low level of complexity displayed in Roman and Byzantine star-and-polygon patterns, their later versions in Norman Italy (eleventh and twelfth centuries), and in Cosmatesque ornament (early twelfth to early thirteenth century).[115] In that sense, Umayyad geometric ornament can be seen as an offshoot of the Late Antique tradition, despite its deliberate divergences from and transformations of classical precedents. However, the later Islamic additions mentioned above are hardly insignificant modifications explicable by a model of evolutionary continuity. Moreover, the exclusive focus on planar geometric designs ignores intimate connections between two- and three-dimensional patterns, such as projections of *muqarnas* and stellate vaults, whose geometries are almost invariably worked out together in design scrolls and, to a lesser degree, in the *Anonymous Compendium*. Equally debatable is the assertion that "the argument from academic geometry fails" because patterns with "unprecedented nonsextal and nonoctal symmetries" required "no new geometric knowledge, only a desire to achieve novel results" and thus do not constitute "a new class of decoration."[116]

First of all, constructive "practical geometry" is not the same as "academic geometry." Secondly, recent mathematical studies have argued that some highly complex Islamic patterns displaying five- and ten-fold symmetries are quasiperiodic (i.e., tiling the plane in a deterministic way that does not repeat), or closely approximate quasiperiodicity, first understood in the West from the 1970s onward.[117] As one mathematician has pointed out, the derivation of "irregular tiles from supplementary shapes to compositional elements in their own right" gave Islamic artists "flexibility to assemble the tiles in novel ways and led to a new category of designs"; in his view, this was an innovation of Seljuk Iran and Anatolia particularly during the twelfth and thirteenth centuries.[118] Whether those designs constitute a "new class of decoration" may be a matter of definition and taxonomy, but two- and three-dimensional star-and-polygon patterns were certainly not immune to "new geometric knowledge."

Successive moments in the development of a pervasive tradition of constructive geometry can be discerned in Abu'l-Wafa''s late tenth-century *Geometrical Constructions*, which lacks ornamental patterns; the *Anonymous Compendium*, which is provisionally datable to the 1300s and survives in a modified mid-fifteenth-century copy; and the subsequent *Topkapı Scroll* (late fifteenth to early sixteenth century). The *Anonymous Compendium*, annotated with terse instructions typical of "how-to manuals," addresses advanced practitioners already familiar with basic procedures such as those outlined in Abu'l-Wafa''s practical geometry treatise and others, including the works of Abu Bakr. It condenses star-and-polygon patterns into repeat-units through an intense investigation of their underlying properties, such as similarity, complementarity, and ratios/proportions. The omission of written instructions toward the end of this document brings the repeat-units of those textless patterns close to their counterparts in the *Topkapı Scroll* (figs. 14, 15a–b). This and subsequent design scrolls catalogued the same mode of geometric ornament by means of relatively more standardized, textless repeat-units, which became integrated into visual and oral workshop traditions.[119]

Although several compositions in the *Anonymous Compendium* find parallels in the textless *Topkapı Scroll*, many of them have been eliminated altogether, especially preliminary procedures and "dividing and assembling" constructions. The scroll is neither a "how-to manual," nor simply a pattern book, since its inked and uninked construction lines make visually explicit the generative geometry that underpins its repeat-units. Most of its planar geometric patterns have solid or dotted construction lines in red, orange, or black ink, which delineate generative polygonal grids (fig. 16a–b). In the *Anonymous Compendium,* with the exception of patterns structured by the "four-legged" square, inked generative grids composed of contiguous polygons only appear in two cases: the aforementioned repeat-units for complex heptagonal designs (figs. 10a and 17). Hence, the "polygonal technique" was known but not articulated graphically in this document, whose written commentaries and constructions are dominated by the "point-joining" technique.[120] By contrast, polygonal grids abound in the *Topkapı Scroll* as a way to provide shorthand workshop formulas. Yet codification was complemented by the invention of new experimental patterns, which, like those of the *Anonymous Compendium,* do not find counterparts in extant monuments.

I have acknowledged that the inked generative grids with contiguous polygons found in the *Topkapı Scroll* may have been intended as templates. Nevertheless, some of its patterns were generated by stellate grid systems more complex than those described by Ernst Hanbury Hankin.[121] He discovered traces of templates that functioned as construction lines faintly scratched on the plastered walls of two late sixteenth-century Mughal bathhouses in the Fatehpur Sikri Palace near Agra, consisting in one case of contiguous triangles, and, in the other, of simple polygons. The final pattern was created by joining the midpoints of each polygon with lines that crossed one another at certain angles.

The more intricate generative grids of the *Topkapı Scroll* include not only the "four-legged" square scheme featured in the *Anonymous Compendium* (fig. 18a–d), but also interlocking stars and polygons (fig. 19a–b). I referred to the latter as "double-layered" patterns because a second layer of proportionally interrelated patterns of different scale is superimposed on these star-and-polygon compositions.[122] Having no counterpart in the *Anonymous Compendium*, those patterns display considerable visual complexity, and are more advanced than former double-layered designs, such as the one seen in an ornamental vault panel at the so-called Abbasid Palace in Baghdad (ca. 1184–1230). The latter double-layered pattern, combining octagonal and pentagonal stars with six-pointed stellate forms, nevertheless foreshadows subsequent versions in that the vertices of its large-scale motifs coincide with the centers and edges of stars in the small-scale star-and-polygon composition (fig. 20a).[123] Not surprisingly, the same building, which is a monument with extravagant geometric decoration in carved brick and terracotta, boasts a unique *muqarnas*-vaulted arcade that reminds us of what has been lost in Baghdad. The *muqarnas* vaulting of a recess in that palace, which frames a flat ceiling decorated with interlocking star-and-polygon designs, comes close to the orthogonal *muqarnas* projections depicted in the *Anonymous Compendium*. This vaulted recess once again demonstrates the intimate relationship between two- and three-dimensional geometrical patterns (fig. 20b).

The closest extant parallels to the elaborate double-layered patterns of the *Topkapı Scroll* appear on fifteenth-century Timurid-Turkmen monuments from western and central Iran, corroborating its proposed date and provenance.[124] In those examples, the superimposed star-and-polygon patterns in differing scales are rendered in the polychrome mosaic-tiling tech-

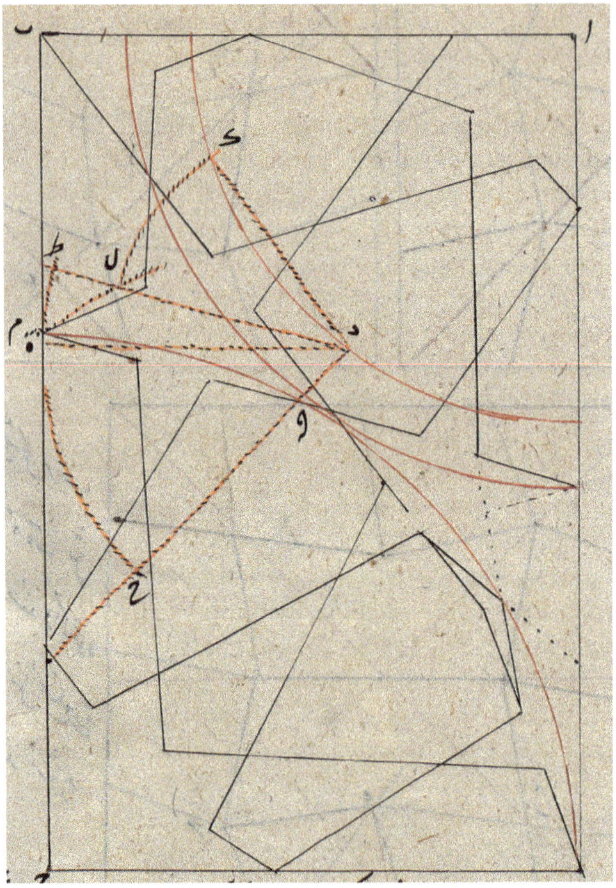

Fig. 14. Rectangular repeat-unit of an incomplete construction with pentagonal, decagonal, and dodecagonal stars. From the *Anonymous Compendium*. Paris, Bibliothèque nationale de France, Ms. Persan 169, fol. 195v [54]. (Photo: Bibliothèque nationale de France)

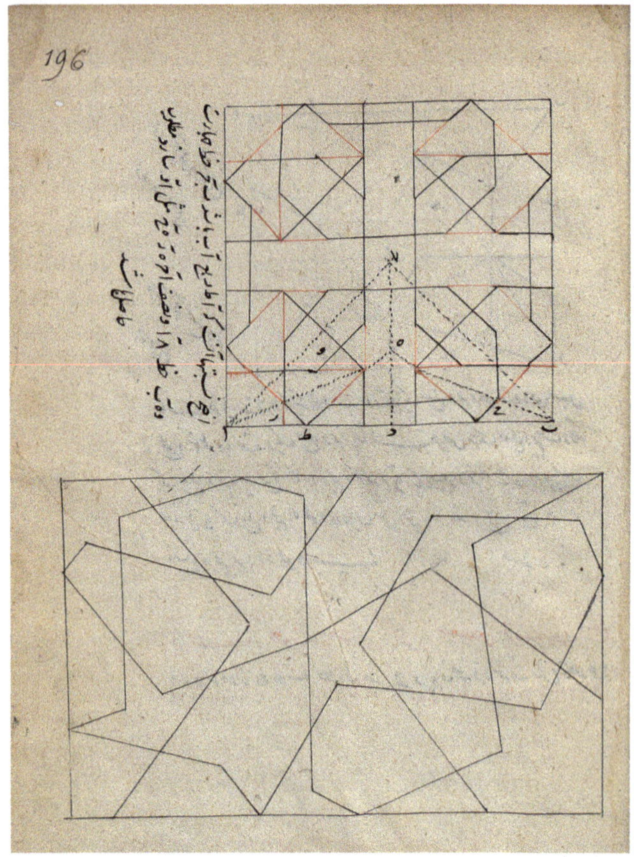

Fig. 15a. Bottom: Rectangular repeat-unit with decagonal and dodecagonal stars; top: square repeat-unit with a central cross-axis of eight arms and quarter-octagonal stars at each corner. From the *Anonymous Compendium*. Paris, Bibliothèque nationale de France, Ms. Persan 169, fol. 196r [55, 56]. (Photo: Bibliothèque nationale de France)

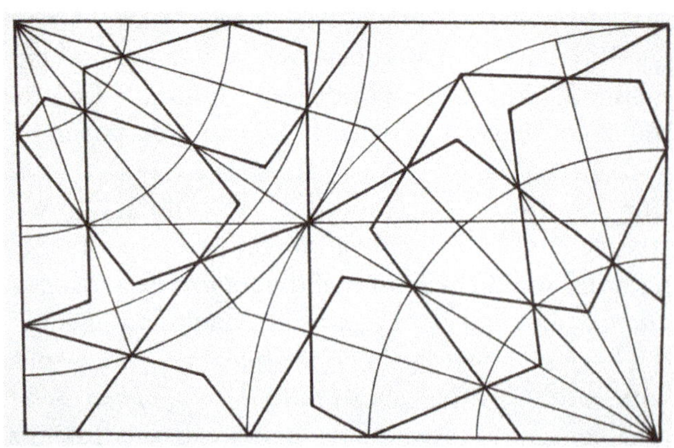

Fig. 15b. Uninked generative grid of construction lines in figure 15a (bottom). (Drawing: Elizabeth Dean Hermann, in Necipoğlu 1995, p. 147, fig. 109b)

nique, which not only allowed for but also required much greater precision than cut brick. The increased exactitude of this technique, mediated by drawings on paper, was developed substantially in the post-Mongol era and may have contributed to the diminished use of strapwork, which disguises uneven crossings by means of ribbons passing over and under each other. The former taste for linear interweaving thus gave way to *girih* patterns with clearly bounded interlocking compartments.

Despite notable differences, the patterns of the *Topkapı Scroll* have features in common with those of the *Anonymous Compendium*, such as inked or uninked incised oblique lines demarcating dynamic axes of symmetry that proportionally cut the edges and surfaces of

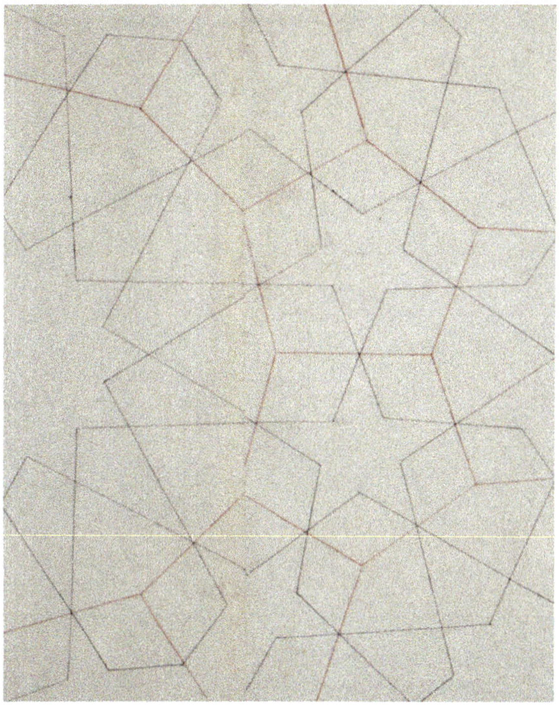

Fig. 16a. Rectangular repeat-unit with pentagonal, decagonal, and dodecagonal stars. From the *Topkapı Scroll*. Istanbul, Topkapı Palace Library, Ms. H. 1956. (Photo: Necipoğlu 1995, p. 310, cat. 44)

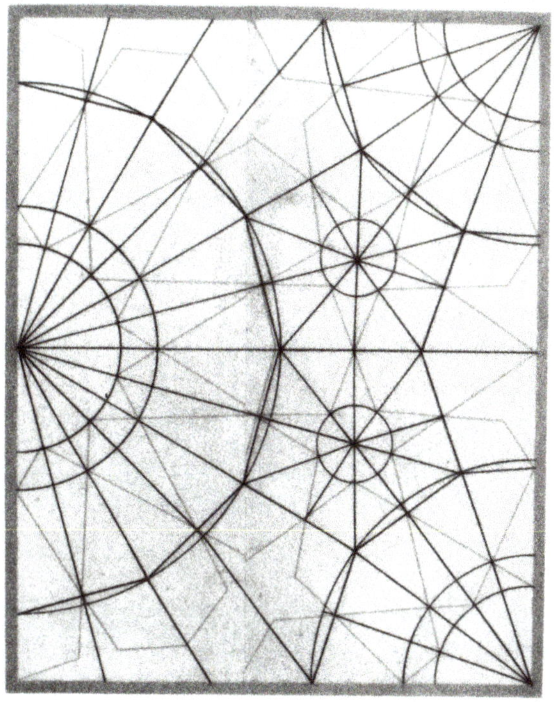

Fig. 16b. Uninked generative grid of construction lines in fig. 16a. (Drawing: Elizabeth Dean Hermann, in Necipoğlu 1995, p. 256, no. 44)

their repeat-units. Both documents also share uninked concentric circular construction lines subdivided by equidistant radii corresponding to particular angles, which emanate from star centers. By comparing the ornamental repeat-units of these two sources it is possible to discern the increasing elaborateness of generative construction lines in the *Topkapı Scroll*, in accordance with its function as a practice-oriented workshop document lacking textual instructions: (1) rectangular repeat-units with decagonal and dodecagonal stars, whose underlying grid of contiguous polygons is delineated with red ink only in the scroll (figs. 14, 15a–b, and 16a–b);

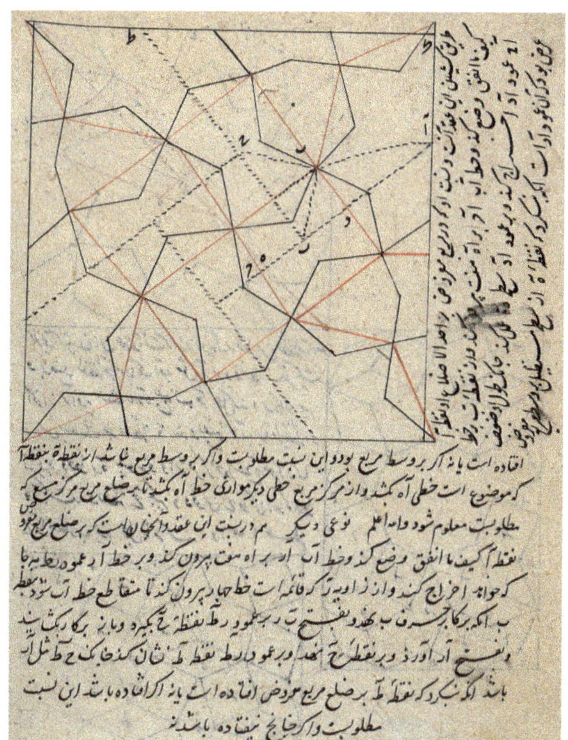

→

Fig. 17. Square repeat-unit with four central double-pentagons, heptagonal stars, and hexagons, generated by a rotated "four-legged" square drawn in black dotted lines. From the *Anonymous Compendium*. Paris, Bibliothèque nationale de France, Ms. Persan 169, fol. 194v [48]. (Photo: Bibliothèque nationale de France)

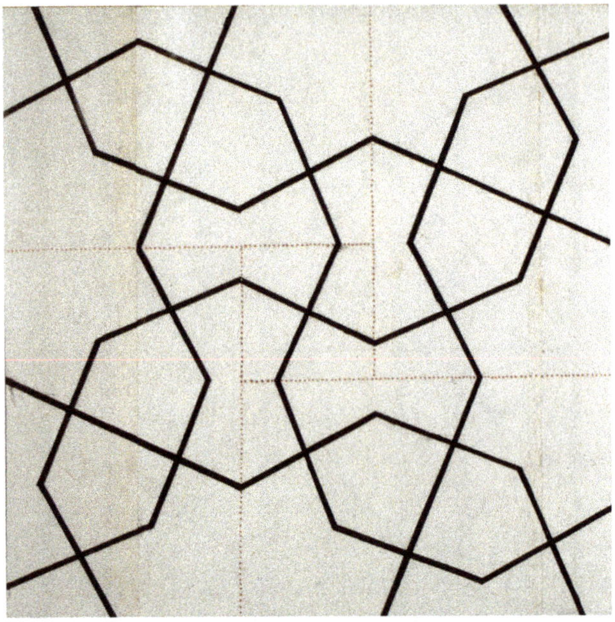

Fig. 18a. Square repeat-unit with four central double-pentagons and octagonal stars, generated by a central "four-legged" square drawn in red dotted lines. From the *Topkapı Scroll*. Istanbul, Topkapı Palace Library, Ms. H. 1956. (Photo: Necipoğlu 1995, p. 320, cat. 61)

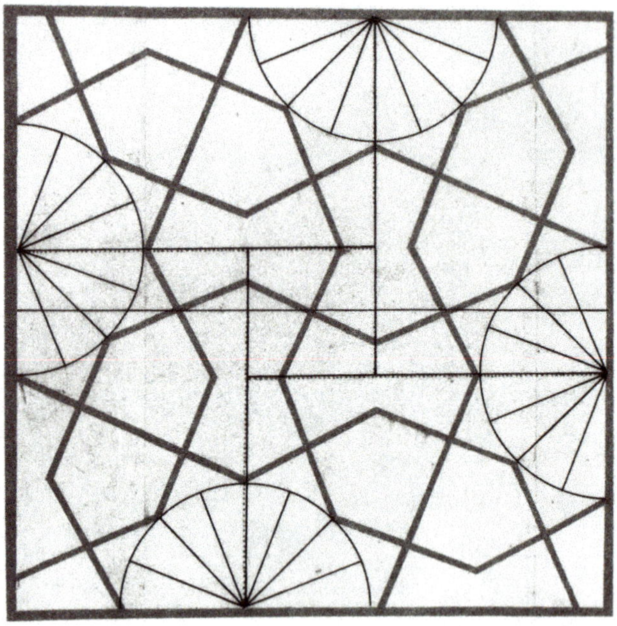

Fig. 18b. Uninked generative grid of construction lines in fig. 18a. (Drawing: Elizabeth Dean Hermann, in Necipoğlu 1995, p. 263, no. 61)

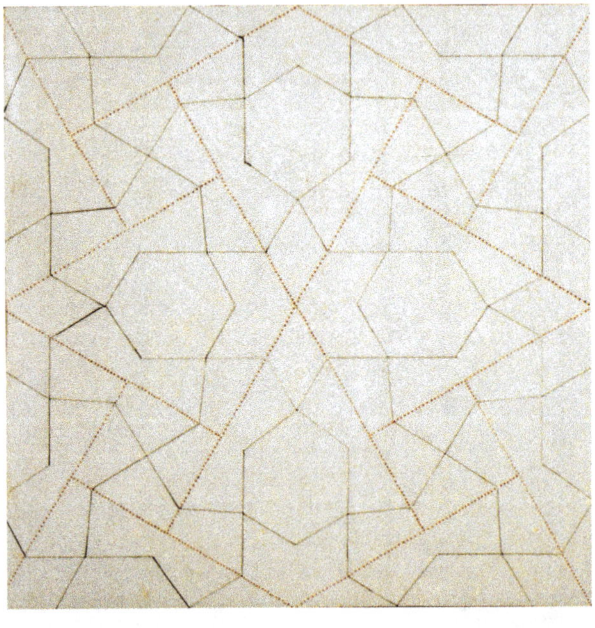

Fig. 18c. Square repeat-unit with hexagonal stars, hexagons, and double-pentagons, generated by four rotated "four-legged" squares drawn in red dotted lines. From the *Topkapı Scroll*. Istanbul, Topkapı Palace Library, Ms. H. 1956. (Photo: Necipoğlu 1995, p. 319, cat. 59)

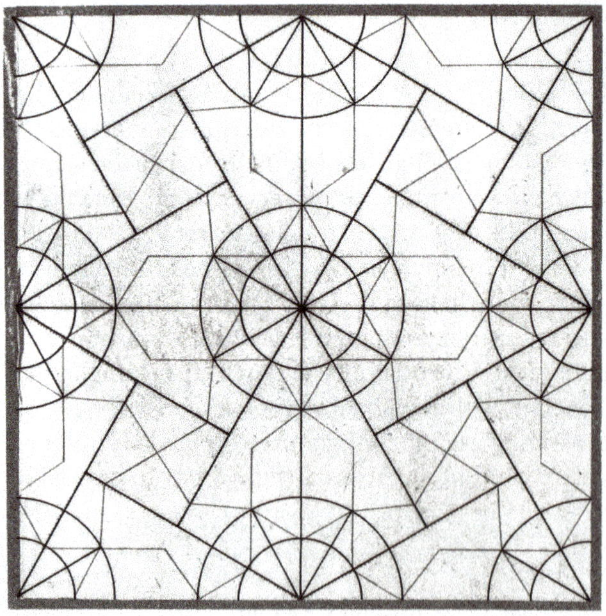

Fig. 18d. Uninked generative grid of construction lines in fig. 18c. (Drawing: Elizabeth Dean Hermann, in Necipoğlu 1995, p. 262, no. 59)

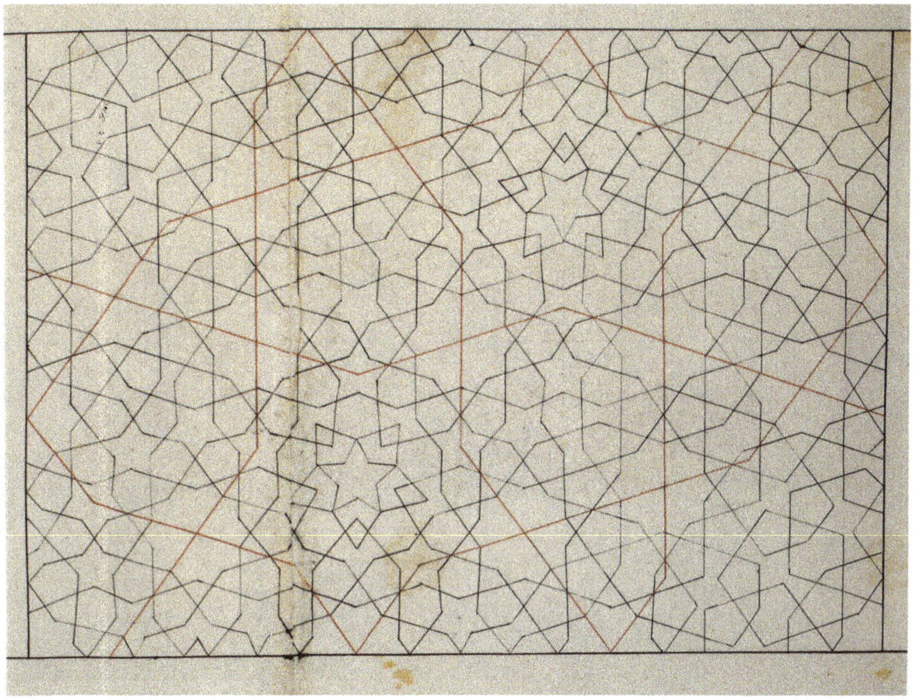

Fig. 19a. Rectangular repeat-unit with double-layered pattern comprising pentagonal and decagonal stars. From the *Topkapı Scroll*. Istanbul, Topkapı Palace Library, Ms. H. 1956. (Photo: Necipoğlu 1995, p. 300, cat. 28)

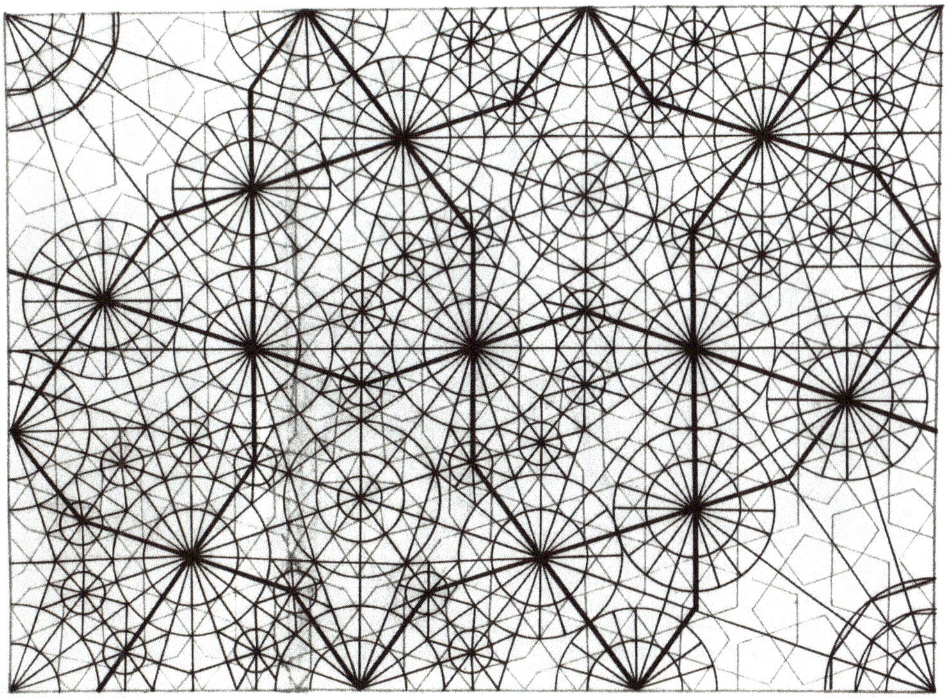

Fig. 19b. Uninked generative grid of construction lines in fig. 19a. (Drawing: Elizabeth Dean Hermann, in Necipoğlu 1995, no. 28, p. 249)

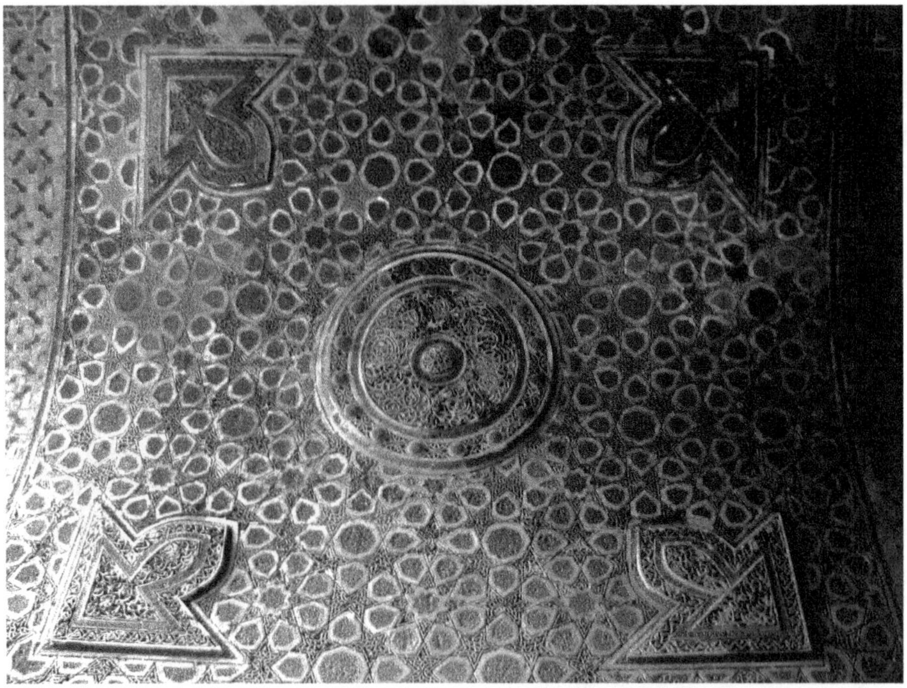

Fig. 20a. Carved brick vault panel with double-layered pattern comprising octagonal and pentagonal stars and six-pointed stellate forms, overlaid with a central roundel and four arrow-shaped medallions. Courtyard iwan of the Abbasid Palace (ca. 1184–1230) in Baghdad. (Photo: Public domain)

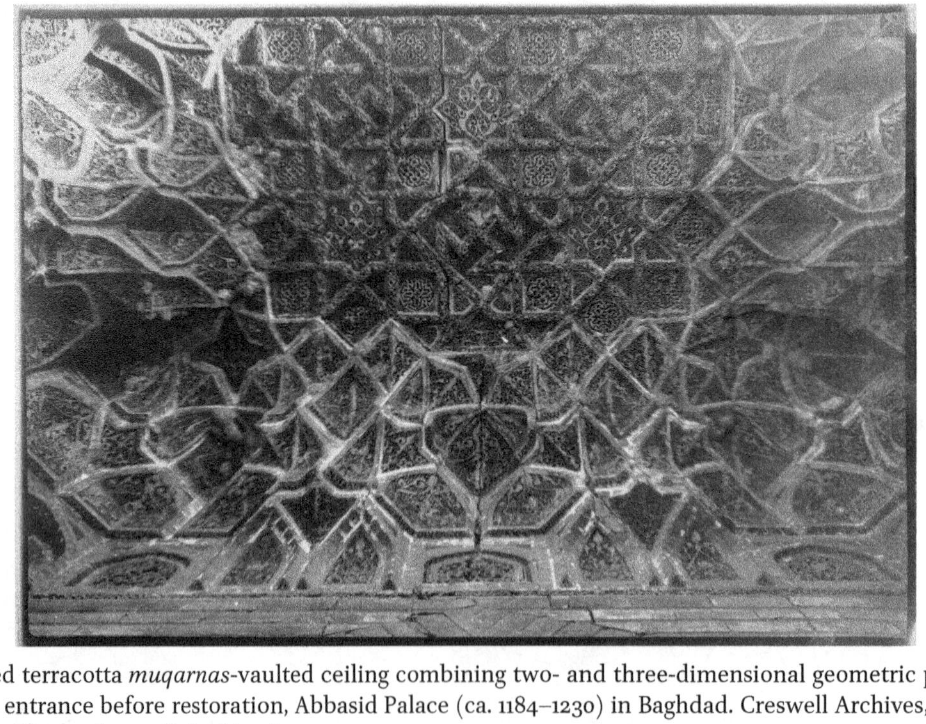

Fig. 20b. Carved terracotta *muqarnas*-vaulted ceiling combining two- and three-dimensional geometric patterns. Recess facing the west entrance before restoration, Abbasid Palace (ca. 1184–1230) in Baghdad. Creswell Archives, EA.CA.6286 161 C127 3Ab(i)Re3. (Photo: Creswell Archives)

(2) square repeat-units with four central, interlocking double-pentagons generated by an underlying scheme of "four-legged" squares drawn in dotted lines, which are combined in one case with heptagonal stars (fig. 17), in another case with octagonal stars (fig. 18a–b), and, in yet another variant, with hexagonal stars (fig. 18c–d); (3) rectangular repeat-units that juxtapose rotating swastikas with pentagonal and decagonal stars in one example (fig. 21a–b), and with dodecagonal stars in another (fig. 22a–b).[125]

Some of the uninked construction lines seen in the repeat-units of these two documents are not taken into consideration in contemporary studies that explore the design methods of Islamic patterns. Such studies tend to devise their own preferred techniques for constructing the same patterns.[126] Since there are generally several ways in which to construct a pattern, the discrepancy between proposed modern design methodologies and the historical documents is understandable. Those methods are ultimately oriented toward contemporary applications, aimed at formulating standard recipes or even computer software programs for revitalizing traditional Islamic geometric patterns.[127] Hence, they do not consistently replicate historical construction procedures recorded in the primary sources, even when they claim to do so. The *Anonymous Compendium* and the *Topkapı Scroll* demonstrate that polygonal templates applied as guidelines on walls would have been derived from patterns with more complex construction lines that were initially worked out on paper or an alternative surface. In other words, there is a difference between transferring a design on a wall and creating it anew by differing methods suitable to particular circumstances. The ability of unskilled artisans to reproduce or modify decorative patterns assembled in written sources such as the *Anonymous Compendium*, in textless workshop collections such as the *Topkapı Scroll*, or in historical buildings and artifacts, did not necessarily require a comprehension of how master designers commanding higher levels of skill had constructed them in the first place. Besides differing levels of competence, then, one must recognize that neither the process of designing a pattern nor that of transferring it to a surface followed a single universal method.

Since the "polygonal method" is documented in Hankin's report, the *Topkapı Scroll*, and the *Anonymous Compendium* (though in just a few ornamental repeat-units), its proponents have assumed that this was the main historical technique for designing geometric patterns, which simply required the use of a ruler and compass.[128] However, the primary sources do mention other professional drafting tools, such as set squares, the moving ruler, and versions of the T-square that facilitated mechanical drawing procedures, especially those involving subtle proportions approximating conic sections. The renowned Maghribi scholar Ibn Khaldun (d. 1406) testifies to the usefulness of conic sections, which to him was "apparent in practical crafts that have to do with bodies, such as carpentry and architecture," and in making mechanical devices.[129] This does not, of course, mean that all patterns were constructed with such advanced tools. Yet the T-square was used, at least once, in designing the star-and-polygon pattern of the bronze door produced by the mechanical engineer al-Jazari in Diyarbakır (fig. 13a–c). Versions of this drafting tool were most likely used in other times and places as well, as is implied by the somewhat simpler variant depicted in the *Anonymous Compendium* (fig. 12). Therefore, the common supposition that the easier "polygonal method" must necessarily have been preferred by medieval designers of Islamic geometric ornament is debatable. This opinion was not shared, for instance, by the great theorist of ornament, Owen Jones (d. 1874):

> As with proportion, we think that those proportions will be the most beautiful which it will be most difficult for the eye to detect....in the best periods of art all moldings and ornaments were founded on curves of a higher order, such as the conic sections; whilst, when art declined, circles and compass-work were much more dominant.[130]

What have recently been termed "polygonal templates" by Jay Bonner, or "*girih* tiles" by Peter J. Lu and Paul J. Steinhardt, were used in certain cases for transferring geometric patterns onto given surfaces.[131] However, I would argue that the square and rectangular repeat-units for ornamental designs in the *Anonymous Compendium* were themselves intended as templates. Notably, the centers of multiple star fragments generally overlap with the vertices and edges of these abridged quadrangular repeat-units. When applied to a surface having the same proportion as that of a square or rectangular repeat-unit, this would ensure a perfect fit

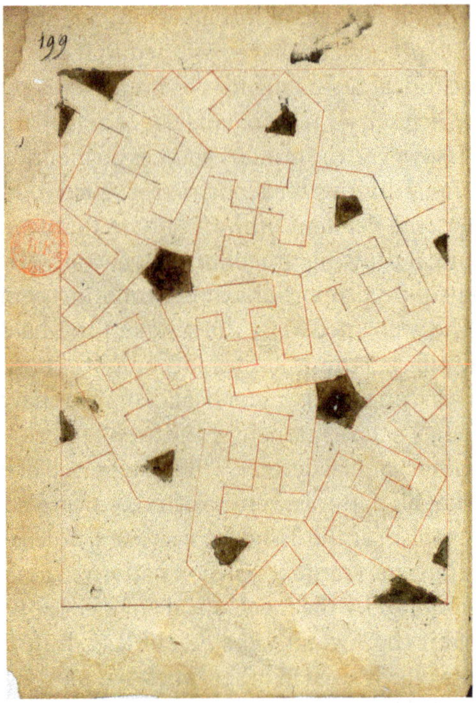

Fig. 21a. Rectangular repeat-unit with rotating swastikas, kites, and pentagonal and decagonal stars. From the *Anonymous Compendium.* Paris, Bibliothèque nationale de France, Ms. Persan 169, fol. 199r [61]. (Photo: Bibliothèque nationale de France)

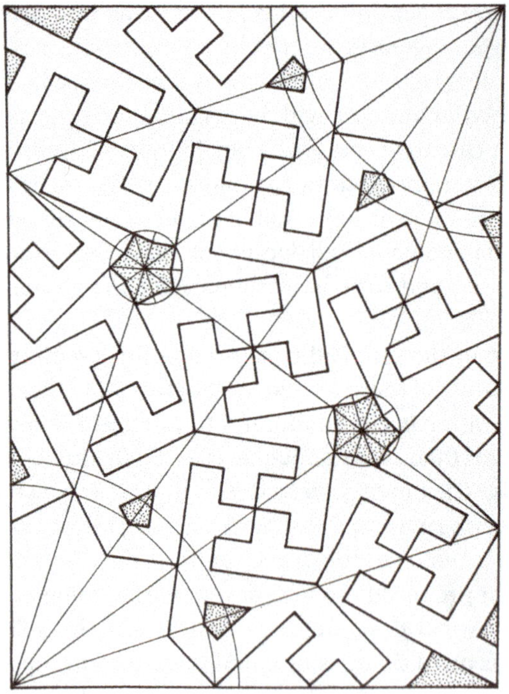

Fig. 21b. Uninked generative grid of construction lines in fig. 21a. (Drawing: Elizabeth Dean Hermann, in Necipoğlu 1995, p. 149, fig. 113b)

between the pattern and its support. Some rectangular repeat-units, for example, feature quarter-stars at opposite corners (figs. 14, 15a [bottom], and 21a). One of the square repeat-units, on the other hand, has a slightly rotated octagonal star at the center, with quarter-hexagonal stars occupying the four corners, while each of its edges is lined up with a half-composite polygon that fuses two pentagons and a heptagon (fig. 23 [top]). The slight tilting of the octagonal star yields a pattern with dynamic symmetry, in which the sides of all the interlocking figures are equal. As Özdural has shown, an untilted octagon would have produced a static design with non-equilateral composite polygons, a comparison testifying to the skill of its designer in subtly integrating stars with differing rotational symmetries (pl. 45.2a–b).

The strict economy of design observed in the *Anonymous Compendium*'s condensed repeat-units, which reduce ornamental patterns to their lowest common denominator, is not always followed in the *Topkapı Scroll*, which includes some fuller compositions, visually displaying the designer's bravura. Nevertheless, both documents shed light on the historical use of repeat-units as templates. Ornamental patterns that are hardly imaginable from their abbreviated repeat-units in the *Anonymous Compendium* may have initially been invented by diverse methods and even copied from earlier manuals and buildings.[132] In addition to serving as templates for present and future use, these repeat-units, whose associated texts clarify proportional relationships, could guide advanced masters in developing new innovative compositions.[133]

The ambition of some studies to impose a single design methodology upon the vast repertory of Islamic geometric ornament is an anachronism contradicted by the multiplicity of techniques recorded in the *Anonymous Compendium* and, to a lesser degree, in more standardized design scrolls. This diversity was recognized long ago by Chorbachi, who acknowledged the need to

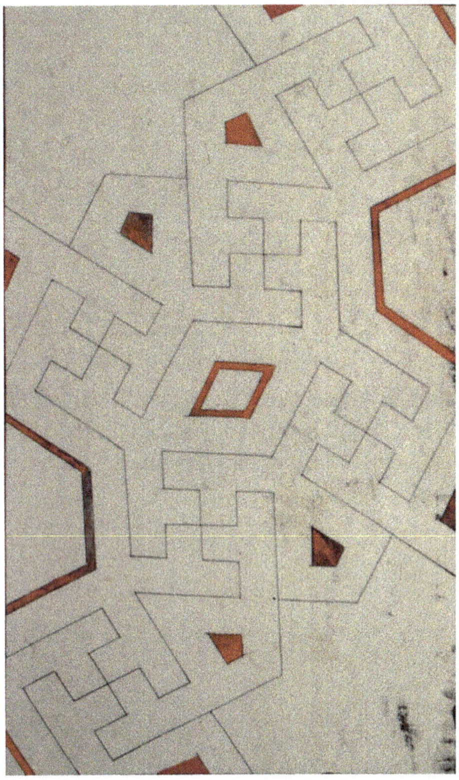

Fig. 22a. Rectangular repeat-unit with rotating swastikas, dodecagonal stars, hexagons, kites, and lozenges. From the *Topkapı Scroll*. Istanbul, Topkapı Palace Library, Ms. H. 1956. (Photo: Necipoğlu 1995, p. 325, cat. 70)

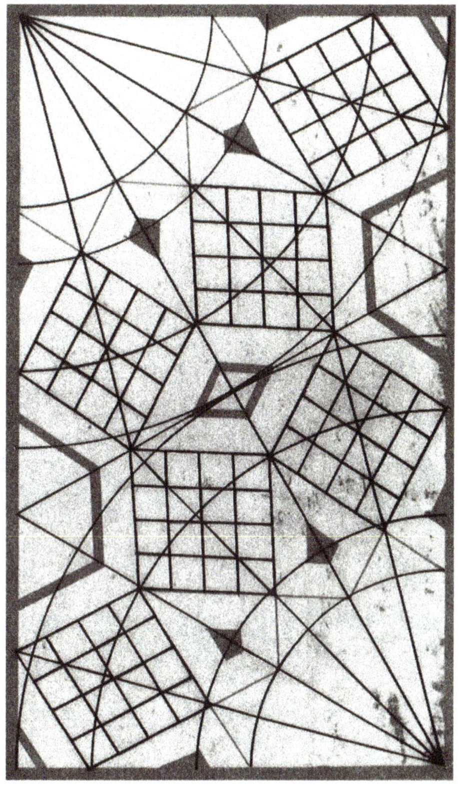

Fig. 22b. Uninked generative grid of construction lines in fig. 22a. (Drawing: Elizabeth Dean Hermann, in Necipoğlu 1995, p. 266, no. 70)

"look beyond the symmetry patterns to the actual process of design and to recognize the significant aspects and issues of geometry it involves."[134] The predominant focus of recent studies on contemporary interests, such as quasiperiodic tilings and symmetry groups, has had "limited relevance to the aesthetic complexity of Islamic geometric design," as one researcher put it, since "this tells us little about the properties of the design other than its symmetry group."[135]

Taking a closer look at the inked or uninked construction lines and actual motifs of ornamental patterns in the *Anonymous Compendium* and the *Topkapı Scroll* unveils historical clues for dating that are not apparent if one focuses on the similar symmetry principles of their generative polygonal grids. A comparison of motifs shows that their design repertories are consistent with the respective dates of compilation I proposed for both documents.

Most importantly, the former document does not include the arch-net or radial *muqarnas* vault projections so characteristic of Timurid-Turkmen architecture that abound in the *Topkapı Scroll*.[136] The archaic orthogonal *muqarnas* vault projections illustrated in the *Anonymous Compendium* are also compatible with its earlier date (fig. 23 [middle and bottom]), when compared with the more intricate drafting conventions used for distinguishing corbelled spatial tiers in the radial *muqarnas* patterns of the *Topkapı Scroll* (fig. 24a–b). Moreover, the geometrical constructions in the *Anonymous Compendium* are not only marked with letters but also drawn in red and black ink, in the manner of medieval scientific treatises, setting them apart from those of the *Topkapı*

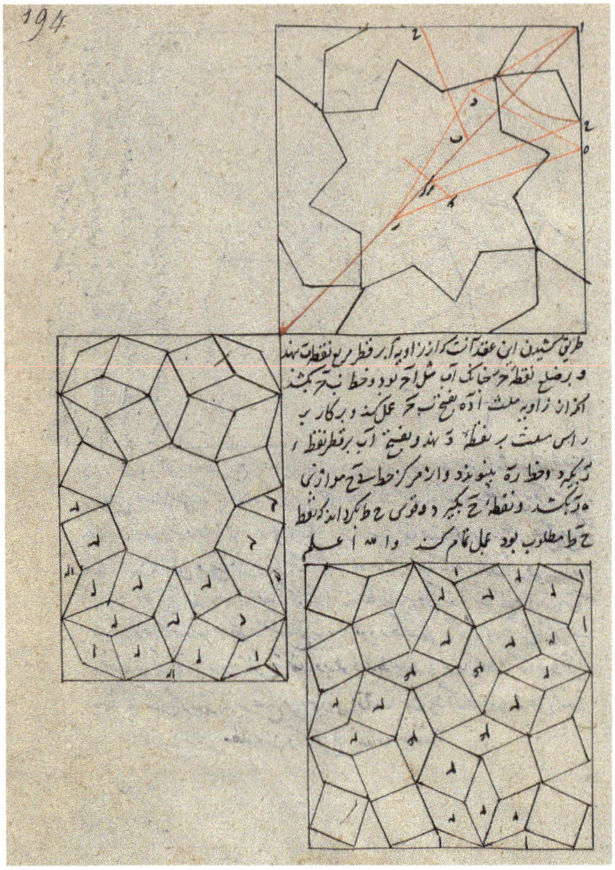

Fig. 23. Top: square repeat-unit with hexagonal and octagonal stars and composite polygons; middle and bottom: orthogonal projections of a rectangular and a square *muqarnas* vault. From the *Anonymous Compendium*. Paris, Bibliothèque nationale de France, Ms. Persan 169, fol. 194r [45–47]. (Photo: Bibliothèque nationale de France)

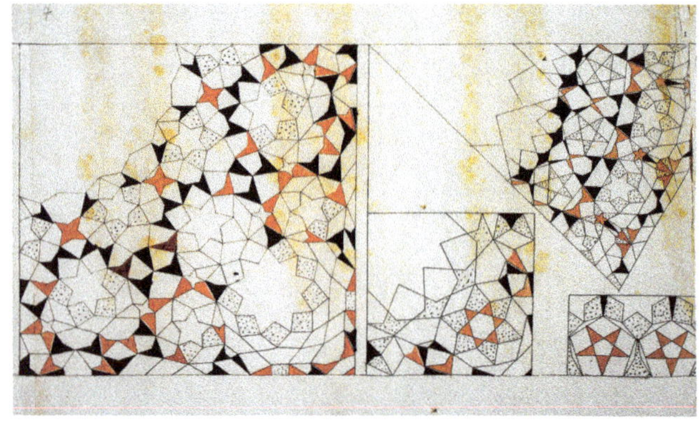

Fig. 24a. Repeat-units of radial *muqarnas* vaults. From the *Topkapı Scroll*. Istanbul, Topkapı Palace Library, Ms. H. 1956. (Photo: Necipoğlu 1995, cat. 16, 17, p. 294)

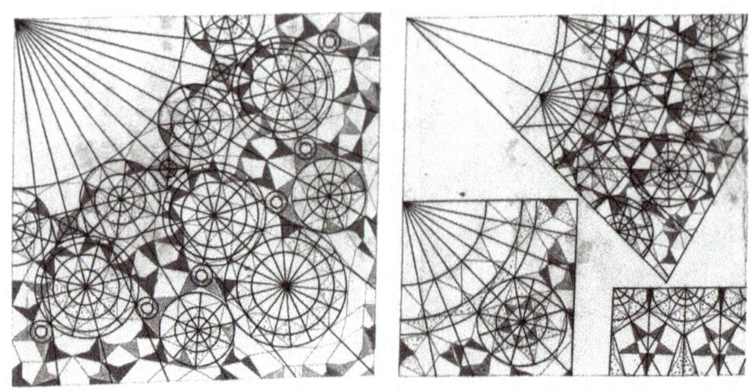

Fig. 24b. Uninked generative grid of construction lines in fig. 24a. (Drawing: Elizabeth Dean Hermann, in Necipoğlu 1995, p. 245, nos. 16 and 17)

Scroll and its successors, which lack lettering and feature multiple colors.

Some of the patterns in the scroll, particularly its "double-layered" repeat-units with highly complex five- and ten-fold symmetries, have lately attracted the attention of contemporary designers, mathematicians, mathematical physicists, and crystallographers (fig. 19a–b). Recent studies, which regard these patterns as innovative experiments with quasiperiodic tilings and self-similarity, have triggered a lively debate revolving around two issues: whether they can be classified using such modern categories, and the level of mathematical awareness or lack thereof possessed by their creators. However, in-depth research using a wider range of written primary sources to seek historicized answers to these questions has not been forthcoming, largely due to the trend toward mathematizing ornamental patterns and the tendency to underestimate the interdisciplinary potential of the subject. Ornamental geometric patterns, situated at the intersection of history of science and cultural artifact, have yet to be analyzed in studies of cognitive history as historical constructs and practices of knowledge.[137] The *Anonymous Compendium* itself compels us to ask whether the highly complex *Topkapı Scroll* patterns could have been intuitively invented by master builder-architects without any direct or indirect input of mathematical knowledge, which I find rather unlikely. With respect to the *Anonymous Compendium*, the textual, visual, and circumstantial evidence that we have considered reveals a reciprocal exchange between the arts and sciences in the process of making technical drawings and the use of mechanical instruments. It would therefore not be farfetched to propose that unrecorded exchanges between theoretical and practical knowledge, such as this one, left their invisible imprint not only on the *Topkapı Scroll*, but also on successive iterations of comparable geometric patterns in different times and places.

CONCLUSION: INDIVIDUALS MEDIATING BETWEEN ART AND SCIENCE

To summarize, the *Anonymous Compendium* is the only known medieval Islamic primary source with instructions on how to construct some challenging geometrical figures, including repeat-units for composite ornamental patterns. Correct and approximate constructions are juxtaposed, sometimes offering alternative methods, suggesting that this document was initially put together in a milieu where individuals with varying degrees of mathematical knowledge and artisanal expertise commingled, or at least became aware of each other's differing yet partly intersecting geometrical construction methods. This was a setting in which geometer-astronomers with a practice-oriented proclivity, such as Abu Bakr b. al-Khalil al-Tajir al-Rasadi, apparently came into contact with practitioners of ornamental geometry.

Rather than a "collaboration," it might be more appropriate to characterize the unknown circumstances of that encounter as an "actor-network," within which the flow of knowledge was not unidirectional but bilateral, or even multidirectional. In this particular context, practical geometry, with its applications in both surveying and the constructive geometry of artisan-designers and master builder-architects provided a middle ground and meeting point for experts in these subfields of the "mixed" mathematical sciences. It is difficult to classify the *Anonymous Compendium* precisely because of the "mixed" character of the problems compiled in it, which overlap with those in related areas such as surveying and mechanics (especially the construction of mechanical drafting tools). Although this document is informed by practices in surveying (*misāḥa*), that is, the science-cum-art of measurement, which fuses arithmetical (numerical) and geometrical methods, it differs methodologically in its minimization of numerical computation. As such, the primarily visual language of its constructive geometry problems comes close to that of Abu'l-Wafa''s *Geometrical Constructions*. The diagrammatic drawings in both of these sources omit numbers that correspond to the measurement of dimensions and angles, unlike surveying manuals, whose drawings are typically marked with numerals to be used in calculating and dividing areas.

The present chapter has drawn attention to previously unnoted ways in which the geometrical constructions of the *Anonymous Compendium* resonate with those of the two treatises in the Paris Codex that contain extracts from Abu Bakr's works. It was meant not merely for theoretical scientists dabbling in mathematical puzzles as purely intellectual exercises, but rather a varied spectrum of scholar-practitioners, artisan-designers, and perhaps master builder-architects experienced in constructive geometry. This deduction is buttressed by the omission of proofs and of detailed textual instructions in the *Anonymous Compendium*, its technical language, its persistent attention to close approximations, and its recourse to shortcuts by means of advanced professional tools. These features hardly conform to conventions typical of treatises on theoretical geometry, just as the *Anonymous Compendium*'s composite interlocking figures are uncharacteristic of standard practical geometry or surveying manuals. While some of its highly complicated constructions would only have been comprehensible to experts with advanced skills, others providing basic preliminaries were presumably unnecessary for "masters" such as Abu Bakr and his associates, suggesting that its intended audience(s) included less experienced disciples and novices as well.

Embodying a pedagogical concern, posed in terms of questions and answers in its "problem-solution" (*mas'ala-javāb*) and "section-example" (*faṣl-miṣāl*) format, the *Anonymous Compendium* can be interpreted as the collective "inscription" of a heterogeneous "actor-network" involving a series of mediations and translations. It was perhaps a written record of oral teaching methods, similar to the explanations a master might have given to associates and disciples. Or else, the format in which its problems are modeled may have been part of a conventional device, also used in other technical manuals such as those comprising extracts from Abu Bakr's works in the Paris Codex.

Having been recorded on paper, the contents of the *Anonymous Compendium* traveled across time and space. The mid-fifteenth-century copy preserved in the Paris Codex, a collection with a strong emphasis on applied mathematics, intimates that its contents were of interest to such mathematician-astronomers as Kubanani Yazdi, whether the codex was copied by him or not. The context(s) and locale(s) in which the *Anonymous Compendium* was initially put together, probably sometime in the 1300s, will remain opaque unless future research unveils new clues about the web of actors surrounding its compilation and subsequent transcription in the Paris Codex. Given the present state of our knowledge, it is equally difficult to surmise whether other sources like it existed. Despite such uncertainties, the *Anonymous Compendium*'s textual and graphic inscriptions document instances of exchange between the realms of theory and practice, generally assumed to have been separate.

We have already encountered some individuals who mediated those realms, including geometer-astronomers and engineers such as Abu'l-Wafa', known as *al-muhandis*; Abu Bakr b. al-Khalil al-Tajir, known as *al-raṣadī*; Ibn al-Haytham; and Ibn al-Razzaz al-Jazari. Another geometer-engineer whose skills combined theory and practice was Mu'ayyad al-Din Abu'l Fadl Muhammad b. ʿAbd al-Karim b. ʿAbd al-Rahman al-Harithy (d. ca. 1202–3), also known as *al-muhandis*. This Damascene master carpenter-cum-stonecutter studied Euclid's *Elements* to improve his skills in the art of joinery by mastering the details of geometry (*al-handasa*) and their application. He subsequently immersed himself in the *Almagest*, Ptolemy's treatise on mathematical astronomy, and tabulated astronomical tables. He also studied with the famed Iranian mathematician-astronomer Sharaf al-Din al-Tusi (d. 1213–14), who spent some time in Damascus.[138]

According to his biography, Abu'l Fadl b. ʿAbd al-Karim *al-muhandis* produced most of the woodwork doors for the hospital of the Zangid ruler Nur al-Din (r. 1146–74) in Damascus. He also renovated the mechanical clock of the city's Great Mosque and wrote books on medicine upon becoming a doctor. One of the hospital's extant woodwork doors bears his signature and the date 549 (1154–55). It is encased in copper sheets fixed with decorative brass nails, and its linear strapwork pattern is composed of pentagonal stars, rosettes inscribed with hexagonal stars, and six-pointed stellate forms, joined together with heptagons, hexagons, and filler units.[139] This sophisticated composition has been identified as a variant of an earlier pattern that adorns one of the tympanum arches at the Seljuk-period North Dome in Isfahan (1088–89).[140] The hospital itself features an Iraqi-type red brick conical *muqarnas* dome and a matching *muqarnas*-hooded portal in carved stone on its main façade, which is framed by a band of interlaced geometric strapwork (fig. 25a–d). Founded by a Zangid ruler who was the son of a former Seljuk *atabeg* of Mosul, this monument is a perfect example of interrelated two-and-three-dimensional geometric designs in concert with one another and with their architectural support.[141]

Sharaf al-Din al-Tusi, the Iranian teacher of the multitalented Damascene *muhandis* who produced that door, also trained a number of celebrated polymaths. His best student was Ibn Yunus (d. 1242), the previously mentioned commentator (in Arabic) on Abu'l-Wafa''s practical geometry treatise. Ibn Yunus, in turn, became the teacher of Nasir al-Din Tusi (d. ca. 1274), the first director of the Maragha observatory, where the pre-Mongol scientific heritage was preserved and advanced further. (It housed a vast collection of manuscripts saved from the Mongol sack of Baghdad in 1258 and elsewhere, and remained active from 1259 to the end of the thirteenth century.) It is therefore significant that the commentary of Ibn Yunus was copied at Maragha in 680 (1281–82).[142]

The *Anonymous Compendium* provides a striking parallel. Its advancements also build upon an already established tradition of practical and constructive geometry, to the extent that it puts forth new solutions to some older construction problems. Like Abu'l-Wafa''s *Geometrical Constructions*, this source makes a distinction between the methods of geometers and those of artisans, yet their respective authors attempt to bridge that

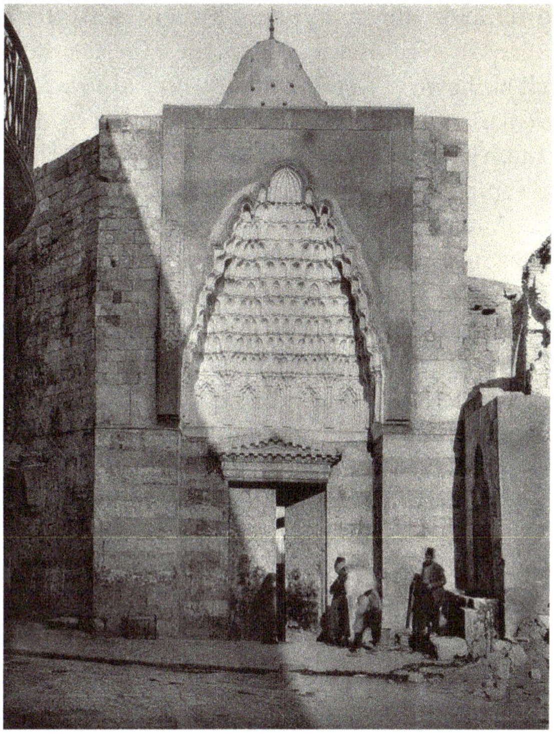

Fig. 25a. Façade of the Hospital of Nur al-Din b. Zangi, Damascus. Undated photograph. (Photo: courtesy of the Harvard Fine Arts Library, Visual Collections)

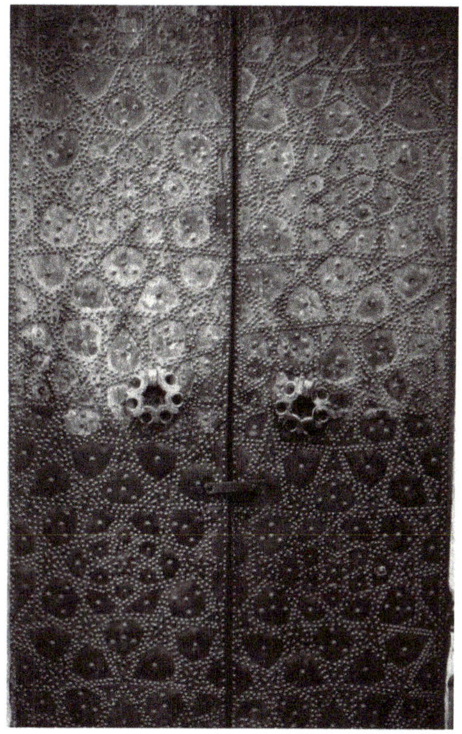

Fig 25b. Wooden door sheathed in copper and ornamented with brass nails, dated 549 (1154–55). Hospital of Nur al-Din b. Zangi, Damascus. (Photo: courtesy of the Yasser Tabbaa Archive, Aga Khan Documentation Center at MIT)

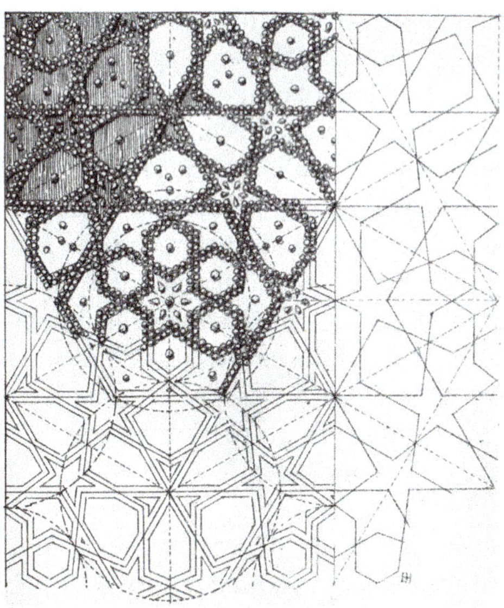

Fig 25c. Drawing of the strapwork pattern combining pentagonal and hexagonal stars with hexagons, heptagons and filler units. Hospital of Nur al-Din b. Zangi, Damascus. (Photo: courtesy of Yasser Tabbaa)

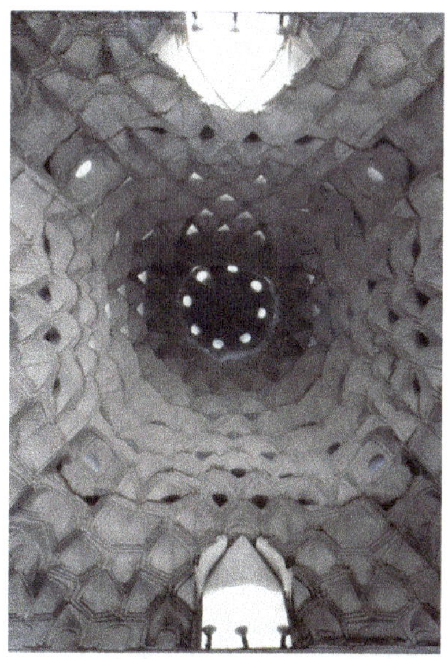

Fig. 25d. Interior view of the *muqarnas* dome. Hospital of Nur al-Din b. Zangi, Damascus. (Photo: courtesy of the Yasser Tabbaa Archive, Aga Khan Documentation Center at MIT)

divide by finding a scientifically sound middle ground. Abu'l-Wafa' and the author(s) of at least some constructions in the *Anonymous Compendium* were practice-oriented geometers, who appear to have been well informed about the approximation methods preferred by artisan-designers and surveyors. In addition, both works address some difficult problems related to the proportional subdivision of triangles, squares, and quadrangles involving square numbers.

To the list of individuals who bridged theory and practice, one might also add the makers of high-precision instruments such as the "perfect compass" for drawing conic sections, and the astrolabe for astronomical observations (pls. 36.3–4).[143] Such devices seem to have inspired the invention of the simpler T-square, with its perpendicular alidade, versions of which are described and illustrated in both the *Anonymous Compendium* and al-Jazari's treatise. One of several scientists involved in the making of instruments was Abu Mahmud Hamid b. al-Khidr al-Khujandi (ca. 940–1000), a metalworker, designer of astronomical devices, and observational astronomer, who mingled theoretical knowledge with manual skill. Trained in Greek geometry and astronomy, he helped build an observatory with a huge sextant in Rayy under Buyid patronage, and made one of the most beautiful astrolabes intended for use in Baghdad; it bears his signature and the date 374 (984–85).[144]

In connection with makers of scientific instruments, Saliba mentions the autobiography of the mathematician-astronomer Ibrahim b. Sinan b. Thabit b. Qurra (908–46). Active in Baghdad, he wrote a book about gnomons on spherical surfaces, dictating a version of it to a "technically clever" artisan in a different language that befits artisans who work with their hands.[145] This echoes Abu'l-Wafa''s observations concerning the diverging methods of geometer-astronomers and artisans, while at the same time testifying to points of contact between them. Saliba also cites a related remark made by the mathematician-astronomer Abu Rayhan al-Biruni (d. ca. 1048): "Some of the artisans favor the use of arithmetic and prefer it over craft methods...as we have found with all the makers of astrolabes and instruments." Yet Biruni's work on calculating the prayer direction (*qibla*) of cities did provide an alternative approximate method for architects and artisans because "they cannot work out the precise amounts which we have derived." In other words, they were not particularly capable of or concerned with precision, unlike those who "wish to use elaborate methods." This too is in keeping with the remark of Abu'l-Wafa', who in 998 collaborated with al-Biruni in observing a lunar eclipse from Baghdad and Kath (in Khwarezm), respectively.[146] Al-Biruni's statement goes a long way in substantiating the argument in my book and this chapter that artisanal-architectural practice relied primarily on constructive geometry methods, ranging from exact to approximate, in which arithmetical computation was of secondary importance, aided by tabulated numerical tables if required. Al-Biruni once again reminds us of gradations in the level of mathematical skills commanded by practitioners, who cannot be lumped into a single undifferentiated category labelled "artisan."

Some "technically clever" artisans and architects were not only literate but even harbored literary pretensions that found an outlet in composing poetry. One of them was the Timurid architect Ustad Qavam al-Din b. Zayn al-Din Shirazi (fl. 1410–1438 or 1440), who was accomplished in engineering-geometry (*muhandisī*), architecture (*miʿmārī*), and drawing-design (*ṭarrāḥī*). It has been convincingly postulated that exceptional masters like him probably had their own crew of decorators and "followed through on every stage of the work, from the working drawings to the design of the vaults and the working out of all types of decoration." Qavam al-Din's biographers eulogized him as "the greatest architect of the age," who was also an "expert in astronomy." When he presented an astronomical almanac that he had calculated to his Timurid patron, Shahrukh (r. 1405–47), the ruler allegedly recited the following line of poetry: "You did the work of the earth so well that you took up the heavens too."[147]

Treatises on ratio and proportion written by such medieval mathematician-astronomers as Abu ʿAbdallah Muhammad b. Isa b. Ahmad al-Mahani (d. ca. 884), Abu'l-Wafa' al-Buzjani, Ibn al-Haytham, and ʿUmar Khayyam have yet to be studied in themselves and in relation to the *Anonymous Compendium*. They point to a wider cultural phenomenon at the intersection of the arts and mathematical sciences, with implied connections to aesthetic philosophy and metaphysics (cosmology).[148] The manifestly geometric ornamental patterns of the *Anonymous Compendium* visually evoke an unarticulated concept of beauty relying on the harmony of proportionally interrelated similar and complementary figures. The planar and spatial versions of such patterns can be seen as an embodiment of the aesthetics of proportion and transcendence. Since the professional training of Islamic geometer-astronomers, artisan-designers, and master builder-architects was connected to the epistemological-metaphysical background of the mathematical sciences, it may not be unreasonable to suggest that star-and-polygon patterns potentially resonated with cosmological notions of order and absolute beauty. Such a resonance certainly permeates Kubanani Yazdi's postface to the Persian translation of the *Geometrical Constructions,* capturing his conviction that geometry is a source of wisdom and truth capable of unveiling the wonders of the divine creation. He thus completed his translation by citing the following Koranic verses: "Those who contemplate the creation of the heavens and the earth" (3:191); and "He to whom wisdom (*ḥikma*) is granted, truly he has received abundant good." (2:269).[149]

Nor is it a coincidence that Johannes Kepler (d. 1630), one of the first scholars in early modern Europe to systematically investigate regular and semiregular tessellations, was an astronomer who regarded the geometry of harmonic figures as the source of all order and beauty in the cosmos. Drawing upon earlier sources that were also familiar in the Islamic context, Kepler, in the first book of his *Harmonice Mundi libri V* (Linz, 1619), quoted a statement by the Greek Neoplatonist philosopher Proclus (d. 485) concerning the importance of mathematics. In his commentary on Euclid's *Elements*, Proclus wrote: "It reveals the orderliness of the ratios according to which the universe is constructed, and the proportion that binds together all that is in the cosmos." The divine order of creation could thus be comprehended by human wisdom through the arts of "measurement" (*miṣāha*) and geometry, a common notion in Christiandom and Islamdom alike.[150]

Just as some clever surveyors, architects and master artisans could become experts in geometry, several Iraqi geometers are known to have participated as engineers in construction and irrigation projects, besides writing technical manuals: the Banu Musa brothers (ninth century), al-Karaji (d. ca. 1029), and Ibn al-Haytham (d. ca. 1040). The latter two were the authors of now-lost treatises classified in encyclopedias of knowledge under the "Science of the Construction of Edifices" (*ʿilm ʿuqūd al-abniyya*), an applied subfield of the science of geometry concerning the erection of castles, houses, and bridges, as well as diverting rivers, building water channels, and creating hydraulic works. Ibn al-Haytham's treatise is described in a list of his own writings as a work on the drafting of canals and buildings, with illustrations in which he applied "all the geometric shapes, until I arrived in this to the shapes of the three conic sections—the ellipse, hyperbola, and ellipse."[151] But the involvement of scientists and the use of scientific treatises in artisanal/architectural practice was relatively rare in the medieval Islamic and Christian lands, where most practitioners mastered geometrical design methods in which arithmetic played a relatively subordinate role and measured drawings were exceptional.[152]

Potential locales for fruitful exchanges between artisan-designers, master builder-architects, and geometer-astronomers included major construction sites, observatories, royal or elite courts, artisanal workshops, and wide-reaching circles of learning, both institutional and private. In 901, during the construction of his palace at Baghdad, the Abbasid caliph al-Muʿtadid Bi'llah (r. 892–902) designated an area where masters could teach all the crafts (*kull ṣināʿat*) and various branches of the theoretical and technical sciences (*al-ʿulūm al-naẓariyya wa al-ʿamaliyya*), providing it with generous financial support.[153] The spatial proximity of such court-sponsored activities would have facilitated the cross-fertilization of theory and practice across disciplines in the very locus of the Abbasid caliphal palace (*Dār al-Khilāfa*) in Baghdad, an aspiration that was apparently short-lived as the caliph died soon thereafter.[154]

In his account of the rebuilding of the ruinous city of Baghdad by the Buyid amir ʿAdud al-Dawla, the historian and philosopher Miskawayh (932–1030) mentions this ruler's palace, where "a place was set apart for men of special attainments and distinguished philosophers" near his own apartment, so that in that chamber such persons could hold discussions undisturbed by the "unenlightened and the vulgar." In order to revive those studies, ample allowances were given to learned men, including literati, physicians, astrologers, arithmeticians, and engineers. Hence, "the young were encouraged to study and the old to instruct, talent had free scope, and there was a brisk market for ability."[155]

In 988 ʿAdud al-Dawla's eldest son and successor, Sharaf al-Dawla (r. 982–89), had an observatory built in the royal palace (*Dār al-Mamlaka*) of Baghdad. There Abu'l-Wafa' *al-muhandis* was among the luminaries who collaborated with the chief geometer-astronomer Abu Sahl al-Quhi (d. 1000).[156] Sharaf al-Dawla was succeeded by his younger brother, Baha' al-Dawla, to whom Abu'l-Wafa' dedicated his practical geometry treatise. As I have conjectured above, one of the gatherings (*majālis*) mentioned in that treatise, where artisans and surveyors interacted with geometers, may have taken place within the Buyid palace in Baghdad, in the presence of his royal patron. In letters that he sent to his father in Kashan, the mathematician-astronomer Jamshid al-Kashi described similar learned discussions in gatherings at which the Timurid ruler Ulugh Beg was present. Through the generosity of this royal patron, who had been studying mathematics for over a decade, five hundred students began to learn this science in twelve locations in Samarqand, the site of an observatory he had founded in 1420 on the model of its predecessor in Maragha.[157] It is not surprising, then, that a deluxe Arabic manuscript of Abu'l-Wafa''s *Geometrical Constructions* was made around 1447–49 for this ruler's library.

In conclusion, it is overly skeptical to insist that the many documented intersections of theory and practice in the broadly defined domains of the mathematical sciences and aesthetic philosophy had little or no impact on artisanal-architectural practice, or vice versa. In conjunction with the complexity of ornamental geometries in historical artifacts and buildings, the *Anonymous Compendium* and *Topkapı Scroll* are important documents providing sufficient evidence that such exchanges must have occurred. The Paris Codex, which contains a Persian translation of Abu'l-Wafa''s *Geometrical Constructions* and the *Anonymous Compendium*, therefore deserves closer examination as a collection that in itself attests to the pervasiveness of these intersections. Demarcations between the visual arts and mathematical sciences did not deter certain individuals from acquiring cross-disciplinary skills and engaging with technologies in related areas. Some scholars may find it difficult to imagine the extent and fluidity of such crossings, given the clear-cut demarcation of modern disciplinary boundaries. I have argued here in favor of some instances of a two-way exchange between geometer-astronomers and practitioners of ornamental geometry, which cannot be reduced to a chicken-or-egg proposition. I also emphasized the often underestimated agency of royal and elite patrons, overshadowed by a preference to privilege artisanal autonomy. These patrons sponsored some of the site-specific occasions and locations in which creative exchanges took place. "Actor networks" such as these certainly transcend the binary of "top down" or "bottom up" paradigms of cultural production.[158] It is heartening to observe a mounting attentiveness to the multifaceted interfaces between the arts and sciences in current research on the visual cultures of pre-modern Europe.[159] Collaborations in the Islamic field between historians of science and philosophy, mathematicians, physicists, and art and architectural historians would likewise open up a vast arena of interdisciplinary inquiry, whose largely untapped primary sources are still waiting to be discovered and scrutinized.

Aga Khan Program for Islamic Architecture,
Department of History of Art and Architecture,
Harvard University, Cambridge, Mass.

NOTES

Author's note: I thank Melis Taner for her unstinting help as my research assistant; Wheeler M. Thackston for generously checking some of my translations; and Farshid Emami, Elaheh Kheirandish, and Peter J. Lu for their valuable comments on a draft of this chapter. Special thanks go to Karen A. Leal, whose meticulous editing enhanced not only my own chapter but the volume as a whole.

1. Paris, Bibliothèque nationale de France, Ms. Persan 169, item 24, fols. 180r–199r. See catalogue entries in Blochet 1912, no. 772, 41–47; and Richard 1989, 183–87.
2. Bulatov 1988 (first ed. 1978), and Chorbachi 1989a. Chorbachi independently discovered this manuscript in 1971 (161n3, 163n7). See also Chorbachi 1989b; and Chorbachi and Loeb 1992.
3. Publications that engage with the *Anonymous Compendium* include, in chronological order: Holod 1986; Holod 1988;

Golombek and Wilber 1988; Necipoğlu 1992; Necipoğlu 1995; Özdural 1995; Özdural 1996; Özdural 1998; Özdural 2000; Özdural 2002; Hogendijk 2012; and Bier 2012 (also see her collected articles at http://works.bepress.com/carol_bier/). Debated issues will be referred to throughout this chapter.

4. Necipoğlu 1995, 123, 133, 146–50 (figs. 108–14), 168–73; see especially two chapters, titled "Theory and Praxis: Uses of Practical Geometry" and "Manuals of Practical Geometry and the Scroll Tradition," 131–81. At the time I wrote my book, Chorbachi's dissertation was not accessible for consultation by readers. I have much benefited from reading it while preparing the present chapter. In her articles she translated the heading of the *Anonymous Compendium* as "On the Interlocking of Similar and Congruent Figures," and "On Interlocking Similar and Congruent Figures." I adopted the second version in Necipoğlu 1995, 133, but also referred to this source with the shorter modern title assigned to it in Richard 1989, 183–87: *Aʿmāl va ashkāl* (Constructions and Figures).

5. See Necipoğlu 1995, 32–39, where I explained that a sixteenth century date and Safavid provenance are less likely than a late fifteenth-century Timurid-Turkmen provenance, probably in or near Tabriz. Subsequent art historical studies have generally concurred with my attribution; examples include Roxburgh 2005, where the scroll is ascribed to late fifteenth-century Iran or Central Asia (433, cat. no. 223; 257–59, fig. 223; and overleaf). See also Golombek and Koch 2017. The Topkapı Scroll has recently been attributed to Tabriz in the second half of the fifteenth century on the basis of identical patterns appearing in the tilework decoration of monuments built for the Qaraqoyunlu and Aqqoyunlu Turkmen dynasties, around the region between Tabriz and Yazd: "Sans doute réalisé à Tabriz, au cours de la seconde moitié du XVe siècle (vers 1450–1460?), ce rouleau constitue la seule compilation de modèles décoratifs conserves pour la période qui nous interesse." See Aube 2017, esp. 268–72, citation at 72.

6. See, for example, Saliba 1999, and Allen 2004. Exceptions include Berggren 2007, 515–675, who deals with the mathematics of practitioners (such as surveyors, builders, artisans in geometric design, tax and treasury officials, and some merchants), which was part of an oral tradition promoting a close relationship between theory and practice, unlike the classical Greek mathematical tradition. Other exceptions are Kheirandish, Bier, and Yousefi 2008, and studies by Bier, Chorbachi, Kheirandish, Özdural, and Hogendijk, cited in this chapter.

7. See in particular Blair and Bloom 2006, 18, 25–27; and Blair and Bloom 2003, 171. In the latter article the authors criticize the use of primary written sources in my 1995 book on the *Topkapı Scroll*: "There is no reason to believe that the literate milieus that produced texts were identical to those that produced works of art." These milieus were obviously not identical, but they did intersect, as Blair herself admits in a later article on the Iranian potter Abu Zayd Kashani (turn of the thirteenth century): "We sometimes envision medieval craftsmen as anonymous, impoverished, and illiterate laborers. The case of Abu Zayd shows us that they were anything but. As potter, poet, scholar, and scribe, he created works of art that were designed to appeal to his contemporaries, who were as visually and literarily sophisticated as he was." See Blair 2008, 155–76.

8. Richard 1989, 183–87. See also the previous catalogue entry by Blochet 1912, no. 772, 41–47.

9. Ms. Persan 169, item 23, fols. 141v–179v. Discussed in Necipoğlu 1995, 133–39, figs. 99–103. I have now adopted the date for Abu'l-Wafa''s Arabic treatise proposed by Elaheh Kheirandish in her chapter in the present volume, titled "An Early Tradition in Practical Geometry: The Telling Lines of Unique Arabic and Persian Sources." The date ca. 993–98 is suggested in Raynaud 2012. For an Arabic edition by Ṣāliḥ Aḥmad al-ʿAlī, see Abū'l-Wafāʾ al-Būzjānī 1979. This and other problematic editions of the Arabic and Persian manuscripts are discussed in Kheirandish's chapter.

10. Abu'l-Wafa''s work (whether the original Arabic text or the Persian translation) will henceforth be referred to throughout this volume as the *Geometrical Constructions*.

11. His full name was Abu Ishaq Shaykhzadah b. Khadim al-Burhani al-Kubanani; his *nisba* relates him to Ku[h]banan, between Yazd and Kirman. On the postface and Kubanani Yazdi, see, in the present volume, chapter 4, Alpay Özdural's "Preliminaries"; the introduction by Gülru Necipoğlu, "In Memory of Alpay Özdural and His Unrealized Book Project"; and chapter 2, Kheirandish, "An Early Tradition in Practical Geometry." Kheirandish, who refers to Kubanani Yazdi as a "co-translator," partially translates his postface and discusses his life and works, using previously untapped primary sources. From a colophon she has established the date "post-1483" for his scholarly activity. See also Munzavī 1969 (1348), 1:143, 152, 318, 368; Rūḥ al-Amīnī 1990 (1369); and Karāmatī and Nigahbān 2009.

12. In Blochet 1912, 46–47, the geometrical drawings accompanying the Persian translation of Abu'l-Wafa''s treatise and the *Anonymous Compendium*, as well as the whole codex are attributed to Kubanani Yazdi. This hypothesis is repeated in Özdural's unpublished book without reference to Blochet. Kheirandish's chapter affirms the likelihood of that hypothesis. On the alternative dates proposed for the copying of the Paris Codex as a whole, which range from the fifteenth century (Alpay Özdural and Priscilla Soucek) to the sixteenth (Francis Richard) and seventeenth centuries (Edgar Blochet), see the introduction to the present volume: Necipoğlu, "In Memory of Alpay Özdural."

13. The Arabic codex inscribed with that verse and likely owned by Kubanani (Paris, Bibliothèque nationale de France, Ms. 4821) is discussed in Rāshid 2013, 571–76. The annotation on fol. 8r is cited in 572–73.

14. See Bulatov 1988. For the suggestion that Abu'l-Wafa''s treatise itself was written by a student of his from class notes, see Woepcke 1855.

15. According to Chorbachi there is "no reason whatsoever to connect these two works"; the *Anonymous Compendium*, dating from ca. 1200, remains both "anonymous and unique": Chorbachi 1989a, 160. She ascribed the latter to

Iraq sometime before 1180–1230, on the basis of a pentagonal star depicted in it (fol. 180r [1]), which also appears in a decorative panel at the so-called Abbasid Palace in Baghdad, built approximately between those dates (see my fig. 20a): Chorbachi and Loeb 1992, 752–54. Chorbachi noted a revived interest in Abu'l-Wafa''s practical geometry treatise at that time in Iraq: in Mosul, a commentary was written on it by Kamal al-Din Musa b. Yunus, known as Ibn Man'a. On the Ilkhanid provenance and date proposed by Özdural, see his "Preliminaries" (chapter 4 in the present volume), and my summary of his book manuscript in "In Memory of Alpay Özdural."

16. Bulatov 1988 regarded the *Anonymous Compendium* as a textbook explicating the basic methods of applied geometry and geometric ornament, dating it to the eleventh century. According to Chorbachi 1989a, this primary source furnishes evidence for a "direct meeting between the artisan-designers and geometrician-scientists," in which the former enlisted the help of the latter as teachers who established "practical geometry as the basis for artistic precision, ensuring high quality of geometric designs." Özdural interpreted the same source as a product of the "collaboration" between artisans and mathematicians: see in this volume Necipoğlu, "In Memory of Alpay Özdural," and Özdural, "Preliminaries." Jay Bonner, a design consultant specializing in Islamic architectural ornament, concurs with Özdural's views in his forthcoming book: "One is led to conclude that this treatise was the work of multiple individuals with variable levels of both mathematical proficiency and geometric design skills." I thank him for having shared with me the second chapter, titled "Geometric Diversity and Design Classification," of his forthcoming book. Hogendijk suggests that the *Anonymous Compendium* would have been used under the guidance of a competent teacher and was "probably the work of one or more craftsmen because it agrees with most of what Abu'l-Wafā' says about their methodology." He recognizes that "we cannot exclude the possibility that a few mathematicians were also involved in the design and construction of geometric ornaments": Hogendijk 2012, 38–39, 42.

17. I thank Wheeler Thackston for pointing out that the spelling of the second title shows that the writer did not have a good command of Arabic. See Ms. Persan 169, item 16, fols. 108r–118v; and item 19, fols. 124v–138v. I have briefly discussed these two treatises, which I examined in Paris: see Necipoğlu 1995, 168–69. Later on, excerpts from the *Anonymous Compendium* and other works of the Paris Codex were posted on Mandragore, the library's website for manuscripts: http://mandragore.bnf.fr/jsp/rechercheExperte.jsp (under "*cote*," search for "Persan 169").

18. On the differentiation between "constructive geometry" and "practical geometry" in the Latin West in the Middle Ages, see Shelby 1972, and Shelby 1977. While works in both fields omitted proofs, the former had little use for arithmetic, and its "mind set" differed from that of computational practical geometry. On the use of geometric design methods in Gothic architecture, see also Bucher 1972, and Bork 2011.

19. See J. Sesiano, s.v. "Wafḳ," *Encyclopaedia of Islam, New Edition* (Leiden: Brill, 1954–2004). Interestingly Sesiano illustrates the same magic square in fig. 10. Euclid is cited in "Constructions by Abu Bakr" on fol. 126v and Ptolemy on fol. 129v of the Paris Codex.

20. The marginal gloss on fol. 135v is also mentioned in Richard 1989, 187; and, in the present volume, by Özdural, "Preliminaries" (chapter 4), and Kheirandish, "An Early Tradition in Practical Geometry" (chapter 2). Al-Jurjani was the teacher of the Ottoman mathematician-astronomer Qadizada al-Rumi (ca. 1364–1436), who became the tutor of the Timurid prince Ulugh Beg (1394–1449) in Samarqand and eventually the head of that city's royal observatory, which was modelled on the Mongol-Ilkhanid observatory in Maragha: See Bāqirī 1997; Kennedy 1960.

21. In Necipoğlu 1995, 168–70, I suggested that parts of the *Anonymous Compendium* may have been modified while it was being copied, most likely by Kubanani Yazdi in the fifteenth century. Accepting Bulatov's eleventh-century dating of the *Anonymous Compendium*, Golombek and Wilber 1988, 1:159, hypothesized that some designs might have been added in the Timurid period.

22. See Necipoğlu, "In Memory of Alpay Özdural," and Özdural, "Preliminaries," in the present volume. See also Özdural 2000, 172n4: "According to the information kindly provided by Mohammed Bagheri and Mehran Akhbarifar (both of Tehran), Abū Bakr al-Khalīl is cited as the father of the copyist of a manuscript in Mashhad. Since the manuscript was copied in 1327–28, Abū Bakr al-Khalīl must have been active around 1300."

23. For the two treatises on surveying in Arabic and Persian, copied in 728 (1327–28) in the hand of 'Ali b. al-Khalil al-Tajir and bound together in the same codex (Mashhad, Āstān-i Quds-i Rażavī Library, Ms. 5429), see, in the present volume, Kheirandish, "An Early Tradition in Practical Geometry." She explains that Özdural's two informants (see n. 22 above) had mentioned the father-and-son relationship to him as a conjecture, one that still remains unconfirmed. Kheirandish also discusses the authorship and contents of the surveying manuscript in Mashhad.

24. Ibn al-Nadīm 1970, 1:667–68.

25. On this codex (Saint Petersburg, Library of the Oriental Institute, Ms. B 2139), comprising mathematical works in Arabic and the name of the copyist, see Rāshid 2013, 504. Rashid does not mention Abu Bakr's works in the Paris Codex (Ms. Persan 169).

26. I am assuming that Abu Bakr b. al-Khalil al-Tajir was not the son of Abu Bakr al-Khalil al-Tajir, which would seem rather farfetched. See Kheirandish's chapter in the present volume and n. 90 below for the suggestion of a possible connection between the *Anonymous Compendium* and the environs of Yazd.

27. Rāshid 2013, 504.

28. On Ibn al-Haytham, see Ibn al-Haytham 1989 (ed. Sabra).

29. Discussed in Rāshid 2013, 504.

30. Trained in the religious, mathematical and medical sciences, 'Imad al-Din 'Abdallah b. Muhammad b. 'Abd al-Razzaq b. al-Khawwam al-Baghdadi was a preceptor of the

sons of ministers and princes in Isfahan around 1277; by the beginning of the fourteenth century he taught Shafiʿi *fiqh* at the Dar al-Dhahab in Baghdad. On his life and works, see Rozenfel'd and İhsanoğlu 2003, 231, no. 657; and Abdeljaouad and Hedfi 1986, 159–77, esp. 159. I thank Kheirandish for bringing to my attention the following source, which gives additional information on Ibn al-Khawwam's presence in Isfahan: Qurbānī 1986–87 (1365), 23–24.

31. For Ibn al-Haytham's treatises on the regular heptagon and his "Completion of the Conics," see Hogendijk 1985; and Rāshid 1979. Özdural argued, rather dubiously, that the related pattern in the *Anonymous Compendium* corresponds to the definition of a special right-angled triangle offered by ʿUmar Khayyam, and that the scribe "attributes it wrongly to Ibn al-Haytham": see Özdural 1998, 702. In "Analyses," the second chapter in his unpublished book (pp. 150–51, 267), the presumed misattribution is unconvincingly explained as a scribal error, due to the "similarity between the pronunciations of 'Khayyam' and 'Haytham'," because Özdural believed that none of Ibn al-Haytham's treatises addressed this type of triangle. For a different view, see chapter 3 in this volume, by Jan Hogendijk, "A Mathematical Classification of the Contents of an Anonymous Persian Compendium on Decorative Patterns," where he suggests that both Ibn al-Haytham and ʿUmar Khayyam may have written treatises on that triangle. See Chorbachi 1989a, 272–73, for two possibly relevant works by Ibn al-Haytham: "A Passage in Which Lemmas Are Laid Down for the Construction of [Conic] Sections by Mechanical Means" (Florence, Biblioteca Medicea-Laurenziana, Ms. Or. 152, fols. 97v–100r), and "An Epistle on the Properties of the Triangle in Respect of Its Height," copied in Mosul in 632 (1234) and published in Hyderabad in 1947 (see Ibn al-Haytham 1947). Chorbachi explains that the second treatise describes a right-angled triangle whose "base plus its height is equal to the height of the original triangle (x:y:√3)": when illustrated, this shows a conic shape inscribed in an equilateral triangle (fig. 7, p. 591).
32. Discussed in Rāshid 2013, 470–568, esp. 470–72.
33. Discussed in ibid., 471–503, 513–64.
34. Cited from ibid., 559. The extract, without proofs, is translated on pp. 559–64: "Account of the Procedures for Measurement Set Out in This Treatise."
35. See Rāshid 2013, chapter on "Practical Geometry: Measurement," 469–568. On the "mixed" sciences, see also Kheirandish 2006, and Kheirandish 2013. She explains that the "mixed sciences" are generally associated with such mathematical disciplines as astronomy, harmonics, optics, and mechanics because of their strong natural/physical component. Coined recently, this term was initially applied to astronomy, a field sharing mathematical-natural sciences such as optics and mechanics. I am using the term here in a broad sense, with reference to "mixed" subfields of mathematics such as surveying, mechanics, engineering, and construction.
36. On al-Farabi's classification in the *Iḥṣāʾ al-ʿulūm* and the "mixed" sciences, see Rāshid 2013, the chapter on "Practical Geometry: Measurement," 469–568, esp. 469. See also an English translation of al-Farabi's discussion of the science of mechanical devices (*ḥiyal*) in Saliba 1985, 145–46. Al-Farabi's and al-ʿAmiri's classifications of mathematical sciences and their discussions of geometry are cited in Necipoğlu 1995, 132–33.
37. See chapter 3 ("A Mathematical Classification") in the present volume, where Hogendijk examines the marginal small pentagon (fig. 6).
38. Because Özdural interprets the term "*ustādān*" as "artisans" rather than "masters," and assumes that they were seeking help from the geometers, constituting a separate group, his conclusion that the two groups were participating in a collaborative working session is questionable.
39. See Özdural's interpretation of this construction in his unpublished book (pls. 30 and 31 in the present volume), and Hogendijk's analysis (chapter 3, "A Mathematical Classification," figure 11), in the present volume.
40. In the *Anonymous Compendium*, some phrases use numerals and others have the number written out; the quotations above correspond to the way they appear in the original document. See the glossary in this volume for the angle measurements corresponding to these lines, mentioned in Ms. Persan 169, fols. 193v [44]; 194v [48]; 195v [54]; 196r [55]; and 190v [35].
41. Hogendijk, "A Mathematical Classification" (chapter 3 in this volume), classifies patterns constituting variants of kite-shaped quadrangles. For set squares, see the glossary in this volume.
42. The pattern in Zuzan is discussed in Bonner's forthcoming book, chapter 2, fig. 112a–d. I found the similar pattern in Sivas in an illustration of Mülayim 1982, 136, pl. 22, fig. 136. The pattern in Mashhad is cited and redrawn by Özdural. Bonner argues that since the same pattern can be more easily generated by an underlying tessellation of squares and triangles, Abu Bakr and his associates were either unfamiliar with the less complex approach or "cared not to reveal it."
43. For these terms and related designs, see Necipoğlu 1995, 170–71. Versions of this decorative pattern in later sixteenth- and seventeenth-century monuments of Isfahan are analyzed mathematically in Cromwell and Beltrami 2011, where it is assumed to have been inspired by mathematical diagrams (p. 85).
44. Chorbachi 1989b, 460–62, fig. 6.3 on 762; and 764, fig. 8 on 766.
45. These woodwork examples are found in the Afyon Ulu Cami (1272), the Beyşehir Eşrefoğlu Camii (1297–99), and the Hamid Camii in Çorum (fourteenth century). Their patterns are redrawn in a plate (Tafel 16) in Schneider 1980, 96 (nos. 180–82), 176.
46. Unlike Özdural, Bonner finds the tessellation generated by this particular repeat-unit "unsatisfactory" as a decorative pattern (fol. 185v [18], pl. 16.3) because the hexagonal stars and octagons are irregular. He observes that an idealized version of the same pattern with regular hexagonal stars and octagons was used as a generative tessellation for an ornamental design in the entry portal of the Sultan Hasan complex in Cairo (1356–63). See Bonner forthcoming, figs.

70a–c, and 413a–b. "Special triangles-kite shapes" in the *Anonymous Compendium* are analyzed in Chorbachi 1989a, 267–76; see also Hogendijk, "A Mathematical Classification" (chapter 3 in the present volume).

47. See chapter 10 in Abū'l-Wafā' al-Būzjānī 1979; and the Persian and French translation in Abū'l-Wafā' al-Būzjānī 2010, esp. 217–29.

48. Özdural writes that the inclusion of the two diagrams in that treatise "appears as a natural product of the discussions between mathematicians and artisans, rather than belonging to a work of mensuration." See Özdural 2000, 182, 186–88 (figs. 9A, 9B); and "Analyses," the second chapter in his unpublished book, pp. 61–62. For another interpretation of these two patterns, see chapter 2 in this volume: Kheirandish, "An Early Tradition in Practical Geometry."

49. Hogendijk also suggests the possibility that the *Anonymous Compendium* may contain sections copied from earlier sources: see Hogendijk 2012, and his chapter in the present volume. A similar conjecture was made earlier in Chorbachi 1989a, 163, who suggested that it could be the "fragment of a previously complete text." Elsewhere she argued that its errors show that it was copied and "thus there must have been another manuscript like it": See Chorbachi 1989b, 777. In chapter 2 of his forthcoming book, Bonner hypothesizes that the *Anonymous Compendium* was likely produced sequentially, and quite possibly over several centuries, from the time of the Seljuks around the twelfth-century to that of the Khwarazmshahs and Timurids.

50. See Raynaud 2012, where he observes that two-thirds of the geometric constructions in Abu'l-Wafa''s treatise appear in Renaissance works on practical geometry, though not necessarily based on it directly.

51. Cited and discussed in Özdural, "Preliminaries" (chapter 4 in the present volume). See Özdural 2000, 173–85, and 193–200, for the translation and the Arabic text (Istanbul, Süleymaniye Library, Ms. Ayasofya 2753, fols. 52–53); also published in Berggren 2007, 611–15. The same passage was translated and interpreted much earlier in Chorbachi 1989b, 766–67, as referring to an "assembly of scientist-geometricians and artisans" in which "the two groups arrived at different methodologies for the construction of one square from the sum of three squares." See also Kheirandish's transcription and translation of the whole passage in her chapter in this volume.

52. "Cut-and-paste" constructions are defined by Jens Hayrup as denoting "methods in which two planes are shown to be equal by dividing each of them into collections of pair-wise-congruent figures": cited in Berggren 2007, 517n3. The modern term "cut-and-paste" is adopted in Hogendijk 2012 and his chapter in the present volume ("A Mathematical Classification"), and in Özdural's unpublished book.

53. See translation and Arabic text in Özdural 2000, 173–85, 193–200.

54. Bergé 1979, 171, 173–74, 203–6.

55. Discussed in Necipoğlu 1995, 157–58: *Kitāb al-Manāzil fī mā yaḥtāju ilayhi al-kuttāb wa al-ʿummāl min ʿilm al-ḥisāb* (Book of the Stations on What the Secretaries and Administrators Need of the Science of Arithmetic). See Saidan 1971.

56. The tenth century was "one of the most active periods of geometrical research in medieval Islam," during which Buyid kings and viziers supported the work of celebrated geometers: Berggren 2003.

57. Rāshid 2013, 300; and Rāshid 2005, 1–4. Despite evidence to the contrary, Terry Allen argues that "geometry was not necessarily part of a cultured man's education," in Allen 1988, 36.

58. On some of these works, see Rāshid 1979; Hogendijk 1981; Hogendijk 1984; and Hogendijk 1985. For treatises on constructing a regular heptagon and a nonagon, and on inscribing the pentagon in a square, see Berggren 2007, 517, 552, 580–87.

59. Discussed in Necipoğlu 1995, 104–8, fig. 93; and Tabbaa 2001, 44–48, figs. 14 and 15.

60. See Hogendijk 2008, 121; Hogendijk 2012, and his chapter in the present volume. He suggests that the "heptagonal tiling" may have been invented by ʿUmar Khayyam (d. 1131), who lived in Isfahan at that time. This hypothesis echoes that of Özdural, according to whom the same poet-mathematician contributed to the design of the pentagonal stellate pattern that decorates the inner surface of the brick cupola crowning the North Dome in Isfahan. Özdural identified the special right-angled triangle that generated this stellate pattern, issuing from a central pentagon, as the one on which ʿUmar Khayyam wrote a treatise in response to a question posed at a meeting. Although the author of the treatise only remarks on the high status of that meeting and of the one who posed the question, Özdural assumed that this was a *conversazione* between mathematicians and artisans/architects in Isfahan. In 1073 the Seljuk sultan Malikshah invited ʿUmar Khayyam to the city and put him in charge of a new observatory there: See Özdural 1995, 60–63, and Özdural 1998, 702–3.

61. Another tympanum panel, comprising hexagonal and pentagonal stars with regular hexagons, has been identified as a version of a pattern featuring only hexagonal stars, whose repeat-unit is included in the *Anonymous Compendium* (fol. 193r [43]). See Bonner's forthcoming book, chapter 2, fig. 309a–b, where he describes these two patterns as "nearly identical," since he focuses on their similar generative polygonal grids, rather than on the final patterns produced by those grids, which are less alike.

62. I am grateful to Farshid Emami for bringing this pattern and its archival photograph to my attention. Similar patterns with terracotta units and painted plaster inlays, which sequentially combine stars with differing numbers of points, appear on the mihrab of the Seljuk-period Friday Mosque (1105) in Barsian near Isfahan. Its patterns featuring pentagonal, hexagonal, heptagonal, and octagonal stars come close to those of the nearby North Dome of the Friday Mosque in Isfahan (1088–89).

63. On Fatimid examples of woodwork cited as evidence against the dissemination of geometric ornament from neighboring Abbasid and Seljuk realms (the earliest being

a minbar commissioned by the vizier Badr al-Jamali in 1091–92 for a shrine in Ascalon, and the rest dating from the twelfth century), see Bloom 2009; and O'Kane 2013. O'Kane even claims that "the Fatimids were leaders in the development and use of geometric ornamentation, embracing it before any other dynasty," thus underestimating the disappearance of relevant objects and monuments from the Buyid and Seljuk periods in and near Baghdad. Tabbaa argues more reasonably that even though some geometric woodwork was known in Fatimid Egypt, it only became widespread under the Ayyubids: Tabbaa 2014, 188–215; and Korn 2004.

64. For the reference to making "geometry manifest," see Bier 2016, "Geometry Made Manifest: Reorienting the Historiography of Ornament on the Iranian Plateau and Beyond," which she kindly shared with me. The likely origin of conical *muqarnas* domes is Iraq, the earliest extant example being the Imam Dur mausoleum near Samarra, which dates from 1088 and is coeval with the North Dome in Isfahan: see Tabbaa 2001, 103–36.

65. See Necipoğlu 1995, 92–126; Tabbaa 2001, 73–136; Tabbaa 2008; and Balbale 2017. Balbale disagrees with Jonathan Bloom's hypothesis that a wooden Almoravid minbar (begun in Cordoba in 1137 and completed ca. 1147) exemplifies a Cordoban tradition. Instead, she connects its geometric strapwork patterns, featuring octagonal stars and irregular polygons, to Almoravid traditions in North Africa, inspired by the Eastern Abbasid tradition and expressing political allegiance to the caliphate in Baghdad. See Bloom 1998, 3–30, esp. 21–25. On the a priori assertion that the use of geometric strapwork ornament in the Shiʿi Fatimid context is sufficient evidence for the absence of any context-specific associations, see Bloom 2009, 135; and O'Kane 2013, 324–25. See also Bloom 1998, 29n62, for the statement that "[t]he presence of geometric strapwork designs on woodwork ordered by Shiʿite and Sunni patrons alike should put to rest Necipoğlu's thesis...that contemporary patrons saw in such patterns any ideological content." His statement distorts the argument of my 1995 book, according to which this mode of geometric ornament was initially associated with allegiance by the Seljuk and Seljuk successor states to the Abbasid caliphate in Bagdad, rather than with sectarianism per se, and that its associations were adapted to new contexts as it spread further in time and space.

66. See Tabbaa 2017, in which he divides the late Baghdad caliphate into two periods: the period of Seljuk domination and Sunni revival (1055–ca. 1150), and the period of the resurgent Abbasid caliphate and Sunni ecumenism (ca. 1150–1258). In his chronicle, Miskawayh (ca. 932–1030) describes the measures taken by the Twelver Shiʿi Buyid amir ʿAdud al-Dawla (r. 949–83) against religious factionalism: Ismaʿili (Sevener Shiʿi) Fatimid missionaries were expelled and their activity in Baghdad curtailed. Sunnis and Shiʿis thereafter mingled together in their visits to shrines and oratories due to suppressed factionalism: see Donohue 2003, 80–81. For a nuanced interpretation of the blurring of sectarian boundaries in shrines, see Mulder 2006; Mulder 2014.

67. For "actor-network" theory, see Latour 2005.

68. There is no reason to assume, however, that the earliest extant manuscript necessarily has the "correct" version of the original title. The title of the very humble ca. 600 (1203) manuscript in Cairo (Dār al-Kutub al-Miṣriyya, Ms. 260 [31024], 1: 1r–58v) may even have been coined by or for the artisans who used it in woodwork. Perhaps deriving from it, a later Arabic manuscript in Cairo (Dār al-Kutub al-Miṣriyya, Ms. 366 [44795], 46 fols.) is dated 831 (1427) and bears the same title. See Kheirandish's chapter in the present volume ("An Early Tradition in Practical Geometry"), where she discusses the five known Arabic manuscripts and published editions. However, a systematic comparative study of these manuscripts has not been attempted thus far.

69. The two "early" Persian translations, with no date or place, are titled *Īn Kitāb-i Ustād Abū al-Wafā' Muḥammad ibn Muḥammad al-Būzjānī Dar Ānch/ Ānchih ṣunnāʿ va muḥtarifa bi ān muḥtāj bāshand az aʿmāl-i handasa va īn kitābrā Kitāb-i Nijārat khwānand*: Mashhad, Āstān-i Quds-i Rażavī Library, Ms. 5264 (twelfth-thirteenth centuries?), and Tehran University, Central Library, Ms. 2876 (eleventh-twelfth centuries?). See Kheirandish's chapter on these manuscripts. For Āqāyānī-Chāvushī's Persian edition and French translation of the Tehran manuscript, without comparison to its counterpart in Mashhad, see Abū'l-Wafā' al-Būzjānī 2010.

70. Istanbul, Süleymaniye Library, Ms. Ayasofya 2753.

71. See the abbreviated Persian version of this preface (Tehran University, Central Library, Ms. 2876) in Abū'l-Wafā' al-Būzjānī 2010, 37–38, 123. I am grateful to Himmet Taşkömür for the following transcription and transliteration of the Arabic preface: Istanbul, Süleymaniye Library, Ms. Ayasofya 2753, fol. 1v: هذا كتاب أبي الوفاء محمد بن محمد بن محمد البوزجاني فيما يحتاج إليه الصانع من أعمال الهندسة قد امتثلت ما رسمه مولانا الملك شاهانشاه السيد الأجلّ المنصور بهاء الدولة وضياء الملة وغياث الأمة -أطال الله بقاه وأدام علاه وقدرته وسلطانه- من إثبات المعاني التي كان يتذاكر بحضرته العالية من الأعمال الهندسية التي يكثر استعمالها عند الصناع مجرداً من العلل والبراهين ليسهل على الصناع تناوله ويقرب عليهم طريقته وجعلت الكتاب ثلثة عشر باباً

Transliteration: *Hādhā kitābu Abī al-Wafā' Muḥammad bin Muḥammad al-Būzjānī fī-mā yaḥtāju ilayhi al-ṣāniʿu min aʿmāl al-handasa. Qad imtathaltu mā-rasamahu mawlānā al-maliku shāhinshāh al-sayyid al-ajall al-manṣūr bahā' al-dawla wa ẓiyā' al-milla wa ghiyāth al-umma - aṭāla allāhu baqāhu wa adāma ʿalāhu wa qudratahu wa sulṭānahu- min ithbāt al-maʿānī allatī kāna yatadhākaru bi-ḥaẓratihi al-ʿāliyati min al-aʿmāl al-handasiyyati allatī yakthuru ʾstiʿmāluhā ʿinda al-ṣunnāʿi mujarradan min al-ʿilal wa al-barāhīn li-yusahhila ʿalā al-ṣunnāʿi tanawula, wa yuqar-*

riba ʿalayhim ṭarīqatahu wa jaʿaltu al-kitāba thalāthata ʿashara bāban.

72. In her chapter in the present volume, Kheirandish provides her own translation and interpretation of the Arabic preface in Ms. Ayasofya 2753. While it can be argued that the treatise addresses the author's patron and courtly circles, rather than artisans, who were unlikely to be addressed in textual form, I see little reason to doubt the stated intention of the preface. Some master artisans and surveyors were certainly literate to some degree; moreover, the geometrical constructions themselves provide sufficient visual information in most cases.
73. Ibn Khaldūn 1967, 2:360–61. Cited and discussed along with other sources that assigned the epithet "*al-najjār*" to Euclid and Apollonius of Perga, in Necipoğlu 1995, 153–54.
74. See Kheirandish's chapter in the present volume for these manuscripts.
75. Ibn Yunus indirectly communicated with the Latin West via the answers he gave to the scientific questions posed by Emperor Frederick II to Arabic scholars, and through his students, ʿAlam al-Din and Theodore of Antioch: see Raynaud 2012, 25; and Burnett 1995. The commentary by Ibn Yunus (Mashhad, Āstān-i Quds-i Rażavī Library, Ms. 5357) is discussed in Chorbachi 1989a, 6, 148–51, 447. She suggests that this work was probably dedicated to Salah al-Din in 1187–93. According to Chorbachi, the date is not easily legible on the manuscript that was copied in Maragha either in 608 (1211–12) or 680 (1281–82). Kheirandish adopts the latter date based on close examination and the library catalogue.
76. On Ibn Yunus's commentary, see Raynaud 2012, 25–28; and Chorbachi 1989a, 148–51. The Arabic commentary in the margins of Ulugh Beg's manuscript is ascribed to Abu'l-Qasim al-Andijani and dated to the fifteenth century in Chorbachi 1989a, 154–56, 270. A more likely attribution was recently suggested by Rushdi Rashid, who has identified the commentator as Ahmad b. Ahmad b. Jaʿfar al-Ghundijani, a geometer who lived in the second half of the eleventh or the beginning of the twelfth century: see Rāshid 2005, 13n36. On p. 142, Rashid also mentions a treatise on burning mirrors by al-Ghundijani, copied in 690 (1291). Muhammad Baqir Yazdi's much later commentary is discussed in Kheirandish's chapter, along with that of Ibn Yunus.
77. On these differences, see Raynaud 2012, 25–28; and Chorbachi 1989a, 148–51, who cites the reference to woodworkers.
78. Cited from the translation and Arabic text provided in Özdural 2000, 173–85, 193–99. For the contents of this work, see also its Persian version, which is accompanied by a French translation in Abū'l-Wafā' al-Būzjānī 2010.
79. For the observation that artisans were not always informed about or receptive to the advice of mathematicians such as Abu'l-Wafa', and that they continued to make errors thereafter, see Özdural 1995, 66–67.
80. Saliba 1999, 643.
81. See Berggren 2007, 611–20: he presents Abu'l-Wafa''s practical geometry treatise, the *Anonymous Compendium*, and the *Topkapı Scroll* as primary sources for the "mathematics of practitioners," where figures are more informative than texts. On constructive geometry in the Latin West, see n. 18 above. The inappropriate term "academic geometry" is used in another critical review of my book: see Allen 2004 (which is titled "Islamic Art and the Argument from Academic Geometry").
82. See Allen 2004, and Saliba 1999.
83. For the translation and Arabic text, see Özdural 2000, 173, 194. A new translation is provided in the chapter in the present volume by Kheirandish.
84. Concerning Abu'l-Wafa''s account of gatherings, Chorbachi observed long ago that "[i]t was the artisans and the architects who called for the services of science and scientists to assist them in solving the design problems that they were facing." See Chorbachi 1989b, 788.
85. Discussed in Chorbachi 1989a, 130–39; Abū'l-Wafā' al-Būzjānī 2010, 99–100, 126–27, 132–33.
86. Discussed in Chorbachi, 1989a, 129–30, 140–41; Abū'l-Wafā' al-Būzjānī 2010, 22–28, 105–7, 152–68.
87. For a discussion of the problem of constructing a square brick, see the Persian text and its French translation in Abū'l-Wafā' al-Būzjānī 2010, 229, and other related problems in 221–43. See also Chorbachi 1989a, 132–36, 142–44; Berggren 2007, 612–16.
88. Abū'l-Wafā' al-Būzjānī 2010, 105–113, 221–49.
89. Ibid., 221–33.
90. In her chapter in this volume, Kheirandish points out the similarity between the term *tadākhul*, used in the *Anonymous Compendium*'s heading, and that of a Persian text on surveying copied in 728 (1327–28) by ʿAli b. al-Khalil al-Tajir: *Fī al-Saṭḥ al-mutadākhil* (On Inscribed Surfaces). She explains that the latter is an addendum to the Persian translation of an Arabic treatise on surveying preserved in the same codex in Mashhad. She attributes both the translation and the addendum to Abu al-Fath al-Yazdi (ca. 1300), probably a contemporary of the copyist, ʿAli b. al-Khalil al-Tajir. Noting the author's connection with Yazd, she points out that Muhammad Baqir Yazdi, the seventeenth-century commentator on Abu'l-Wafa''s *Geometrical Constructions* in Persian, was also from Yazd, as was Kubanani Yazdi who completed the Persian translation of Abu'l-Wafa''s treatise preserved in the Paris Codex.
91. Blochet 1912, 42; Richard 1989, 184: see Ms. Persan 169, item 5, fols. 37v–41v. The Arabic root "*dakhala*" has a range of meanings, including the entering, inscribing, or insertion of one thing into another; to enter, join, mesh, or interlock. In Persian and Arabic mathematical texts *tadākhul* refers to a pair of numbers in which one is "contained" in the other a number of times. In other words, two numbers are "*mutadākhil*" if one is a divisor of the other; the text in the Paris Codex uses the pair "4, 20" as an example.
92. Chorbachi 1989a, 318. Unlike Bulatov, who mathematically analyzed nearly all the geometrical constructions in the *Anonymous Compendium*, Chorbachi focused on ten constructions but based her astute analyses on the manuscript's incised construction lines, which were invisible on the microfilm used by Bulatov.

93. For a detailed analysis of this page, see Chorbachi and Loeb 1992, 283–305; see also Chorbachi 1989a, 253–54. Bier observes that this construction is "not only a self-contained design," but "may also serve as a generative form to yield a planar pattern of overlapping decagons," such as a striking decagonal-pentagonal pattern in cut brick at the decagonal tomb tower in Maragha (northwestern Iran), the Gunbad-i Qabud, built in 593 (1196–97): see Bier 2012, 267–68.
94. On Gothic geometrical design strategies, see Bork 2011, 33.
95. For a different view, see chapter 2 in Bonner's forthcoming book: "the large portion of this treatise dedicated to geometric dissections does not appear to be of any practical use to artists working with geometric design. Similarly many of the instructions are of questionable relevance to artists." For a classification of constructions on kites, see Hogendijk's chapter in the present volume.
96. See Chorbachi 1989a, 239–45, 547–53. She connects this pattern with its versions on the minarets of the Friday Mosques in Damghan and Sava. See also fig. 145 in chapter 2 of Bonner's forthcoming book. He compares the same pattern with the two examples mentioned by Chorbachi, as well as a design on the mihrab of the Friday Mosque in Gulpaygan, and on a minaret in Daulatabad, Afghanistan. Unaware of Chorbachi's analysis, Bonner states that this pattern "has not been identified as a well-known 4-fold pattern in previous studies." Özdural's redrawing and expansion of the pattern shows that he was not aware of its connection with the historical examples cited by Chorbachi and Bonner (pls. 55.1–2 in the present volume). For some of the Anatolian examples in carved stone, see Schneider 1980, Tafel 28, 121, 183: Alay Han, portal (1180–1200); Divriği, Sitte Melik Türbesi, portal (thirteenth century); Sivas, Ulu Cami, mihrab (thirteenth century); Konya, Beşare Bey Mescidi, mihrab (1213–19).
97. Özdural proposes that the fifteenth-century copyist may have added the constructions on the last two pages of the *Anonymous Compendium.* I previously pointed out that the last pattern, which may have been added in the fifteenth century, is nevertheless in keeping with rotating swastika compositions seen in earlier Anatolian Seljuk monuments: see Necipoğlu 1995, 168–70, 180n113. On the possibility that the last page was a mid-fifteenth-century Timurid addition, see "Preliminaries," the chapter by Özdural in this volume. The mosaic tile dado panel at the Friday Mosque in Yazd is illustrated, together with another Timurid version featuring dodecagonal stars, in Necipoğlu 1995, 35, figs. 59–60. Variants of the swastika pattern with octagonal stars in Seljuk Anatolia and Kharraqan are illustrated in Mülayim 1982, 192 (pl. 78, fig. 83); 232 (pl. 118, fig. 124); 344 (pl. 228, fig. 270). The *Topkapı Scroll* has a swastika pattern with octagonal stars and another version with dodecagonal stars (see fig. 18a–d in the present essay).
98. In his forthcoming book, Bonner regards this pattern as aesthetically inadequate because he assumes that it is a finished ornamental design. He compares it with the successful combination of pentagonal, hexagonal, heptagonal, and octagonal stars on the mihrab of the Seljuk-period Friday Mosque (1105) in Barsian near Isfahan (fig. 332).
99. The term *ʿaqd* appears in fols. 185v [17, 18], 187v [26, 27], 188v–189r [30, 31], 190r [34], 190v [35], 191r–191v [36], 193v [44], 194r [45], and 194v [48].
100. Necipoğlu 1995, 172–73.
101. Based on Özdural's identification in "Analyses" (chapter 2 of his unpublished book), 78–87, 178–79, Hogendijk's chapter in the present volume classifies those three constructions as *muqarnas* vaults. The *muqarnas* design on a wall of the Armenian monastery (Astvatsankal Monastery, Armenia) subsequently disappeared: see Ghazarian and Ousterhout, 2001, 141–54; figs. 11–13. On pp. 148–50 are measured drawings and reconstruction diagrams of the now-lost drawing. For primary sources on the *muqarnas*, see Necipoğlu 1995, 4–27, 44–50, 159–60, 179, 350, 353, 359, and 173, as well as Al-Asad's appendix on "The Muqarnas: A Geometric Analysis," 349–59. Also see Dold-Samplonius 1992; Dold-Samplonius 1996; and Dold-Samplonius 2003.
102. See Hogendijk, "A Mathematical Classification," chapter 3 in the present volume, figs. 8–10.
103. On these translated sources, see Hogendijk 1986; Hogendijk 1993; and Necipoğlu 1995, 138–39, 177nn18–19. Sections from Book II of Euclid's *Elements* are in Ms. Persan 169, item 13, fols. 102v–104v.
104. This numerical table recalls two purely theoretical treatises by Ibn al-Haytham, in which he calculated the angles of all possible triangles for constructing a regular heptagon, solved by the intersection of two conic sections, a parabola and a hyperbola. This parallel has been suggested by Özdural (pl. 13.2). For Ibn al-Haytham's two treatises on the heptagon, see Rāshid 2013, 360–65, 383–84.
105. On these set squares and design principles, see Chorbachi 1989a, 207–10, 259, 265; and Özdural, "Synthesis," the third chapter in his unpublished book, p. 273. This construction is analyzed in Hogendijk's chapter ("A Mathematical Classification") in the present volume (fig. 7).
106. Unlike Özdural, I prefer to interpret "*kātib*" literally, as a reference to the scribe or copyist. In "Analyses" and "Synthesis," the second and third chapters of his unpublished book (pp. 156–60, 236, 258–59, 271–72), he proposes to read this as a name, "*Kātib* or *Kātib*[*ī*]" that has been spelled incorrectly. Özdural is inclined to identify him as ʿAli b. Ahmad Katibi, the designer of a luster-tile mihrab in the shrine of Imamzada Yahya at Varamin (1305), in support of what he sees as the "intimate link between theory and praxis." He states that this person was well-versed in conic sections and familiar with the mathematical sciences, and that he probably invented the T-square described in this passage, an instrument about the capacities of which the ignorant copyist expressed disbelief. On p. 236n177 of his unpublished book, Özdural explains that he interpreted "*kātib*" as a reference to the scribe/copyist in Özdural 1996, 192, 196, but then leaned towards reading *Kātib* or *Kātib*[*ī*] as the name of a particular person. In his chapter in the

present volume, Hogendijk also thinks that this may refer to "someone called Katib."

107. Discussed in Necipoğlu 1995, 141, 145, figs. 107a–b, where I only cited the 1974 English translation of al-Jazari's treatise by Donald Hill (see Jazarī 1974, 191–98, 267–68). For the Arabic edition by Aḥmad Yūsuf al-Ḥasan, see Jazarī 1979, 469–85. The palace is referred to as being located in "Āmid" (modern Diyarbakır) and the description of the tool appears right after that of the bronze door, thereby implying the connection between the two.

108. See Jazarī 1979, 469; translated in Jazarī 1974, 193, and Tabbaa 2001, 97.

109. See Hill's 1974 translation of al-Jazarī's treatise: Jazarī 1974, 191–95. Discussed in Necipoğlu 1995, 145, 151–52; and in Tabbaa 2001, 96–99.

110. For a different view, see chapter 2 in Bonner's forthcoming book, where he questions the *Anonymous Compendium*'s overall relevance to the history of Islamic geometric design. He assumes that "the geometric idiom was approaching full maturity, and the needs for artists to have direct mathematical input would have been less of an aesthetic imperative than during the earlier time of al-Buzjani." Bonner thereby underestimates continuing problems arising from the application of patterns to large-scale surfaces and to non-rectilinear areas with curved arch profiles. For *girih* patterns adapted to the curvature of arch spandrels in the *Topkapı Scroll*, see Necipoğlu 1995, 305, cat. 35; 307, cat. 39

111. See Necipoğlu 1995, 145. In her dissertation, which was not accessible when I wrote my 1995 book, Chorbachi translates *ʿaqd* as "unit of design," "repeat-unit," or "fundamental region": see Chorbachi 1989a, 164, 168. She provides a range of meanings for this word, which includes terms referring to the calculation of hand and finger joints (finger reckoning): e.g., "knit, knot, tie, to put together, to write, lock, to arch, vault." She notes that "all such meanings connect with the concept of tying, making a unit, as in unit of design, or an area within which the geometrical constructions of the sought pattern are carried": Chorbachi 1989a, 276–77.

112. For the incorrect assumption that *girih* is a post-nineteenth-century term, see Allen 2004; Bier 2012, 269; and Milwright, 2012. On sixteenth- and seventeenth-century Persian primary sources that often mention the term *girih* (and its variants) as a specific mode of ornament, see Necipoğlu 1995, 112–114; Necipoğlu 2016, 136–37. For the *girih* mode in present-day Iran (pronounced as *gereh* or *gireh*), see Shiʿrbāf 1993–200 (1372–97); Firishta-nizhād 1977 (1356); and Khānlarī 197?. On Central Asian masters of the *girih* mode, see Baklanov 1944; Baklanov 1947; and Gaganov 1958.

113. On geometric ornament in Seljuk Anatolia, see Schneider 1980; and Mülayim 1982.

114. On the differing regional tradition of geometric ornament in Nasrid Granada, see Makovicky 2011. The type of *muqarnas* that prevailed in the Islamic West and Sicily also remained distinct from examples in the Eastern Islamic lands: see Garofalo 2010. The proliferation of floral ornament in the early modern period is discussed in Necipoğlu 2016.

115. See Allen 2004. The low level of complexity in classical Roman, Byzantine, and Cosmatesque geometric patterns, in comparison with their Islamic counterparts, is discussed in Cromwell preprint 2013a, 34–36. See also Pajares-Ayuela 2001, esp. 146–95, 247–70.

116. See Allen 2004.

117. While acknowledging their complexity, not all authors agree on how to classify these patterns. See Bonner 2003; Lu and Steinhardt 2007; Saltzman 2008; Makovicky 2008; Bodner 2009; Cromwell 2009a; Cromwell 2009b; Cromwell 2010; Bodner 2012; Bonner 2012; Sarhangi 2012; and Bier 2015.

118. Cromwell 2009b, 41.

119. The earliest surviving example is the *Topkapı Scroll*. Chronologically next in line are the Tashkent scrolls, ascribed to sixteenth- and seventeenth-century master builder-architects of Bukhara in Uzbek Central Asia. These and other known scrolls are discussed in Necipoğlu 1995, "The Scroll Tradition," 3–57, and "Geometric Design in Timurid-Turkmen Architectural Practice," 48–67. For unpublished scrolls owned by master builders in Arab villages of Iraq, which Chorbachi examined in 1981 along with the Mirza Akbar scrolls from Qajar Iran, see Chorbachi 1989a, 262–63, fig. 22 at 608. A nineteenth-century Iranian scroll with twenty-four patterns was published in Khānlarī 197?.

120. These terms are used in Bonner's forthcoming book, chapter 2. See n. 121 below for the "polygonal technique" first noted by Hankin. "Point joining" refers to joining with lines critical points from which major and minor proportions are derived; see n. 92 above.

121. Necipoğlu 1995, 49–50: "In the Topkapı scroll polygonal grids or construction lines … define grid networks with polygons in contact … It would be easier to transfer such polygons onto wall surfaces than the multiple concentric circles used in generating the patterns on paper. Not all star-and-polygon patterns in the Topkapı scroll, however, are accompanied by construction lines defining polygons in contact. Those generated by more complex un-inked radial grid systems based on dynamic axes of symmetry use grids composed of stars and polygons in contact, which demonstrates that Hankin's formula was not universally applicable." The scroll's repeat-units were "generated by three types of grid systems: squared, triangulated, and composite radial (polygonal)": 31–32. See my catalogue of patterns and drawings of uninked grids, the mathematical analysis of which I left to mathematicians, on 231–39, 240–83.

122. See Necipoğlu 1995, "Catalog: Pattern Types and Drawings in the Topkapı Scroll," 231–39. See Lu and Steinhardt 2007, who demonstrate the complexity of fig. 19a–b, in which overlapping patterns at two different scales are generated by the same "*girih*-tile" shapes. Their argument for the quasicrystallinity of some Islamic patterns with five- and ten-fold symmetries is based on ideas illustrated in the *Topkapı Scroll*, where double-layered designs reflect-

ing sophisticated mathematics show how the large tiles divide into small ones. The "advanced nature" of seven "2-level patterns" in the *Topkapı Scroll* (cat. 28, 29, 31, 32, 34, 38, 49) is also recognized in Cromwell 2013a, 1–46, esp. 39–40: "In fact, they display a high degree of ingenuity and imagination, and are unconventional in some respects." But he unconvincingly speculates that, "given the advanced nature of the designs, a Safavid date, possibly quite late, seems more appropriate" (p. 40). Cromwell cites examples of two-layered patterns from late Safavid and Qajar monuments, which became rather "formulaic and routine to produce," but none of those late patterns are found in the *Topkapı Scroll*, and their "formulaic" character differs from their counterparts in the scroll. More importantly, none of the *Topkapı Scroll* patterns appears on any known monument; therefore Cromwell's alternative suggestion makes more sense: "Their constructions extend the methods we have seen in our analyses of surviving mosaics and they may be the result of someone exploring possibilities rather than a record of best practice" (p. 39). Nonetheless, the fact that the scroll patterns do not appear in the few surviving Timurid-Turkmen monuments of western and central Iran, a region prone to earthquakes, might also be an accident of survival.

123. See also the two-layered tympanum patterns in the North Dome (1088–99) of the Friday Mosque in Isfahan (fig. 10b). Another relatively early example of a two-layered geometrical pattern in carved brick is seen at the Gunbad-i Qabud in Maragha, dated 593 (1196–97), analyzed and illustrated in Lu and Steinhardt 2007, 1108 (figs. 2B–D), and Makovicky 2008. For the compelling argument that the so-called Abbasid Palace is not a madrasa, as some have proposed, but rather a highly original palace built ca. 1184 by the late-Abbasid caliph al-Nasir (r. 1180–25) and completed under al-Mustansir (r. 1228–42), see Tabbaa 2017.

124. Extant examples are illustrated in Necipoğlu 1995, 3–57. The Timurid-Turkmen provenance I have suggested for the *Topkapı Scroll* has been accepted in the art historical scholarship: see n. 5 above. An exception is Cromwell, who speculates that the scroll's "2-level designs" might date from the eighteenth or nineteenth centuries (see n. 122 above). However, there are no "2-level designs" in the Qajar-period Mirza Akbar scrolls or in any other late scroll from Iran and Central Asia that I have examined. For reasons why the *Topkapı Scroll* was not compiled at a later date, see Necipoğlu 1995, 34: "this is unlikely given the striking difference between its sophisticated drafting conventions and those of nineteenth- and twentieth-century scrolls. The fragmentary documents that compose the Topkapı scroll have been damaged over time, and this also seems to suggest an early date." The scroll was kept at the imperial treasury of the Topkapı Palace as a prized item, as indicated by its inventory number (H[azine] 1956), and was subsequently repaired. The type of paper, too, suggests a late fifteenth- or early sixteenth-century date, as do the types of patterns and kufic inscription panels compiled in the *Topkapı Scroll*: see Necipoğlu 1995, 29–39, esp. 38–39, 56n73.

125. The *Topkapı Scroll* also has a version of the swastika pattern with octagonal stars (cat. 67), which was common in monuments of Seljuk Anatolia and elsewhere: see n. 97 above.

126. This is also noted in Cromwell 2013b: "in discussions of Islamic patterns there is a tendency to focus on geometrical properties and to over-mathematise the design process," and "when circles are used they are not exactly as in the scrolls."

127. A computer program such as this has been developed: see Kaplan 2000; and Kaplan 2005.

128. For the "polygonal" method, see chapter 2 in Bonner's forthcoming book, and his articles cited above. Examples of studies that assume that only simple ruler-and-compass constructions were used by Islamic artisans include Bonner's forthcoming book; Broug 2013; Bloom 1998, 27. See also n. 139 below for Bloom's dismissal of advanced geometrical knowledge among woodworkers, even when the evidence suggests otherwise in particular cases.

129. Ibn Khaldūn 1967, 3:132.

130. Jones 1982, 69 (originally published in London in 1856).

131. See Bonner 2003; Bonner 2012; Lu and Steinhardt 2007. In the latter study, the "*girih*-tiles" correspond exactly to the dotted lines of the *Topkapı Scroll* patterns, as mentioned above in n. 122.

132. This is also acknowledged in Chorbachi 1989a, 73, 261. She observes that the patterns in the *Anonymous Compendium* may have been arrived at in several possible ways, and that the mechanics of the drawing process are not always clear from the illustrations, some of which were possibly copied from another model. For a summary of the different methods of designing star-and-polygon patterns, see Sarhangi 2012, who suggests that the combined use of ruler-and-compass constructions and *girih* tiles (i.e., the "modularity technique") accounts for the invention of more elaborate star-and-polygon patterns comprising ten-pointed star polygons, which would be "difficult if not impossible to execute using compass and straightedge alone." See Lu and Steinhardt 2007 for the use of edge-to-edge *girih* tiles, which was more convenient and quicker than employing a ruler and compass. An overview of differing design methods is provided in Bier 2015.

133. In his forthcoming book, Bonner hypothesizes that the "point-joining" instructions "may be the result of artists consulting with mathematicians to produce step-by-step instructions for specific patterns for their future use in this prescribed manner." While this is possible, I am less convinced by his statement that only the polygonal technique is "predisposed to the creation of new designs."

134. See Chorbachi 1989a, 319–20. For pragmatic reasons she adopted "an appropriate scientific language and notation system to classify Islamic geometric patterns according to their point group symmetry."

135. See Saltzman 2008, 153–54: "Imputing nascent group theory or nascent theory of quasi-crystals to medieval artists and mathematicians simply cannot be justified on the basis of

the historical record." He concludes: "Ultimately, attempts to find precursors of contemporary mathematical thought in the cultural production of medieval Islam, or any other period for that matter, fall prey to the *presentist fallacy*—the fallacy of reading the present into the past."

136. Later drawings in the Tashkent scrolls from Uzbek Central Asia and in the Mirza Akbar scrolls from Qajar Iran are discussed and illustrated in Necipoğlu 1995, 7–17.

137. The debate was triggered by Lu and Steinhardt 2007, with its wide media coverage; see articles cited in n. 117 above. Cromwell considers it anachronistic to apply modern mathematical concepts, such as crystallographic groups and quasiperiodic tilings to medieval Islamic ornamental patterns: see Cromwell 2013a; and Cromwell 2009a, 55: "Islamic artists were certainly familiar with generating designs by applying reflection, rotation and translation to repeat a template. They probably had an intuitive understanding of the crystallographic restriction and a feeling that global 5-fold and 10-fold rotation centres are somehow incompatible with periodicity. They did have the tools available to construct quasiperiodic designs but not the theoretical framework to appreciate the possibility or significance of doing so." See also Hogendijk 2012, who states that although *girih* tiles in the *Topkapı Scroll* "have drawn recent attention because they can be used to define aperiodic tilings, in the absence of textual evidence, it is impossible to say whether the craftsmen had an intuitive notion of aperiodicity." In his chapter in the present volume, Hogendijk likewise remarks that reading modern mathematical concepts such as aperiodic tilings or group theory into medieval texts and buildings is "problematic anyway." In contrast, Makovicky argues that the language of crystallography can be applied to Islamic patterns "without ascribing our modern definitions to the artists themselves." He does not agree that fourteenth-century artists of the Alhambra were "unable to have their own definitions of the reflection operation and kinds of rotational symmetry they used, even if we lack written records on this subject. Their geometric genius is clearly set in stone": see Makovicky 2011, 3. On the need for studies in cognitive history in the field of Greek mathematics, see Netz 1999.

138. See Ibn Abī Uṣaybiʿa 1998, 621–22; Wiedemann 1970, 114n1; Mayer 1958, 53. Subsequently discussed in Necipoğlu 1995, 152; Tabbaa 2001, 88–90, fig. 38 at 90; Bloom 2009, 132–34, 139–40; Allen 1999, "Chapter Eleven, Architects and Patrons."

139. While Bloom is correct in stating that not all woodworkers studied Euclid, he is not convincing in his attempt to disprove that Abu'l Fadl b. Abd al-Karim's knowledge of geometry informed his woodwork. He assumes that "[e]ven the most complex designs.... were probably made by combining set patterns rather than by complex geometric analysis": Bloom 2009, 134.

140. For a comparison of the two patterns in Damascus and Isfahan, see chapter 2 in Bonner's forthcoming book. He additionally compares both patterns to a third version in the *Anonymous Compendium* (fol. 193r [43]), which is based on a comparable generative polygonal tessellation, though its motifs differ significantly (fig. 309a–c).

141. On the wooden door, *muqarnas* portal, and two *muqarnas* domes of Nur al-Din's hospital, madrasa, and mausoleum complex in the Damascus complex, see Tabbaa 2001, 119–21.

142. See also Kheirandish's chapter in the present volume on this point. The manuscript was copied in Maragha by Husayn b. ʿAmr b. Ahmad al-Mardini. On Ibn Yunus's teacher, see Rāshid 2005, 692. On the transfer of manuscripts rescued from Baghdad and other Iraqi and Syrian towns to the Maragha observatory, see Saliba 2007, 244. The Maragha observatory was succeeded by other Ilkhanid observatories, including those of Ghazan Khan in Tabriz (ca. 1300) and Öljeytü in Sultaniya (ca. 1310): see Mozaffari and Zotti 2012, and Mozaffari and Zotti 2013. Damascus too was a major center for the study of astronomy.

143. The perfect compass, which could draw straight lines, the circumferences of circles, and conic sections, was probably invented by Abu Sahl al-Quhi (ca. 940–1000), who flourished in Baghdad and wrote a treatise on it. Other treatises on the perfect compass were written by al-Sijzi (ca. 945–ca. 1020), al-Biruni, and Ibn Yunus's student al-Abhari: see Rāshid 2005, 629–848. In addition to its use in theoretical mathematics, the perfect compass had technological applications in the construction of astrolabes and sundials, for which conics were essential.

144. This famous astrolabe is now in the Museum of Islamic Art in Doha, Qatar: see King 2005, 503–17; Hogendijk 2012, 42.

145. See Saliba 1999, 641–42, where he supplements this information with a comment that attributes a somewhat crude assumption to my 1995 book on the *Topkapı Scroll*: "This situation should, at least, warn against the assumption that artisans of medieval Islam, in particular, could on their own render and modify mathematical concepts simply because mathematical concepts were available in society."

146. Saliba 1999, 642. Quoting Biruni, Saliba writes: "Contrary to what Necipoğlu has argued in the book under review, there were apparently some artisans, especially astrolabe and instrument makers known to Bīrūnī, who preferred exact arithmetical measurements over the approximative artisanal ones apparently used by masons and architects and which were usually frowned upon by scientists rather than being inspired by them." This criticism is unwarranted because my book primarily focused on builders and designers of geometrical ornament, who generally preferred the approximation methods of practical geometry, rather than on the makers of astronomical instruments, who obviously had to use precise methods. On the jointly observed lunar eclipse, see Rāshid 2005, 997.

147. Golombek and Wilber 1988, 1:117, 189–93. See Thackston 1989, 159, where he translates the biographical accounts of Dawlatshah al-Samarqandi and Khwandamir. On the Timurid architect Qavam al-Din and a family of seventeenth-century Mughal architects descended from Ustad Ahmad Lahori, whose members studied mathematics and

geometry, some of them writing treatises on these subjects, see also Necipoğlu 1995, 5, 50–53, 155–58.

148. Mathematical treatises on proportion and the Neoplatonic epistle on ratio/proportion written by the Brethren of Purity (Ikhwan al-Safa') are cited in Necipoğlu 1995, 180n129. The philosophical and aesthetic dimensions of geometric patterns are discussed in chapters in Part V: "The Aesthetics of Proportion and Light," and "Geometric Abstraction and the Psychology of Visual Perception," 185–215. The treatises of al-Mahani are titled *Risāla fī al-nisba* (Epistle on Proportions), and *Risāla fī mushkil min amr al-nisba* (Epistle on Complex Proportions); that of Abu'l Wafa' is *Risāla fī al-nisāb wa al-taʿrīfāt* (Epistle on Ratio and Proportion). On those works and others by Ibn al-Haytham and ʿUmar Khayyam, see Sezgin 1974, 100, 260–62, 324.

149. Ms. Persan 169, fols. 179v–180r. I would like to thank Maria Subtelny for sharing with me her translation of this part of the postface, which she had prepared at the request of Özdural, who did not cite it. She notes that at the end of his postface, in a passage that recalls the Neoplatonic epistles by the Brethren of Purity (Ikhwan al-Safa'; see n. 148 above), Abu Ishaq Kubanani Yazdi encouraged the learned men (*fuẓalā'*) of his day to take a close look at the intricate problems discussed in Abu'l Wafa''s *Geometrical Constructions*: "so that they might appreciate the degree to which human capability (*qābiliyat-i bashar*) has grasped the workings of the earth and the extent to which [human] thought regarding the basic principles (*basā'iṭ*) [governing] the heavens and their understanding (*żabṭ-i ān*) has reached for 'Those who contemplate the creation of the heavens and the earth' (Q 3:191). 'He to whom wisdom (*ḥikma*) is granted, truly he has received abundant good.' (Q 2:269). It is hoped that the seeker after Truth (*ṭālib-i ḥaqq*) will receive his full portion of the sciences of Certainty/Proof (*ʿulūm-i yaqīnī*) (i.e., geometry), which are the special nourishment of the spirit (*rūḥ*) and the everlasting ornament of the soul (*jān*)." Subtelny explains that *ʿulūm-i yaqīnī* is usually used in a technical sense in Sufism to refer to intuitive certainty, as in *ʿilm-i yaqīn*, which denotes knowledge of certainty through proof, but *yaqīniyyāt* also means proofs or demonstrations in geometry: "It must have been used in both these meanings here, but with the emphasis on the mathematical sciences. I believe the point he is making is that there is no conflict between spiritual truth and scientific truth." For cosmological notions of beauty, see Necipoğlu 1995, "Part 5: Geometry and Aesthetic Theory," 183–226.

150. The quotation from Kepler is cited from Endress 2003, 122; see also Necipoğlu 1995, 162–64, 163 (fig. 118), 179n93.

151. These two lost treatises are mentioned as representative of that science in the *Irshād al-qāṣid ilā asnā al-maqāṣid* by the Mamluk scholar Ibn al-Akfani (d. 1348–49); this was one of the models for the *Kitāb Miftāḥ al-saʿāda wa misbāḥ al-siyāda fī mawḍūʿāt al-ʿulūm* (Key of Happiness) by the Ottoman scholar Taşköprizade (d. 1561). See Ibn al-Akfānī 1978, 81–82; and Ṭāşköprizāde 1911–37 (1328–56), 1:311–12; discussed in Necipoğlu 1995, 140–41, 145. I have cited here Chorbachi's translation of the description of Ibn al-Haytham's lost work on the science of building (*ʿilm ʿuqūd al-abniya*), recorded in a list of his own works (composed before the end of October 1038), which was copied in the biographies of physicians by Aḥmad b. al-Qāsim b. Abī Uṣaybiʿa (d. ca. 1269–70). See Chorbachi 1989a, 274–75, cited from Ibn Abī Uṣaybiʿa 1965, 2:93. In Saliba's opinion, the term *ʿuqūd* in "*ʿuqūd al-abniya*" (Science of Erection of Structures) is predominantly concerned with vaulting (*ʿaqd*), since the dictionary specifies that "*ʿaqada al-binā* means to 'vault the structure'." However, sources mentioned above refer to diverse construction types. Although Saliba believes that al-Karaji's treatise might be the same as his extant treatise, titled *Inbāṭ al-miyāh al-khafiyya* (Extraction of Hidden Waters), both works by Karaji are listed under separate titles by Taşköprizade, who classifies them under different sciences: see Saliba 1999, 642–43.

152. For a comparison of architectural-artisanal practice and extant medieval drawings in the Latin West and medieval Islamic lands, see Necipoğlu 1995, 3–57, 131–81, esp. 41–53.

153. Maqrīzī 1987, 2:362–63; cited in Necipoğlu 1995, 123.

154. See Chorbachi, 1989a, 114–15, on the addendum to Miskawayh's chronicle according to which the caliph's early death meant that his wish was left unfulfilled.

155. Ibn Miskawayh 1920–21, 5:443–48, at 447.

156. Rāshid 2005, 18, 997.

157. See Bāqirī 1997, 241–46.

158. Contextualizing exchanges between geometer-astronomers, artisans, surveyors, and master builder-architects requires multiple perspectives and a consideration of different kinds of agency, including that of scientists, practitioners, and their patrons, as well as the semi-autonomous agency of inanimate artifacts, technologies, and design traditions. On human and nonhuman agency in the arts and sciences, see Gell 2010, and Latour 2005.

159. See, for example, Galison and Jones 1998; Turner 2006; Stafford 2007; and Payne 2015.

BIBLIOGRAPHY

Abdeljaouad, Mahdi and Hmida Hedfi. 1988. "Vers une étude des aspects historiques et mathématiques des problèmes ouverts d'Ibn Khawwam (XIIIe. s.)." In *Histoire des mathématiques arabes, Actes du 1er colloque maghrébin sur l'histoire des mathématiques arabes, Alger, 1–3 décembre 1986*, 159–77. Algiers: Maison du Livre.

Abū'l-Wafā' al-Būzjānī. 2010. *Ketâb al-Nejârat: Sur ce qui est indispensable aux artisans dans les constructions géométriques.* Persian edition with French translation and commentary by Ja'far Āqāyānī-Chāvushī and Bernard Vitrac. Bibliothèque iranienne 71. Tehran: Institut français de recherche en Iran (IFRI).

———. 1990–91 (1369). *Handasa-i Īrānī: Kitāb-i Tijārat: Fī mā yaḥtāju ilayhi al-'ummāl wa-al-ṣunnā' min al-ashkāl al-handasiyya, yā, Kārburd-i handasa dar 'amal.* Edited and adapted (to modern Persian) by Sayyid 'Alī Riżā Jaẕbī. Tehran: Surūsh.

———. 1979. *Mā yaḥtāju ilayhi al-ṣāni' min 'ilm al-handasa.* Edited by Ṣāliḥ Aḥmad al-'Alī. [Baghdad]: Jāmi'at Baghdād, Markaz Iḥyā' al-Turāth al-'Ilmī al-'Arabī.

Allen, Terry. 2004. "Islamic Art and the Argument from Academic Geometry." Occidental, Calif.: Solipsist Press. http://www.sonic.net/~tallen/palmtree/academicgeometry.htm#dOe1 127

———. 1999. "Ayyubid Architecture." Occidental, Calif.: Solipsist Press. http://www.sonic.net/~tallen/palmtree/ayyarch/

———. 1988. *Five Essays on Islamic Art.* Sebastopol, Calif.: Solipsist Press.

Aube, Sandra. 2017. *Le céramique dans l'architecture en Iran au XVe siècle: Les arts qarâ quyûnlûs et âq quyûnlûs.* Paris: Presses de l'université Paris-Sorbonne.

Baklanov, Nikolai B. 1947. "Gerikh: Geometricheskii ornament Srednei Azii i metodi ego postroeniia." *Sovetskaia Arkheologiia* 9:101–20.

———. 1944. "Arkhitekturnye chertezhi uzbekskogo mastera XVI veka." *Soobshcheniia Instituta istorii i teorii arkitektury, Akademii arkhitektury SSSR* 4:1–21.

Balbale, Abigail. 2017. "Bridging Seas of Sand and Water: The Berber Dynasties of the Islamic Far West." In *A Companion to Islamic Art and Architecture*, edited by Barry Flood and Gülru Necipoğlu, 2 vols. Blackwell Companions to Art History. Malden, Mass.: Wiley-Blackwell.

Bāqirī, Muḥammad. 1997. "A Newly Found Letter of Al-Kāshī on Scientific Life in Samarkand." *Historia Mathematica* 24:241–56.

Bergé, Marc. 1979. *Pour un humanisme vécu: Abū Ḥayyān al-Tawḥīdī.* Damascus: Institut français de Damas.

Berggren, J. L. 2007. "Mathematics in Medieval Islam." In *The Mathematics of Egypt, Mesopotamia, China, India, and Islam: A Sourcebook*, edited by Victor J. Katz et al., 515–675. Princeton, N.J.: Princeton University Press.

———. 2003. "Tenth-Century Mathematics through the Eyes of Abū Sahl al-Kūhī." In *The Enterprise of Science in Islam: New Perspectives*, edited by Jan P. Hogendijk and Abdelhamid I. Sabra, 177–96. Cambridge, Mass.: MIT Press.

Bier, Carol. 2015. "Geometry in Islamic Art." In *Encyclopaedia of the History of Science, Technology, and Medicine in Non-Western Cultures*, edited by Helene Selin. Dordrecht: Springer.

———. 2016. "Geometry Made Manifest: Reorienting the Historiography of Ornament on the Iranian Plateau and Beyond." In *Historiography of Persian Architecture*, edited by Mohammad Gharipour, 41–79. New York and London: Routledge.

———. 2012. "The Decagonal Tomb Tower at Maragha and Its Architectural Context: Lines of Mathematical Thought." *Nexus Network Journal* 14, no. 2: 251–73.

Blair, Sheila S. 2008. "A Brief Biography of Abu Zayd." *Muqarnas* 25:155–76.

Blair, Sheila S., and Jonathan Bloom. 2006. "Ornament in Islamic Art." In *Cosmophilia: Islamic Art from the David Collection, Copenhagen*, 9–30. Chestnut Hill, Mass.: McMullen Museum of Art.

———. 2003. "The Mirage of Islamic Art: Reflections on the Study of an Unwieldy Field." *Art Bulletin* 85, no. 1: 152–84.

Blochet, Edgar. 1905–34. *Catalogue des manuscripts persans.* 4 vols. Paris: Imprimerie nationale.

Bloom, Jonathan. 2009. "Woodwork in Syria, Palestine and Egypt during the 12th and 13th Centuries." In *Ayyubid Jerusalem: The Holy City in Context 1187–1250*, edited by Robert Hillenbrand and Sylvia Auld, 129–46. London: Altajir Trust.

———. 1998. "The Minbar from the Kutubiyya Mosque." In *The Minbar from the Kutubiyya Mosque*, edited by Jonathan Bloom et al., 3–30. New York: Metropolitan Museum of Art.

Bodner, B. Lynn. 2012. "From Sultaniyeh to Tashkent Scrolls: Euclidean Constructions of Two Nine- and Twelve-Pointed Interlocking Star Polygon Designs." *Nexus Network Journal* 14:307–32.

———. 2009. "The Eleven-Pointed Star Polygon Design of the Topkapı Scroll." In *Bridges Banff 2009: Mathematics, Music,*

Art, Architecture, Culture, edited by Craig S. Kaplan and Reza Sarhangi, 147–54. [Hertfordshire, Eng.]: Tarquin Publications.

Bonner, Jay. Forthcoming. *Islamic Geometric Patterns: Their Historical Development and Traditional Design Methodologies*. New York: Springer-Verlag.

———. 2012. "Creating Non-Systematic Islamic Geometric Patterns with Complex Combinations of Star Forms." In *Proceedings of Bridges 2012: Mathematics, Music, Art, Architecture, Culture*, edited by Robert Bosch, Douglas McKenna, and Reza Sarhangi, 593–98. Phoenix, Ariz.: Tessellations Publishing.

———. 2003. "Three Traditions of Self-Similarity in Fourteenth and Fifteenth Century Islamic Geometric Ornament." In *Meeting Alhambra: ISAMA-Bridges 2003 Conference Proceedings: Mathematical Connections in Art, Music and Science*, edited by Reza Sarhangi and C. Sequin, 1–12. Granada: University of Granada.

Bork, Robert. 2011. *The Geometry of Creation: Architectural Drawing and the Dynamics of Gothic Design*. Farnham: Ashgate.

Broug, Eric. 2013. *Islamic Geometric Design*. London: Thames & Hudson.

Bucher, François. 1972. "Medieval Architectural Design Methods, 800–1560." *Gesta* 11, no. 2: 37–51.

Bulatov, Mitkhat S. 1988. *Geometricheskaia garmonizatsiia v arkhitekture Srednei Azii IX–XV vv*. Rev. ed. Moscow: "Nauka." Original edition Moscow, 1978.

Burnett, Charles. 1995. "Master Theodore, Frederick II's Philosopher." In *Federico II e le nuove culture: Atti del XXXI Convegno storico internazionale, Todi, 9–12 ottobre 1994*, 225–85. Spoleto: Centro italiano di studi sull'alto medioevo.

Charette, François. 2006. "The Locales of Islamic Astronomical Instrumentation." *History of Science* 44, no. 2: 123–38.

Chorbachi, Wasma'a K. 1989a. "Beyond the Symmetries of Islamic Geometric Patterns: The Science of Practical Geometry and the Process of Islamic Design." PhD diss., Harvard University.

———. 1989b. "In the Tower of Babel: Beyond Symmetry in Islamic Design." *Computers and Mathematics, with Applications* 17:751–89.

Chorbachi, Wasma'a K., and Arthur L. Loeb. 1992. "An Islamic Pentagonal Seal (from Scientific Manuscripts of the Geometry of Design)." In *Fivefold Symmetry*, edited by István Hargittai, 283–305. Singapore: World Scientific.

Cromwell, Peter R. Preprint, 2013a. "Modularity and Hierarchy in Persian Geometric Ornament." http://www.liv.ac.uk/~spmr02/islamic/cromwell_persian.pdf

———. Preprint, 2013b. "On Irregular Stars in Islamic Geometric Patterns." http://www.liv.ac.uk/~spmr02/islamic/cromwell_irregular_stars.pdf

———. 2010. "Islamic Geometric Designs from the Topkapı Scroll II: A Modular Design System." *Journal of Mathematics and the Arts* 4, no. 3: 119–36.

———. 2009a. "Islamic Geometric Designs from the Topkapı Scroll I: Unusual Arrangements of Stars." *Journal of Mathematics and the Arts* 4, no. 2: 73–85.

———. 2009b. "The Search for Quasi-Periodicity in Islamic 5-fold Ornament." *The Mathematical Intelligencer* 31, no. 1: 36–56.

Cromwell, Peter R., and Elisabetta Beltrami. 2011. "The Whirling Kites of Isfahan: Geometric Variations on a Theme." *The Mathematical Intelligencer* 33, no. 3: 84–93.

Dawlatshāh al-Samarqandī. 1901. *Tadhkīrat al-shuʿarāʾ* (*"Memoirs of the Poets"*). Edited by Edward G. Browne. London: Luzac.

Dold-Samplonius, Yvonne. 2003. "Calculating Surface Areas and Volumes in Islamic Architecture." In *The Enterprise of Science in Islam: New Perspectives*, edited by Jan P. Hogendijk and Abdelhamid I. Sabra, 235–65. Cambridge, Mass.: The MIT Press.

———. 1996. "How al-Kashi Measures the *Muqarnas*: A Second Look." In *Mathematische Probleme im Mittelalter: Der lateinische und arabische Sprachbereich*, edited by Menso Folkerts, 56–90. Wiesbaden: Harrassowitz Verlag.

———. 1992. "Practical Arabic Mathematics: Measuring the Muqarnas by al-Kāshī." *Centaurus* 35:193–242.

Donohue, John J. 2003. *The Buwayhid Dynasty in Iraq 334 H. /945 to 403 H. /1012: Shaping Institutions for the Future*. Islamic History and Civilization 44. Leiden: Brill.

Endress, Gerhard. 2003. "Mathematics and Philosophy in Medieval Islam." In *The Enterprise of Science in Islam: New Perspectives*, edited by Jan P. Hogendijk and Abdelhamid I. Sabra, 121–76. Cambridge, Mass.: MIT Press.

Firishta-nizhād, Murtażā, ed. 1977 (1356). *Girih-sāzī va girih-i chīnī dar hunar-i miʿmārī-i Īrān*. Tehran: Anjuman-i Ās̱ār-i Millī.

Gaganov, G. A. 1958. "Geometricheskii ornament Srednei Azii." *Arkhitekturnoe nasledstvo* 11:181–208.

Galison, Peter, and Carolyn A. Jones, eds. 1998. *Picturing Science, Producing Art*. New York: Routledge.

Garofalo, Vincenza. 2010. "A Methodology for Studying Muqarnas: The Extant Examples in Palermo." *Muqarnas* 27:357–406.

Gell, Alfred. 2010. *Art and Agency: An Anthropological Theory*. Oxford: Clarendon Press.

Ghazarian, Armen, and Robert Ousterhout. 2001. "A Muqarnas Drawing from Thirteenth-Century Armenia and the Use of Architectural Drawings during the Middle Ages." *Muqarnas* 18:141–54.

Golombek, Lisa, and Donald Wilber. 1988. *The Timurid Architecture of Iran and Turan*. 2 vols. Princeton, N.J.: Princeton University Press.

Golombek, Lisa, and Ebba Koch. 2017. "In Search of the Timurid Roots of Mughal Architecture." In *A Companion to Islamic Art and Architecture*, edited by Barry Flood and Gülru Necipoğlu, 2 vols. Blackwell Companions to Art History. Malden, Mass.: Wiley-Blackwell.

Hogendijk, Jan P. 2012. "Mathematics and Geometric Ornamentation in the Medieval Islamic World." *Newsletter of the European Mathematical Society* 86:37–43.

———. 2008. "Ancient and Modern Secrets of Isfahan." *Nieuw Archief voor Wiskunde*. 5th ser., 9, no. 2: 121.

———. 1993. "The Arabic Version of Euclid's 'On Divisions.'" In *Vestigia Mathematica: Studies in Medieval and Early Modern Mathematics in Honour of H. L. L. Busard*, edited by Menso Folkerts and Jan P. Hogendijk, 143–62. Amsterdam: Rodopi.

———. 1986. "Arabic Traces of Lost Works of Apollonius." *Archive for History of Exact Sciences* 35:187–253.

———. 1985. *Ibn al-Haytham's Completion of the Conics*. New York: Springer-Verlag.

———. 1984. "Greek and Arabic Constructions of the Regular Heptagon." *Archive for History of Exact Sciences* 30:197–330.

———. 1981. "How Trisections of the Angle were Transmitted from Greek to Islamic Geometry." *Historia Mathematica* 8:417–38.

Holod, Renata. 1988. "Text, Plan, and Building: On the Transmission of Architectural Knowledge." In *Theories and Principles of Design in the Architecture of Islamic Societies: A Symposium Held by the Aga Khan Program for Islamic Architecture at Harvard University and the Massachusetts Institute of Technology, Cambridge, Massachusetts, November 6–8, 1987*, edited by Margaret Bentley Ševčenko, 1–12. Cambridge, Mass.: Aga Khan Program for Islamic Architecture.

———. 1986. "Defining an Art of Architecture." In *Architecture Education in the Islamic World: Proceedings of Seminar Ten in the Series Architectural Transformations in the Islamic World Held in Granada, Spain, April 21–25, 1986*, edited by Ahmet Evin, 1–32. Singapore: Concept Media for the Aga Khan Award for Architecture.

Ibn Abī Uṣaybiʿa, Aḥmad ibn al-Qāsim. 1998. *ʿUyūn al-anbāʾ fī ṭabaqāt al-aṭibbā*. Edited by Muḥammad Bāsil ʿUyūn al-Sūd. Beirut: Dār al-Kutub al-ʿIlmiyya.

———. 1965. *ʿUyūn al-anbāʾ fī ṭabaqāt al-aṭibbā*, Edited by Nizār Riḍā. Beirut: Dār Maktabat al-Ḥayāt.

Ibn al-Akfānī, Muḥammad ibn Ibrahīm. 1978. *Irshād al-qāṣid ilā asnā al-maqāṣid*. Edited by ʿAbd al-Laṭīf Muḥammad al-ʿAbd. Cairo: Maktabat al-Anjlū al-Miṣriyya.

Ibn al-Haytham, Abū ʿAlī al-Ḥasan ibn al-Ḥasan. 1989. *The Optics of Ibn al-Haytham, Books I–III: On Direct Vision = Kitāb al-Manāzir*. Translated and edited by Abdelhamid I. Sabra. 2 vols. London: Warburg Institute, University of London.

———. 1947. *Majmūʿ al-rasāʾil Ibn al-Haytham*. Hyderabad: Maṭbaʿat al-Maʿārif ʿUthmāniyya.

Ibn al-Nadīm, Muḥammad ibn Isḥāq. 1970. *The* Fihrist *of al-Nadīm: A Tenth-Century Survey of Muslim Culture*. Edited and translated by Bayard Dodge. 2 vols. New York: Columbia University Press.

Ibn Khaldūn. 1967. *The Muqaddimah: An Introduction to History*. Translated by Franz Rosenthal. 3 vols. Princeton, N.J.: Princeton University Press.

Ibn Miskawayh, Aḥmad ibn Muḥammad. 1920–21. *Tajārib al-umam, Books 5 and 6 = The Concluding Portion of* The Experiences of the Nations. Arabic text edited by H. F. Amedroz. Translated by D. S. Margoliouth. Oxford: Basil Blackwell.

Jazarī, Ismāʿīl ibn al-Razzāz. 1979. *Al-Jāmiʿ bayna al-ʿilm wa-al-ʿamal al-nāfiʿ fī ṣināʿat al-ḥiyal = A Compendium on the Theory and Practice of the Mechanical Arts*. Edited by Aḥmad Yūsuf al-Ḥasan. Aleppo: Maʿhad al-Turāth al-ʿIlmī al-ʿArabī, Jāmiʿat Ḥalab.

———. 1974. *The Book of Knowledge of Ingenious Mechanical Devices = Fī maʿrifat al-ḥiyal al-handasiyya*. Translated by Donald R. Hill. Dordrecht: Reidel.

Jones, Owen. 1982. *The Grammar of Ornament*. London : Day & Son, 1856. Reprint, New York: Van Nostrand Reinhold.

Kaplan, C. S. 2005. "Islamic Star Patterns from Polygons in Contact." *Proceedings of Graphics Interface 2005*, ed. Glenn E. Healey et al., 177–86. Victoria, British Columbia: A. K. Peters / CRC Press.

———. 2000. "Computer Generated Islamic Star Patterns." *Proceedings of Bridges 2000: Mathematical Connections in Art, Music, and Science*, edited by Reza Sarhangi, 105–112. Winfield, Kans.: Southwestern College.

Karāmatī, Yūnus, and Farzīn Nigahbān. 2009. *Encyclopedia Islamica*, s.v. "Abū Isḥāq al-Kūbunānī." Brill: London.

Kennedy, Edward S. 1960. "A Letter of Jamshīd al-Kāshī to His Father: Scientific Research and Personalities at a Fifteenth Century Court." *Orientalia* 29:191–213.

Khānlarī, Parvīz Nātil, ed. 197?. *Girih-sāzī dar hunar-i miʿmārī-yi Īrān*. With tile patterns by Muḥammad ʿAlī Fuṣhatī and ʿAbd al-ʿĀlī Fuṣhatī, photographed by Karāmat Raʿnā Ḥusaynī. Tehran: Bunyād-i Farhang-i Īrān.

Kheirandish, Elaheh. 2013. "The Mixed Mathematical Sciences: Optics and Mechanics in the Middle Ages." In *The Cambridge History of Science*, vol. 2, *Medieval Science*, edited by David C. Lindberg and Michael H. Shank, 84–108. Cambridge: Cambridge University Press.

———. 2006. "Organizing Scientific Knowledge: The 'Mixed' Sciences in Early Classifications." In *Organizing Knowledge: Encyclopaedic Activities in the Pre-Eighteenth Century Islamic World*, edited by Gerhard Endress, 135–54. Leiden: Brill.

Kheirandish, Elaheh, Carol Bier, and Najm al-Din Yousefi, eds. 2008. "Sciences, Crafts, and the Production of Knowledge: Iran and Eastern Islamic Lands (ca. 184–1153 AH/800–1740 CE)." Special issue, *Iranian Studies* 41, no. 4: 433–600.

Khvānd Mīr, Ghiyās̱ al-Dīn. 1954. *Tārīkh-i ḥabīb al-siyar fī akhbār afrād bashar*. Edited by Jalāl al-Dīn Humāʾī. 4 vols. Tehran: Kitābkhāna-i Khayyām.

King, David A. 2004–5. *In Synchrony with the Heavens: Studies in Astronomical Timekeeping and Instrumentation in Medieval Islamic Civilization*. Vol. 2, *Instruments of Mass Calculation*. Leiden: Brill.

Korn, Lorenz. 2004. *Ayyubidische Architektur in Ägypten und Syrien: Bautätigkeit im Kontext von Politik und Gessellschaft 564–658/1169–1260*. Heidelberg: Heidelberger Orientverlag.

Latour, Bruno. 2005. *Reassembling the Social: An Introduction to Actor-Network Theory*. Oxford: Oxford University Press.

Lu, Peter J., and Paul J. Steinhardt. 2007. "Decagonal and Quasi-Crystalline Tilings in Medieval Islamic Architecture." *Science* 315:1106–10.

Makovicky, Emil. 2011. *Crystallographer's Alhambra: Beauty of Symmetry—Symmetry of Beauty*. Copenhagen: University of Copenhagen; Granada: University of Granada.

———. 2008. "Another Look at the Blue Tomb of Marāgha, a Site of the First Quasicrystalline Islamic Pattern." *Symmetry: Culture and Science* 19, nos. 2–3: 127–51.

Maqrīzī, Aḥmad ibn ʿAlī. 1987. *Mawāʿiẓ wa-al-iʿtibār fī dhikr al-khiṭāṭ wa- al-āthār*. 2 vols. Cairo: Maktabat al-Thaqāfa al-Dīniyya.

Mayer, L. A. 1958. *Islamic Woodcarvers and Their Works*. Geneva: A. Kundig.

Milwright, Marcus. 2012. "Gereh-Sāzī ii. Architecture." *Encyclopaedia Iranica*. Online edition: http://www.iranicaonline.org/articles/gereh-sazi.

Mozaffari, S. Mohammad, and Georg Zotti. 2013. "The Observational Instruments at the Maragha Observatory after AD 1300." *Suhayl: International Journal for the History of the Exact and Natural Sciences in Islamic Civilisation* 12:45–179.

———. 2012. "Ghāzān Khān's Astronomical Innovations at Marāgha Observatory." *Journal of the American Oriental Society* 132, no. 3: 395–425.

Mülayim, Selçuk. 1982. *Anadolu Türk Mimarisinde Geometrik Süslemeler: Selçuklu Çağı*. Ankara: Kültür ve Turizm Bakanlığı Yayınları.

Mulder, Stephennie. 2014. *The Shrines of the ʿAlids in Medieval Syria: Sunnis, Shiʿis and the Architecture of Coexistence*. Edinburgh: Edinburgh University Press.

———. 2006. "The Mausoleum of Imam Shafiʿi." *Muqarnas* 23:15–46.

Munzavī, Aḥmad. 1969 (1348). *Fihrist-i nuskhahā-yi khaṭṭ-i Fārsī*. 6 vols. Tehran: Muʿassasa-i Farhangī-i Minṭaqaʿī.

Necipoğlu, Gülru. 2016. "Early Modern Floral: The Agency of Ornament in Ottoman and Safavid Visual Cultures." In *Histories of Ornament: From Global to Local*, edited by Gülru Necipoğlu and Alina Payne, 132–55. Princeton, N.J.: Princeton University Press.

———. 2000 (1379). *Handasa va tazyīn dar miʿmārī-yi Islāmī: Ṭūmār-i Tūpqāpī = The Topkapı Scroll: Geometry and Ornament in Islamic Architecture; Topkapı Palace Museum Library MS H. 1956*. Translated into Persian by Mihrdād Qayyūmī Bīdhindī. Tehran: Rawzana.

———. 1995. *The Topkapı Scroll: Geometry and Ornament in Islamic Architecture; Topkapı Palace Museum Library MS H. 1956*. Santa Monica, Calif.: Getty Center for the History of Art and the Humanities. (Digitized Copy Available in the Getty's Virtual Library.)

———. 1992. "Geometric Design in Timurid/Turkmen Architectural Practice: Thoughts on a Recently Discovered Scroll and Its Late Gothic Parallels." In *Timurid Art and Culture: Iran and Central Asia in the Fifteenth Century*, edited by Lisa Golombek and Maria Subtelny, 48–66. Leiden: Brill.

Netz, Reviel. 1999. *The Shaping of Deduction in Greek Mathematics: A Study in Cognitive History*. Cambridge: Cambridge University Press.

O'Kane, Bernard. 2013. "A Tale of Two Minbars: Woodwork in Syria and Egypt on the Eve of the Ayyubids." In *Ferdowsi, the Mongols and the History of Iran: Art, Literature and Culture from Early Islam to Qajar Persia: Studies in Honour of Charles Melville*, edited by Robert Hillenbrand, A. C. S. Peacock, and Firuza Abdullaeva, 316–28. London: I. B. Tauris.

Özdural, Alpay. 2002. "The Use of Cubic Equations in Islamic Art and Architecture." In *Nexus IV: Architecture and Mathematics*, edited by José Francisco Rodrigues and Kim Williams, 165–79. Fucecchio, Italy: Kim Williams Books.

———. 2000. "Mathematics and Arts: Connections between Theory and Practice in the Medieval Islamic World." *Historia Mathematica* 27, no. 2: 171–201.

———. 1998. "A Mathematical Sonata for Architecture: Omar Khayyam and the Friday Mosque of Isfahan." *Technology and Culture* 39, no. 4: 699–715.

———. 1996. "On Interlocking Similar or Corresponding Figures and Ornamental Patterns of Cubic Equations." *Muqarnas* 13:191–211.

———. 1995. "Omar Khayyam, Mathematicians and *Conversazioni* with Artisans." *Journal of the Society of Architectural Historians* 54, no. 1: 54–71.

Pajares-Ayuela, Paloma. 2001. *Cosmatesque Ornament: Flat Polychrome Geometric Patterns in Architecture*. Translated by Maria Fleming Alvarez. New York: W.W. Norton & Company.

Payne, Alina, ed. 2015. *Vision and Its Instruments: Art, Science and Technology in Early Modern Europe*. University Park, Pa.: The Pennsylvania State University Press.

Pope, Arthur U., and Phyllis Ackerman, eds. 1938. *A Survey of Persian Art from Prehistoric Times to the Present*. 6 vols. London: Oxford University Press.

Qurbānī, Abū al-Qāsim. 1986–87 (1365). *Zindagīnāma-i rīyāżīdānān-i dawra-i Islāmī az ṣada-i sivvum tā ṣada-i yāzdahum-i Hijrī*. Tehran: Markaz-i Nashr-i Dānishgāhī.

Rāshid, Rushdī. 2013. *Ibn al-Haytham's Theory of Conics, Geometrical Constructions and Practical Geometry*. Translated by J. V. Field. Vol. 3 of *A History of Arabic Sciences and Mathematics*, edited by Nader El-Bizri, 4 vols. London: Routledge.

———. 2005. *Geometry and Dioptrics in Classical Islam*. London: Al-Furqān Islamic Heritage Foundation.

———. 1979. "La construction de l'heptagone régulier par Ibn al-Haytham." *Journal for the History of Arabic Science* 3:309–87.

Raynaud, Dominique. 2012. "Abū al-Wafā' Latinus? A Study of Method." *Historia Mathematica* 39:34–83.

Richard, Francis D. 1989. *Catalogue des manuscrits persans*. Vol. 1, *Ancien fonds*. Paris: Bibliothèque nationale de France.

Rosenthal, Franz. 1975. *The Classical Heritage in Islam*. Translated by Emile and Jenny Marmorstein. London: Routledge and Kegan Paul.

Roxburgh, David, ed. 2005. *Turks: A Journey of a Thousand Years, 600–1600*. London: Royal Academy of Arts.

Rozenfel'd, Boris. A., and Ekmeleddin İhsanoğlu. 2003. *Mathematicians, Astronomers, and Other Scholars of Islamic Civilization and Their Works (7th–19th c.)*. Istanbul: Research Center for Islamic History, Art and Culture (IRCICA).

Rūdhrāwarī, Abū Shujāʿ. 1920–21. *Dhayl Tajārib al-umam = Continuation of The Experiences of the Nations*. Vols. 3 and 6 of *The Eclipse of the ʿAbbasid Caliphate: Original Chronicles of the Fourth Islamic Century*, edited and translated by H. F. Amedroz and D. S. Margoliouth. Oxford: Basil Blackwell.

Rūḥ al-Amīnī, Maḥmūd. 1990 (1369). "Abū Isḥāq Kūbanānī, riyāżīdān va adīb-i qarn-i nuhum-i Hijrī." In *Nashriyya-i Dānishkada-i Adabiyyat va ʿUlūm-i Insānī*, 42–49. Kerman: Dānishgāh-i Shahīd Bāhunar-i Kirmān.

Saidan, A. S., ed. 1971. *Arabic Arithmetic: The Arithmetic of Abū al-Wafā' al-Būzjānī, 10th Century Mss Or. 103 Leiden and 42 M Cairo*. In *Tārīkh ʿilm al-ḥisāb al-ʿArabī*. Amman: Jamʿiyyat Aʿmāl al-Maṭābiʿ al-Taʿāwuniyya.

Saliba, George. 2007. *Islamic Science and the Making of the European Renaissance*. Cambridge, Mass.: The MIT Press.

———. 1999. "Artisans and Mathematicians in Medieval Islam." Review of *The Topkapı Scroll: Geometry and Ornament in Islamic Architecture* by Gülru Necipoğlu. *Journal of the American Oriental Society* 119, no. 4: 637–45.

———. 1985. "The Function of Mechanical Devices in Medieval Islamic Society." *Annals of the New York Academy of Sciences* 441:141–52.

Saltzman, Peter. 2008. "Quasi-Periodicity in Islamic Geometric Design." In *Nexus VII: Architecture and Mathematics*, edited by Kim Williams, 153–68. Turin: Kim Williams Books.

Sarhangi, Reza. 2012. "Interlocking Star Polygons in Persian Architecture: The Special Case of the Decagram in Mosaic Designs." *Nexus Network Journal* 14: 345–72.

Sarkis, Hashim. 2016. "Inscription: On the Surface of Exchange between Writing, Ornament, and Tectonic in Contemporary Architecture." In *Histories of Ornament: From Global to Local*, edited by Gülru Necipoğlu and Alina Payne, 34–43. Princeton, N.J.: Princeton University Press.

Schneider, Gerd. 1980. *Geometrische Bauornamente der Seldschuken in Kleinasien*. Wiesbaden: Ludwig Reichert Verlag.

Sezgin, Fuat. 1974. *Geschichte des arabischen Schrifttums. Band 5: Mathematik bis ca. 430 H.* Leiden: E.J. Brill.

Shelby, Lon. 1977. *Gothic Design Techniques: The Fifteenth-Century Design Booklets of Mathes Roriczer and Hanns Schuttermayer*. Carbondale: Southern Illinois University Press.

———. 1972. "The Geometrical Knowledge of Mediaeval Master Masons." *Speculum* 47, no. 3: 395–421.

Shiʿrbāf, Aṣghar. 1993–2000 (1372–79). *Girih va kārbandī*. 2 vols. Tehran: Sāzmān-i Mīrāṣ-i Farhangī-i Kishvar.

Stafford, Barbara Maria. 2007. *Echo Objects: The Cognitive Work of Images*. Chicago: University of Chicago Press.

Tabbaa, Yasser. 2017. "The Resurgence of the Baghdad Caliphate." In *A Companion to Islamic Art and Architecture*, edited by Barry Flood and Gülru Necipoğlu, 2 vols. Blackwell Companions to Art History. Malden, Mass.: Wiley-Blackwell.

———. 2014. "Originality and Innovation in Syrian Woodwork of the Twelfth- and Thirteenth-Centuries." In *Material Evidence and Narrative Sources: Interdisciplinary Studies of the History of the Muslim Middle East*, edited by Daniella Talmon-Heller and Katia Cytryn-Silverman, 188–215. Leiden: Brill.

———. 2008. "Andalusian Roots and Abbasid Homage in the Qubbat al-Barudiyyin in Marrakech." *Muqarnas* 25:133–46.

———. 2001. *The Transformation of Islamic Art during the Sunni Revival*. Seattle: University of Washington Press.

Ṭāşköprīzāde, Aḥmed ibn Muṣṭafā. 1911–37 (1328–56). *Kitāb Miftāḥ al-saʿāda wa-miṣbāḥ al-siyāda fī mawḍūʿāt al-ʿulūm*. 2 vols. Hyderabad: Maṭbaʿat Dāʾirat al-Maʿārif al-Nizāmiyya.

Thackston, Wheeler M. 1989. *A Century of Princes: Sources on Timurid History and Art*. Cambridge, Mass: The Aga Khan Program for Islamic Architecture at Harvard University and MIT.

Turner, Mark, ed. 2006. *The Artful Mind: Cognitive Science and the Riddle of Human Creativity*. Oxford: Oxford University Press.

Wiedemann, Eilhard. 1970. *Aufsätze zur arabischen Wissenschaftsgeschichte*. Vol. 1. New York: G. Olms.

Woepcke, Franz. 1855. "Recherches sur l'histoire des sciences mathématiques chez les Orientaux d'après des traités inédits arabes et persans. Deuxième article. Analyse et extrait d'un recueil de constructions géométriques par Aboûl Wefâ." *Journal Asiatique*, ser. 5, no. 5: 218–56, 309–59.

BRILL

ELAHEH KHEIRANDISH

AN EARLY TRADITION IN PRACTICAL GEOMETRY: THE TELLING LINES OF UNIQUE ARABIC AND PERSIAN SOURCES

This chapter focuses on the subject of practical geometry through an examination of the text and textual tradition of the *Kitāb Aʿmāl al-handasa* (Book of Geometrical Constructions [hereafter *Geometrical Constructions*]) of Abu'l-Wafa' al-Buzjani (d. ca. 998 [see Appendix I]),[1] a work composed in Arabic in the last decade of the tenth century, whose dates have here been further narrowed down through a detailed examination of the written primary sources.[2] I treat both the Arabic and Persian traditions of Abu'l-Wafa''s *Geometrical Constructions* and the related aspects of treatises on surveying. The principal aim is to complement the other chapters in this volume, which are primarily focused on a single Persian document transmitted under the Arabic title *Fī tadākhul al-ashkāl al-mutashābiha aw al-mutawāfiqa* (*On Similar and Complementary Interlocking Figures*), a work referred to throughout this *Supplement to Muqarnas* as the *Anonymous Compendium* in order to highlight its non-uniform structure, as well as its anonymous and undated content.[3] This text is part of Ms. Persan 169, notable for the unique items contained between its covers and hereafter referred to as the Paris Codex, after its location in the Bibliothèque nationale de France.[4] The present chapter examines several items in the Paris Codex, as well as additional related works, in an attempt to provide further historical context for the rare and complex *Anonymous Compendium*, which is treated from different perspectives in each chapter.[5]

My secondary aim is to supplement relevant publications on practical geometry, a subject historically designated as *handasa ʿamalī* in both Arabic and Persian, and on surveying (*misāḥa*), that have been found to be limited and/or problematic in various ways. The text of Abu'l-Wafa''s *Geometrical Constructions*, which can be classified most accurately as a treatise on practical geometry, has been poorly represented in the few published works devoted to it. Within the Arabic tradition, there is a published facsimile reproduction, as well as a "critical" edition, but even between them these do not account for all the known manuscripts of Abu'l-Wafa''s book.[6] The Persian tradition of the *Geometrical Constructions* is even more concerning in light of the problematic way the texts forming that tradition have been treated and frequently cited. For example, the Persian texts of the *Anonymous Compendium*, along with a late translation of Abu'l-Wafa''s *Geometrical Constructions*, ahistorically "rewritten into modern Persian," appear in a volume misleadingly titled *Handasa-i Īrānī* (Iranian Geometry), with the subtitle *al-Ashkāl al-handasiyya* (Geometrical Propositions/Figures).[7] There is also a so-called edition that curiously disregards one of only two known manuscript variants of an early Persian translation of Abu'l-Wafa''s text, with its rare features and annotations.[8]

With respect to the closely related field of surveying, various aspects of which can be classified under the term *misāḥa*, as in the "science of measurement" or mensuration (*ʿilm al-misāḥa*), the situation is different. Publications on that subject, such as the Arabic texts and translations published in Rushdi Rashid's *Ibn al-Haytham's Theory of Conics, Geometrical Constructions and Practical Geometry*,[9] focus on aspects of "practical geometry" and "geometrical constructions" other than those analyzed here. These include a surveying text titled *Uṣūl al-misāḥa* (Principles of Measurement) by Ibn al-Haytham (d. post ca. 1040), who is named in the *Anonymous Compendium* and is best known for his *Kitāb al-Manāẓir* (Book of Optics), which had a wide impact in

 | DOI 10.1163/9789004315204_004

Europe, where it was known by the Latin titles *De aspectibus, Perspectiva*, and *Prospettiva*.[10]

As will be argued and demonstrated in this chapter, the Arabic and Persian traditions of Abu'l-Wafa''s *Geometrical Constructions* (*Aʿmal al-handasa*), the main subjects of the present study, were closely related to specific textual traditions within practical geometry and surveying. Most scientific works from Arab and Persian lands of the Islamic Middle Ages were composed primarily in Arabic, the standard scientific language of the time. Known sources in the Persian tradition of Abu'l-Wafa''s *Geometrical Constructions*, however, exceed those in Arabic, in number, range, and date, including relatively early translations and unexpectedly late compositions.[11] The original Arabic text is among the earliest works to be examined here, as part of larger subjects such as theoretical aspects of both geometry and surveying that go beyond ornamental geometry.[12]

The historical sources presented in this chapter are texts in Arabic and Persian dating from the tenth to the seventeenth centuries, most of which are known through unique manuscripts that are discussed here for the first time. These and other sources were selected for the important revelations they offer regarding the Arabic and Persian traditions of practical geometry through a set of "telling lines." Lines of text in both prose and verse are part of a selection that further allows for reconstructions of these traditions, which extend to the *Anonymous Compendium*.

Abu'l-Wafa''s *Geometrical Constructions* is closely related, on the one hand, to theoretical and practical aspects of ancient Greek geometry, and, on the other hand, to the field of surveying in both its geometrical and numerical traditions. More specifically, the methods of "geometrical construction" described in Abu'l-Wafa''s book, which are explicitly associated with both "geometricians" (*muhandisīn*) and "craftsmen" (*ṣunnāʿ*), are visible "form[s]" (*ṣūra*) of proof (*ithbāt*) as distinct from textual illustrations (*mithāl*) or demonstrations (*burhān*). His constructions reveal close correspondences to procedures such as "applications of areas," in the tradition of Euclid's *Elements of Geometry*, without citing them specifically. These methods, involving the equality of the areas of surfaces, as in Book II of Euclid's *Elements*, are, significantly, different from those involving the determination of magnitudes such as distances, as in the surveying propositions of Euclid's *Optics*.[13]

In this chapter, the forms of historical evidence uncovered in the textual tradition of Abu'l-Wafa''s *Geometrical Constructions*, and in related works by him and others in both Arabic and Persian, are all drawn directly from primary sources and supplied with linguistic analyses to offer new perspectives on both the practitioners and practices under discussion. Questions involving the identification and dates of individuals associated with the *Anonymous Compendium* are detailed in the conclusion.[14]

What follows presents all known major texts in the Arabic and Persian traditions of Abu'l-Wafa''s *Geometrical Constructions*, as well as a few other related works on surveying, along with other primary sources, from manuscripts and coins to tilings and woodwork, with reference to specific types of "lines": "Lines of Text" refer to lines or passages in a text; "Lines of Verse," to verses or hemistiches in a text; "Lines of Craft," to lines or passages in a text found on crafted productions; and "Lines of Proof," to lines or passages in a text corresponding to geometrical illustrations.[15] These subheadings are meant to encompass themes ranging from the historical and changing meanings of various methods and practices, to the profiles and relations of authors and practitioners.

I. A TEXTUAL AND CONTEXTUAL OVERVIEW OF THE *BOOK OF GEOMETRICAL CONSTRUCTIONS* (APPENDIX I)

Abu'l-Wafa''s *Geometrical Constructions* provides important evidence for the practices of contemporary geometers and craftsmen. This becomes immediately clear from the content and language of an especially revealing passage describing the presence of both geometers and craftsmen at a gathering in which geometrical problems were discussed. In this passage, the demonstrations (*barāhīn*) of geometers (*muhandisīn*) are distinguished from the constructions (*aʿmāl*) of craftsmen (*ṣunnāʿ*), as are specific errors (*khaṭāʾ*) and mistakes (*ghalaṭ*) made by the latter in their treatment of the same geometrical problems.[16] One can trace the transformation and evolution of these terms and concepts in the Arabic and Persian traditions of the *Geometrical Constructions*, which include textual variants and translations, as well as commentaries and other texts often treated as the same work or as originating in the same time and place. These transformations in usage have often been misunderstood in the secondary literature, leading to errors in interpretation, which were then perpetuated. This section is therefore primarily concerned with identifying, and distinguishing between, the various versions of the *Geometrical Constructions* and related texts in the Arabic and Persian traditions, and with making clear the historical context within which each was written.

Most importantly, the largely understudied Arabic and Persian traditions of Abu'l-Wafa''s *Geometrical Constructions* raise important historical questions about the sciences and crafts under examination in these works, ranging from the theoretical to the practical and, above all, the often overlooked fine lines that distinguish the sciences and crafts from one another. General questions about the historical meanings of "science" and "arts/crafts"—the *ʿilm* and *ṣināʿa* (or *ṣanʿa*) of Arabic and Persian, corresponding to the *epistēmē* and *technē* of ancient Greek—have been addressed by me elsewhere, as have the historical contexts of both theoretical and practical aspects of geometry (*handasa*) in related fields, from subdivisions of geometry, such as optics (*manāẓir*) and mechanics (*ḥiyal*), to surveying (*misāḥa*)—itself called both a "science" and "art/craft," and employed in titles such as *al-Misāḥāt al-manāẓiriyya* (Optical Surveying).[17]

Distinctions between divisions and subdivisions within the sciences and arts/crafts are also documented in periods both before and after the *Geometrical Constructions*. An early Arabic composition by Qusta ibn Luqa (d. ca. 912–13), *Aṣnāf al-ṣinaʿāt* (Classes of Crafts),[18] is focused on the arts/crafts. This is distinct from the more common topic of the sciences (*ʿulūm*), treated in the slightly later *Iḥṣāʾ al-ʿulūm* (Enumeration of the Sciences) by Abu Nasr al-Farabi (d. ca. 950), a figure directly associated with the tradition of the *Geometrical Constructions* through a variant of that text (1.1.v) carrying his name, with a date as early as 321 (933).[19] Following Farabi, distinctions in classifications may be found in works ranging from al-Katib al-Khwarizmi's *Miftāḥ al-ʿulūm* (Key to the Sciences, composed after ca. 977), in Arabic, to the Persian classification work by Fakhr al-Din al-Razi (d. ca. 1209), *Jāmiʿ al-ʿulūm* (Compendium of the Sciences [=*Kitāb Sittīnī*, for its sixty sections]). The latter reflects more recognizable distinctions, such as "production" (*sākhtan*) versus "knowledge" (*shinākhtan*) with reference to specific sciences.[20]

The early terminology of Ibn Luqa's classification of crafts, as well as Farabi's classification of the sciences, both covered from different angles elsewhere,[21] readily captures a sense of the practices and practitioners involved in practical geometry shortly before Abu'l-Wafa was active. In Ibn Luqa's classification of the four crafts, it is instructive to consider the distinction between crafts that are "productive" (*fiʿliyya*), using both hands and instruments (*ālāt*), and crafts that are "operative" (*ʿamaliyya*), which may or may not involve an instrument: the examples given for productive crafts include joinery (*nijāra*) and construction (*banāʾ*)—the former term, as we shall see, was later used in the title of Abu'l-Wafa''s text. The distinction that Ibn Luqa draws between the "productive" and "operative" crafts is that the former remain with the cease (*bāṭil*) of the agent/maker of the craft (*ṣāniʿ*), as opposed to the latter, which does not. As such, productive crafts well fit the joinery and building constructions associated with geometrical constructions, as do dancing and fencing in the "operative" crafts. Similarly, fine distinctions may be discerned in the treatment in Farabi's *Enumeration of the Sciences* of geometry (*handasa*)—the second of the seven "mathematical sciences," after arithmetic and before optics, astronomy, music, mechanics, and geography. For instance, in Farabi's text, the scope of practical (*ʿamaliyya*) geometry, which is atypically treated before theoretical geometry (*naẓariyya*), includes surveying (*misāḥa*). Practical geometry itself is described as "looking into lines and surfaces, within a wooden body (*khashab*) if it is used by a joiner (*najjār*), within a metal body (*ḥadīd*) if it is used by a metal worker (*ḥaddād*), within a wall (*ḥāʾiṭ*) if it is used by a builder (*bannāʾ*), or within surfaces (*suṭūḥ*) of lands (*arḍ*) and fields (*mazāriʿ*) if it is used by a surveyor (*māsiḥ*)."[22]

The documented practices specific to the tradition of Abu'l-Wafa''s *Geometrical Constructions* go beyond methods such as the demonstrations (*burhān*) of the sciences (*ʿulūm*) and constructions (*aʿmāl*) of the crafts (*ṣinaʿāt*), the latter being a form of scientific craft itself, with its own textually transmitted traditions. Fine distinctions in his text find clear expression in the proofs (*ithbāt*) of geometry, from the visible form (*ṣūra*) of figures in visual illustrations (*mithāl*), to the procedural "way/method" (*ṭarīq*) of crafts in "constructions" (*aʿmāl*), as distinct from logical "reasoning" (*ʿilla*) in "geometrical" demonstrations (*barāhīn*).[23] The important point raised by Gülru Necipoğlu in her chapter in this volume, "Ornamental Geometries: A Persian Compendium at the Intersection of the Visual Arts and Mathematical Sciences," about the visual transmission of some working methods given the scarcity of written sources in fact finds support in a historical source as early as Abu'l-Wafa''s *Geometrical Constructions*, which prominently features visual methods, as she observes. My own emphasis here is that the visible "form[s]" (*ṣūra*), some verbalized next to figures throughout that text, function as a prominent kind of visual "proof" (*ithbāt*) for the "concepts" (*maʿnā*) behind the geometrical "constructions" (*aʿmāl*). These elementary "proofs" capture visually equivalent or superimposed areas as distinct from methods of "demonstrations" (*barāhīn*) or "reasonings" (*ʿilal*), which are explicitly left out of the text.

In the closely related field of surveying (*misāḥa*), we find distinctions similar to those seen in practical geometry between "demonstrations" and "constructions," as described above. Here the distinctions involve measurements (*misāḥa* [the same term as that used to designate "surveying" as a whole]) and calculations (*ḥisāb*). The term *misāḥa* can refer to the measurement of lengths, surfaces, and volumes of geometrical objects, and to determinations (*maʿrifa/wijdān*) of magnitudes (*abʿād*), such as heights (*irtifāʿ*) or depths (*ʿumq*), with or "without measurement" (*ghayru misāḥa*). The latter, namely methods not involving measurement, also include aspects of surveying specific to physical objects, the "elevation" (*aʿmidat*) of mountains (*jibāl*) and "height" (*irtifāʿ*) of "upright structures" (*al-ashkhāṣ al-qā'ima*), most based on proportional relations.[24]

Surveying is represented as the measurement (*misāḥa*) of surfaces (*musaṭṭaḥāt*) and volumes (*mukaʿabāt*), with all its attendant subdivisions, in the twelfth-century Persian classification work *Jāmiʿ al-ʿulūm* (Compendium of the Sciences) by Fakhr al-Din Razi.[25] The difference between land surveying (*misāḥa*) and the measurement (*misāḥa*) of distant bodies (*ajrām al-baʿīda*) becomes clear through the subdivision of geometry into topics such as optics (*manāẓir*), as found in the later classification work (in Arabic), *Irshād al-qāṣid* (Guidance of the Messenger), by Ibn al-Akfani (d. ca. 1348).[26] Among all the various meanings that "*misāḥa*" encompasses, the most relevant in terms of the present discussion is what I refer to as "area surveying," when it involves surfaces, from lands to buildings, and "distance surveying," regarding distances both near and far. The latter is treated in this chapter in fewer cases including an item in the Paris Codex.

Area surveying differs from distance surveying in that it involves surfaces, rather than lines or solids, which have their own geometrical and numerical traditions. As mentioned above, the geometrical methods of construction in Abu'l-Wafa''s *Geometrical Constructions*, and in related works by him and others in both Arabic and Persian, reveal particularly close correspondences to procedures such as "applications of areas," in the tradition of Book II of Euclid's *Elements*. These are distinct from methods such as the "determination of magnitudes," in the tradition of Euclid's *Optics*, and its surveying propositions, involving other historical distinctions and expressions, and have more direct relevance to the rest of this volume.[27]

Returning to the practices of geometry, historical distinctions between proofs (*ithbāt*) and visible forms (*ṣūra*) are also identifiable in the major works spanning the Arabic and Persian traditions of Abu'l-Wafa''s *Geometrical Constructions*, through expressions largely common to both languages as preserved mostly in their known unique manuscripts. Methods such as proofs involving visible forms of geometrical subjects appear not only in Abu'l-Wafa''s *Geometrical Constructions* but also in a thirteenth-century commentary on it by Kamal al-Din ibn Yunus (ca.1156–1242), also in Arabic.[28] Later expressions include "verifiable sciences" (*ʿulūm yaqīnī*) or "rejections" (*radd*) of "reasonings" (*ʿilal*) in the Persian translation of the *Geometrical Constructions* "completed" in the fifteenth century by Kubanani Yazdi, who composed a postface of that translation in the Paris Codex, as we shall see below.[29] Likewise, a seventeenth-century Persian commentary on Abu'l-Wafa''s text by Muhammad Baqir Yazdi distinguishes between methods such as demonstration (*burhān*) for the rejection (*buṭlān*) of constructions within land surveying, those in accordance with (*munṭabiq*) rational (*nāṭiqa*) demonstrations (*barāhīn*) of geometry, actualization (*bi-fiʿl dar-āvardan*), and the use of compasses with more than one opening (*fatḥ*).[30]

The practitioners themselves must also be considered, including the geometer (*muhandis*), craftsman (*ṣāniʿ*), surveyor (*massāḥ*), and calculator (*ḥāsib*). These designations occasionally overlap, as seen in joint expressions such as "geometer-calculator" (*al-muhandis al-ḥāsib*) and "scholar-practitioner" (*ʿālim-i ʿāmil*), respectively used by Abu'l-Wafa' and the aforementioned Kubanani Yazdi (the fifteenth-century co-translator of the *Geometrical Constructions* from Arabic into Persian).[31] In addition to overlaps within one and the same figure, we must also consider the connections between different individuals, which ranged from indirect to direct contact or communication, to even finer sorts of interactions among various figures at different times. In earlier periods, it seems more appropriate to describe the occasional meetings between geometers and craftsmen as instances of "joint participation" rather than "connection" or "collaboration."[32] Beyond the "gathering" with "geometers" and "craftsmen" reported by Abu'l-Wafa', written documents from early periods only rarely refer to occasions of direct interaction across science and craft disciplines. The few known cases of such contact include a scientific figure addressing a craftsman, as reported by Ibrahim ibn Sinan (ca. 908–46), as well as by Abu Rayhan al-Biruni (d. ca. 1048).[33] Kubanani, who is credited with a commentary on a work by Ibrahim ibn Sinan,[34] names a figure as "master-geometrician," which is itself a rare example of overlaps between the activities of geometers and craftsmen, in his "completion" of the Persian translation of Abu'l-Wafa''s *Geometrical Constructions*. There are more known examples of recorded exchanges between scientists than among figures who were versed in sciences and crafts: these include a pre-arranged joint observation of the lunar eclipse of 997, described by Biruni, which involved Abu'l-Wafa' and Biruni, and a report by Muhammad Baqir Yazdi in the seventeenth century of his meeting with a fellow mathematician regarding scientific works from Euclid to Abu'l-Wafa''s *Geometrical Constructions*.[35] Later periods also have limited, though to a lesser degree after the fifteenth century, written documentation of cases of direct interaction. The scarcity of written documentation for the representation of crafts next to the sciences may reflect limitations or omissions of bibliographical sources for various reasons. This can be contrasted to increasing evidence for such representations in textual primary sources, especially those in Persian and Turkish after the fifteenth century.[36]

II. THE ARABIC TRADITION OF THE *BOOK OF GEOMETRICAL CONSTRUCTIONS* (APPENDIX II)

The earliest primary sources within the Arabic tradition of Abu'l-Wafa''s *Book of Geometrical Constructions* (Appendix I, 1.1) include related compositions of that author (1.2–1.5), and other texts, such as a few in surveying (e.g., 2.2). The corpus of works that form the Arabic and Persian traditions of his *Geometrical Constructions* (Appendix I, nos. 1–3 [Arabic], and 4–10 [Persian]) may be distinguished as "dependent" as opposed to "related" compositions (Appendix I, nos. 3, 4, 5, 9, and 10; and nos. 2, 6–8, respectively). The terms "Arabic" and "Persian" refer to the languages in which the works under examination were written, rather than the regions where they were produced or the ethnicities of those involved. A "dependent" composition refers to a direct translation or commentary, as distinct from a "related" composition or tradition. The textual selections briefly discussed here and more fully referenced below may be placed in only a rough chronological order, since a few of the sources are undated and anonymous, the most important and complex of these being the aforementioned Persian *Anonymous Compendium* (Appendix I, no. 8), which is central to the present volume.

In its collective form this work cannot be assigned to a single author, date, or place. Its non-uniform structure shields it against any blanket conjectures about possible dates, though the content of none of its sixty-one constructions may be dated to a period after the fifteenth century—the likely date of the text's transcription. The other textual selections to be discussed here (see Appendix I) fall rather evenly around it in terms of chronology. Those from the tenth to fourteenth centuries (nos. 1–3 in Arabic; nos. 4–7 in Persian), i.e., those that precede the *Anonymous Compendium*, include the earliest related translations and compositions; the works that succeed it (nos. 9 and 10 in Persian), from the fifteenth and seventeenth centuries, offer further revelations and reconstructions related to our anonymous document.

The *Geometrical Constructions* and other related works of Abu'l-Wafa' (1.1–1.5) have elements in common with the *Anonymous Compendium* in terms of the geometrical figures that are described in each. So do the dependent commentaries on Abu'l-Wafa''s *Geometrical Constructions* in both Arabic and Persian (3.1; 10.1), though they stress the need to add geometrical demonstrations and thus differ from Abu'l Wafa''s explicit preference for visible forms (*ṣūra*) of proofs (*ithbāt*) in his *Geometrical Constructions*.

As indicated above, the field of surveying is closely related to the *Geometrical Constructions* (1.1). This includes Abu'l-Wafa''s own related works involving geometrical figures and areas (1.2–1.5), some of which use the term *misāḥa* in their very titles or section headings.[37] Surveying is also addressed in Arabic works composed shortly after Abu'l-Wafa''s time, such as *al-Misāḥa* (2.2 [Measurement]) by Abu Mansur al-Baghdadi (d. ca. 1037), and another work that has long been attributed to him under the title *al-Īḍāḥ ʿan ʿuṣūl-i ṣināʿat al-massāḥ* (2.1 [Clarifications of the Principles of the Art/Craft of Surveying]). However, this last work significantly bears the name of a certain "Abū al-Fatḥ al-Yazdī [or Mardī/Murādī]."[38] The latter author is so far unknown, but both his name and date are derivable, and his measurements and calculations in surveying offer important clues to questions posed in this chapter and volume.

In the Arabic tradition we find just one known commentary on Abu'l-Wafa''s *Geometrical Constructions*: it is by the aforementioned "Kamāl al-Dīn ibn Yūnus" and entitled *Sharḥ al-Aʿmāl al-handasiyya* (Commentary on the Geometrical Constructions) (3.1).[39] The Persian tradition, by contrast, was much more prolific. We find early translation(s) of Abu'l-Wafa''s *Geometrical Constructions* (4.1 and 5.1), hypothetically assumed to be copies from the eleventh-twelfth and twelfth-thirteenth centuries, related works in surveying from the twelfth to fourteenth centuries (6.1 and 7.1), another fifteenth-century Persian translation (9.1), and a Persian commentary (10.1) from the seventeenth century. Only parts of the *Anonymous Compendium* (8.1) may be estimated as dating somewhere between the fourteenth and fifteenth centuries, based on the respective citations and transcriptions of that text (7.1, 9.1).

The Arabic and Persian traditions of Abu'l-Wafa''s *Geometrical Constructions* and the related works selected here for what they reveal beyond issues of dating have been given limited attention in the secondary literature, despite the significance of some previously overlooked and presently discovered historical sources.

The Author: Abu'l-Wafa' Muḥammad ibn Muḥammad al-Buzjani

As his *nisba*, Buzjani, indicates, "Abu-l-Wafā' Muḥammad ibn Muḥammad al-Būzjānī" (Appendix I, no. 1), was born in Buzjan, in Khurasan in northeastern Iran. Various primary sources describe him as a scientific figure of some significance. The biobibliographical work closest to his own time, the well-known catalogue (*Fihrist*) of the bookseller and author Ibn al-Nadim (ca. 995), lists his long name along with details on both his life and works. The *Fihrist* includes the place and date of his birth, down to the day and month, even noting that it was the time of the new moon. Then it describes his works, his education and his teachers, and names his teachers' teachers and their subjects. The author discusses Abu'l-Wafa''s early move to Iraq, and provides a list of titles from "among his books."[40] Another early source, ʿAli Bayhaqi (d. ca. 1176–77), calls Abu'l-Wafa' a "scholar/philosopher" (*ḥakīm*), but later biographers such as Ibn al-Qifti (d. ca. 1248) and Ibn Khallikan (ca. 1211–82) lack any such descriptions. Safadi (ca. 1297–1363) and Hajji Khalifa (i.e., Katip Çelebi [d. ca. 1657]) identify him as an arithmetician (*ḥāsib*) and geometer (*muhandis*), respectively.[41] Occasionally the overlapping form "*al-muhandis al-ḥāsib*" is used.[42]

Although Ibn al-Nadim's *Fihrist* provides a list of Abu'l-Wafa''s arithmetical and astronomical titles, it does not mention his *Book of Geometrical Constructions*, nor do any biobibliographical sources up to and including Kâtip Çelebi (Hajji Khalifa). However, manuscript copies and variants of this and related works do include the names Abu'l-Wafa', al-Buzjani, or both, in titles or incipits, indicating that historically his authorship of the book was considered authentic. His name and compositions are mentioned in other primary sources besides biobibliographies; these include works in Arabic by authors of scientific texts who lived around or after the period in which Abu'l-Wafa' was active. Among these, the earliest and most important are the writings of his aforementioned younger associate, Abu Rayhan al-Biruni (d. ca. 1048), a figure originating from Khwarazm, in the outer parts of northeast Iran, who similarly wrote in Arabic, but, unlike Abu'l-Wafa', traveled widely.

Two works by Biruni, *Qānūn al-Masʿūdī* (Masʿudi's Canon) and *Taḥdīd al-amākin* (Determination of Coordinates [of Cities]), contain evidence of Abu'l-Wafa''s scientific activities in addition to Biruni's report of a lunar eclipse, simultaneously observed by them from two different cities. These tracts not only detail Abu'l-Wafa''s astronomical observations but also contain the names of his local patrons.[43] Biruni describes observations Abu'l-Wafa' made at the "Bāb al-Tibn" in Baghdad circa 974–76, as well as later ones made at an observatory in the same city around 988. Reports go beyond the ones mentioned in Biruni's own typically long and resourceful dedications.

While the latter had a diverse pool of patrons, throughout his life Abu'l-Wafa''s activities were sponsored by powerful members of the Buyid dynasty,[44] ranging from ʿIzz al-Dawla (r. ca. 967–68) of Baghdad, who supported his astronomical observations, to Sharaf al-Dawla (r. ca. 987–89) in Kirman, who founded an observatory in the "royal garden" in Baghdad. Between these two, both temporally and geographically, was ʿAdud al-Dawla (r. ca. 979-83), who ruled in Fars, as well as Baghdad. ʿAdud al-Dawla is mentioned as an important patron of Abu'l-Wafa' by the contemporary biobibliographer Ibn al-Nadim; his name also appears in the dedication of one of Abu'l-Wafa''s important arithmetical works (1.3, discussed below). But, from the standpoint of the present study, Abu'l-Wafā''s most important patron was Baha' al-Dawla, who ruled from Baghdad to Fars and Kirman (ca. 989–1012), and was the sponsor of his *Geometrical Constructions* (1.1), as well as a related text described below (1.2). In his dedications, Abu'l-Wafa' listed the long chain of names and titles associated with Baha' al-Dawla. This list proved to be an aid in further narrowing down the date of composition of the work most central to this chapter.

Abu'l-Wafa''s Geometrical Constructions *and related works by him (Appendix I, 1.1–1.5)*

Abu'l-Wafa''s *Geometrical Constructions*, whose full title is *Fī mā yaḥtāju ilayhi al-ṣāniʿ min aʿmāl al-handasa* (On What Is Needed by the Craftsman from Geometrical Constructions [1.1]), provides information about practices such as "geometrical constructions," and practitioners such as "craftsmen" (to be discussed later). A related composition is similarly titled: *Kitāb Fī mā yaḥtāju ilayhi al-kuttāb wa al-ʿummāl min ʿilm al-ḥisāb* (On What Is Needed by Secretaries and Administrators from the Science of Arithmetic [1.3]). This work is also known under a shorter alternative title as *Kitāb al-Manāzil* (Book of Stations) and through more than one manuscript copy.[45] Of the three unpublished compositions by Abu'l-Wafa' bearing various geometrical titles (1.2, 1.4, and 1.5), the one closest in time to his *Geometrical Constructions* is titled *Fī jamʿ aḍlāʿ al-murabbaʿāt wa al-mukaʿabāt wa akhdh tafāḍulihā* (On the Addition of the Sides of Quadrangles and Cubes and Their [Magnitude] Differences [1.2]). This is a work that, though dedicated to the patron of Abu'l-Wafa''s *Geometrical Constructions*, Baha' al-Dawla, names the astronomer-astrologer "Abū Bishr al-Ḥasan Ibn Sahl al-Munajjim" as the one asking for demonstrations (*barāhīn*) [of geometrical constructions].[46] Another work, titled *Jawāb … al-Būzjānī ʿammā saʾalahu al-Faqīh Abū ʿAlī al-Ḥasan ibn al-Ḥārith al-Ḥubūbī* (Buzjani's Answer to Questions Put to Him by the Jurist al-Hububi [1.4]), about "proving a rule for calculating an area," is, by contrast, directed specifically to the jurist himself.[47] Yet another work, titled *Masʾalat Arshimīdis fi misāḥat al-muthallath* (The Problem of Archimedes on Measuring a Triangle [1.5]), names only the ancient Greek mathematician Archimedes without referring to a patron.[48] All these works are in need of full and in-depth studies.

Only a few of the works mentioned above may be ordered chronologically based on the patrons cited in them. The one dedicated to the earlier ruler ʿAdud al-Dawla (1.3), dates before the two works dedicated to his son Baha' al-Dawla (1.1, 1.2), leaving open the chronology of the other two, which lack a named patron or datable figure (1.4, 1.5).

The manuscript tradition of Abu'l-Wafa''s Book of Geometrical Constructions (*Appendix I, 1.1.i–v*)

Abu'l-Wafa''s *Geometrical Constructions* provides useful information about the subjects under discussion. But the textual variations of its manuscript copies and variants, which are crucial for understanding the text's intended meanings, await a proper critical edition. This work is known through three copies, one in Istanbul (1.1.i) and two in Cairo (1.1.ii–iii), which, at least in the case of one of the first two accessible to me, can be considered "manuscript copies," that is, duplicate versions of the same text, despite notable variations in their respective titles (see below). I call the two documents in Milan and Uppsala (1.1.iv–v) "manuscript variants" as versions that may be from different manuscript traditions; their relation to the first three, and to each other, remains more difficult to determine, given their particular features and defects in parts of the text where critical information resides.[49]

The secondary literature has not been comprehensive in its documentation of this work, or its important textual variations. The known manuscripts copies and variants of the text are either partially or incorrectly documented in catalogues and reference works, including the facsimile and critical editions of the *Geometrical Constructions*.[50] The facsimile reproduction includes only the Istanbul manuscript (1.1.i), which it treats as a "unique copy," and does not mention any of the other manuscript copies or variants, or the much earlier published edition; it also refers to the Arabic Milan variant as a Persian version.[51] The previously published "edition," is, in turn, based on only the manuscript in Istanbul and the two in Cairo (1.1.i–iii); the versions in Milan and Uppsala (1.1.iv–v) have not been taken into account.

The telling lines of some Arabic sources

1. The Book of Geometrical Constructions (*Appendices I and II, 1.1*)

Abu'l-Wafa''s *Geometrical Constructions* is among the earliest Arabic sources to offer a sense of the current practices and practitioners of his time, particularly when it is considered in connection with his related compositions (1.2–1.5). It is helpful to consider the work's title, as well as its opening lines, which explain the aims of the work and provide information on the date of composition. We will then turn to some key passages in order to examine the details of specific constructions and identify cases associated with "geometricians" and "craftsmen."

Title: The Book of Geometrical Constructions (*Appendix II, 1.1a*)

The title of Abu'l-Wafa''s *Geometrical Constructions* survives in two distinct forms through its most complete manuscript copies in Arabic (1.1.i–ii). The manuscript in Istanbul (1.1.i) is dated circa 1447–49, based on the dates of the Timurid ruler Ulugh Beg, whose seal appears on the cover page. Originally from the royal library in Samarqand (fig. 1a), it bears the title *Kitāb Abī al-Wafāʾ Fī mā yaḥtāju ilayhi al-ṣāniʿ min aʿmāl al-handasa* (Book of Abu al-Wafa' on What Is Needed by the Craftsman from Geometrical Constructions). This title is different from the earlier of the two manuscript copies known in Cairo (1.1.ii), dated 1203, which has the variant form of *Kitāb al-Nijāra fī ʿamal al-masṭara wa al-birkār wa al-kūniyā taṣnīf Abī al-Wafāʾ Muḥammad ibn Muḥammad al Būzjānī* (Book of Joinery on Constructions with Ruler, Compass, and Set Square composed by Abu al-Wafa' ... al Buzjani).[52] The title thus names the craft of joinery, not mentioned in the text itself, and tools that are introduced early in the text.[53]

The phrase *aʿmāl al-handasa* (geometrical constructions) in Abu'l Wafa''s title is critical for understanding the workings of the Arabic text. While the text includes "constructions" by both geometers (*muhandisīn*) and craftsmen (*ṣunnāʿ*), in the title the expression is used with specific reference to the needs of the latter group. For this reason, some modern interpretations and trans-

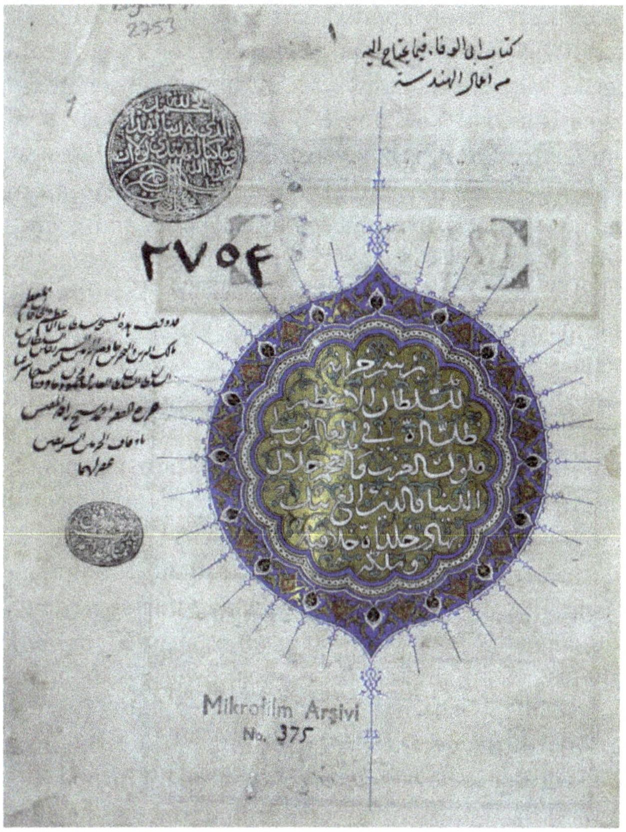

Fig. 1a. The cover of the Arabic manuscript of Abu'l-Wafa''s *Geometrical Constructions*, stamped with the seal of Ulugh Beg's royal library in Samarqand, dated circa 1447–49. Istanbul, Süleymaniye Library, Ms. Ayasofya 2753, fol. 1r. (Photo: Süleymaniye Library)

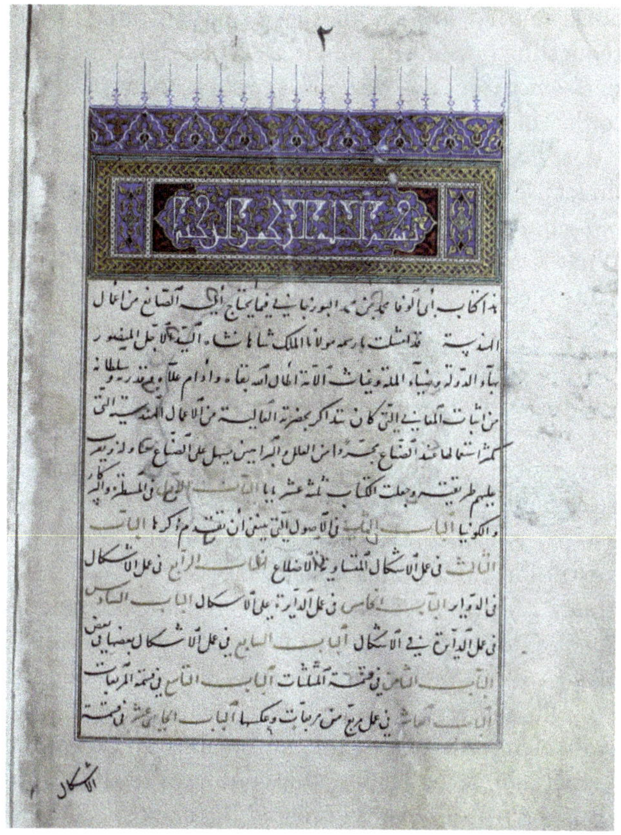

Fig. 1b. The opening page of the Arabic manuscript of Abu'l-Wafa''s *Geometrical Constructions*, stamped with the seal of Ulugh Beg's royal library in Samarqand, dated circa 1447–49. Istanbul, Süleymaniye Library, Ms. Ayasofya 2753, fol. 1v. (Photo: Süleymaniye Library)

lations regard the craftsmen as the targeted audience of that composition. It is useful, then, to consider the language of the opening lines of the *Book of Geometrical Constructions* in light of this phrase and the distinctions made about it within the text itself.

Opening Lines (*Appendix II, 1.1b*)

> I have complied with the order of our Master, Monarch, King of Kings, the Highest, [Divinely-assisted], Victorious, Noble Baha' al-Dawla, Light of the Nation, Aid of the Community, may God prolong his life and preserve his high standing, power, and reign, regarding the proof[54] of concepts that were mentioned to his High Excellency involving geometrical constructions most used by craftsmen, abstracted from reasonings and demonstrations, so that it might be easy for craftsmen to grasp and bring its way closer to them, and I set this book in thirteen chapters …

As mentioned earlier, the Istanbul manuscript preserves the rich opening lines of Abu'l-Wafa''s text (fig. 1b). The author is quite explicit about having followed the order of his patron, Baha' al-Dawla: he literally "complied with his order" (*qad imtathaltu mā rasamahu*) [in composing the text]. This is in contrast to merely naming that patron next to the one who requested the text, as Abu'l-Wafa' does in another work of his titled *Fī jamʿ al-aḍlāʿ al-murabbaʿāt wa al-mukaʿabāt*, which was dedicated to the same patron but requested by someone else, the aforementioned astronomer-astrologer "Abū Bishr al-Ḥasan Ibn Sahl al-Munajjim" (1.2). The same is true for a work dedicated to a different ruling patron, ʿAdud al-Dawla (1.3), and for another work dedicated to the requesting jurist al-Hububi (1.4). While the dedication in *Geometrical Constructions* is explicit about both the "order" of Baha' al-Dawla and the fact that he was alive, it does not state that the text itself was composed for the direct use of the craftsmen mentioned in the title (1.1: *Fī mā yaḥtāju ilayhi al-ṣāni' min a'māl al-handasa* [On What Is Needed by the Craftsman from Geometrical Constructions]). Nevertheless, the intention to address the needs of practitioners is clearly mentioned in the foreword of the Arabic manuscript in Istanbul, on account of which Abu'l-Wafa says "he is complying with the order of his Buyid patron, Baha' al-Dawla, regarding the proof of concepts that were mentioned to that High Excellency involving geometrical constructions most used by craftsmen, abstracted from reasonings and demonstrations, so that it might be easy for craftsmen to grasp." The exact wording of these opening lines alone is not enough, and is inconclusive as to whether Abu'l-Wafa is directing his book to craftsmen, who were not typically addressed in textual form, though some master craftsmen were literate, thereby creating an exception to that rule.[55] Neither does the wording by itself suggest that Abu'l-Wafa's patron ordered the book specifically for the benefit of craftsmen; only that elementary forms of "proof" (*ithbāt*), those without reasonings (*'ilal*) or demonstrations (*burhān*) for the better grasp of craftsmen (*ṣunna'*), had been "mentioned to" that patron, whether by Abu'l-Wafa himself or by others, though the text is not clear enough here to rule out other readings (see Appendix, II, 1.1b, and n. 54 above). Still, the book is the only one of its kind from an early period to include a discussion of craftsmen and their geometrical constructions in its title, opening lines, and key passages, to which I will return shortly.

Particularly revealing are the opening dedicatory lines naming the patron of the text, Baha' al-Dawla, within a chain of his titles (fig. 1b).[56] Based on these titles it has been estimated that the text was composed after the year 383 (998), when this Buyid ruler was presumed to have assumed those titles,[57] and before the author, Abu'l-Wafa', reportedly died.[58] Yet dated coins that have not been previously studied in connection with this text bear the exact same succession of titles for Baha' al-Dawla as are seen in Abu'l-Wafa''s dedication, allowing us to more closely approximate the date of this manuscript (fig. 1c–d).[59] The date 381 (991–92) on an early coin (fig. 1c) suggests that the text could have been composed as early as 991;[60] a date close to 387–89 (997–98) on a later coin (fig. 1d) is still consistent with the latest possible date of composition. This is earlier than the date of ca. 993–98 recently proposed by Raynaud, and it similarly challenges Woepcke's previous argument that the *Geometrical Constructions* was likely a posthumous composition by a student of Abu'l-Wafa'.[61]

The dedication of the *Geometrical Consructions* is followed by a long two-part statement introducing the text's subject and aim. The prominent placement of this statement after the dedication requires us to consider it more carefully. According to the first part, the subject of the text is the "proof" (*ithbāt*) of "concepts" (*ma'ānī*) that were "mentioned to that High Excellency," Baha' al-Dawla, regarding the geometrical constructions (*a'māl al-handasiyya*) that are most used by craftsmen (*ṣunnā'*). The use of the singular term *ithbāt*, which is the main subject of the sentence under discussion, shows that the "proof" of the geometrical concepts in the "constructions most used by craftsmen" is central to the text (see Appendix II, 1.1b). Abstracted from strict geometrical "reasonings" and "demonstrations," such a visually comprehensible proof would justify the validity of the concepts (*ma'ānī*) behind those constructions, in accordance with their "mention" as being "easier for craftsmen to comprehend."

As noted above, the term "*ṣūra*" ([visible] form), a common historical counterpart to "*ma'nā*" ([rational] concept), used for "figures" throughout the text, is a visual illustration, as distinct from a "*mithāl*" ([model] illustration), which refers to a textual example within geometrical propositions.[62] The visible "proof" of drawn figures is justified despite being "abstracted" from more advanced forms of "demonstration," and includes methods such as the "application of areas," as well as the superposition of figures in and beyond geometry.

From the cases that appear later in Abu'l-Wafa''s *Geometrical Constructions* it becomes clear that the methods intended for the geometrical construction(s) (*a'māl*) and procedure(s) (*ṭarīq*) used in the crafts, as distinct from the geometrical demonstration(s) (*burhān*) and reasoning(s) (*'illa*) of the sciences, include those formu-

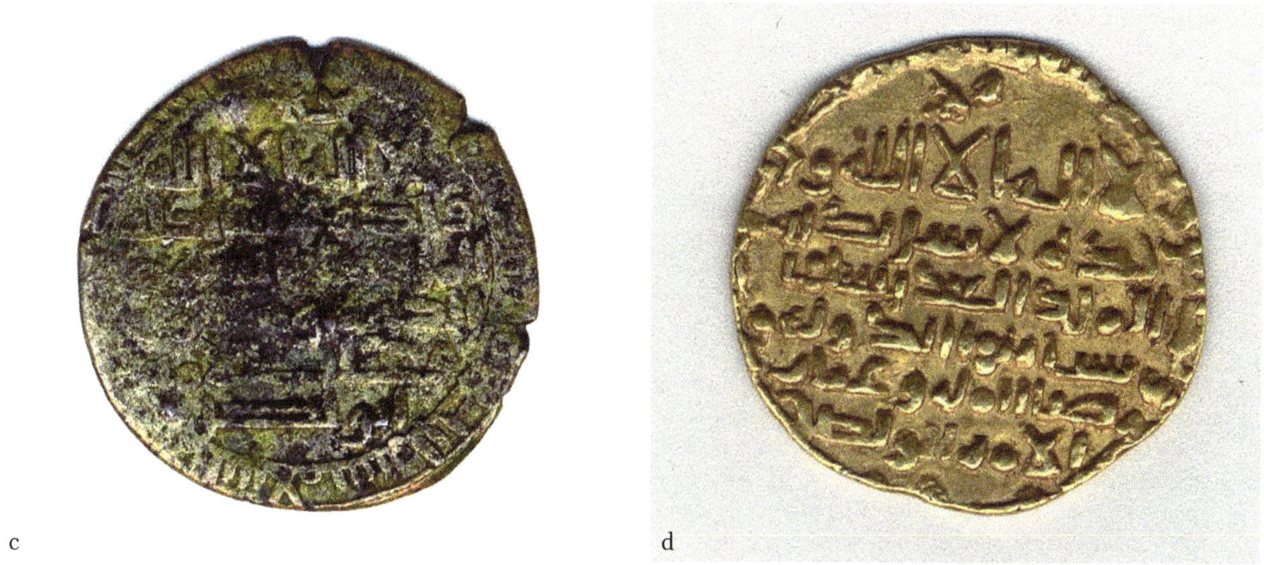

Fig. 1c–d. Buyid coins dated (c) 381 (991), Madinat al-Salam (Baghdad), and (d) 387–89 (997–999), Suq al-Ahwaz, with the exact name and titles of Baha' al-Dawla, the named patron of Abu'l-Wafa''s *Book of Geometrical Constructions*, as given in the book's dedication. London, British Museum: (c) Silver dirham, BMC 682, Reg. no. OR 9404; (d) gold dinar, BMC 668, Reg. no. 1850,0823.1. (Photos: © Trustees of the British Museum)

lated and applied in Book II of Euclid's *Elements* and used in disciplines from algebra to surveying. It must be noted that Abu'l-Wafa' wrote now-lost works within the geometrical traditions of Euclid, the arithmetical tradition of Diophantus (ca. third century), and the algebraic tradition of Muhammad ibn Musa al-Khwarizmi (d. ca. 850). His *Geometrical Constructions* is notable for preserving methods, some with specific reference to surveyors (*māsiḥ*) and to craftsmen (*ṣunnaʿ*).[63] Whether the patron, geometers, surveyors, or, less commonly, craftsmen, the intended audiences of this work are never explicitly stated, either in the text's opening lines, where the subjects and aims of the text are offered, or in the procedures and methods that follow.

Key Passages

The following selection of four passages in Abu'l-Wafa''s *Geometrical Constructions* sheds some light on the methods of geometers and craftsmen in different terms than in the text's title and opening lines (Appendix II, 1.1a–b). These all appear in Chapter Ten ("On the Division of Quadrangles and Their Assemblage"), in a section titled "On the assemblage and division of squares not composed of two squares" [i.e., "square numbers" having square roots] (Appendix II, 1.1c–f: passage labels and titles are added). The first two passages (1.1c–d) are on the mistakes (*ghalaṭ*) of geometers and craftsmen (1.1c), and a gathering with craftsmen and geometers (1.1d); the last two passages (1.1e–f) are on geometrical constructions by a craftsman (1.1e) and a geometer (1.1f).

Passage 1: On the mistakes of geometers and craftsmen (Appendix II, 1.1c)

A group of geometers (*muhandisīn*) and craftsmen (*ṣunnāʿ*) have been mistaken in the matter of these quadrangles (*murabbaʿ*) and their assemblage (*ta'līf*), geometers because of their little training in construction (*ʿamal*), and craftsmen because they lack knowledge of demonstrations (*barāhīn*).[64] That is, since the geometer does not have training in construction, it is difficult for him to approximate what is for him correct by demonstration through lines [i.e., geometrical proofs] in the way it is required by the craftsman. But the aim of the craftsman is what is approximated by construction, and [for him] correctness is apparent by what he sees through sense and observation, and he is not con-

> cerned with demonstrations through lines. [While] for the geometer, if the demonstration of something is established by way of assumption, he is not concerned with correctness through observation, if it is not correct [through demonstration] … No doubt, what the craftsman sees is taken from what is already (*awwal*) worked by the geometer and established for its correctness. But the craftsman and surveyor (*māsiḥ*) rather take the choice part of something, and do not think about the ways by which its correctness is justified. And because of that, mistake (*ghalaṭ*) and error (*khaṭā'*) falls upon it.[65]
>
> As for the geometer, he already knows the correctness of what he wants by means of demonstrations, since he is the one who has derived the concepts on which a craftsman and the surveyor base their work. And [even] if the skillful among these geometers are asked about something in dissecting figures or multiplying lines, they are perplexed and need a long [time] to think, sometimes this occurs and is accessible to them, other times it is difficult and they do not find its construction.[66]

Passage 2: On a gathering with craftsmen and geometers (Appendix II, 1.1d)

> I was present at a gathering (*ba'ḍ al-majālis*) in which a group (*jamā'a*) of craftsmen and geometers [were present], and they were asked about the construction of a square from three squares. The geometer derived a line whose square is equal [in area] to three squares easily, but one of the craftsmen was not satisfied by his construction. The craftsman wanted to divide these squares into parts from which one square can be assembled, as we described for two squares and five squares. The craftsmen presented a number of ways, for some of which demonstration could be established and others which were incorrect, or [cases] where that for which demonstration could be established corresponded with correctness [so] closely in appearance that whoever looked at it thought it is correct. But we present these ways so that the correct [one] may be known from the false [one] and one who looks into this concept will not make a mistake by accepting what is false. God Willing.[67]

The two passages from Abu'l-Wafa''s *Geometrical Constructions* cited above are primarily celebrated for the author's unique discussion of a gathering where the differing methods of craftsmen (*ṣunnā'*) and geometricians (*muhandisīn*) are presented, and the "mistakes" (*ghalaṭ*) in the approximation methods of the former are contrasted with the more accurate methods of the latter. But these and the two passages that follow (1.1e–f) are also discussed here with a view to placing their occurrence in Abu'l-Wafa's *Geometrical Constructions* in the wider context of other texts by him and others. The first of these four passages on the mistakes of craftsmen and geometers (1.1c) recalls a passage in an earlier work by Abu'l-Wafa' himself, this time as part of a section titled "Surveying Constructions" (*A'māl al-misāḥāt*) in his aforementioned arithmetical text *'Ilm al-ḥisāb* (Science of Arithmetic) (1.3), which was dedicated to Baha' al-Dawla's father, 'Adud al-Dawla. There, he points to the "grave errors" (*fāḥish al-khaṭā'*) of not craftsmen but "secretaries" (*kuttāb*), administrators (*'ummāl*), and "surveyors" (*massāḥ*) in measuring [areas] (*misāḥa*) of geometrical figures.[68]

The second passage (1.1d) is also notable for reasons beyond its report of a "gathering" with a "group" (*jamā'a*) of geometers and craftsmen, in one such gathering (*ba'ḍ al-majālis*), in light of the corresponding singular term "*fīhi*" (in it). This and the two passages that follow (1.1e–f) represent parallel methods by craftsmen and geometers, respectively, for the construction of a square from three smaller squares. Those included here as the first of two constructions representing craftsmen, and a demonstration representing geometers, can be contextualized with reference to works outside Abu'l-Wafa''s own: what is identifiable in all the constructions presented is a method used but not cited. That method, known as the "application of areas" through the correspondence of equal areas, is explicated in propositions found in Book II of Euclid's *Elements*.[69] The textual expressions, however, are as close to those propositions as the methods they involve, from the chapter title on dividing (*qisma*) and assembling (*ta'līf*) "quadrangular figures" (*murabba'āt*) (see Appendix II, 1.1), to sections with constructions for "composing" (*tarkīb*) various figures and their combinations (e.g., 1.1e).

Let us now turn to the two passages from the *Geometrical Constructions* that specifically refer to the methods of craftsmen and geometers with corresponding illustrations (1.1e–f), and their comparability to the Euclidean method of the "application of areas." The constructions presented in these passages, which have apparent counterparts in arithmetic and surveying, involve the equality of the areas of a square and three smaller

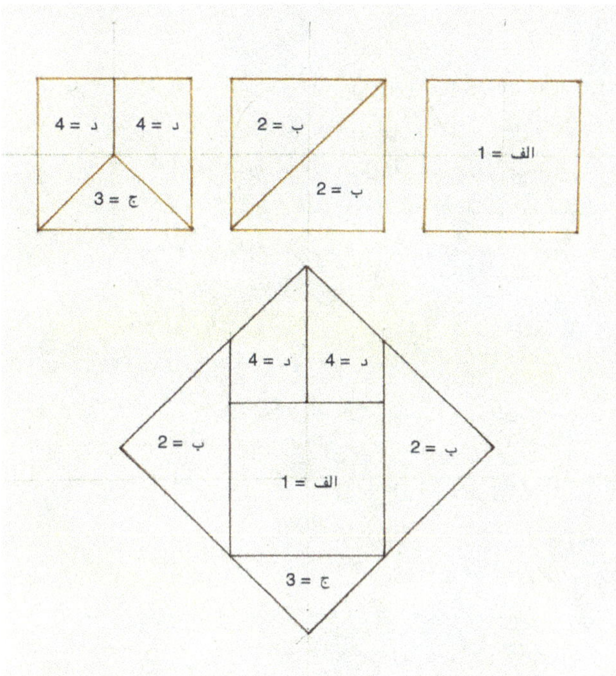

Fig. 1e. Drawing of a construction by a craftsman in Abu'l-Wafa''s *Geometrical Constructions*, labeled here to show the correspondence in areas between the three small squares above and the encompassing square below. (Drawing: Dr. Katya Nosyreva, Prince's School of Traditional Arts in London, 2015; model: manuscript illustration in Qurbānī and Shaykhān 1992, facs. 54)

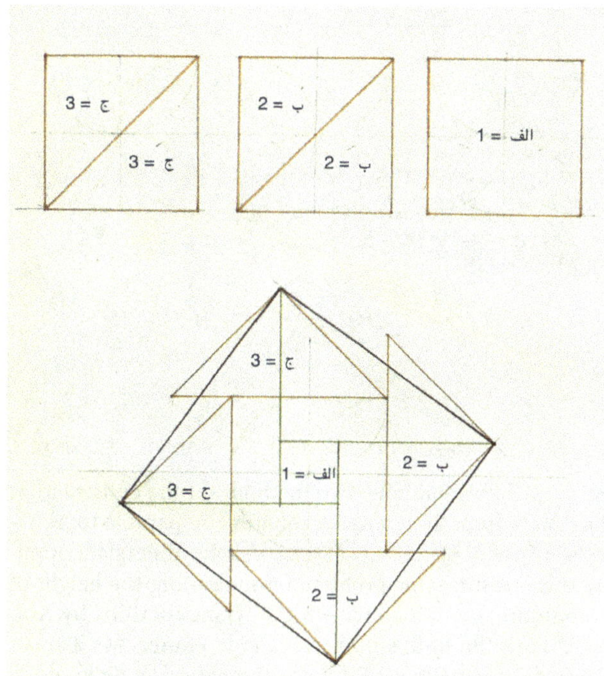

Fig. 1f. Drawing of a construction by a geometer in Abu'l-Wafa''s *Geometrical Constructions*, labeled here to show the correspondence in areas between the three small squares above and the encompassing figures below. (Drawing: Dr. Katya Nosyreva, Prince's School of Traditional Arts in London, 2015; model: manuscript illustrations in Qurbānī and Shaykhān 1992, facs. 57)

squares assembled in it, here labeled to mark area correspondences.

Passage 3: On the construction of a craftsman (Appendix II, 1.1e, fig. 1e)

... One of the craftsmen (*ṣunnāʿ*) placed one of the [three] squares in the middle [الف = 1] and bisected the second [square] by means of the diagonal, and placed it [i.e., two triangles] on both sides of the [first] square [= 2]. He [then] drew from the center of the third [square] two straight lines to two of its angles, not on one diagonal, and he [also] drew a line from it [i.e., the same center] to the midpoint of the side opposite the triangle, which is produced by [those] two lines. Then the [third] square is divided into two deviated [squares, i.e., trapezoids] [= 4] and a triangle [= 3]. Then he put the triangle [= 3] below the first square, and the two trapezoids [د = 4] above it, and he joined the two longer sides of a trapezoid in the middle. This led him to a [larger encompassing] square, as in the illustration of this figure (*ṣura*). Abu'l-Wafā' said, "But this figure which he constructed is in the imagination, and someone who has no training in the art /craft (*ṣināʿa*) or in geometry may consider it correct ... This is [not correct] because if we make the side of the square composed of three squares, [each] approximately ten cubits for the learner, [then] the side of the [encompassing] square is by approximation seventeen and one-third cubits. But the side of the square [constructed] is seventeen and half of one-seventh cubits [i.e., one fourteenth], and there is a great [magnitude (of)] difference between these ... So we have explained the falsity of that division and composition (*tarkīb*) ... Some people have divided these squares with a kind of division more clearly false than the first division ....[70]

Passage 4: On the correct construction of geometers (Appendix II, 1.1f, fig. 1f)

As for the division of squares in the correct way as required by demonstration, it will be clear according to the way we

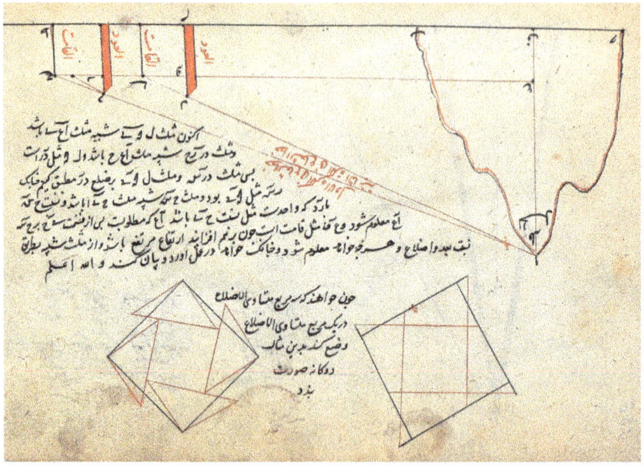

Fig. 2a. A folio containing two methods for the construction of a square from three smaller squares comparable to Abu'l-Wafa''s *Book of Geometrical Constructions* in Persian; opposite this is a surveying problem for measuring the height of a mountain, contained within the "Constructions by Abu Bakr." Paris, Bibliothèque nationale de France, Ms. Persan 169, 19, fol. 136v. (Photo: Bibliothèque nationale de France)

> now mention. That is, we bisect two of the squares along their diagonals [ب=2, ج=3], and add each of those [half squares: ب=2, ب=2, ج=3, ج=3] to one of the [four] sides of the third square [in the middle]. Then we place from [each] triangle, the angle which is half a right angle at [such an] angle to the square [الف=1] that its diagonal [falls on] the side of that square. And part of the triangles at the other angle [of the triangle] remains for us. Then we join the middle [points] of the right angles of [each] triangle by straight lines. That becomes the side of the desired square ... and this is its figure. If the geometer (*muhandis*) is asked for the construction of a square from less or more [number of] squares, then he finds a line whose square is equal [in area] to three squares, and he is not concerned with how the squares are cut. That is to say, if he is asked for the construction of a square from three [smaller] squares, he will draw the diagonal of one of the squares. He then erects at one of the endpoints of the diagonal a line perpendicular to it and equal to the side of the square. He then joins its endpoint with the [other] endpoint of the diagonal by a straight line, and that is the side of the square composed of three squares.[71]

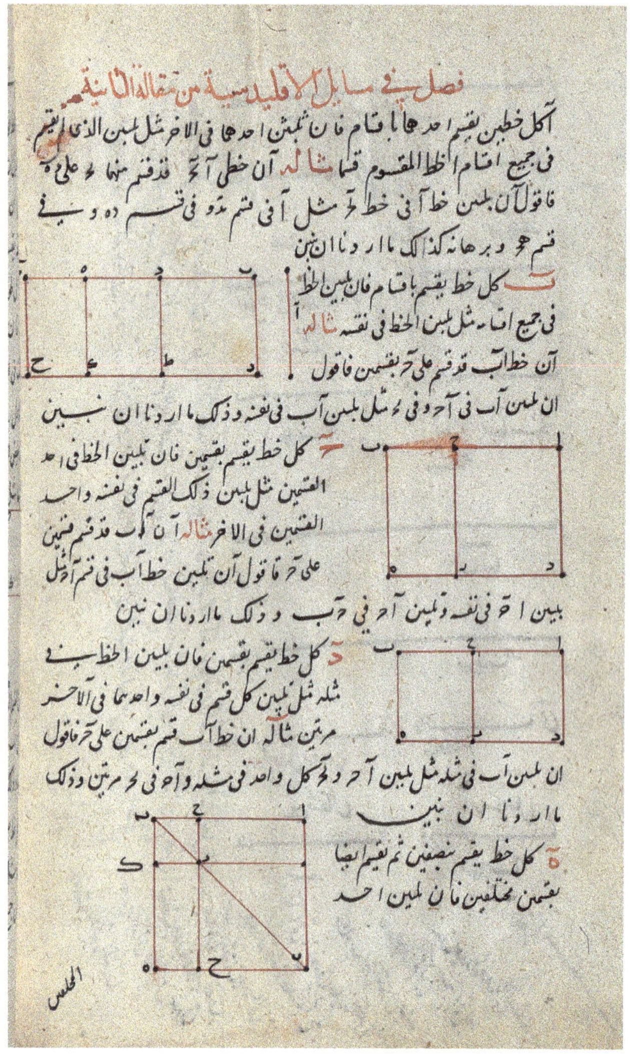

Fig. 2b. A folio titled "Chapter on Euclidean Problems from the Second Book [of Euclid's *Elements of Geometry*]": bottom illustration: Book II, Prop. 4. Paris, Bibliothèque nationale de France, Ms. Persan 169, 13, fol. 102v. (Photo: Bibliothèque nationale de France)

Most significant from the standpoint of the *Anonymous Compendium* are constructions in Abu'l-Wafa''s *Geometrical Constructions* that are comparable to those in it and in other parts of the Paris Codex. The most obvious are two methods for constructing a square from three squares on a single folio in a work of the Paris Codex titled *Lil-Aʿmāl Abī Bakr al-Khalīl al-Tājir al-Raṣadī* (Of the Constructions by Abu Bakr al-Khalil al-Tajir al-Rasadi) (fig. 2a). These two constructions correspond to Abu'l-Wafa''s examples of methods by a craftsman and geometer (Qurbānī and Shaykhān 1992, facs. 55 and 57). The fact that they face a surveying problem on that same folio is an indication of the close association between practical geometry and surveying. A much simpler case is Abu'l-Wafa''s construction of a square from two smaller squares and two rectangles (facs. fol. 59) immediately following the three square constructions (facs. 55 and 57) in a form that corresponds to one of the propositions in excerpts from Book II of *Euclid's Elements*. Significantly, this same proposition is included in the Paris Codex as part of its second book, the only one included from its mutiple volumes (fig. 2b).

Most notable among the constructions of Abu'l-Wafa''s *Geometrical Constructions* is the simple corresponding construction involving unequal quadrangles: there two squares and two rectangles form a larger encompassing square, one that corresponds to the fourth proposition in Book II of Euclid's *Elements* (facs. 59, top right illustration), supplied here with labels to mark area correspondences to a larger encompassing square (fig. 3a). Next to Abu'l-Wafa''s construction and its accompanying figure is the special case of one central and four peripheral triangles placed within a larger square equal in area to their sum (facs. 59, top left illustration), based on the same method of application of areas. The four peripheral triangles are here extended to four corner quadrilateral figures, each made up of two facing right-angled triangles, similarly labeled for marking the corresponding areas (fig. 3b). A close model to the latter is a popular ornamental square pattern, comprising a central square with four such peripheral quadrilaterals or kites (*turunj*), each composed of two congruent right-angled triangles, used individually or as a repeat-unit (fig. 3c–d). The latter construction is historically documented by a wooden door from the Imam Ibrahim Mosque at Mosul, dated 498 (1104), drawn here as a panel with repeat patterns (fig. 4a), and the former by a wall tiling that I photographed in the north iwan of the Friday Mosque of Isfahan, dated 1112 (1700–1701) (fig. 4b).[72]

Another notable example that finds a parallel in the *Anonymous Compendium* is a five-pointed star at the center of a polygon at the end of Abu'l-Wafa''s *Geometrical Constructions*, where the term "*mukhammas*" corresponds to a pentagonal star as well as a five-sided polygon (Qurbānī and Shaykhān 1992, facs. 68–69: fig. 5a). The chronologically later *Anonymous Compendium*, which comprises interlocking figures, has many constructions comparable to those in Abu'l-Wafa''s *Geometrical Constructions*, consisting of central and peripheral motifs. This is the case on its opening page, where a central "pentagonal" (*mukhammas*) star (*khātim*, lit. ending [piece]) is illustrated, surrounded by five irregular hexagons that fit within an encompassing larger decagon (fol. 180r [1], drawn here with labels to mark area correspondences [fig. 5b]).

Among other comparable constructions in that source are those involving figures with different numbers of sides. One of these is a simple construction of a square at the center of an octagon, surrounded by four irregular pentagonal figures, with a short text and useful illustration elucidating basic area identities behind the construction, here based on the so-called "Pythagorean theorem" in Book I of Euclid's *Elements* (fol. 181r [4]). But it is the much simpler, and methodologically similar version of a construction in Abu'l-Wafa''s *Geometrical Constructions,* involving unequal "quadrangles," that is behind the previously mentioned ornamental pattern deployed on the wooden door in Mosul (fig. 4a) and the wall tiling in Isfahan (fig. 4b), so far without identifiable historical links to Abu'l-Wafa'. In both cases, a central square has four peripheral kites (*turunj*), made up of two adjacent congruent right-angled triangles. In the Isfahan tiling, inscriptions consisting of Persian verses visually form an interlocking geometrical figure (fig. 4b). On this quadrangular construction, comparable to central-peripheral models in the texts of Abu'l-Wafa' and the *Anonymous Compendium* (figs. 4b and 5b), the name of the craftsman behind the "construction" (*ʿamal*) forms a square at the center, and four lines of verse form the corner kites. The inscription outlining the central square of the tiling reads: "Construction of Ibn Muḥammad Mu'min Muḥammad Amīn."[73] The "lines of verse" delineating the four corner kites, moving clockwise from the lower left, read:

> When they rolled up the letter of our accusation
> Took and weighed it against the balance of actions
>
> Our guilt was more than everyone else, and yet
> They forgave us through the kindness of ʿAlī.[74]

A note must be added here about "lines of text" that appear all around the margins of the Istanbul manuscript

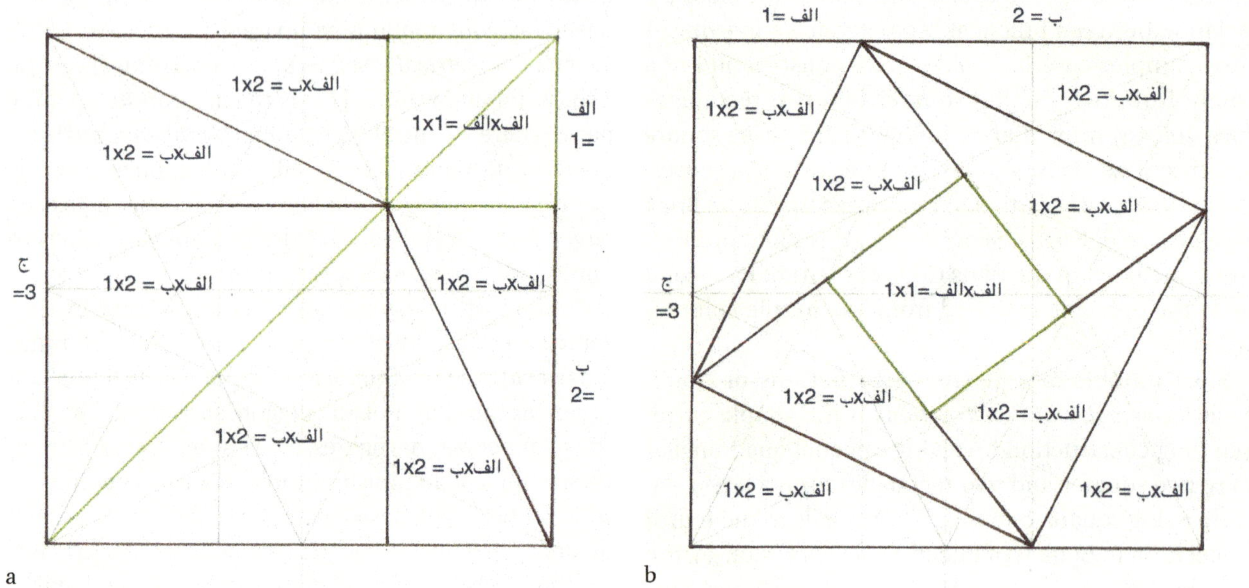

Fig. 3a–b. a) Drawing of the Fourth Proposition in Book II of Euclid's *Elements* on the problem of a square equal in area to two smaller squares and two rectangles; b) Drawing of the construction of a central square and four peripheral figures based on Abu'l-Wafa''s *Geometrical Constructions*. Drawings are labeled to mark area correspondences. (Drawings: Dr. Katya Nosyreva, Prince's School of Traditional Arts in London, 2015; models: manuscript illustrations in Qurbānī and Shaykhān 1992, facs. 59: top right illustration [a]; top left illustration adjusted [b])

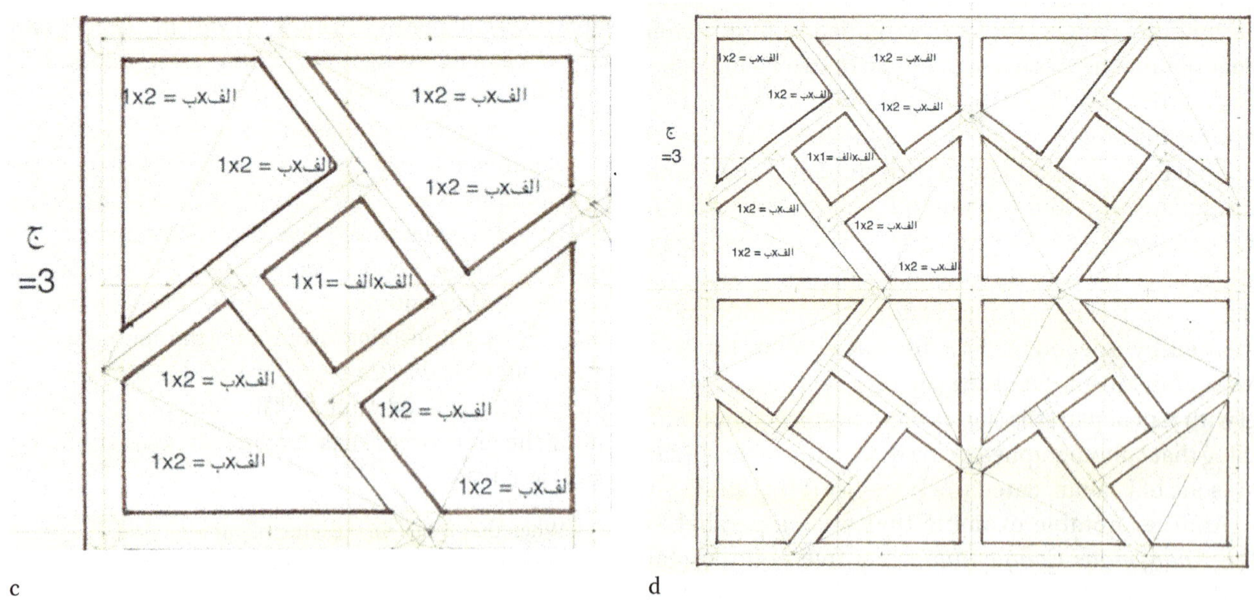

Fig. 3c–d. Drawings of (c) individual figure and (d) repeat patterns for the construction of a central square and four peripheral figures based on Abu'l-Wafa''s *Geometrical Constructions*. Drawings are labeled to mark area correspondences. (Drawings: Dr. Katya Nosyreva, Prince's School of Traditional Arts in London, 2015; models: manuscript illustrations in Qurbānī and Shaykhān 1992, facs. 59: top left illustration adjusted [c]; adjusted illustration as a repeat pattern [d])

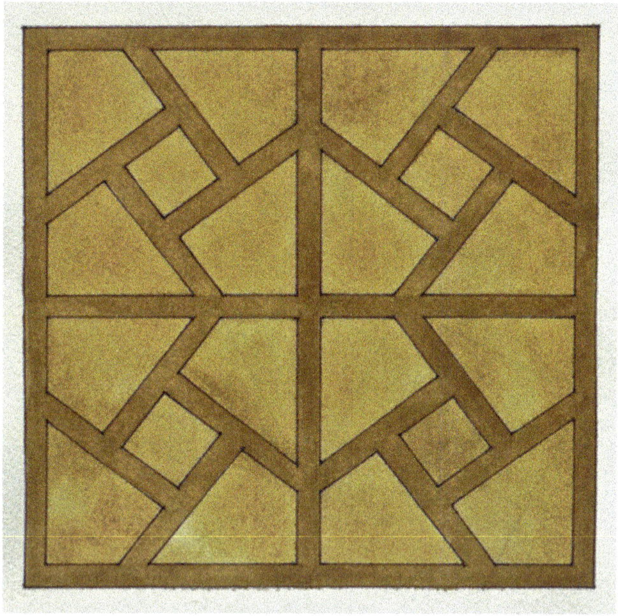

a

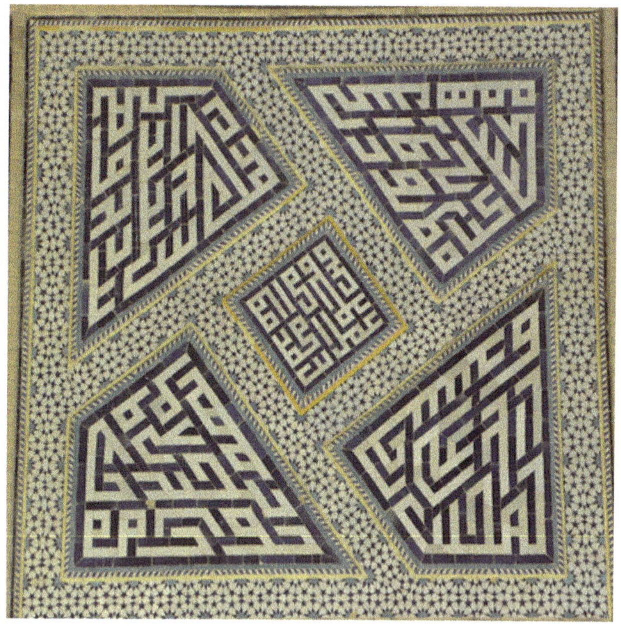

b

Fig. 4a–b. a) Drawing of a geometrical construction on a door of the Imam Ibrahim Mosque in Mosul, dated 498 (1104). (Drawing: Dr. Katya Nosyreva, Prince's School of Traditional Arts in London, 2015); b) Wall tiling, north iwan of the Friday Mosque in Isfahan, dated 1112 (1700–1701). (Photo: Elaheh Kheirandish)

of Abul-Wafa''s *Geometrical Constructions*. Among these marginal notes, all seemingly in the same hand, are a number of commentaries starting with "*burhān-i ẕālik*," providing the "demonstrations" left out by Abu'l-Wafa', which, on closer inspection, are associated with a name that requires further identification and elucidation. That name, which may be read as "Ghundijānī," is identifiable with a figure from the late eleventh-early twelfth century. The long first name of this figure corresponds exactly with the name of the author of an undated surveying text titled *Kitāb Fī qismat al-arāḍī* (Book on the Division of Lands).[75] Regardless of such name or date associations, the marginal commentaries on Abu'l-Wafa''s division of quadrangular figures, which is only one of a few cases to bear the name "Ghundijānī" (Qurbānī and Shaykhān 1992, facs. 59), indicate a correspondence between geometrical constructions and land surveying. The same may be said of the aforementioned single folio in the Paris Codex where two side-by-side constructions comparable to those of a craftsman and geometer in Abu'l-Wafa''s text (facs. 55 and 57) appear next to a surveying problem on the opposite side of the same folio (fig. 2a).

The problem in Euclidean geometry, which involves the equality of the areas of a square divisible into two smaller equal squares and rectangles (fig. 2b), has its own counterparts in land surveying. It also has associations with traditions within arithmetic, such as those referred to by the problematic expressions "geometric algebra" or "algebraic identity." While the geometrical methods of application of areas correspond closely to arithmetical identities later called "algebraic," the use of such expressions has been a subject of debate when applied to a pre-algebraic age.[76]

Abu'l-Wafa''s report on the construction methods of craftsmen (*ṣunnāʿ*) and geometers (*muhandisīn*) that were presented at a gathering, though a rarity in itself, especially in the early medieval sources, calls for its own forms of caution with respect to the use of these terms. A historical report on the specific methods of craftsmen and geometers, side by side, does not in and of itself establish any form of historical "relation" between them

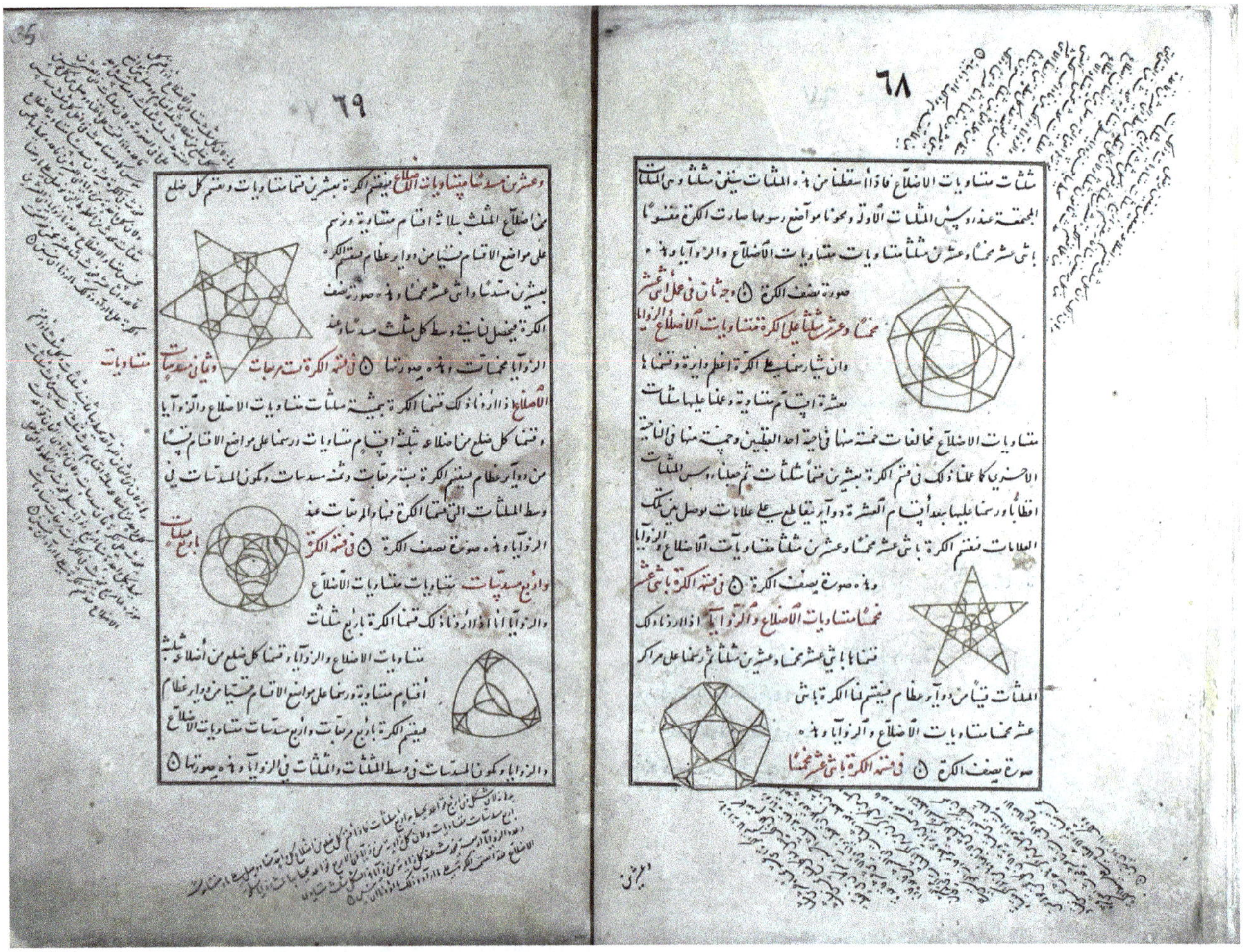

Fig. 5a. Double page of the last folio of the Arabic manuscript of Abu'l-Wafa''s *Geometrical Constructions*, stamped with the seal of Ulugh Beg's royal library in Samarqand. Istanbul, Süleymaniye Library, Ms. Ayasofya 2753, fols. 34v–35r. (Photo: Süleymaniye Library)

beyond documenting the mere presence of some practitioners in a gathering. The closest "relation" that may be deduced from these important passages is the one between practices, such as the application of areas in constructions by craftsmen and geometers, and indeed surveyors who are also named—as are parallel methods within the arithmetical and geometrical traditions.

A text fittingly titled *Fī Taṣḥīḥ al-masā'il al-jabr bi al-barāhīn al-handasiyya* (On the Verification of Algebraic Problems with Geometrical Demonstrations),[77] by Thabit ibn Qurra (d. ca. 901), a predecessor of Abu'l-Wafa', highlights the historical correspondence between methods of "applications of areas" within distinct geometrical and algebraic traditions. In that text, the correspondence appears in the form of verifications (*taṣḥīḥ*) of specific numerical problems through generalized geometrical methods. Abu'l-Wafa''s methods for quadrangular figures of various sizes also find parallels in the much later, yet still surprisingly elementary, methods seen in the *Anonymous Compendium*, such as the visual depictions of the correspondence of areas within the "interlocking figures" mentioned earlier (figs. 8a, 5b).

In other fields and periods, visual "forms" of proof (*ṣūra*) without geometrical "demonstrations" were also used, as in optical treatises in both Arabic and Persian by figures such as Nasir al-Din al-Tusi (d. ca. 1274), the

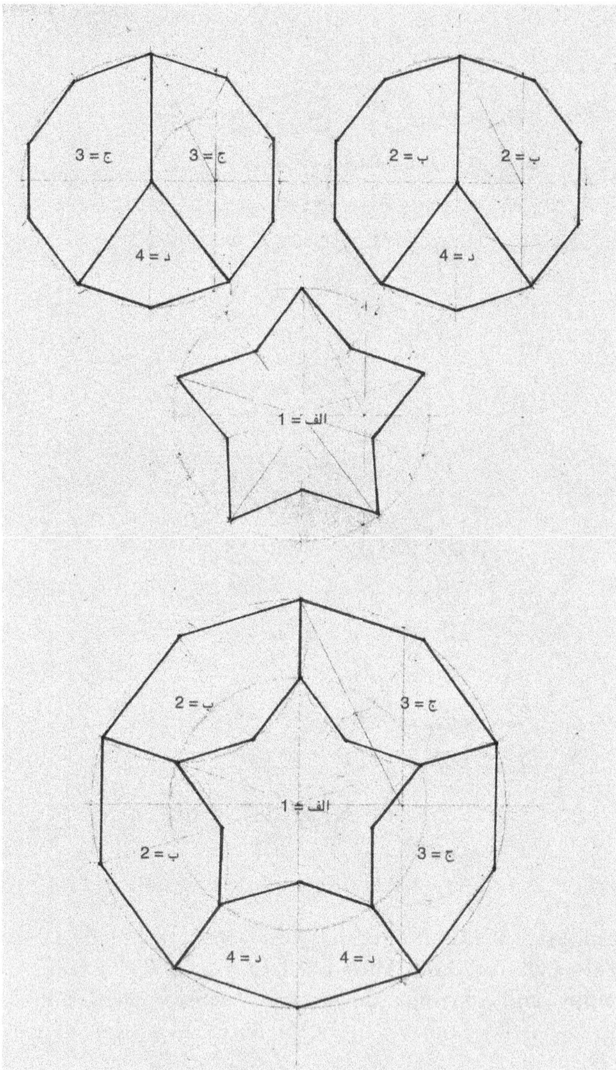

Fig. 5b. Drawing of the opening construction from the *Anonymous Compendium*, labeled here to mark area correspondences between two decagons and a ten-sided star above, and a larger encompassing decagon below. (Drawing: Dr. Katya Nosyreva, Prince's School of Traditional Arts in London, 2015; model: Paris, Bibliothèque nationale de France, Ms. Persan 169, 24, fol. 180r)

well-known mathematician and director of the Maragha Observatory in northwest Iran.[78] Also notable are the related field of surveying (*misāḥa*) and its geometrical and numerical traditions. One early author in this field was Thabit ibn Qurra; another was Yaʿqub al-Kindi (d. ca. 870), to whom are attributed work(s) titled *Īḍāḥ* (Clarification), on the determination of magnitudes such as the height of mountains, the depth of wells, and the width of rivers.[79] This same title is used in another surveying text by a different author, which will be treated next as a way to address questions of identification and dating related to the *Anonymous Compendium*.

2. *Surveying texts* (*Appendix I, 2.1–2.2*)

There is only one known manuscript of the Arabic text on surveying titled *Kitāb al-Īḍāḥ ʿan ʿuṣūl ṣināʿat al-massāḥ* (Book on the Clarification of the Principles of the Craft of Surveying) (2.1).[80] This text, which involves both measurements and calculations in surveying, opens with the word "clarification" (*Īḍāḥ*) in its title, a word that appears in the title of other surveying texts, as noted above. But there are features of this work that are quite uncommon: the text has a short supplement appended to it as a kind of addendum, which is presented as the work of the same author. These are transcribed in the same hand with two successive colophons on a double folio (fig. 6a). The colophons offer unique information relevant to the *Anonymous Compendium* and the traditions under discussion, starting with the immediate issues of identification of related individuals and dating. Two otherwise unknown authors and transcribers are mentioned in the folios containing those colophons (fig. 6a, facs. 84–85). The name of the author is given in the first colophon (fig. 6a, facs. 84) as "Abū'l-Fatḥ al-Yazdī (more likely than the alternative reading Abū'l-Fatḥ al-Mardī/Murādī),"[81] directly after the stated "completion of the book" (*tammat kitāb*) and the full title of the *Kitāb al-Īḍāḥ*. This raises questions regarding the long-time attribution of the work to the well-known eleventh-century author of another surveying work: *Al-Misāḥa*, by Abu Mansur al-Baghdadi (d. ca. 1037) (2.2).[82] Indeed the authorship of the *Kitāb al-Īḍāḥ* has long required clarification, and for this we must turn to another related work, which is included in the same manuscript volume as the *Kitāb al-Īḍāḥ*. This is a Persian translation by Abu'l-Fath Isfahani (fl. ca. 1095–1119) of an Arabic surveying treatise titled *Dar ʿIlm-i misāḥat* (On the Science of Surveying) (6.1).[83] According to Isfahani, Abu Mansur al-Baghdadi was the author of the Arabic work behind his rendering into Persian. Isfahani does not specify any

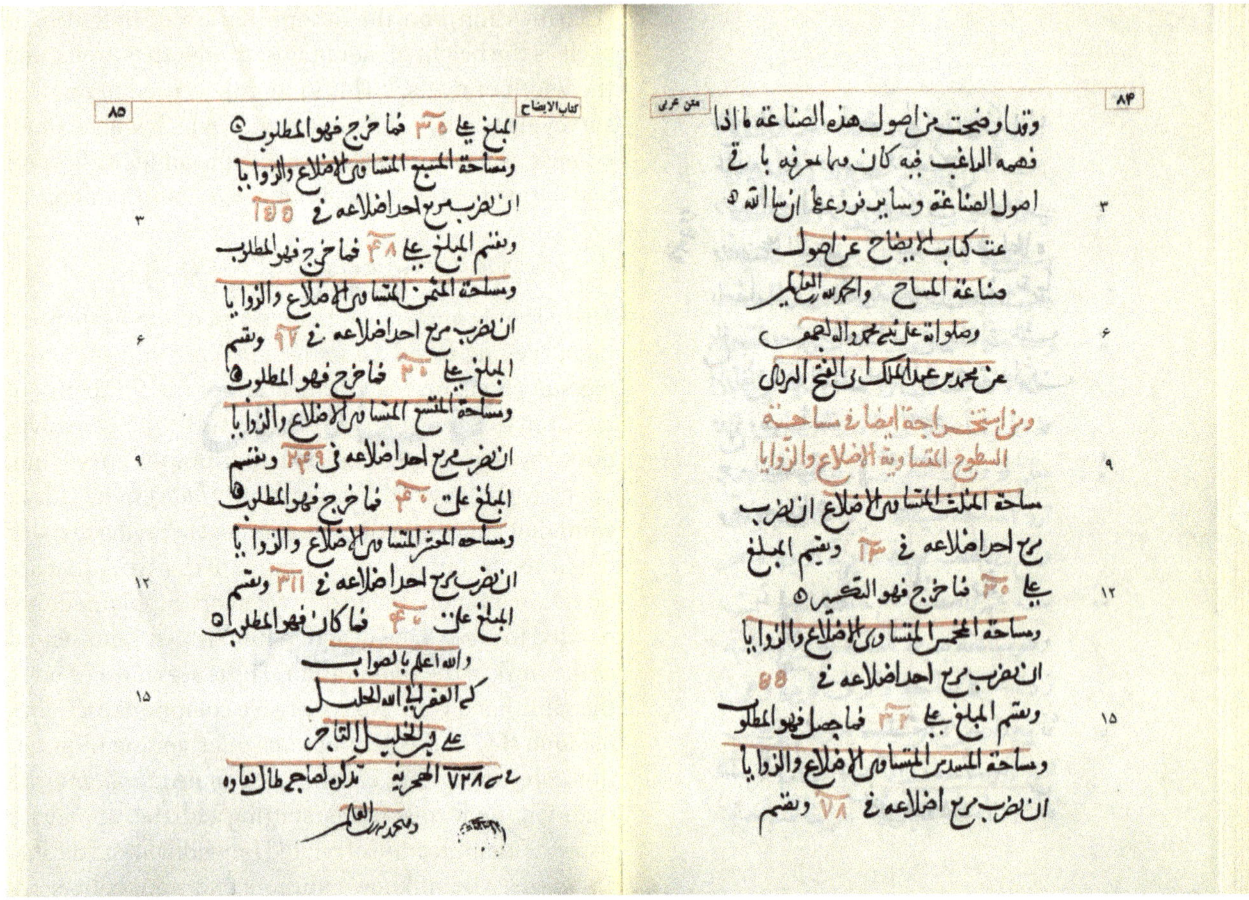

Fig. 6a. Double page from the "Book of Clarifications of the Principles of the Craft of [Land] Surveyors" attributed to Abu Mansur al-Baghdadi (d. ca. 1037), colophon name: "Abū al-Fatḥ al-Yazdī" (or Mardī/Murādī) (fl. ca. 1327–28?) (facs. 84): A surveying supplement indicating the same author (fol. 84v, red script), and containing the name and date of a transcription in the same hand as "ʿAlī ibn al-Khalīl al-Tajir," 728 (1327–28) (facs. 84–85). Arabic manuscript: Mashhad, Āstān-i Quds-i Rażavī Library, Ms. 5429. (Photo: Gulchīn Maʿānī 1969, facs. 84–85)

title for the original Arabic treatise that he attributes to Baghdadi and may well have meant the author's *Fī ʿilm al-misāḥa*, rather than the *Kitāb al-Īḍāḥ*, which was subsequently assumed to have been written by Baghdadi. In light of this, and the fact that instead of Baghdadi another author, Abu'l-Fath al-Yazdi, is cited in the colophon of the *Kitāb al-Īḍāḥ* itself, we may now put the issue of that text's authorship to rest.[84]

In other words, we are probably dealing with a work by an otherwise unknown author with the *nisba* "al-Yazdī." The long-held attribution of the text *Īḍāḥ* to Baghdadi can now be summarized as having been due to the following overlooked "lines of text": (a) the closing lines of the unique manuscript of the Arabic text *Īḍāḥ* (2.1), where the name of the Arabic author can be read as "Yazdī," with no mention of Baghdadi or any other author; (b) the opening lines of a Persian surveying text (6.1), where the Persian translator, Isfahani, gives the name "Baghdadī" as the author of the Arabic text he translated, but not its original Arabic title—possibly *al-Misāḥa* (2.2); and, more recently, (c) the misleading joint publication of the Arabic text of *Īḍāḥ* (2.1) and the Persian translation of a surveying text by Baghdadi (6.1). The latter two texts were bound in the same volume and treated as if the Arabic *Īḍāḥ* had been authored by Baghdadi and translated into Persian by Isfahani.[85]

Now that issues of the authorship of the *Kitāb al-Īḍāḥ* and the possible title of the Persian text bound with it have been set aside, let us turn to other standing problems of identity and dating, which may benefit from related "lines of text." Immediately after Abu'l-Fath al-Yazdi is named in the first colophon (fig. 6a [fol. 84v]) as the author of the book reported to be "completed" (*tammat kitāb*), followed by the full title of *Kitāb al-Īḍāḥ,* there are two lines marked in red. These introduce the title of a short addendum of only two pages in a wording that seems to attribute it to the same author, whose dates may be deciphered from the second colophon. The second title given in the two lines in red as *Fi Misāḥa al-Suṭūḥ al-mutasāwiyya al-aḍlāʿ wa al-zawāyā* (On the Measure of Equilateral and Equiangular Surfaces) is preceded by the critical, and almost invisible set of words "also among his derivations" (*min istikhrājihi ayḍan*). From the word "*ayḍan*" we may infer that Yazdi was believed to have authored not only the text of *Kitāb al-Īḍāḥ* but also the short addendum on surveying that immediately follows it. The "lines of text" that follow by the same reported author are also revealing for the content of this short addendum to the text of the *Kitāb al-Īḍāḥ*. Devoted to the subject of the measurement of surface areas of regular polygons, with sides numbering anywhere from three to ten, and numerical values, the text may by itself be regarded as important as part of the understudied subject of "area surveying" in its geometrical and numerical traditions. But the colophon remains a particularly important part of this text for the leads that it provides on issues of identification and dating.

A copyist is named in what seems like a second colophon for both Arabic texts in the same hand (fig. 6a [fol. 85r]): the name appears as "ʿAlī ibn al-Khalīl al-Tājir," next to the transcription date of "728" (1327–28). This figure has possible name and date associations with constructions (*aʿmāl*) in the *Anonymous Compendium* (8.1) and elsewhere in the Paris Codex (7.1). The common part of the last name "al-Khalīl al-Tājir," if considered part of the same family name, allows us to approximate a date for the figure named Abu Bakr, based on the one named ʿAli (see Part III. The Persian Tradition of the *Geometrical Constructions*: 7.1).[86]

The same colophon concludes with a reference to the text's owner (*ṣāḥib*). This most likely meant al-Yazdi as the author of the *Kitāb al-Īḍāḥ* and its addendum, which leads us to other deductions. From the wish for his life to be prolonged (*ṭāla baqāʾahu*) we may infer that the author was living at the time; this therefore places the dates of both the author and the texts on surveying within the same timeframe of 728 (1327–28). If the author of the surveying text(s) is indeed "al-Yazdī," named after the city of Yazd in central Iran, and if he has any relation to the well-wishing transcriber, who signed his name as "ʿAlī ibn al-Khalīl al-Tājir," this in itself would be significant, with larger implications for related datings and identifications of names relevant to the other chapters in this volume.[87]

3. *An Arabic commentary* (*Appendix I, 3.1*)

In contrast to compositions in geometry or surveying within the Arabic tradition related to Abu'l-Wafa''s *Geometrical Constructions*, the unpublished commentary by Kamal al-Din ibn Yunus (no. 3) directly refers to that work. Known through a unique manuscript titled *Sharḥ al-Aʿmāl al-handasiyya li Abī al-Wafāʾ* (Commentary on the *Geometrical Constructions of Abū'l-Wafāʾ* [3.1]), this commentary is a largely unexamined work. Interestingly, the reference to "*Aʿmāl al-handasiyya*" in its title differs from that of the known Arabic manuscript copies of Abu'l-Wafa''s *Geometrical Constructions,* where "*Aʿmāl al-handasa*" is used.[88] The commentary is treated here for the valuable clues it offers: besides a name and date for the copyist, the colophon of this commentary atypically names a specific place (i.e., Maragha), where manuscripts related to the *Geometrical Constructions* were transcribed.

The closing lines of Ibn Yunus's commentary furnish certain hints about the workings of the tradition beyond tenth-century Baghdad, where Abu'l-Wafa' is believed to have composed the *Geometrical Constructions*. Written in Arabic by an author with the *nisba* al-Mawsili, after nearby Mosul in northern Iraq, this commentary was possibly composed not far from Baghdad, though about two centuries after Abu'l-Wafa' lived. An introductory section contains the full name of the important patron of the text, that is, the Ayyubid ruler "Ṣultān … Ṣalāḥ al-Dīn" (r. 1168–98), whose name occasionally appears alongside that of the author, Kamal al-Din ibn Yunus.[89] Just before the colophon, one finds the name of the copyist, i.e., "Ḥusayn ibn … Aḥmad Mārdinī," and the date of copying, 680 (1281).[90] It is the less commonly reported place of transcription, here given as "Maragha," a city in northwestern Iran, that stands out most for its various associations.

Maragha has long been noted as the site of the foundation of a rich compound, including a school, an observatory, and a library.[91] Beyond the construction of this famous Ilkhanid scientific center under the directorship of Nasir al-Din al-Tusi, who was a student of the author of the commentary under discussion, Ibn Yunus, the city was associated with various types of geometrical constructions, ranging from monuments such as the decagonal tomb tower known as the Gunbad-i Qabud, dated 593 (1196–97), to ornamental constructions, such as the "decagonal/pentagonal pattern" decorating that structure.[92] Relevant texts were also transcribed in Maragha shortly before and immediately after the transcription of Ibn Yunus's commentary, including two in 688 (1289). One of these is an unpublished early Arabic text titled *Ashkāl mutasāwī al-aḍlāʿ* (Equilateral Figures), which names Abu'l-Wafa''s early Buyid patron, ʿAdud al-Dawla; the only known manuscript of this work was reportedly copied in that year (688 [1289]) in the Maragha Observatory (*bi-Marāgha al-raṣad*).[93] The other work is an unpublished Arabic manuscript titled *Madkhal Babūs fī al-ḥiyal* (Introduction to Mechanics by Pappus [of Alexandria]), whose colophon mentions that it was transcribed at the "Yāqawiyya" School at Maragha in that same year (1289); this text has some overlaps with both the text of Abu'l-Wafa' and its twelfth-century Arabic commentary by Ibn Yunus, as discussed below.[94]

In the closing lines of his commentary, Ibn Yunus returns to the aim (*gharaḍ*) of the work, which is spelled out mainly in terms of the manuscript tradition of Abu'l-Wafa''s *Geometrical Constructions*. Ibn Yunus hopes to "facilitate the difficult ways" (*tashīl ṭuruq mushkila*) of that work, "follow what is acceptable or good in its way and correct what goes (*tanāwal*) into defect (*fisāda*) of its manuscript copies (*nusakh*)," and explain the "points of error" (*mawāḍiʿ al-khaṭāʾ*) in its proofs (*ithbāt*), and validate them (*ḥaqqaqna*).[95] The objectives in the closing lines also recall propositions in mechanics (*ḥiyal*), with specific reference to a screw (*ḥafr lawlabī*) used by carpenters (*najjārūn*). This is a device treated in Pappus's *Introduction to Mechanics*, an anonymous Arabic translation of which was transcribed in Maragha according to one of the two known manuscripts of it mentioned above. The same colophon notably documents a much earlier transcription by the important tenth-century geometer "al-Sijzī": this has occasioned the suggestion that Sijzi's copy was available to Abu'l-Wafa', whose *Geometrical Constructions* has a construction using a compass with a single opening that is comparable to parts of Pappus's *Mechanics* surviving only in Arabic.[96] Another link to the tradition of the *Geometrical Constructions* is through one of five mechanical devices or "simple machines." This is a lifting device known as a compound pulley (*bakra*),[97] comparable to a "construction" (*ʿamal*) associated with the aforementioned treatise in the Paris Codex titled "Of the Constructions by Abu Bakr al-Khalil al-Tajir al-Rasadi" (7.1), and hence with the *Anonymous Compendium* as well, where his name is cited (8.1). As already noted, the unique Arabic manuscript of Ibn Yunus's commentary on the *Geometrical Constructions*, which, like another Arabic text titled *Equal-Sided Figures*, was copied in Maragha (though it is now kept in Mashhad), requires further study. The cover page of the unique manuscript of this commentary states that it was "lost" and "transferred to Isfahan." We learn from its library title page that it was endowed by the "late Shaykh Baha'i," namely, the Safavid capital's famed scholar and statesman, Baha' al-Din ʿAmili (d. ca. 1621).[98]

III. THE PERSIAN TRADITION OF THE *BOOK OF GEOMETRICAL CONSTRUCTIONS* (APPENDIX I, 4.10): LINES OF TEXT, LINES OF VERSE

The Persian tradition of the *Geometrical Constructions* has elements both distinct from and in common with the Arabic tradition. In terms of distinctions, immediately noticeable is the profile of the figures behind the known compositions. Whereas the dedications in the Arabic works reveal the major role of ruling patrons in the production of those manuscripts, we do not find a similar level of royal involvement in the Persian tradition as it is preserved. Rulers figure prominently in the Arabic texts of Abu'l-Wafa' dedicated to Baha' al-Dawla (1.1 and 1.2) and ʿAdud al-Dawla (1.3), as well as in the thirteenth-century Arabic commentary on his *Geometrical Constructions* by Ibn Yunus (3.1) naming the Ayyubid sultan, Salah al-Din. A manuscript of Abu'l-Wafa''s own text under that title was made for Ulugh Beg's library, now in Istanbul (1.1.i). The other known Arabic manuscripts (1.1.ii–1.1.v) are not royal, and their respective standard conditions suggest that they were disseminated beyond elite circles. But while this can be verified for at least the earlier of the two known copies in Cairo (l.1.ii–1.1.iii), which uses the term "joinery" (*nijāra*) in its title (1.1.ii), the variants in Europe (l.1.iv–1.1.v) have defective openings and closing sections that prevent knowledge of any signs of a commission where such specifications are most expected.

By contrast, in none of the texts known within the Persian tradition, which are mostly complete, are there references to recognizable courtly or high-profiled figures. The two anonymous Persian translation(s) (Appendix I, 4.1 and 5.1), hypothetically considered to be earlier than the one in the Paris Codex completed by Kubanani Yazdi and datable to the fifteenth century (9.1), only partially repeat the dedication in Abu'l-Wafa''s Arabic text as preserved in the Istanbul manuscript; Baha' al-Dawla is mentioned in the opening lines of each. The presumably later Persian translation in the Paris Codex similarly lacks reference to any ruling patron in association with that translation, although various other persons are named (a requester, and two translators) after a colophon signed "Kūbanānī Yazdī," who was one of those translators. This work and the seventeenth-century Persian commentary on Abu'l-Wafa''s *Geometrical Constructions* (10.1) refer to specific individuals and their exact roles in the production of these texts. In the fifteenth-century translation (9.1) there are persons who initiate, request, and complete that translation. The individual who requested the seventeenth-century commentary (10.1), despite being well versed in that and related works, is mentioned, but not named. In this regard, both late Persian texts share some features with two compositions by Abu'l-Wafa' himself, who was asked by a mathematician named Ibn Sahl al-Munajjim (1.2), and a jurist named al-Hububi (1.4), respectively, to write those aforementioned treatises. In other words, overlaps and distinctions between the Persian and Arabic traditions of Abu'l-Wafa''s *Geometrical Constructions* range from problems of transmission to documentation.

The telling lines of some Persian sources

Anonymous Translation(s) of the Book of Geometrical Constructions *(Appendix I, 4.1 and 5.1)*

The earliest Persian translation(s) of Abu'l-Wafa''s *Geometrical Constructions* are anonymous works, itemized separately here given the absence of studies on whether they are two independent translations, or one and the same. One of them (4.1) is known through a manuscript now in Mashhad; it is called "unique" (*bī tā*) in the library's manuscript catalogue.[99] As mentioned above, the Tehran manuscript contains an anonymous Persian translation with comparable features (5.1).[100] The two manuscripts, with the conjectured transcription dates of "twelfth-thirteenth century" and "eleventh-twelfth century" in their respective library catalogue entries,[101] are respectively titled *Dar Ānch* and *Dar Ānchih ṣunnāʿ wa muḥtarifa bi ān muḥtāj bāshand az aʿmāl-i handasa* (On What Is Needed by Craftsmen and Artificers from Geometrical Constructions). They both have the ending *"va īn kitābrā Kitāb-i Nijārat khwānand"* (and They Call this Book, the Book of Joinery). These two anonymous translation(s) are distinct from the aforementioned fifteenth-century Persian translation "completed" (*itmām*) by Kubanani Yazdi (9.1),[102] whose related works (9.2–9.5) are central to the discussions that follow. This late Persian translation, with the shorter title of *Tarjuma-i Aʿmāl-i handasiyya* (*Translation of Geometrical Constructions*), is in the same Paris Codex as the *Anonymous Compendium* (8.1) and the "Constructions by Abu Bakr" (7.1).[103] The latter work, like the translation completed by Kubanani Yazdi, may have been transcribed by him as part of the Paris Codex, as we shall see in the concluding section of this chapter.

The anonymous Persian translations of the *Geometrical Constructions* in Mashhad and Tehran have been itemized here as two texts (4.1 and 5.1), in case they were translated by different anonymous translators. They are numbered as they are because of the apparently earlier linguistic style of the Mashhad manuscript, compared to that of its counterpart in Tehran.[104] These manuscripts share many aspects, including their title and the erroneous transmission of the Buyid patron Baha' al-Dawla's epithet "*Manṣūr*" (Victorious) as part of his name (that epithet is correctly cited only in the Arabic manuscript in Istanbul, indicating its closer connection to the original text submitted to Baha' al-Dawla).[105] Nevertheless, several "lines of text" are unique to each of the Mashhad and Tehran manuscripts in Persian.

Annotations

The manuscript in Mashhad (4.1), which seems to represent an early stage in the Persian translations of Abu'l-Wafa''s *Geometrical Constructions*, is a notable copy of poor quality, with some unique features. Besides the reported "script," "stamp," and signature," at the back of its first folio is a reference to "Shaykh Baha'i," namely Baha' al-Din ʿAmili, who endowed the manuscript.[106] A thus far unrecorded annotation on its front folio may provide some leads once it is deciphered. This unclear transcription appearing on the top left corner of the opening folio, which is not noted in any catalogue or the secondary literature, may be read as "Abū Isḥāq," or "Abū'l-Fatḥ," each name presenting different possible leads. The reading "Abū Isḥāq" may be a reference to the first name of Kubanani, who completed the incomplete Persian translation of the same work in the Paris Codex (9.1). The more likely reading, "Abū'l-Fatḥ," corresponds to the first names of at least two figures with direct links to the field of surveying, already discussed above, i.e., the first name of the author of the Arabic surveying text *Kitāb al-Īḍāḥ*, Abu'l-Fath al-Yazdi (2.1), and of Abu'l-Fath Isfahani (6.1), who translated into Persian a text on surveying titled *Dar ʿIlm-i Misāḥat*, both of them bound together in a codex in Mashhad. If the annotation is intended for any of those names, rather than for Abu Ishaq Kubanani or another figure, it leads to a direct connection between Abu'l-Wafa''s *Geometrical Constructions* in Persian and the field of surveying.

The Persian translation of a surveying text (Appendix I, 6.1)

The reported Persian "translation" (*naql*) of an Arabic text on surveying offers features that reveal further connections between practices and practitioners in "area surveying" in particular. This is a work datable from the late eleventh to early twelfth centuries, based on the dates of its named translator, Abu'l-Fath Isfahani (no. 6.). As we have already seen, the Persian text was long considered a translation of an Arabic surveying text titled *al-Īḍāḥ* (2.1) attributed to the Persian translator Abu'l-Fath Isfahani, which named the original Arabic author as Abu Mansur al-Baghdadi without specifying a text title. The Persian translation may indeed be based on another Arabic text by Baghdadi with the similar title of *al-Misāḥa* (Surveying) (2.2).[107] Particularly notable are the closing sections of the Persian "translation" (6.1), which include the aforementioned supplement added in Persian bearing the title "*Fī al-Saṭḥ* [i.e., *Suṭūḥ*] *al-mutadākhil*" (On Inscribed Surfaces [Within One Another]) (fig. 6b). This title is very similar to the heading of the *Anonymous Compendium: Fī tadākhul al-ashkāl al-mutashābiha aw al-mutawāfiqa* (8.1). The surface areas involved in both cases, whether described as "inscribed" within one another, as in the datable surveying text, or "interlocking," in the sense of lacking spaces between them, in the undated *Anonymous Compendium*, seem related to old, and apparently close, traditions of area surveying, a branch with its own variations in the related literature.

Distinctions between measurement in surveying (*misāḥa*) involving surfaces (*musaṭṭaḥāt*) as opposed to measuring volumes (*mukaʿabāt*) have already been noted above in works such as the Persian classification of Fakhr al-Din Razi (ca. 1178–79), which is from a period rather close to that of the Persian surveying text by Abu al-Fath Isfahani (6.1). In the same way, distinctions between methods of area surveying specific to land (*arḍ*) have been discussed in the section on "inscribed surfaces" in this work. These methods were contrasted with those applicable to architectural constructions (*banāʾ*), listed under the science of "erection of structures" (*ʿuqūd al-abniyya*) as one of ten subdivisions (*furūʿ*) of geometry (*handasa*) in the later Arabic classification work titled *Irshād al-qāṣid* (Guidance of the Messenger) by Ibn al-Akfani (d. ca. 1348), next to surveying (*misāḥa*). Thus, "area surveying" with respect to the division of land was different from the surveying of structures or buildings, especially in later periods, where the role of practical geometry (*handasa*) and mensuration (*misāḥa*) in the training of architects (*miʿmār*) was more explicitly stated.[108]

The applications of the "inscribed areas" in the brief supplement with that title in the Persian surveying text

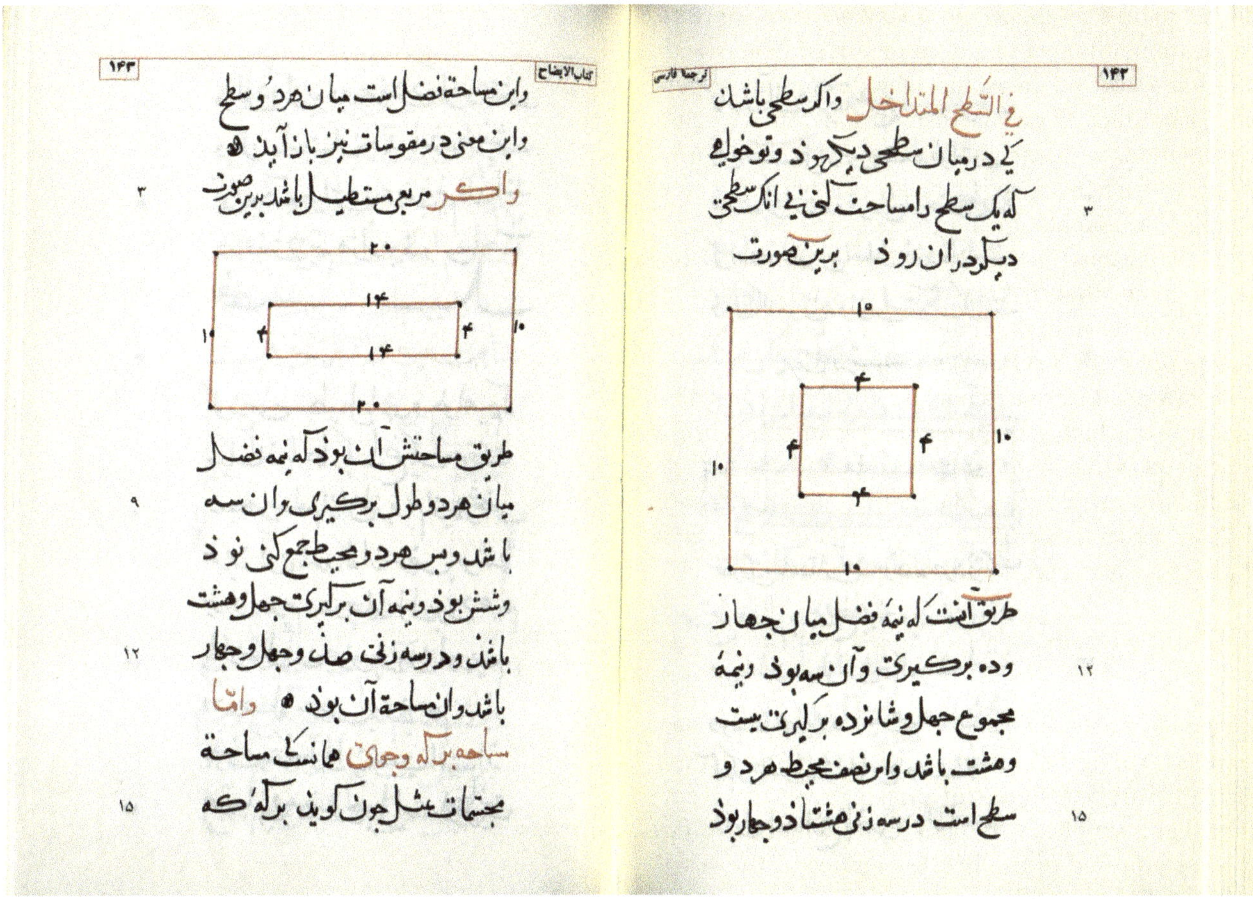

Fig. 6b. Double page from a Persian translation, "Treatise on the Science of Surveying" by Abu'l-Fath Isfahani (fl. ca. 1095–1119), of an Arabic text by Abu Mansur al-Baghdadi: this section, titled "Inscribed Surfaces," is in the same hand and datable to 728 (1327–28). Persian manuscript: Mashhad, Āstān-i Quds-i Rażavī Library, Ms. 5462. (Photo: Gulchīn Maʿānī 1969, facs. 124–125)

(fig. 6b) are not limited to finding areas of figures such as a square or rectangle. As such, it differs from the Arabic surveying text bound with it. I have already discussed the possible author, Abu'l-Fath al-Yazdi and the signed transcriber, ʿAli ibn al-Khalil al-Tajir, of this work and its addendum, which is dated 728 (1327–28) (fig. 6a). In the latter, the measurement (*misāḥa*) of surface areas is limited to regular polygons of three to ten sides that are given numerical values, except for quadrangular areas, pointing to possible applications outside land surveying that are more standard for square and rectangular areas. In the Persian supplement, "inscribed areas" refer to figures falling inside one another, where the use of such explicit expressions presents a clear contrast to other treatments of inscribed figures, including Abu'l-Wafa''s. Notably, this involves measurements (*misāḥa*) of the difference (*faḍl*) between the areas of those surfaces, when one is subtracted from the other. The only two illustrated examples in this short supplement are limited to pairs of rectangles and squares inscribed within one another (fig. 6b). Such measurements (*misāḥa*), involving abstract numerical calculations regardless of any actual cubit measurements, are followed by concepts (*maʿānī*) applicable to curved figures (*muqawwisāt*), as distinct from physical bodies (*mujassamāt*) such as ponds and streams (*barka va jūy*). Further study is required with regard to any direct relation between the "inscribed surfaces" in this Persian surveying text (6.1) and the "interlocking figures" of the *Anonymous Compendium* (8.1), which are not limited to angled or curved figures with different sides and areas, but include composite patterns assembled in various inscribed and interlocking combinations with one another. "Interlocking" figures, as distinct from "inscribed" surfaces, may be used for the complex ornamental geometric patterns in two- and three-dimensions (including *muqarnas* projections) discussed in the other chapters of the present volume.

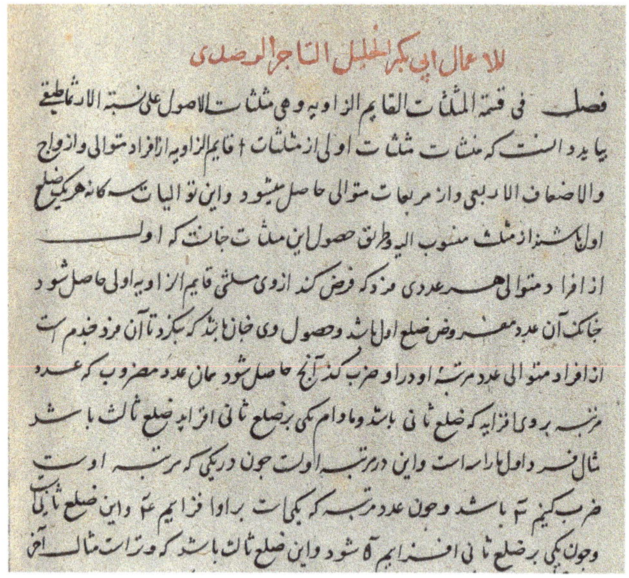

Fig. 7a. The opening page of the "Constructions by Abu Bakr." Paris, Bibliothèque nationale de France, Ms. Persan 169, 19, fols. 124v. (Photo: Bibliothèque nationale de France)

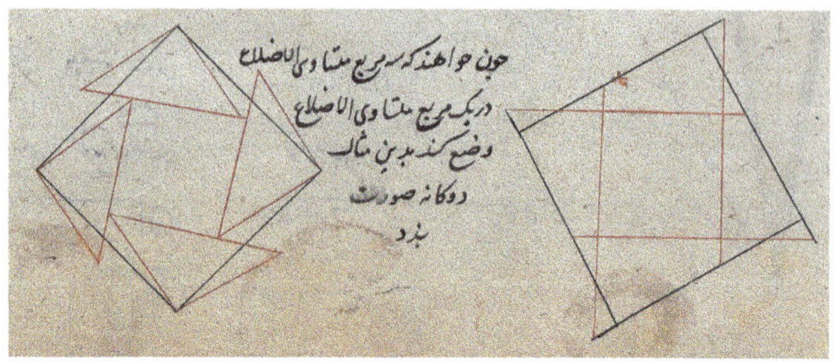

Fig. 7b. Detail of a folio with two illustrations and Persian script referring to alternative construction methods comparable to Abu'l-Wafa's constructions for a craftsman (right) and a geometer (left), contained within the "Constructions by Abu Bakr." Paris, Bibliothèque nationale de France, Ms. Persan 169, 19, fol. 136v. (Photo: Bibliothèque nationale de France)

With reference to the *Anonymous Compendium*, the Persian surveying text *Dar ʿIlm-i misāḥat* (6.1) is also notable for what it lacks, and not just what it contains. The Persian text, which, though missing a colophon, is in the same hand as that in the Arabic text bound with it (2.1), would have carried the same name "ʿAlī ibn al-Khalīl al-Tājir," for its transcription, and the date "728" (1327–28), or close to it, occasioning other examinations.

Two Persian texts from the Paris Codex (Appendix I, 7.1 and 8.1)

Any name or date associated with the *Anonymous Compendium* (8.1, fig. 8a), a work with the unprecedented feature of various complex interlocking geometrical patterns that is supplied with both text and illustrated figures, as opposed to the textless figures of the scroll tradition,[109] is particularly important. But a text that is unique, anonymous, undated, and non-uniform presents certain challenges. Since the citation of names or works associated with the *Anonymous Compendium* can have an impact on how its date is estimated, the forms of such citations and associations and variations also matter.

Most relevant is the citation of the name "Abū Bakr al-Khalīl al-Tājir" in association with some "constructions" (*aʿmāl*) in *Lil-Aʿmāl Abī Bakr al-Khalīl al-Tājir al-Raṣadī* (7.1, fig. 7a) in the Paris Codex, a work containing two constructions on a folio next to a surveying problem (fig. 2a) that correspond to Abu'l-Wafa''s constructions by a craftsman and a geometer (fig. 7b). The same name form, "Abū Bakr al-Khalīl al-Tājir," as given in the title of those "Constructions," also appears in the *Anonymous Compendium* (fig. 8b).[110] The importance of that citation for dating the *Anonymous Compendium* has been noted, since that name seems to overlap with that of the copyist of the surveying text discussed above: "ʿAlī ibn al-Khalīl al-Tajir" (6.1, fig. 6a, fol. 85r). In scholarly discussions in this volume, there has been speculation over whether these two were father and son, or brothers, which could affect the dating of the *Anonymous Compendium*. Regarding date approximations for the *Anonymous Compendium*, Özdural has suggested a father-son relationship, approximating that date to ca. 1300. Necipoğlu has noted citations of a variant form of one of

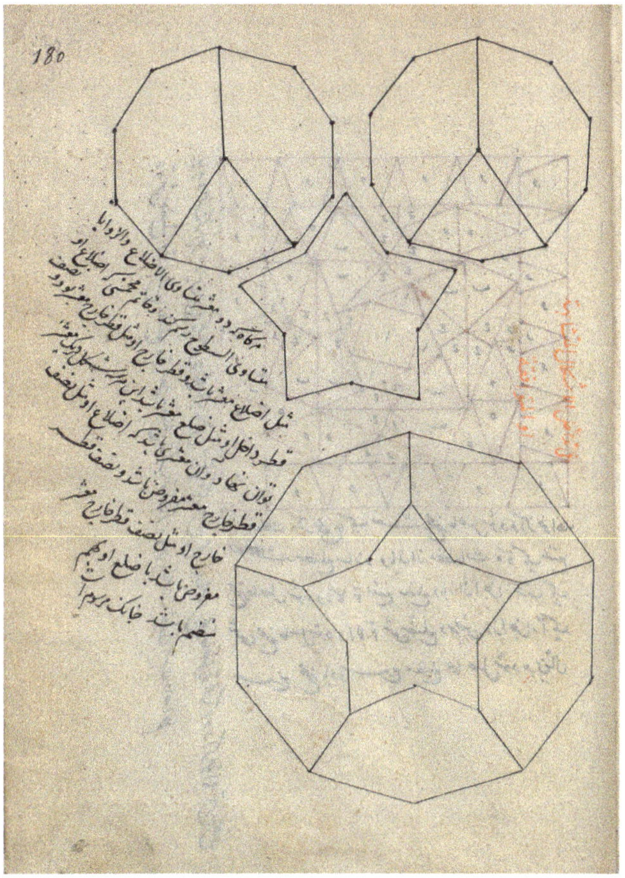

Fig. 8a. The first page of the *Anonymous Compendium*, with the title in the right margin. Paris, Bibliothèque nationale de France, Ms. Persan 169, 24, fol. 180r. (Photo: Bibliothèque nationale de France)

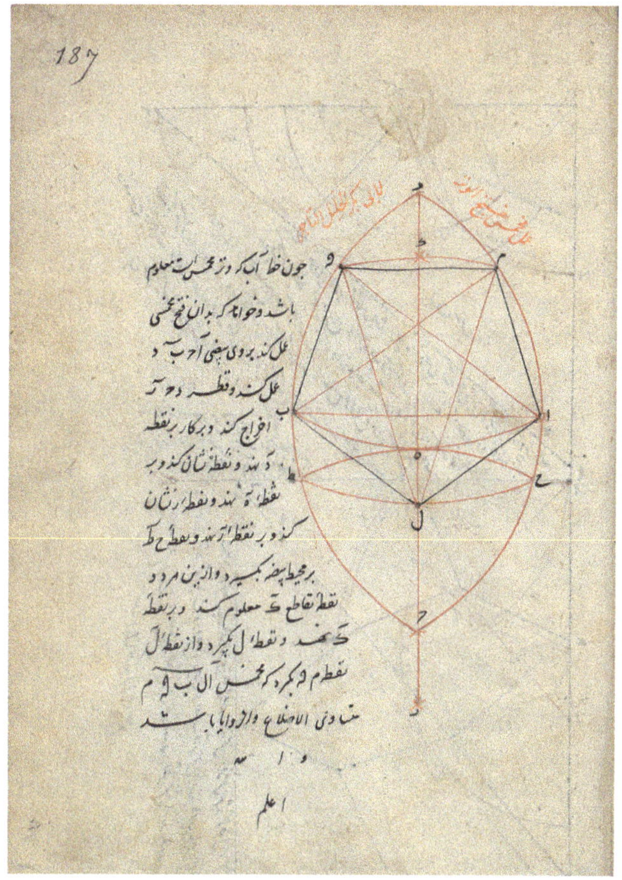

Fig. 8b. Folio from the *Anonymous Compendium* with a citation of the name "Abū Bakr al-Khalīl al-Tājir." Paris, Bibliothèque nationale de France, Ms. Persan 169, 24, fol. 187r. (Photo: Bibliothèque nationale de France)

those names as "Abū Bakr ibn al-Khalīl al-Tājir," with an extra "ibn" (son) in both the Paris Codex and in a codex now located in Saint Petersburg, and considers the possibility that they may have been brothers, with the resulting shift of date approximations (ca. 1300 or 1300s).[111] The occurrence of the form "Abū Bakr ibn al-Khalīl al-Tājir" in the Paris Codex, where he is referred to as the "pride of the geometers" (*mufakhkhar al-muhandisīn*) on folio 118v,[112] and, outside of it, as a copyist of the Saint Petersburg Codex,[113] is behind the position adopted by Necipoğlu to add a bracketed "[b.]" to the wording of the "Constructions by Abu Bakr" (7.1, fig. 7a) and to citations of his name in the *Anonymous Compendium* (8.1, fig. 8b).

The various names cited and their direct association with the constructions above, their slightly different forms in the *Anonymous Compendium*, as well as in other parts of—and beyond—the Paris Codex, are not insignificant. The form "Abū Bakr al-Khalīl al-Tājir al-Raṣadī," which appears within the codex in the context of some "constructions" (*aʿmāl*) associated with him (7.1, fig. 7a), is a longer version of his name as cited in several places in the *Anonymous Compendium*. But the same may not be said about the variation "Abū Bakr ibn al-Khalīl al-Tājir," with an added "ibn" (son [of]) before the last name, as if he were the son of someone named "Khalīl al-Tājir." This involves more than an extra few letters in one name. Hence, the form "Abū Bakr ibn al-Khalīl al-Tājir" may itself be not a variation, but the same name as "Abū Bakr al-Khalīl al-Tājir," the "ibn" of which could have dropped through transmission.[114]

It has been noted that besides Ibn-i Haytham, whose own compositions help us to establish the lower limits for dating that part of the *Anonymous Compendium* to circa 1020s, when he was first active, the only other name that appears in this work is "Abū Bakr al-Khalīl al-Tājir." We find the latter in two forms, in two contexts, as discussed in more detail in Necipoğlu's chapter. The name "Abū Bakr al-Khalīl al-Tājir" first occurs in exactly that form in the label accompanying the construction of a regular pentagon (*ʿamal-i mukhammas*), using a compass with a single opening, that is attributed to him (fol. 187r [25]). It next occurs in the abbreviated form "Abū Bakr al-Khalīl," when he is described as subjecting a different construction to multiple "testing" (*imtiḥān*), which this time involved quadrangular areas and different procedures (fol. 189r [31]).[115]

This quadrangular construction, involving both squares, right triangles, and their derivatives, is a more complicated version of the "application of areas" method, with a square at the center surrounded by four compound triangular areas, called kites (*turunj*), all inscribed within a larger square of equal area. As such, it is comparable to the square constructions with central-peripheral figures in both Abu'l-Wafa''s *Geometrical Constructions* (1.1) and in the *Anonymous Compendium* (8.1), as discussed before (see Part II above). On the other hand, we should also note the reported testings (*imtiḥān*) for proportional ratios (*nisbat*) behind such constructions as performed by "masters" (*ustādān*): for example, "Abū Bakr al-Khalīl" is specifically named in this part of the text in relation to both a "construction" (*ʿamal*) and its "multiple" (*chand vajh*) testing.

The terminology of testing (*imtiḥān*) in that construction is distinct from other methodological terms used throughout the *Anonymous Compendium*, ranging from ones as elementary as "actualization" (*fiʿl*) of potentiality (*quwwa*) through superposition (*taqrīr*) (fol. 182v), to others as advanced as "[comparative] examination" (*iʿtibār*) (fol. 184v), representing an early form of experiment. The latter term was identified as a crucial part of "experimentation" by Ibn al-Haytham himself, who is cited elsewhere in the *Anonymous Compendium* in a different context (fol. 191r), and in his Latin translation of the term *iʿtibār* as *experimentare*.[116] Such methodological terms and concepts, however, seem more basic in the *Anonymous Compendium*, and cannot by themselves suggest specific dates for parts of the text.

Another independent citation useful for dating at least the "Constructions by Abu Bakr" in the Paris Codex (7.1), and, through it, the *Anonymous Compendium* (8.1), appears in a marginal note containing the name "Sayyid Sharīf" on one of the "constructions" (*aʿmāl*) of Abu Bakr (fol. 135v). The reading of that name as "al-Sayyid al-Sharīf [al-Jurjānī]" (1339–1413) is consistent with the glosses of Jurjani "on most books by the ancients and moderns," as reported by the sixteenth-century historian Khwandamir.[117] If Jurjani, a native of Iran whose life and travels can be roughly dated, is the figure meant by "Sayyid Sharif" in the marginal note on that particular "construction" in the Paris Codex (as first suggested quite reasonably by Francis Richard), his own dates can also establish an upper limit for the date of "Constructions by Abu Bakr," as well as parts of the *Anonymous Compendium*. That upper limit of 1413 would concern original compositions, rather than later transcriptions or compilations of the Paris Codex, some of which may be examined in connection with the figure discussed next, namely Abu Ishaq Kubanani.[118]

The Persian Translation of the Book of Geometrical Constructions, *completed by Kubanani (Appendices I and III, 9.1)*

The treatise in the Paris Codex titled *Tarjuma-i Kitāb-i Abī al-Wafā' Muḥammad ibn Muḥammad al-Būzjānī dar aʿmāl al-handasiyya* (Translation of the Book of Abu al-Wafa' … al-Buzjani on Geometrical Constructions) may be considered a late translation of the Arabic text in comparison with the seemingly two earlier ones discussed above (i.e., Appendix I, 4.1 and 5.1). It is not clear whether the text is based on the Arabic original from about five hundred years earlier, or on an earlier Persian translation closer to its own time.

The author signing the work as "Abū Isḥāq ibn ʿAbdallāh Kūbanānī Yazdī" (9.1, fig. 9a) explicitly states that the Persian translation was "in abbreviated form" (*bar sabīl-i ikhtiṣār*), and that he "completed" (*itmām*) a "previous" (*sābiq*) "translation" (*tarjuma*) by "reconstructing" (*banā'*) and "writing" (*taḥrīr*) it in the course of two months. As such, the resulting work may be characterized as an abbreviated Persian "co-translation" and "reconstruction" of Abu'l-Wafa''s *Geometrical Constructions*. Thus, it is not a typical translation, as it is called in its title, in the more familiar senses of that term. The text also stands out for lacking any date or place associated with it, as is the case in many other works of Kubanani included in what is believed to be an autograph codex now in Tehran (9.2). Positioned in the Paris Codex just before the *Anonymous Compendium*, the text contains various "telling lines" in both text and verse form, from which selections are presented here (see Passages 1–4 below, and Appendix III, 9.1).

Like those of all the Persian treatises in the Paris Codex, the opening lines of Kubanani's "completion" of the Persian translation of *Geometrical Constructions* do not include a dedication to a ruling patron as do its Arabic counterparts (Appendix I, 1.1–1.3, and 3.1). This is in striking contrast to several of Kubanani's own works, which often bear details on his various commissions within Iran, i.e., those in the 1440s and 1450s in Sari, in the north, where he dealt mostly with astronomical works; in the 1450s and 1460s in Kirman and Jarun (now Hormuz), in the south (9.2), where he worked on arithmetical tracts as well; and in the 1470s and 1480s, in unknown locations, including a work on Euclidean geometry (9.5). In the closing part of the Persian "translation" bearing the name "Kūbanānī Yazdī" (9.1, fig. 9a), we learn that this was a co-translation completing a work "upon the request" of another individual. In this long colophon, hereafter referred to as a "postface," as it comes after rather than before the main text, as is more standard,[119] Kubanani provides information on his role as the author who completed the work, while also referring to the requester, and the initiator of that translation, from their names to their professional profiles, as well as his own addressed audience.

The following selection of passages from this late Persian "translation" of Abu'l-Wafa''s *Geometrical Constructions* is illuminating beyond its long title and opening lines (Appendix III, 9.1a–b): the title presents one of two forms of the Arabic version, that is, *Aʿmāl-i handasiyya* (instead of *Aʿmāl-i handasa*), and the opening lines mention construction tools (a ruler, compass, and set-square), which were part of Abu'l-Wafa''s variant title (1.1.ii), but here form an "introduction" (*muqaddima*). The passages that follow the text's closing lines (9.1c–f), however, are key to uncovering rarely encountered names and profiles, as well as highlighting practices and practitioners, and, through them, terms and expressions closer to the author's time. The first two of those passages (9.1c–d) describes the way in which the book was requested and completed, while the next two (9.1e–f) are more specific regarding its initial translator and later addressees. These passages refer to many relevant works and figures through both versed prose and rhyming poems.[120] The successive lines of text and verse open with a description of the work involved in that composition (Passage 1, fig. 9a), and the figure who requested it (Passage 2, fig. 9b), but despite the guiding rhymes, these opening passages have at times been read or interpreted other than the way in which they were intended.[121]

Passage 1: On the completion of the book (Appendix III, 9.1c, fig. 9a)

This [text] is completed (*tamma*). God Almighty, may Praise be to Him, gave this impoverished mortal subject, Abu Ishaq ibn ʿAbdallah Kubanani Yazdi, the power to facilitate (*taysīr*) the passages (*ʿibar*) and solve the problems (*ishkāl*) of the propositions/figures (*ashkāl*) of this wondrous (*ʿujāb*) book (*kitāb*).[122]

Passage 2: On the request for the book (Appendix III, 9.1d, fig. 9b)

[The Completion was] according to the request (*iltimās*) of the teacher of mankind (*ustād-i bashar*), and "unique in the art" (*yagāna bar hunar*), who in the work station of fate and destiny (*kārkhāna-i qażā va qadar*) ... hemistich:

Until the [heavenly] wheel has a circular path around this clay
And until swift-handed laborers place brick over brick
[No] celestial beings will see such a master geometer (*ustād-ī muhandis*)
on the surface of the earth
[Nor] will the bird of the fast grasp of Euclid [reach] the heights of his innovative works (*badāyiʿ-i aʿmāl-i ū*), which are all ... hemistich (*miṣraʿ*) ...

In his [i.e., the requester's] studentship and assistantship, [I] placed the finger of obedience on the eyes ... and was

> granted success by God [to carry the order of] Shams al-Din Abu Bakr Shah, the son of the complete master (*ustād-i kāmil*), and wise scholar-practitioner (*ʿālim-i ʿāmil*), Najm al-Din al-Mahmud Shah, son of Zamh al-Hajj wa al-Muʿtamirin Taj al-Din Kaduk, for the reconstruction (*banāʾ*) [of this book], without losing any time, to be written during the course of two months.[123]

The first few lines of Kubanani's postface (9.1c) merit some analysis of the descriptions and terms, however few, as to what the "completion" of the book penned by him amounted to in his own words. Starting with a common self-address as an "impoverished mortal subject ... given power (*quwwa*) by the Almighty," Kubanani introduces a monumental undertaking in the form of two specific tasks: "facilitating (*taysīr*) the passages (*ʿibar*), and solving (*ḥall*) the problems (*ishkāl*) of the propositions/figures (*ashkāl*) of that wondrous (*ʿujāb*) book." While the task of "facilitating passages" suggests problems in textual transmission that are also indicated independently, that of "solving problems" does not make it clear whether the problems (*ishkāl*) specified involved the text's propositions or figures, the term "*ashkāl*" (sing. *shakl*) standing for both items in geometrical texts.

The next passage of the postface (9.1d), describing how the book came to be requested, is longer. In the first part of the passage, Kubanani describes the request (*iltimās*) as having been made by a "teacher of mankind" (*ustād-i bashar*) who is "unique in the art" (*yagāna bar hunar*): the first title was often used for Nasir al-Din al-Tusi, whose *Recension of Euclid's Elements* formed the basis of another work by Kubanani (9.5), and the second expression is a description rhyming with the latter, as well as the next phrase (*kārkhāna-i qażā va qadar*). It is not clear whether Kubanani uses these descriptions merely as rhyming phrases or whether the requester, about whom more follows in the second part of that same passage, actually was "unique in the arts" (*hunar*, i.e., Greek *technē*, Arabic *ṣunʿ*). Some terms, however, should clearly not be taken as literally as they have been, such as "*kārkhāna*" (work station), which, in the context of "fate and destiny" (*qażā va qadar*), has been inaptly interpreted as a "workshop." The same may be said for parts of the so-called hemistich (*miṣraʿ*) that follows, which describes the book's requester as a "master geometer" (*ustād-ī muhandis*) whose unmatchable "innovating works" (*badāyiʿ–i aʿmāl*) were unequaled. Such expressions appear in similarly figurative terms: as hitherto "unseen" (*nādīda*) by the "celestial beings" (*muʾas̲s̲arāt-i aflāk*) on the "surface of the earth" (*saṭḥ-i khāk*), for as long as (*tā*) "the [heavenly] wheel has a circular path" above and "laborers place brick over brick (*khist*)" below.[124]

In the last part of this same passage (9.1d), Kubanani provides some critical information about the requester as well as the request. He begins by describing himself favorably as an obedient student and assistant of the requester (he refers to his "studentship" and "assistantship" [*shāgirdī* and *dastyārī*] and having "fingers of obedience over [his] eyes"). Here we learn what was specifically requested, namely, a textual "reconstruction" (*banāʾ*), and "writing" (*taḥrīr*); this successfully took take place over the course of "two months (*du māh*) ... without losing time." We only then learn that the one who requested this rushed project was named "Shams al-Din Abu Bakr Shah ... son of (ibn) ... Najm al-Din al-Mahmud Shah ..." (fig. 9b).[125] The project was meant to complete a book begun by the latter (Najm al-Din), who was no longer living; the aim was to preserve his work so that it would become known.

The remaining two passages (3 and 4 [fig. 9c–d]), which contain more details and profiles, also require careful analysis.

> *Passage 3: On the initiator of the book (Appendix III, 9.1e, fig. 9c)*
>
> In days of distress ... and with heart sorrowed by and mind suffering from grief over that departed dearest brother, pride of scholars and sages, pole of tying astronomical tables and untying the [puzzles] of the *Almagest*, Shaykh Najm al-Din al-Mahmud, may God bless his soul with a great blessing, [he who] in the vast field of wisdom and scholarship had snapped the ball of speed from the course of time, who had written commentaries on the major books of the *Almagest*, as epistles and excerpts consisting of specific interpolations, and annotations on the Spherics of Menelaus among the "Intermediate Books," in the version of Amir Mansur Nasr ibn ʿIraq, and other kinds of marvels in arts/crafts (*funūn*) of science (*ʿilm*), [and] who passed away in the peak of his youth, becoming the cause of the pressing request of the work of one who had previously translated this book, some of which was destroyed in the maelstrom of time, and his requester wanted that such an effort not remain unrecognized and that this bit of his blessed works be preserved in the world, [so he] initiated (*iqdām*) the completion (*itmām*) of that work.[126]

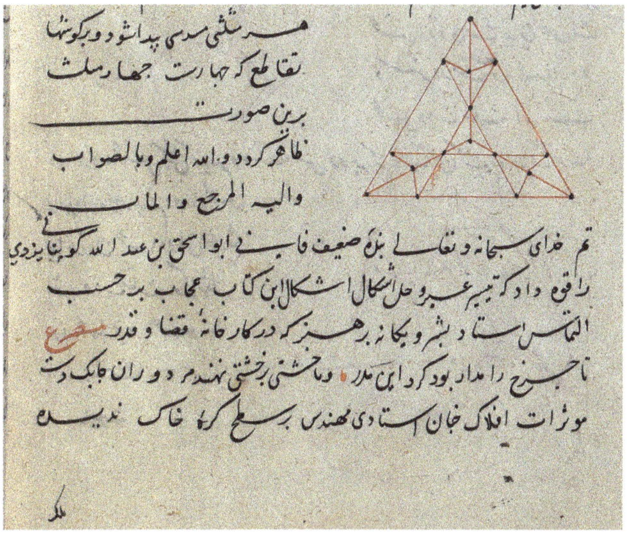

Fig. 9a. The colophon and first folio of the postface in the Persian translation of Abu'l-Wafa''s *Geometrical Constructions* completed by "Abū Isḥāq ibn ʿAbdallāh Kūbanānī Yazdī" with a citation of that name. Paris, Bibliothèque nationale de France, Ms. Persan 169, 23, fols. 178v. (Photo: Bibliothèque nationale de France)

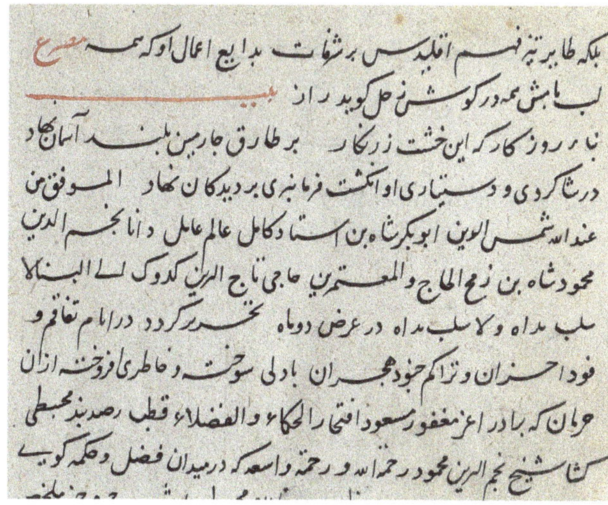

Fig. 9b. The second folio of the postface in the Persian translation of Abu'l-Wafa''s *Geometrical Constructions* completed by "Abū Isḥāq ibn ʿAbdallāh Kūbanānī Yazdī," with the citation of the name "Shams al-Dīn Abū Bakr Shāh" as the book's requester. Paris, Bibliothèque nationale de France, Ms. Persan 169, 23, fol. 179r (top). (Photo: Bibliothèque nationale de France)

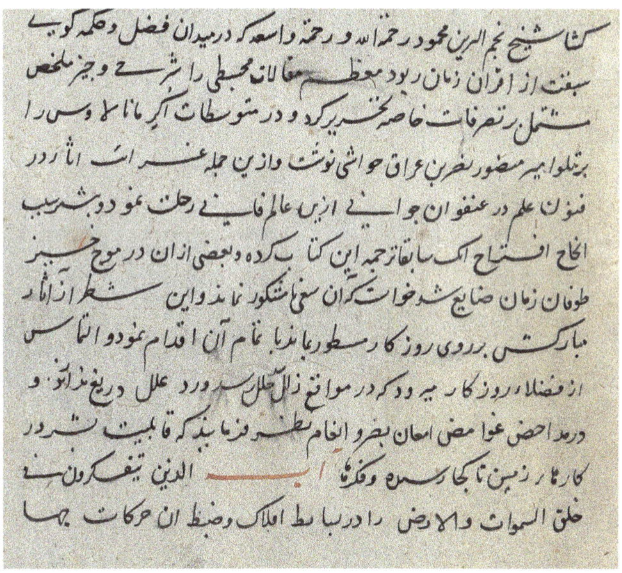

Fig. 9c. The third folio of the postface in the Persian translation of Abu'l-Wafa''s *Geometrical Constructions* completed by "Abū Isḥāq ibn ʿAbdallāh Kūbanānī Yazdī," with the citation of the name "Najm al-Dīn Maḥmūd [Shāh]" as the book's initial translator. Paris, Bibliothèque nationale de France, Ms. Persan 169, 23, fol. 179r (bottom). (Photo: Bibliothèque nationale de France)

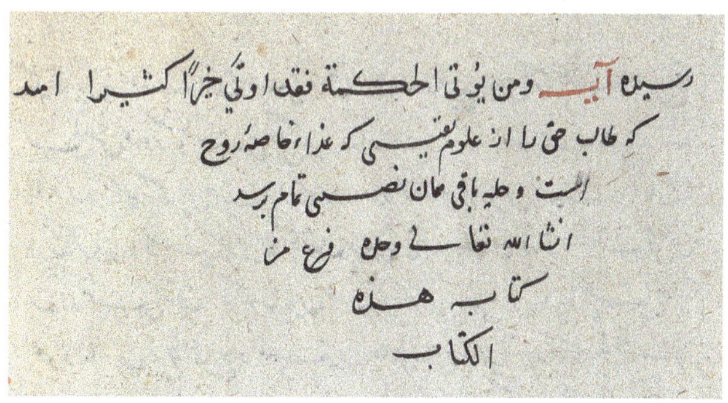

Fig. 9d. The fourth folio of the postface in the Persian translation of Abu'l-Wafa''s *Geometrical Constructions* completed by "Abū Isḥāq ibn ʿAbdallāh Kūbanānī Yazdī," with the citation of the end of the book's transcription. Paris, Bibliothèque nationale de France, Ms. Persan 169, 23, fol. 179v. (Photo: Bibliothèque nationale de France)

Passage 4: On the addressee of the book (Appendix III, 9.1f, fig. 9d)

> So it is requested that the sages of the age, in places of distortions [and] defects, not refrain from rejecting the [text's demonstrative] reasonings (*ʿilal*); and in [the case of] invalidities [and] complexities, employ scrutiny and observation, given how far human capacity has reached in earthly matters. ... [127]

The first of these passages (9.1e) concerns the initial translator of the book, who was introduced in the previous passage (9.1d) as "Najm al-Dīn al-Maḥmūd Shāh, son of Zamh al-Ḥājj wa al-Muʿtamirīn Tāj al-Dīn Kadūk." "Shams al-Dīn Abū Bakr Shāh," the requester of the book, might have been his "son," as indicated by the term "ibn" after "Shams al-Dīn" and before "Najm al-Dīn," as well as a common last name, "Shāh," in that long chain of names (fig. 9b). Earlier, in passage 2 (9.1d), Najm al-Din was described in rhymes as a "complete teacher" (*ustād-i kāmil*), and wise "scholar-practitioner" (*ʿālim-i ʿāmil*). Here, in passage 3 (9.1e), Kubanani also refers to him in rhymes, as the "pride of scholars (*ʿulamā*) and sages (*fużalā*)," and "pole of tying astronomical tables and untying [puzzles] of the Almagest (*Magistī-gushā*)." We should especially note Kubanani's reference to Najm al-Din Mahmud as his "dearest brother" (*barādar-i aʿiz*), for more on which see below. Kubanani also refers to specific works by him that include those within the "Intermediate [Book]s" (*mutawassiṭāt*, i.e., works intermediately between Euclidean geometry and Ptolemaic astronomy), which are "marvels" (*gharāʾib*) in the "arts/crafts of science" (*funūn-i ʿilm*).[128]

In the passage that follows (9.1f), which forms the closing lines of the postface, Kubanani addresses his audience with comparable expressions and rhymes. His plea is that the sages of the time (*fuḍalāʾ-i rūzgār*), not refrain (*darīgh*) from blocking (*sadd*) or rejecting (*radd*) the text's reasonings (*ʿilal*) and insightful scrutiny (*imʿān-i baṣar*), and from bestowing their regard (*inʿām-i naẓar*) upon it.

Some of the information provided in this postface, such as the discussion of textual problems in Passages 1 and 4, may be verified from what is known from the textual tradition of Abu'l-Wafa''s *Geometrical Constructions.* Others, like Kubanani's reference to the named requester as "Sham al-Din Abu Bakr Shah," followed by "son" (*ibn*) of "Najm al-Din Mahmud Shah" (Passage 2), or to the latter as a "brother" (*barādar*) (Passage 3),[129] may be further examined through related works by both Kubanani (9.3) and others (10.1).

In sum, the postface signed by Kubanani (Appendix III, 9.1) documents various problems in transmission, most specifically when Kubanani states that he was asked to "reconstruct" (*banāʾ*) a text (9.1d), an unfinished translation (*tarjuma*) of Abu'l-Wafa''s *Geometrical Constructions* that was apparently finished but no longer complete due to parts that had been dispersed (9.1e). Besides elucidating problems faced by Kubanani, the postface contains important details that allow us to piece together aspects of the longer history behind this work. From its opening lines pointing to the "problems" (*ishkāl*) of "passages" and "propositions/figures" (*ashkāl*) (9.1c) to its closing lines referring to distortions (*zilal*), and defects (*khilal*), a (9.1f), we have further documentation of the poor transmission of the text over time. All these make Kubanani's own work much more than just a "completion" of a translation, one whose striking features include a construction unmistakably close to another within the Paris Codex (fig. 9e–9f; for details, see Part IV).

As for the near contemporaries cited by Kubanani, their profiles and descriptions are no less revealing. The one who requested the text is described as being "unique in the arts," while the initial translator is a "perfect master" next to being a "scholar-practitioner," whose works in the "arts/crafts of science" (*funūn-i ʿilm*) are marvels. Kubanani's postface concludes with yet another revealing expression—about the "verifiable sciences" (*ʿulūm yaqīnī*) and his wish that "the seeker of truth" (*ṭālib-i ḥaqq*) attain a complete (*tamām*) share (*naṣīb*) of them (Appendix III, 9.f). All such descriptions and expressions evoke a simultaneous legacy of sciences (pl.*ʿulūm* [sing. *ʿilm*]) and arts/crafts (*funūn* [sing. *fann*]) that may, in turn, be reconstructed from Kubanani's own works. These include writings beyond his "completion" of the Persian *Geometrical Constructions* (9.1), itself enlightening for a remarkable closing line (in Arabic, as is common in Persian texts as well). Here the end of the "transcription" (*kitābat*) of that "text/book" (*kitāb*) is made explicit in the last two lines containing the respective terms in a particularly close orthography (fig. 9d).

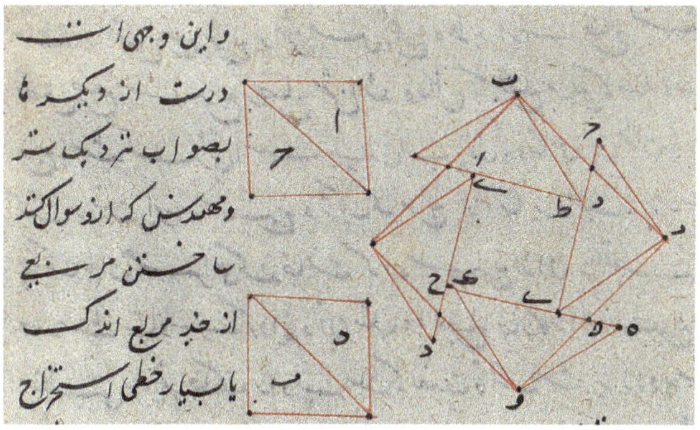

e

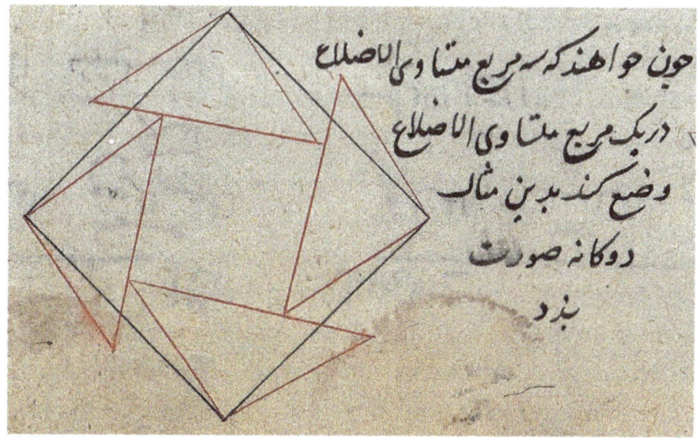

f

Fig. 9e–f. Details of constructions by a geometer of a square from three smaller squares, from illustrations on two different folios in the Paris Codex: e) Kubanani's completion of the Persian translation of Abu'l-Wafa''s *Geometrical Constructions*; f) "Constructions by Abu Bakr." Paris, Bibliothèque nationale de France, Ms. Persan 169, fols. 171v and 136v. (Photos: Bibliothèque nationale de France)

The writings and letters of Abu Ishaq Kubanani (Appendix I, 9.2–9.5)

Various historical connections among practices and practitioners within the traditions of practical geometry and surveying in general, and of Abu'l-Wafa''s *Geometrical Constructions* and area surveying in particular, may be reconstructed through the figure of "Abū Isḥāq ibn ʿAbdallāh Kūbanānī Yazdī" (fl. ca. 1442–post 1483).[130] As we have seen, he was the one who signed the postface to the Persian translation of Abu'l-Wafa''s *Geometrical Constructions,* which also detailed how he "reconstructed" (*banāʾ*) and "completed" (*itmām*) that work (9.1). It has been previously suggested that this figure possibly transcribed that text and others within the Paris Codex, including the *Anonymous Compendium.*[131] This proposition has been raised in the present volume in Necipoğlu's chapter in light of Kubanani's possible ownership of a different codex comprising a collection of mathematical treatises, a proposition discussed in my conclusion through a full examination of all supportable evidence and consistent premises.[132]

Kubanani, for whom new dates and various new works and associations have been identified in the present chapter, has been largely understudied, despite a wealth of material detailing his little-known works (Appendix I, 9.1–9.5).[133] These often recorded the name of the patron or requester of the work in question, and the dates and places of his compositions or transcriptions. The *nisba* Kubanani, which is derived from a place called Kubanan, located between Yazd and Kirman in southern-central Iran, has been misrepresented in occasional references to that individual,[134] including works not verifiably his "compositions."[135] A number of works of historical if not scientific significance have largely been overlooked, including a collection of scientific works in a unique manuscript codex (9.2), a collection of letters compiled by a disciple, Husayn Maybudi (9.3), which cites his other works, along with an excerpt from the measurement (*misāḥa*) of surface areas (9.4), and yet another work on Euclidean geometry (9.5). The latter work proves that he was productive beyond the date known so far, 886 (1481), to after 888 (1483), the date cited in its colophon.

The life and works of Kubanani provide largely untapped and rarely noted information about the subjects under discussion. Because he may have transcribed all the texts contained within the Paris Codex, including the *Anonymous Compendium,* besides completing the Persian translation Abu'l-Wafa''s *Geometrical Constructions,* the time period and place in which he worked are of particular significance. We can glean much from the given dates and places of Kubanani's various works, many of which are considered to be in his own hand. A manuscript codex now in Tehran titled *Majmūʿa-i rasāʾil va kutub* (Collection of Treatises and Books [9.2]), which, according to its library cover, contains his autographed scientific works, documents that he was active in Iran mostly during the mid- to late-fifteenth century, in at least three areas under different patrons, as noted above. But several relevant names and subjects emerge from his little-known letters and fragmented works (*Munshaʾāt*), which were compiled by his disciple, Husayn Maybudi (9.3). For example, the ill health and death of his so-called brother "Shaykh Najm al-Din Mahmud" are noted. This may mean that his completion of the Persian *Geometrical Constructions*, which noted that "Shaykh Najm al-Din Mahmud" was no longer living, came after the parts of his undated letters, which express devastation over his sickness. Kubanani's near contemporary, the mathematician-astronomer ʿAli Qushji, is also mentioned in the letters, though without any reference to his visit to Kirman before joining the Ottoman court in 1472 or his transport of any texts.[136]

Included in the letters of Kubanani compiled by his student (9.3) is a preface to a work on measurement (*misāḥa*) concerning the ratio of the circle's circumference to its diameter (9.4). This ratio is close to the comparable calculations of a well-known contemporary figure, namely, the mathematician-astronomer Ghiyath al-Din Kashi, known for the letters he wrote from Samarqand to his father in Kashan and credited with an early approximation to several digits of the number later known as *pi*.[137]

A versified description of Kubanani himself appearing at the beginning of his writings and letters as arranged by Husayn Maybudi (9.3) also requires evaluation. In that description, cited below, Kubanani is depicted as "the most knowledgeable of the time" (*aʿlam-i zamān*), and "all-knowing of this world" (*ʿallāma-i jahān*); a "single strike of [his] pen" (*yak sarīr-i qalam*) in mathematics opened the "[puzzling] palm of irrational roots" (*jaẕr-i ʿaṣam*); his "rational" [sciences] (*ʿaqlī-ash*) are beyond "measure of reason" (*qiyās-i ʿaql*) and his "transmitted" [sciences] (*naqlī-ash*) surpass "principles of transmission" (*asās-i naql*).[138] These verses praise Kubanani's scientific knowledge in glowing terms (see Appendix III, Verse 2):

Shaykh Abu Ishaq [Kubanani]:
Teacher of scholars on [all] horizons [and]
Highest in philosophy and faith

He is a sage, a philosopher with quick understanding

Heaven calls him the most knowledgeable of the time
Afterlife calls him the all-knowing of this world

In mathematics, his single strike of a pen
Has opened the [puzzling] palm of irrational roots

Whatever has a sign of wisdom in this world
His heart is capable of its power

His rational [sciences] are outside the measure of reason
His transmitted [sciences] are beyond the principles of the transmission

He is like a pouring cloud of bounty on the world's crown
Making the mouth of clever people like an open shell.

Among the works that may be attributed to Kubanani outside the collections and letters bearing his name (9.3) is a text that is directly in the tradition of Euclidean geometry, namely, an *Attachment* (*Ilḥāq*) *to the Recension* of *Euclid's Elements* by Nasir al-Din al-Tusi (9.5).[139] The signed name contained in that manuscript codex, which is now in London, is "Abū Isḥāq," without the last part, "Kubanānī." The content is still identifiable, given the parts of a verse contained in it that are closely comparable to one in his letters (Appendix III, Verse 3).[140]

Still, none of these clues provide enough evidence about when Kubanani "completed" the Persian translation of the *Geometrical Constructions*. Nor does his possible transcription of that text, or of the Paris Codex, lead to any firm conclusions about his involvement in the production of the *Anonymous Compendium*. The transcription of the latter text, as well as of the entire codex in which it is contained, must rather be fully analyzed in light of evidence concerning Kubanani's script found elsewhere, including the author's reported "autograph copy" of his scientific works (9.2), as well as some verses claimed to be in his hand (Appendix III, Verse 4).[141] If the scripts are in fact determined to be in the same hand, the profile of Kubanani could widen in scope from that of a scientific author transcribing his own works to a scribe transcribing the scientific works of others, most likely for his own private use as a mathematician-astronomer. In this regard, the transcription of the Arabic original of Abu'l-Wafa''s *Geometrical Constructions* in Istanbul also merits close examination, as it too dates to the fifteenth century, through the royal stamp of the Timurid ruler Ulug Beg (d. ca. 1449), as discussed above (fig. 1a). Any role Kubanani may have had in the compilation, and not just the transcription, of the *Anonymous Compendium* must be considered in light of the non-uniform nature of that text in terms of both time and place.

The *Anonymous Compendium* has been viewed, on the one hand, as "a compilation of notes taken over a period of time" and, on the other hand, as a work "taken from different sources, probably dating back to different centuries."[142] Due to the "disorganized nature of the folios" forming the text, Özdural suggested that the compendium may have been a series of notes "compiled" to record "discussions between mathematicians and artisans at a succession of joint meetings." He based this conjecture in part on the example of the notes on practical geometry compiled from group discussions by an Ottoman scribe, Caʿfer Efendi, who belonged to the household of the chief court architect Mimar Mehmed Agha. However, that scribe worked in late sixteenth- to early seventeenth-century Istanbul, a time very different from the era in which Kubanani flourished.[143] In other cases, the structure of the text was disregarded, resulting in Jazbi's misleading portrayal of Kubanani as the author of both the *Anonymous Compendium* and the Persian translation of Abu'l-Wafa''s *Geometrical Constructions*. The complexity of the *Anonymous Compendium* is best summed up by Necipoğlu elsewhere in this volume: "the contextual circumstances in which the *Anonymous Compendium* was compiled must remain open questions until more concrete additional evidence is garnered in the future."[144] In my conclusion, I will consider the *Anonymous Compendium* and the Paris Codex as a whole in terms of their composition, transcription, and compilation. This will allow for deductions, based on consistent premises, regarding at least the last two categories.

A Persian commentary on the Book of Geometrical Constructions *(Appendix I, 10.1)*

In comparison with the late Persian translation of Abu'l-Wafa''s *Geometrical Constructions,* completed by Kubanani Yazdi in the fifteenth century (9.1), the late seventeenth-century Persian commentary on that same text by the much better known Mulla Muhammad Baqir Yazdi (10.1) has not received much attention, in part because it was not bound within an important manuscript codex. Preserved in an apparently unique copy in Qum, Iran, the work is not recorded beyond manuscript catalogues and a bibliographical entry under "Buzjani."[145] According to the introduction, this late Persian commentary, titled *Barāhin al-aʿmāl al-handasiyya* (Demonstrations of the *Geometrical Constructions*), was composed by "Muḥammad Bāqir ibn Zayn al-ʿAbidīnī Yazdī ... in the year 1031 (1621)," with the alternative title *Futūḥāt-i ghaybiyya fī al-ʿilm al-handasa* (Hidden Openings in the Science of Geometry).[146] Just like the early and late Persian translations of Abu'l-Wafa''s *Geometrical Constructions* (4.1 and 5.1; 9.1), this much later commentary lacks a dedication naming a ruling patron. It was rather requested by an individual whom Muhammad Baqir-i Yazdi describes as a close associate (*aṣḥāb*), and a "dear brother" (*barādar-i ʿazīz*), although in this case it is clear that there was no familial relation. The detailed explanation of the reason behind the text's composition is comparable to those found in both Abu'l-Wafa''s original Arabic text (1.1) and the thirteenth-century Arabic commentary on it by Ibn Yunus (3.1). And its opening lines, set in the language of the geometrical figures and constructions that form the main subject of both Abu'l-Wafa''s text and the *Anonymous Compendium*, resemble other related texts, such as the surveying text mentioned below.

Muhammad Baqir-i Yazdi's commentary nevertheless possesses some striking features in that introductory section, setting it apart from all of the other texts presented in this chapter. The author begins by praising God and then turns directly to "the single opening (*fatḥ*) of a compass (*pargār*)," creating various (*mukhtalifa*) figures from similar (*mutashābiha*) triangles to geometric shapes of increasing sides: quadrangular, pentagonal, hexagonal, heptagonal, octagonal, nonagonal, and decagonal figures. He also discusses the three tools of geometric constructions mentioned in the opening of Abu'l-Wafa''s Arabic text and in a version of its title (1.1.ii), namely, the set square (*gūnyā*), the ruler (*masṭara*), and the compass (*pargār*). Here the upright (*aʿmada*) trees (*ashjār*) on the plane (*basīṭ*) of the earth (*zamīn*) appear as having also been divinely created. As these descriptions resemble ones found in texts such as an anonymous surveying treatise in the Paris Codex (item 9, fol. 65v [see fig. 10a]), it would be significant if these were works to which Muhammad Baqir-i Yazdi may have had access.

In the introduction, the Safavid author elaborates on why he composed the text that he refers to as *Futūḥāt-i ghaybiyya* (Hidden Openings), presumably in the sense of a compass opening (*fatḥ*) but with obvious mystical and cosmological overtones that are also apparent in his attribution of symbolic meanings to each polygon.[147] Initiating a sequence of three explanations, the author first explains his relation to the one who requested the composition of the text, as well as the nature of the problems in Abu'l-Wafa''s *Geometrical Constructions* that had inspired his requester's questions. Problems in the tradition of Euclid's *Elements* are placed next to those in the tradition of Abu'l-Wafa''s *Geometrical Constructions*. In the second of the three explanations, the author then expresses his own thoughts about the procedures involved in and beyond that book, from constructions (*ʿamal*) and demonstrations (*burhān*), to drawings (*rasm*) and actualization (*fiʿl*). These prompted his associate to ask him to arrange (*tadvīn*) such verbal matters (*madhkurāt*) in a written format, for the benefit of both the elect (*khawāṣṣ*) and ordinary (*ʿawāmm*) people.

The detailed explanations document the transmission and translation of Abu'l-Wafa''s *Geometrical Constructions*, and thus the development of this work's Arabic and Persian traditions. Such shifts, however discrete, start with the title of Abu'l-Wafa''s Arabic text, from *Aʿmāl al-handasa* (Constructions of Geometry) to the *Aʿmāl al-handasiyya* (Geometrical Constructions) of subsequent dependent texts in Arabic and Persian (3.1, 9.1), eventually becoming *Barāhīn al-Aʿmāl al-handasiyya* (Demonstrations of Geometrical Constructions) in the latest version (10.1). Moreover, while Abu'l-Wafa' refers to a group (*jamʿ*) of "geometers and craftsmen" (*muhandisīn wa ṣunnāʿ*) (1.1), these individuals become known as "geometers, craftsmen, and artificers" (*muhandisān, ṣunnāʿ va muḥtarifa*) in the early anonymous Persian translations (4.1 and 5.1), and then as "geometers, artisans, and artificers" in the late translation completed by Kubanani Yazdi, who used Persian equivalents and terms that are closest in meaning to "artisans" (*jamʿī muhandisān, hunar-varān va pisha-kārān*) (9.1). In Muhammad Baqir-i Yazdi's hands, they are referred to as a "group of geometers" (*jamʿī az muhandisīn*) and a "great number of craftsmen" (*kasīrī az ṣunnāʿ*) (10.1, fol. 132v). These shifts also reflect modifications in key statements by Abu'l-Wafa'. The differences in this seventeenth-century Persian commentary by Muhammad Baqir-i Yazdi become reflective of the evolution in the way these practitioners were practicing their craft by Yazdi's time.

According to this much later eye-opening version, "no doubt (*shakkī nist*) all (*jamīʿ*) of the correct (*ṣaḥīḥ*) constructions (*aʿmāl*) that the craftsmen (*ṣunnāʿ*) and surveyors (*massāḥ*) bring to actualization (*fiʿl*) they have learned (*farā girifta*) from the geometers (*muhandisīn*), and demonstration (*burhān*) is established upon (*qā'im*) their correctness (*ṣiḥḥat*)" (10.1). Directly therafter, this much later author describes his own book as "pointing (*ishāra*) to the demonstrations (*barāhīn*) of the constructions (*aʿmāl*)" that were omitted from Abu'l-Wafa''s Arabic text according to the introduction of that book (see Part II above, and Appendix II). Explicit references to the requesters of texts and their role in works may be rare in historical sources, but linguistic shifts such as these, whether explicit or implicit, shed light on the development of practices over time.

IV. TEXTUAL VS. MATERIAL SOURCES: LINES OF CRAFT, LINES OF PROOF

In the previously mentioned section on geometry (*handasa*) in the *Iḥṣā al-ʿulūm* (Enumeration of the Sciences) by Farabi, an author directly associated with the textual tradition of Abu'l-Wafa''s *Geometrical Constructions* (Appendices I and II, 1.1),[148] a distinction is made between "practical geometry, forming lines, surfaces—quadrilateral, circular, and triangular—within a body whose matter belongs to the subject of that practical art/craft (*ṣināʿa al-ʿamaliyya*)," as opposed to "theoretical geometry, looking into lines, surfaces, and bodies by definition and in general." This distinction applies to various subdivisions (*furūʿ*) of geometry both in and beyond that work. Yet, the "lines of craft" of practical geometry and the "lines of proof" of theoretical geometry are not always clearly distinguishable in early geometrical compositions themselves.

Crafts are "produced" (*fiʿliyya*)—which, in the aforementioned language of Qusta ibn Luqa, refers to what remains after the removal of the craftsman (*sāniʿ*); hence, they are not separable from the verification methods involved in their production. With regard to one of the geometrical constructions of the *Anonymous Compendium* (fol. 182v), the language of production carries an even closer association with the Aristotelian concept of "actualization" (*biʾl-fiʿl*) of "potentiality" (*quwwa*), here in the sense of testing a construction against a method of proof, however elementary. A direct correspondence between produced crafts and geometrical constructions that involve actual physical elements is a different matter. The closest one may come to physical settings in a medieval historical text is in the long passage from Abu'l-Wafa''s *Geometrical Constructions* discussed earlier. In addition to practitioners such as geometers, craftsmen, and surveyors, this passage described the constructions of both craftsmen and geometers in relation to the same geometrical problem at a gathering; practitioners and examples of what they practiced thus appear side by side in both the body and figures of that text (Appendix II, 1.1.e–f; figs. 1e and 1f).[149] But the text does not mention any medium for which those constructions were intended.

Just such a correspondence can be detected in the cast-brass door drawn and made by Badiʿ al-Zaman Ibn Razzaz Jazari (fl. ca. 1200) in his *al-Jāmiʿ bayn al-ʿilm waʾl-ʿamal al-nāfiʿ fī ṣināʿat al-ḥiyal* (A Compendium of Theory and Useful Practice in the Mechanical Arts). In Jazari's geometrical templates (*mithāl*) and crafted patterns a direct link is established between craft and method. Both the text and manuscript figure are discussed in further detail by Necipoğlu in her *Topkapı Scroll* and in her chapter in this volume, where some other examples are included.[150]

An entirely different situation is seen in cases where the visual correspondence between "lines of craft" and "lines of proof" comparable to those discussed in Abu'l-Wafa''s *Geometrical Constructions* may not mean an actual historical correspondence through directly verifiable forms of evidence. Take, for example, the aforementioned wooden door from Mosul (ca. 498 [1102]), decorated with carved patterns, which corresponds to geometrical constructions found in Abu'l-Wafa''s practical geometry treatise (fig. 4a).[151] No case can readily be made, however, for an actual connection between the crafting of that door and the proof behind those constructions, even in the case of a work as close in time and place as the Arabic commentary on Abu'l-Wafa''s *Geometrical Constructions* by Kamal al-Din ibn Yunus al-Mawsili (ca. 1156–1242).[152] The same goes for other patterns such as the late Safavid tiling on a wall of the Friday Mosque in Isfahan already considered above (fig. 4b). The geometrical construction of a central square surrounded by four kites after the "*chahār turunj*" of a similar construction in the *Anonymous Compendium* (fols. 188v and 189r [30, 31]) does correspond closely to both the Mosul door and the Isfahan tiling, but once again there is a similar lack of documentation for a historical link. The same is also the case for more complex constructions in the latter work, where an exact correspondence can be demonstrated for other decorative designs in extant monuments or objects. Besides the construction of a five-pointed star within a decagon found on various handicrafts such as inlaid woodwork (*khātam-kārī*), a more complex construction from that text (fol. 192r [37]) was recently superimposed on an exactly corresponding late-eleventh-century geometric pattern from the Seljuk period that decorates a wall in the North Dome of the Friday Mosque in Isfahan. This was described by Jan Hogendijk, who led the process and the teamwork.[153]

By contrast, the close correspondence of practical geometry through the tradition of Abu'l-Wafa''s *Geometrical Constructions* to methods such as the application of areas in the tradition of Book II of Euclid's *Elements* may be documented directly in texts that name physical objects. The short tract in Arabic in the Paris Codex, which contains key propositions from Book II devoted to those methods, is adjacent to corresponding algebraic problems appearing under the heading "*al-Khātam*" and a table of contents naming gems such as the ruby (*laʿl*) and elements such as gold (*ṭalā*) and their weights, with related marginalia in Persian (fol. 102r).

Besides geometrical pieces in inlay work, there are correspondences between the application of areas in arithmetical and mechanical measures of weights and ratios. Correspondences with aspects of surveying are also documented in terms of both the historical figures and works under discussion in this chapter. Individuals such as Abu'l-Wafa' al-Buzjani and Kamal al-Din ibn Yunus in the tenth and thirteenth centuries, as well as Kubanani Yazdi and Muhammad Baqir-i Yazdi in the fifteenth and seventeenth centuries, had direct associations with Euclidean geometry and other ancient sources for methods such as the application of areas. These individuals were apparently familiar with the methods of geometers and craftsmen in a subject such as "area surveying." With respect to Abu'l-Wafa', this is most clear from the celebrated problem of the construction of a quadrilateral (*murabbaʿ*) from three quadrilaterals, which was described earlier. He provides a specific example with two solutions in order to point to the "errors" involved in approximated constructions, which are here reconstructed through the drawings of those solutions with the help of a ruler, compass, and set square (see figs. 1e and 1f). The areas under discussion are marked with *abjad* letters bearing numerical values (fol. 196v [57] of the *Anonymous Compendium* contains these letters) modeled after some constructions in the same work (fol. 180r [1]).

As noted above, such methods are clearly in the tradition of specific propositions in Euclid's *Elements of Geometry*, an important example being from Book II, where the area of a large square equals the area of two smaller squares and rectangles enclosed in it (fig. 2b). The mathematical identity of equal areas allows one to calculate an unknown side from the two known ones, and through it there is a "correspondence" for the historical identification of the methods of geometry and those of "algebra" (*jabr*), a subject with the historical meaning of finding unknown quantities from known ones based on proportions, not symbols or functions.[154]

Significantly, we find evidence for the presence of both Abu'l-Wafa''s constructions and the related Euclidean propositions in the Paris Codex. There we find, in Arabic, the fourth proposition from Book II of Euclid's *Elements* (fig. 2b), often noted for overlaps with simple algebraic identities (fig. 3a). The integration of these Euclidean propositions in a largely Persian codex points to their centrality to the Persian treatises bound with them in that volume. Of particular interest from the standpoint of relations between geometry and craft is the Euclidean text that contains references to building materials such as *talbin* (bricks), a term with linguistic cognates in Syriac and Hebrew.[155] This indicates much older applications for those propositions than comparable ones recorded on the folio just before it in the Paris Codex (fol. 102r) with respect to jewelry making, this time in the form of marginal notes in Persian, the language of the great majority of the texts in that codex.

Directly related to Abu'l-Wafa''s comparable constructions is a single folio from the treatise in the Paris Codex titled "Constructions by Abu Bakr," discussed above, which features two important visual sources as mirror images (fol. 136v [fig. 2a]). The bottom illustrations show two adjacent figures that are comparable to Abu'l-Wafa''s constructions of a geometer (left), and a second craftsman (right), represented there but not part of this chapter's selections (fig. 7b). Here, three squares (*murabbaʿ*, literally equal-sided rectangles) are equal in area to a larger encompassing figure, and the "two ways for this form (*ṣūrat*) are according to [their] illustration (*mithāl*)."[156] No less significant is the left side of the same folio (here appearing on top) for a surveying problem with a figure rotated upward accompanying a text in Persian.[157]

The placement of Abu'l-Wafa''s constructions, based on that of a craftsman and a geometer, on the same folio as a problem concerning the determination of a high elevation (*irtifāʿ-i murtafaʿ*) such as a mountain (*jabal*), to be actualized (*ʿamal*) in the case of constructions and demonstrated (*bayān*) in the case of height determination (fig. 2a), may not have been arbitrary. The author or transcriber of that part of the codex may have merged the constructions of Abu'l-Wafa' that had to do with "area surveying" with a construction involving "distance surveying." The illustrations of the two constructions involving equality of areas are strikingly close to those in the Persian *Geometrical Constructions* co-translated by Kubanani in the Paris Codex. In the case of the geometer, they are almost identical in both detail and orientation, as if one were a duplication of the other (fig. 9e–f). This takes us beyond the issues of compositions, transcriptions, and compilations in the Paris Codex, to the hardly touched-upon subject of its illustrations, at yet another intersection of lines of craft and lines of proof. When we consider that the "constructions" (*aʿmāl*) in that part of the codex titled as such carried the name Abu Bakr al-Khalil al-Tajir, while the above constructions and illustrations, if not the entire codex, could be associated with Abu Ishaq Kubanani, we then return full circle to the issues of identity and dating.

CONCLUSION

As demonstrated in this chapter, the composition, transcription, and compilation of the Paris Codex, and of the *Anonymous Compendium* in particular, have posed vexing challenges for scholars, given the non-uniform structure and dating of both. Leaving out the question of composition, if the Paris Codex was transcribed and/or compiled by someone such as Kubanani Yazdi, the fifteenth-century author who completed the Persian translation of Abu'l-Wafa''s *Geometrical Constructions* in that codex (9.1), and whose name is given in both transcriptions and compilations of other scientific works in both Arabic and Persian (9.2, 9.5), this would be consistent with other findings and indications. Both Edgar Blochet and Francis Richard, the two cataloguers of the Paris Codex, hold that the entire codex was transcribed in the same hand, while Priscilla Soucek has dated the codex to the fifteenth century. Blochet and Özdural further assert that Kubanani transcribed the entire codex.

Özdural held that the *Anonymous Compendium* was not a uniform text, but the product of several meetings in which different individuals participated. Necipoğlu and Hogendijk have suggested the possibility that it may include texts from different authors and periods, not necessarily subscribing to the idea of "meetings," or of "teaching" or "working" sessions as the venue for the compilation. Necipoğlu nevertheless provisionally proposes an initial compilation date in the 1300s, prior to its being copied in a modified compilation during the fifteenth century by or for Kubanani Yazdi in the unique version preserved in the Paris Codex. It is consistent with all these statements, as well as with the research represented in the present chapter, to posit that the Paris Codex was compiled by someone with the profile of Kubanani, if only partially. This does not extend to him the composition of any of the *Anonymous Compendium* beyond what he signs as his completion of Abu'l-Wafa''s *Geometrical Constructions* in Persian. His transcription and/or compilation of any part of the Paris Codex require fuller study of his scripts, works, and codices.

The Tehran Codex (9.2) partly itemized here is similar to the Paris Codex in that it includes scientific texts from different authors and periods. It was catalogued as a *Majmūʿ al-rasāʾil* (Collection of Treatises and Books), which was claimed to be in Kubanani's own hand. According to the title page of the codex, it contains commentaries on works as early as one by Ibrahim ibn Sinan (ca. 908–46), the supposed grandson of Thabit ibn Qurra (d. ca. 901), which was signed by Kubanani with the most complete name and earliest date known for him, i.e., "Abū Isḥāq Shaykh-zāda ibn al-Khādim al-Burhānī al-Kūbanānī, 846 [1442]." The Tehran Codex, which also includes works as late as those authored by Kubanani himself, ends with a commentary on Nasir al-Din Tusi's *Sī Faṣl* (Thirty Chapters), the first item in the Paris Codex. This would suggest that the latter may similarly be in Kubanani's hand.

The aforementioned collection of Kubanani's works and letters arranged by his disciple Husayn Maybudi (*Munshaʾāt* [9.3]) offers a range of other relevant and consistent material. These letters describe Kubanani's involvement in subjects ranging from the rational and traditional sciences in the "Saʿdiyya" School, whose location is not specified, to farm and land issues (*zirāʿat*) in his native land. Two related "prefaces" (*dibācha*) to works by Kubanani are also preserved in the *Munshaʾāt*. One of these, bearing the title *Dar bāb-i [misāḥat-i] muḥiṭ-i dāʾira* (On the Subject of the [Measurement] of the Circumference of the Circle: 9.4 [contained in 9.3]), mentions Archimedes. The second preface is under the title *Risāla dar shaṭranj* (Treatise on Chess), a treatise on a mathematical board game that includes verses by Kubanani in both Persian and Arabic (as in 9.2 and 9.3). Verses appear in Kubanani's completion of Abu'l-Wafa''s *Geometrical Constructions* in the Paris Codex, and in a versed work in that codex attributed to Nasir al-Din Tusi, which could have been added by Kubanani, who included verses in his own works.[158] A related work by Kubanani, in the form of a *Recension* (*Taḥrīr*) of Euclid's *Elements* by Nasir al-Din al-Tusi, is in a codex in London dated 888 (1483) (9.5). The latest among those signed as the colophon of his known works, it appears to be in a

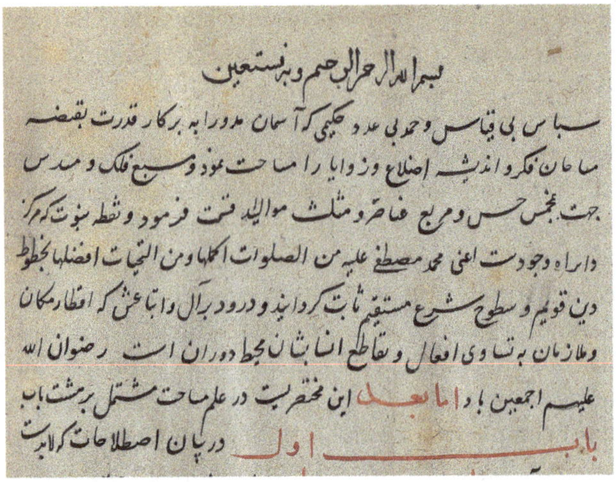

Fig. 10a. Detail of the opening page of an anonymous Persian treatise on surveying. Paris, Bibliothèque nationale de France, Ms. Persan 169, 9, fol. 65v. (Photo: Bibliothèque nationale de France)

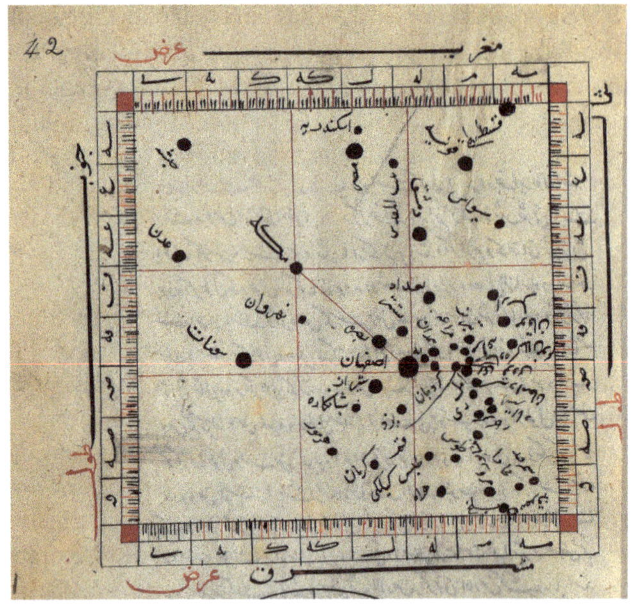

Fig. 10b. Map from a treatise on determining the direction of the qibla. Paris, Bibliothèque nationale de France, Ms. Persan 169, 6, fol. 42r. (Photo: Bibliothèque nationale de France)

hand particularly close to that in which the Paris Codex was copied. While Kubanani atypically leaves out the date when he completed Abu'l-Wafa''s *Geometrical Constructions*, the approximate year, as well as that of the codex containing it, may be closer to the upper limit of the dates so far associated with him. His "completion" of the translation in the Paris Codex can be dated after some of his letters in which he refers to the illness, but not yet the death, of "Najm al-Din Mahmud," the initial translator of Abu'l-Wafa''s *Geometrical Constructions*.

There are items and folios in the Paris Codex that may go beyond Kubanani's own works, as indicated by the way in which they are transcribed and/or compiled: e.g., the Arabic text from the second book of Euclid's *Elements* (fig. 2b), and the previously cited folio from "Constructions by Abu Bakr," which contains Abu'l-Wafa''s constructions as well as a surveying problem, both in Persian (fig. 2a). The two constructions could be considered to be associated with Kubanani as the compiler/transcriber of those parts of the codex because they are closely comparable to the figures of the Persian translation of Abu'l-Wafa''s text completed by him.[159] The list of twenty-five works in the Paris Codex, as identified in the Bibliothèque nationale de France catalogue, is filled with other subjects treated by Kubanani, from a few works on surveying, to those on astronomy, which may have been similarly compiled, though not authored, by someone with his profile. The same may be said for other items (such as a map highlighting sites including Yazd, Kirman, and Hormuz, as well as Sari [fig. 10b]) that coincide with the sites of Kubanani's patrons and compositions. The aforementioned gloss on another page of "Constructions by Abu Bakr" with the name of Sayyid Sharif in the margins of one construction (fol. 135v), if meant to be Jurjani (d. ca.1413), is early enough to allow us to consider Kubanani's involvement as copyist. Among the constructions of Abu Bakr is one related to optics, a subject included in another codex involving Kubanani, with works by two other figures mentioned in the Paris Codex: Mahmud Haravi (ca. 1436) and Sharaf al-Din ʿAli Yazdi (d. 1454). In addition, a name close to "Abū Bakr al-Khalīl al-Tājir al-Raṣadī," which is associated with the "constructions" in that codex, is "Abū Bakr ibn al-Khalīl al-Tājir," the copyist of a text by Ibn al-Khawwam (d. ca. 1325) in the previously mentioned codex in Saint Petersburg. The latter is an author whose work was mentioned as having been commented on by Kubanani, another link to be added to the list of connections between him and the Paris Codex. Additionally, it is through the author of one of those transcribed texts, Ibn Khawwam al-Baghdadi,[160] that a direct link may be made between its copyist, Abu Bakr ibn al-Khalil al-Tajir, and Abu Ishaq Kubanani, a commentator (*shāriḥ*) on that same author in another work.[161] Through that link Kubanani's association with the Paris Codex may be seen to include not only his "completion" of the Persian translation of the *Geometrical Constructions* (9.1), but also his role as a possible copyist and/or compiler of the "constructions" of Abu Bakr, both in the *Anonymous Compendium* (8.1) and elsewhere in the codex, including, *Lil-Aʿmāl Abī Bakr al-Khalīl al-Tājir al-Raṣadī* (7.1). This would be consistent with the suggestion made by both Rashid and Necipoğlu about Kubanani's "possible ownership" of another codex in Paris (Bibliothèque nationale de France, Ms. 4821, fols. 1r–86r), which is a collection of mathematical treatises. It would be even more directly supported by Kubanani's likely ownership of at least the first of those treatises in that codex (fols. 1r–8r), where the script (*khaṭṭ*) ascribed to him appears on the last folio (fol. 8r) of that treatise, titled *Risāla Fī ʿamal ḍilʿ al-musabbaʿ al-mutasāwī al-aḍlāʿ fī al-dāʾira* (On the Construction of the Side of the Regular Heptagon in the Circle).[162]

Significantly, the signature and script with the same form and sequence of his name (Abū Isḥāq ibn ʿAbdallāh al-Kūbanānī, ending with a broken *yā*) does appear on the last folio of "Kūbanānī"'s own completion of the Persian translation of the *Geometrical Constructions* (Ms. Persan 169, fol. 178v [fig. 9a]). As noted above, the closing line of that text's postface, this time in Arabic and in the same hand, is on the next folio (179v [fig. 9d]), stating the end of the book's transcription (*kitābat*), though without naming Kubanani in that exact place. If it did, this could act as conclusive evidence for the premise of Kubanani's transcription of that text and, by extension, others in the Paris Codex, which are all considered to be in the same hand. Still, the one word "transcription" (*kitābat*), clearly written down a page apart from the name of Kubanani in the same text and in the same hand, is the strongest evidence yet in support of such a premise. The more challenging case of Kubanani's possible compilation of any part of the Paris Codex can similarly move forward regardless of the transcription premise.

If we consider the possibility that any parts of the Paris Codex were compiled by someone such as Kubanani, the following may then be added to the previous combination of several consistent working premises: namely, that this codex starts with the *Recension* of Nasir al-Din Tusi's *Sī Faṣl*, a text that is found at the end of Kubanani's collected works in the Tehran Codex (9.2); and that the last three texts of the Paris Codex are Kubanani's own completion of Abu'l-Wafa''s *Geometrical Constructions* in Persian (9.1), followed by copies of the *Anonymous Compendium* (8.1) and what is a late text by his native contemporary Sharaf al-Din ʿAli Yazdi (d. ca. 1454).[163] Rather than constituting conclusive evidence, the above picture remains a working reconstruction, in keeping with the aim of this chapter and volume to provide constructive steps forward for future research.

Aga Khan Program for Islamic Architecture Associate, Department of History of Art and Architecture, Harvard University, Cambridge, Mass.

A note on Appendices I–III (*see next page*)

In Appendix I the texts are listed chronologically under their respective authors and by language (nos. 1–3 are authors writing in Arabic, nos. 4–10, authors writing in Persian; under each author works are listed in order of relevance to this chapter: 1.1–1.5 through 10.1). Full bibliographical information is provided in each case. Manuscripts are organized by lowercase Roman numerals ("i" for unique manuscripts; "ii," "iii," etc., for other copies or textual variants). Hijri dates are followed by the common-era date.

Appendices II and III contain a selection of texts in Arabic and Persian (listed in Appendix I under "Arabic Tradition" and "Persian Tradition," respectively) that are most directly related to this chapter (i.e., 1.1 in Arabic, and 9.1 in Persian). In each case the English translations of titles and added section headings are followed by transcriptions of the original texts in Arabic or Persian, and related notes. The English translations of selected passages in Appendices II and III appear in the main text of this chapter (in Part II: The Arabic Tradition, and Part III: The Persian Tradition). Appendix III concludes with a selection of verses beyond the texts included in Appendix I.

APPENDIX I

List of authors and texts belonging to the Arabic and Persian traditions of the *Geometrical Constructions*) by Abu'l-Wafa' al-Buzjani

Texts

Arabic Tradition

1. Abu'l-Wafā' al-Būzjānī (ca. 328–88 [939/40 –998]).

1.1. *Kitāb Aʿmāl al-handasa* (*Geometrical Constructions*) (ca. 381–88 [991–988]).

Manuscripts:

i. Istanbul: *Kitāb-i Abī al-Wafā' Fī mā yaḥtāju ilayhi al-ṣāniʿ min aʿmāl al-handasa*. Süleymaniye Library, Ms. Ayasofya 2753, dated ca. 851–53 (1447–49).
ii. Cairo: *Kitāb al-Nijāra fī ʿamal al-masṭara wa al-birkār wa al-kūniyā taṣnīf Abī al-Wafā' Muḥammad ibn Muḥammad al-Būzjānī*. Dār al-Kutub al-Miṣriyya, Ms. 260 (31024), 1: folios 1r–58v, dated 600 (1203). (Defective opening. Incomplete.) (See King 1981, vol. 2, p. 844; King 1986, 42.)
iii. Cairo: title unknown. Dār al-Kutub al-Miṣriyya, Ms. 366 (44795), 46 folios, dated 831 (1427). (Defective. Ms. inaccessible.) (See King 1981, vol. 2, p. 844; King 1986, 42.)

Textual Variants

iv. Milan: no title, incomplete (opening chapters of the *Geometrical Constructions*). Biblioteca Ambriosiana, Ms. Arab 68. "Abū al-Wafā" is named on fol. 16b.
v. Uppsala: anonymous, *Kitāb al-ḥiyal al-ruḥāniyya wa al-asrār al-ṭabīʿiyya fī daqā'iq al-ashkāl al-handasiyya*. Universitetbibliotek, Ms. Tornberg 324, Vet 27, folios 1r–60v. (Attributed to Fārābī in opening and closing folios. Unique manuscript?)

1.2. *Fī jamʿ al-aḍlāʿ al-murabbaʿāt wa al-mukaʿabāt*.
i. Mashhad: Āstān-i Quds-i Rażavī Library, Ms. 5521/1, dated 867 (1462). (Unique manuscript?)

1.3. *Kitāb Fī mā yaḥtāju ilayhi al-kuttāb wa al-ʿummāl min ʿilm al-ḥisāb* (also known as *Kitāb al-Manāzil*).
i. Cairo: Dār al-Kutub al-Miṣriyya, Ms. Riyāḍa 42/1, dated 487 (1094).
(Under the title *Kitāb al-Manāzil al-sabʿ*: Rozenfel'd and İhsanoğlu 2003, 96).
ii. Leiden: Bibliotheca Universitatis Or. 103, first three stations (*manāzil*), under the title *Kitāb Fī mā yaḥtāju ilayhi al-kuttāb wa al-ʿummāl min ʿilm al-ḥisāb*.
(Published: Saidan 1971, based on the two manuscripts above)

1.4. *Jawāb Abī al-Wafā' Muḥammad ibn Muḥammad al-Būzjānī ʿammā sa'alahu al-Faqīh Abū ʿAlī al-Ḥasan ibn al-Ḥārith al-Ḥubūbī, wa huwa al-burhān ʿala misāḥat al-muthalathāt min istikhrāj al-ʿamūd wa masqaṭ al-ḥajar*.
i. Damascus: Ẓāhiriyya Library, Ms. 4871, 20: 82a–82b, dated after ca. 555 (1160). (Ragep and Kennedy 1981.)
ii. Oxford: Bodleian Library, Ms. Thurston 3, folios 53r–53v.

1.5. *Mas'alat Arshimīdis fī misāḥat al-muthallath*.
i. Oxford: Bodleian Library, Ms. Thurston 3, folios 55v–58r. (Unique manuscript?)

2. Ibn Ṭāḥir al-Baghdādī (d. ca. 429 [1037]).
(Full name: Abū Manṣūr ʿAbd al-Qādir ibn Ṭāḥir Muḥammad ibn ʿAbd-allāh Baghdādī Tamīmī.)

2.1. *Kitāb al-Iḍāḥ ʿan ʿuṣūl-i ṣināʿat al-massāḥ*. Attributed to Abū Manṣūr Ibn Ṭāḥir al-Baghdādī, but a certain "Abī al-Fatḥ al-Yazdī" (or Mardī/Murādī, possibly ca. 600s/1300s) is named in the colophon and the possible author. The colophon of a text immediately following that in the same hand gives a name and date for its transcription as "ʿAlī ibn al-Khalīl al-Tājir," and "728" (1327–28).
i. Mashhad: Āstān-i Quds-i Rażavī Library, Ms. 5429, 728 (1327–28), based on the Arabic text in the same codex. (Unique manuscript?)

2.2. *Al-Misāḥa*.
i. Istanbul: Süleymaniye Library, Ms. Lâleli 2708/2, folios 1–19. (Unique manuscript?)

3. Kamāl al-Dīn Ibn Yūnus (Mūsā Shāfiʿī) (ca. 551–639 [1156–1242]).

3.1. *Sharḥ al-Aʿmāl al-handasiyya li Abī al-Wafā'*.
i. Mashhad: Āstān-i Quds-i Rażavī Library, Ms. 5357, dated 680 (1282): *Sharḥ al-Aʿmāl al-handasiyya li Abī al-Wafā'*. Copied in Maragha. (Unique manuscript?)

Persian Tradition

4. Anonymous.

4.1. Anonymous Persian translation of *Kitāb Aʿmāl al-handasa* by Abu'l-Wafā' al-Būzjānī, titled *Dar Ānch ṣunnāʿ va muḥtarifa bi ān muḥtāj bāshand az a'māl-i handasa va īn kitābrā Kitāb-i Nijārat khwānand* (referred to as *Kitāb al-Nijāra*).

i. Mashhad: Āstān-i Quds-i Rażavī Library, Ms. 5462, dated 6th–7th centuries (12th–13th centuries)?: see Gulchīn Ma'ānī 1971 (1340), 81. (Manuscript variant?)

5. Anonymous.

5.1. Anonymous Persian translation of *Kitāb A'māl al-handasa* by Abu'l-Wafā' al-Būzjānī, titled *Dar Ānchih ṣunnā' wa muḥtarifa bi ān muḥtāj bāshand az a'māl-i handasa va īn kitābrā Kitāb-i Nijārat khwānand* (referred to as *Kitāb al-Nijārat*).

i. Tehran: Tehran University, Central Library, Ms. 2876: folios 1–70, dated 5th–6th centuries (11th–12th centuries)?: see Dānish Pazhūh 1962 (1340), 1720. Dated 1017 (1608) for ownership. (Manuscript variant?)

6. Abū'l-Fatḥ Iṣfahānī (fl. ca.1095–1119).
(Full name: Abū al-Fatḥ ... Maḥmūd ibn Khalaf ibn Aḥmad ibn Muḥammad Iṣfahānī 'Ajalī.)

6.1. *Dar 'Ilm-i misāḥat.*

i. Mashhad: Āstān-i Quds-i Rażavī Library, Ms. 5462, 728 (1327–28). Same hand as 2.1, where the second colophon was transcribed by 'Alī ibn Khalīl Tājir (ca. 728 [1327–28]). (Unique manuscript?)

7. Abū Bakr al-Khalīl al-Tājir al-Raṣadī (ca. 728 [1327–28]?).

7.1. *Lil-A'māl Abī Bakr al-Khalīl al-Tājir al-Raṣadī.*

i. Paris: Bibliothèque nationale de France, Ms. Persan 169, 19: folios 124v–137v. (Unique manuscript?)

8. Anonymous.

8.1. *Fī tadākhul al-ashkāl al-mutashābiha aw al-mutawāfiqa.*

i. Paris: Bibliothèque nationale de France, Ms. Persan 169, 24: folios 180r–199r.

9. Abū Isḥāq Kūbanānī (fl. ca. 846–88 [1442/43–post 1483]).
(Full name: Abū Isḥāq ibn 'Abdallāh Kūbanānī Yazdī also known as Abū Isḥāq Shaykh-zāda ibn al-Khādim al-Burhānī al-Kūbanānī (or Kūhbanānī).

9.1. *Tarjuma-i Kitāb-i Abī al-Wafā' Muḥammad ibn Muḥammad al-Būzjānī Dar A'māl-i Handasiyya.* Translation completed by Abū Isḥāq Kūbanānī Yazdī.

i. Paris: Bibliothèque nationale de France, Ms. Persan 169, 23: folios 141v–179v. (Unique manuscript?)

9.2. *Majmū'a-i rasā'il va kutub.*

i. Tehran: Tehran University, Central Library, Ms. 2417. (Autograph collection?)

9.3. *Munsha'āt* (Writings [and Letters of "Shaykh Abū Isḥāq [Kubanānī]").
Compiled by Qāḍī Ḥusayn ibn Mu'īn Maybudī Yazdī.

i. Tehran: Malik Library, Ms. 1605, undated.
ii. Tehran: Majlis Library, Ms. 15330, dated 979 (1571–72).
iii. Tehran: Hashimi Private Library, uncatalogued, dated 900 (1494–95), Yazd.

9.4. *Dibācha-i Risāla Dar misāḥat-i dā'ira*, contained in *Munsha'āt*: see 9.3.

9.5. *Ilḥāq-i Abī Isḥāq* (Attachment by Abu Ishaq [of the *Recension of Euclid's Elements*, by Naṣīr al-Dīn al-Ṭūsī]).

i. London: British Library, Ms. Or. 1514, 888 (1483). (Rieu 1894, no. 751.)

10. Muḥammad Bāqir Zayn al-'Abidīnī al-Yazdī (d. ca. 1047 [1638]).

10.1. *Barāhīn al-A'māl al-handasiyya / Futūḥāt al-ghaybiyya fī al-'ilm al-handasa.*

i. Qum: Mar'ashi Najafi Library, Ms. 12936: folios 39v–160r, 1, 1031 (1621). (Unique manuscript?)

APPENDIX II. SELECTIONS IN ARABIC

1. Author

ابوالوفاء محمّد ابن محمّد البوزجانى

Text

كتاب أعمال الهندسة

1.1.i. Manuscript[1]

1.1a. Title

كتاب فيما يحتاج اليه الصّانع من أعمال الهندسة

1.1b. Opening Lines

قد امتثلت ما رسمه مولانا الملك شاهانشاه السيد الأجلّ [المؤيد] المنصور بهاء
الدولة و ضيآء الملة و غياث الأمة، اطال الله بقاه و أدام علاه و قدرته و
سلطانه، من إثبات المعاني التي كان يتذاكر بحضرته العالية من الأعمال الهندسية

1. 1.1.i is the manuscript in Istanbul used in this selection: facs. pages are from Qurbānī and Shaykhān 1992; 1.1.ii–iii are manuscipts in Cairo represented under Variants: readings are from Abū'l-Wafā' al-Būzjānī 1979, ed. al-'Alī.

التي يكثر استعمالها عند الصنّاع مجردًا من العلل والبراهين، يسهل على الصنّاع تناوله ويقرب عليهم طريقته، وجعلت الكتاب ثلثة عشر بابًا.

Key Passages:

Chapter Ten: *On the Division of Quadrangles and Their Assemblage*, Qurbānī and Shaykhān 1992, facs. 46.

الباب العاشر فى قسمة المربّعات و تأليفها ...

On the Assemblage and Division of Squares Not Composed of Two Square [Number]s [i.e., Squares]

فى تركيب المربّعات وقسمتها اذا لم يكن عددها مؤلفة من مربعين

[Passage 1: 1.1c. On the mistakes of geometers and craftsmen: facs. 52–53].

قد غلط جماعة من المهندسين و الصنّاع في أمر هذه المربّعات وتركيبها، أمّا المهندسون لقلة دربتهم بالعمل، وأمّا الصنّاع فلخلوهم من علم البراهين . وذلك أنّ المهندس اذا لم تكن له دربة بالعمل صعب عليه تقريب ما يصّح له بالبراهين الخطوطية علي مايلتمسه الصُّانع . فإنّ الصُّانع غرضه مايقرب عليه العمل و يظهر له صحةً مايراه في الحسّ و المشاهدة ولا يبالي بالبراهين الخطوطية . والمهندس اذا قام له البرهان على الشيء بالتوهّم، لم يبال صحة ذلك بالمشاهدة اذا لم يصح ... لا شكّ أنّ جميع مايراه الصُّانع إنّما هو مأخوذ مما يعمله المهندس أولاً وقام البرهان على صحةً، فإنّ الصُّانع والماسح إنّما يأخذ من الشيء زبدته ولا يفكر فى الوجوه التي ثبتت صحة ذلك به، و لأجل ذلك قد يقع عليه الغلط و الخطاء . فأمّا المهندس فقد علم صحةً ما يريد بالبراهين اذاكان هو المستخرج للمعاني التي عمل عليها الصُّانع والماسح . فإنّ حذاق هولآء المهندسين اذا سئلوا عن شيء من قسمته الأشكال أو شيء من ضرب الخطوط تحيروا فيه و احتاجوا الى فكر طويل ، و ربما سنح لهم هذا و قرب عليهم، و ربما صعب و لم يتأت لهم عمله .

[Passage 2: 1.1d. On a gathering with craftsmen and geometers: facs. 52–53]

و لقد حضرت فى بعض المجالس و فيه جماعة من الصنُّاع و المهندسين و سئلوا عن عمل مربع من ثلثة مربعات. أمّا المهندس فإنّه يستخرج خطّاً يقوي على ثلثة مربعات بسهولة، و لم يرض أحد من الصنُّاع [بما عمله] . فإنّ الصانع يريد أن يقسم تلك المربعات بأقسام يؤلف منها مربعًا واحدًا كما عملنا فى مربعين وخمسة مربعات . وأمّا الصنّاع فأنّهم أوردوا فيها عدّة وجوه قام البرهان على البعض وبطل البعض، الا أنّ ما لم يقم البرهان عليه كان مقاربًا للصحة في المنظر فيخيل لمن ينظر اليه أنه صحيح . ونحن نورد تلك الوجوه ليعلم الصحيح منه من الفاسد، ولا يقع على الناظر فى هذا المعني غلط فى قبول ما هو فاسد، انشاء الله .

[Passage 3: 1.1e. On the construction of a craftsman: facs. 53–55]

... بعض الصنّاع وضع احد المربعات في الوسط وقطع الثاني منها بنصفين بالقطر و وضعه على الجنبتي المربع و اخرجه من وسط الثالث الى الزاويتين منها على غير القطر خطين مستقيمين وخطًا منه الى الوسط الضلع المقابل للمثلث حدث بالخطين . فانقسم المربّع بمنحرفين ومثلث . ثم وضع المثلث أسفل المربّع الاوّل و وضع المنحرفين فوقه وضم الضلعين الاطولين احدهما الى الآخر في الوسط . فصار له مربّعًا على مثل هذه الصورة . قال ابوالوفا أمّا صورة ما عمله فهو فى التخيّل و لمن لايكون دربة بالصناعة والهندسة أنّه يرى صحيح ... وذلک أنّ ضلع المربّع المؤلف من ثلث مربعات اذجعلنا ضلع كل مربّع عشر أذرع تقريبا على المتعلّم هو سبعة عشر ذراعًا وثلث بالتقريب وضلع هذا المربع هو سبع عشر ونصف سبع وبينهم تفاوت كثيرًا ... فقد تبين فساد هذه القسمة والتركيب وقد قسم بعض الناس هذه المربّعات بنوع آخر من القسمة أظهر فساد من القسمة الأولة .

[Passage 4: 1.1f. On the correct construction of geometers: facs. 56–58]

فأمّا قسمة المربّعات على الوجه الصحيح وعلى مايلزم عليه البرهان، فإنّه تبين على الوجه الذي نذكر . وهو أنّا نقسم مربّعين منها بنصفين على قطره، ويضاف كل واحدٍ منها الى الضلع من اضلاع المربّع الثالث . ونجعل الزاويه التي هى نصف قائمة من المثلث على زاويه من الزوايا المرّبع والقطر منه على الضلع . فيفضل لنا من المثلث من عند الزاويه الأخرى بعضه . ثم نوصل بين زوايا المثلث القائمة بخطوط مستقيمه . فيكون ذلک ضلع المربع المطلوب ... هذه صورته . فأمّا المهندس فاذا سئل من عمل مربع من المربّعات قلت او كثرت، فإنّه يوجدک الخط الذي يقوي على تلک المربعات ولا يبالي ينقطع المربعات كيف كانت . وذلک أنه اذا سئل عن عمل مربّع من ثلث مربّعات، فإنّه يوصل قطر أحد مربّعات ويقيم على أحد طرفى القطر خطًا يكون عمودًا عليه مساويًا لضلع المربع . ويوصل من طرفه وبين طرف القطر بخط المستقيم، ويكون ذلک ضلع المؤلف من ثلث مربعات.

Variants:

فيما...الهندسة] كتاب النجارة فى عمل المسطرة والبركار والكونيا تصنيف ابي الوفاء محمد بن محمد البوزجاني: 1.1ii
البوزجاني] + المهندس : العلى.
[المؤيد] المنصور] المؤيد المنصور: العلى.
أدام] + تمكينه و علوه و رفعته و بسطته : العلى.
الهندسية] الهندسة : ق : العلى. .
يسهل] ليسهل : العلى.
لا شكّ ... عليها الصُّانع و الماسح . . .] لا نشكک إنّما يأخذ من الشي زبدته ولا يفكر فى الوجوه التى ثبتت صحّة ذلك به، ولأجل ذلك قد يقع عليه الغلط و الخطاء. فأمّا المهندس فقد علم صحة ما يريد بالبرهان، اذاكان هو المستخرج للمعاني التي يعمل عليها الصانع و الماسح، وإنّما يصعب عليه دربة ما يعمله البرهان الى العمل اذا لم تكن له دربة بما يعمل الصُّانع و الماسح. . . : العلى.

APPENDIX III. SELECTIONS IN PERSIAN

9. Author

ابواسحاق ابن عبدالله کوبنانی یزدی

9.1. Text

ترجمه کتاب اعمال هندسیة

9.1a. Title

ترجمه کتاب ابي الوفاء محمد بن محمد البوزجاني در اعمال هندسية

9.1b. Opening Lines

این ترجمه کتاب ابي الوفاء محمد بن محمد البوزجاني است در اعمال الهندسية
مشتمل بر مقدمه و دوازده باب بر سبیل اختصار، مقدمه در مسطره و
پرگار وگونیا

Key Passages

9.1c–f. Closing Lines

[Passage 1: 9.1c. On the completion of the book]

تم خدای سبحانه وتعالی بنده ضعیف فانی ابواسحاق ابن عبدالله کوبنانی یزدی

را قوه داد که تیسیر عبر وحل اشکال این کتاب عجاب بر حسب التماس
استاد بشر ویگانه بر هنر که در کارخانه قضا و قدر ... مصرع

[Passage 2: 9.1d. On the request for the book]

تا چرخ را مدار بود گرد این مدر
وتا خشتی بر خشتی نهند مزدوران چابک دست
مؤثرات افلاک چنان استادی مهندس بر سطح خاک ندیده
بلکه طایر تیز فهم اقلیدس بر شرفات بدایع اعمال او که همه ... مصرع ...
در شاگردی ودستیاری او انگشت فرمانبری بر دیدگان نهاد الموفق من
عندالله شمس الدین ابوبکر شاه بن استاد کامل عالم عامل دانا نجم الدین محمود
شاه بن زمح الحاج والمعتمرین حاجی تاج الدین کدوک الی البناء لا سلب مداه
در عرض دوماه تحریر گردد.

[Passage 3: 9.1e. On the initiator of the book]

در ایام تفاقم ... بادلی سوخته و خاطری افروخته از آن حرمان که برادر اعز
مغفور مسعود افتخار الحکماء والفضلاء قطب رصد بند مجسطی گشا شیخ نجم
الدین محمود رحمه الله ورحمة واسعة که در میدان فضل وحکمه گویی سبقت
از اقران زمان ربود معظم مقالات مجسطی را شرحی وجیز ملخص مشتمل بر
تصرفات خاصّه تحریر کرد ودر متوسطات اکر مانالاوس را بر تلو امیر منصور
بن عراق حواشی نوشت و از این جمله غرائب آثار در فنون علم، در عنفوان
جوانی از این عالم فانی رحمت نمود وبیشتر سبب انحاح اقتراح آنک سابقًا ترجمه
این کتاب کرده بعضی از آن در موج خیز طوفان زمان ضایع شد خواست که
آن سعی نامشکور نماند واین سطر از آثار مبارکش بر روی روزگار مسطور
بماند باتمام آن اقدام نمود.

[Passage 4: 9.1f. On the addressee of the book]

والتماس از فضلاء روزگار میرود که در مواقع زلل [و] خلل ز سد و رد
علل دریغ ندارند و در مداحض [و] غوامض امعان بصر وانعام نظر فرمایند
که قابلیت بشر در کارهای زمین تا کجا رسیده ... امید که طالب حق را
از علوم یقینی که غذای خاصّه روح است...نصیبی تمام برسد...فرغ من
کتابة هذه الکتاب...

Related Verses

1. From Kubanani's postface to the "Translation" of the *Geometrical Constructions* (Appendix I, 9.1):

... بیت
بناء روزگار که این خشت زرنگار
بر طارق چارمین بلند آسمان نهاد

2. From the *Munsha'āt* (Compilation of Works and Letters) of "Shaykh Abū Isḥāq [Kūbanānī]" (Appendix I, 9.3):

حکیمی فیلسوفی نکته دانیست ...
کاسمان اعلم زمانش خواند
دهر علامه جهانش خواند
در ریاضی به یک سریر قلم
بازکرده است کف جذر اصّم
هرچه در دهر نقش دانائیست
دل او را به آن توانایی است
عقلیش از قیاس عقل برون
نقلیش از اساس نقل فزون
او چو ابر کرم بفرق جهان
زیرکان چون صدف گشاده دهان

3. From the manuscript of *Ilḥāq-i Abī Isḥāq* (Attachment by Abu Ishaq [Kubanani]) (9.5) in 9.3 (quoting another verse):

فکرتش نسخه وجود آمد
رازگردون درو خط الحاق

4. From a manuscript reportedly in Kubanani's hand (Paris, Ms. BNP 4821, fol. 8r: see bibliography):

خوشا اندر شبی روشن چراغی
کتابی و از همه عالم فراغی

NOTES

Author's note: I would like to thank Professor Gülru Necipoğlu, director of the Aga Khan Program for Islamic Architecture and editor of *Muqarnas: An Annual on the Visual Cultures of the Islamic World* and *Supplements to Muqarnas*, and Dr. Karen A. Leal, the managing editor of those series, for their important and constructive work on the present chapter. Professor Necipoğlu and Professor Jan P. Hogendijk both offered expert comments on my essay while preparing their own chapters for this volume, and the works of the late Professor Alpay Özdural were an important starting point for my own analyses and revisions. I am grateful to Wheeler M. Thackston, Professor of the Practice in Persian and Other Near Eastern Languages, retired, Harvard University, for sharing his linguistic expertise and helping me refine some problematic passages in Arabic and Persian. I am also indebted to Adam Gacek, Faculty Lecturer in Arabic Manuscript Studies, retired, McGill University, whose codicological skills helped me in comparing a few scripts. Dr. Hussein Rezaei of the University of Mashhad, Dr. Vahid Qanbarinaniz of the University of Kerman, and Dr. Fariba Afkari of the Central Library at Tehran University provided timely access to important manuscripts in Iran, while Dr. Asadollah Kheirandish was instrumental in facilitating their acquisition and transcription. For sources in Europe, I would like to acknowledge the help of Dr. Dominique Raynaud at the Université Pierre Mendès, Bink Hallum at the British Library, and Vesta Curtis at the British Museum.

Finally, thanks are also due to the staffs at the following institutions: Bibliothèque nationale de France, Paris; Süleymaniye Library and Topkapı Sarayı Library, Istanbul; Astan-i Quds-i Razavi Library, Mashhad; Central Library, Tehran University, and the Majlis Library, Malik Library, and Hashimi Library, all in Tehran; Mar'ashi-Najafi Library, Qum; Biblioteca Ambriosiana Library, Milan; Bodleian Library, Oxford; and, most recently, the British Library and the British Museum in London.

1. Abu'l-Wafā' al-Būzjānī (d. ca. 998 [388]), *Kitāb Fī mā yaḥtāju ilayhi al-ṣāni' min a'māl al-handasa*, abbreviated throughout this chapter as *A'māl al-handasa*, and translated as *Geometrical Constructions* (Appendix I, 1.1; known manuscripts 1.1.i-v), although "Constructions of Geometry" is more exact. This form of the title, which appears as such in the opening lines of the text in Ms. 2753 in Istanbul (Appendix I, 1.1.i) and on the title page of that same manuscript (without "*al-ṣāni'*" [craftsman]), is different from the longer form found in another copy in Cairo (1.1.ii). The final part of the title, "*A'māl al-handasa*," appears in a slightly different form as "*A'māl al-handasiyya*" in the titles of some Arabic and Persian texts in that tradition (Appendix I, 3.1, 9.1, 10.1). An edition of the Arabic text by Abu'l-Wafa', based on a partial representation of extant manuscripts, one in Istanbul and two in Cairo (Appendix I, 1.1.i–iii), is contained in Abū'l-Wafā' al-Būzjānī 1979 (edited by al-'Alī), under the title *Mā yaḥtāju ilayhi al-ṣāni' min 'ilm* [i.e., *a'māl*] *al-handasa li-Abī al-Wafā' Muḥammad ibn Muḥammad ibn Yaḥyā al-Būzjānī*. A facsimile reproduction of the Istanbul manuscript is included in Qurbānī and Shaykhān 1992 (facs. 1–70). Qurbani and Shaykhan mistakenly refer to that manuscript as a unique copy ("*tanhā nuskhah-i 'Arabī-yi mawjūd*") (p. 13). Some illustrated pages of the Istanbul manuscript are reproduced and discussed in Necipoğlu 1995, 134–37. A Russian translation of the manuscript in Istanbul is included in Krasnova 1966, 1:1:42–140, text: 56–130. A partial German translation based on an Arabic variant is included in Suter 1922, 94–109. See Part II and Appendix II in this chapter (under "Title").
2. Details of the newly approximated composition date, with reference to the combined written evidence of the text's dedication and a dated coin, are discussed in Part II, under "Opening lines."
3. The designation of this text as the *Anonymous Compendium* was suggested by the editor of this volume, Gülru Necipoğlu, who uses it in the introduction ("In Memory of Alpay Özdural and His Unrealized Book Project" and in her chapter ("Ornamental Geometries: A Persian Compendium at the Intersection of the Visual Arts and Mathematical Sciences"). It is also used in the present chapter and that of Jan Hogendijk ("A Mathematical Classification of the Contents of an Anonymous Persian Compendium on Decorative Patterns"), who discusses the non-uniform structure of the text, as well as its contents and classification. For an early treatment, see Bulatov 1988, which contains an abbreviated Russian translation of the *Anonymous Compendium* by A. B. Vildanova.
4. The full title of the Persian document, *Fī tadākhul al-ashkāl al-mutashābiha aw al-mutawāfiqa*, comes from a marginal heading in red ink on the opening folio: Paris, Bibliothèque nationale de France, Ms. Persan 169, 24, fols. 180r. Of the two known catalogues of the manuscript, Blochet 1912, 41–47, and Richard 1989, 1:183–87, the former refers to the title as *A'māl va Ashkāl*: Richard 1989, 187. In her unpublished dissertation (Chorbachi 1989a) and articles, Chorbachi translated "*tadākhul*" as "interlocking." The same word was used in Necipoğlu 1995, and in Özdural 1996, 191–211. However, in his unpublished book, the first chapter of which is published as chapter 4 of the present volume, Özdural chose to translate *tadākhul* as "interlocks," as in "On Interlocks of Similar or Complementary Figures." In the present volume, we have returned to translating *tadākhul* as "interlocking," and *aw* (or) as "and," in favor of a better translation into English of the title in the margin: "On Similar and Complementary Interlocking Figures."
5. In the present volume, Necipoğlu and Hogendijk, as well as, posthumously, Özdural, offer their latest viewpoints on this unusual document. In former studies, the *Anonymous Compendium* has been treated from the perspective of the "Science of Practical Geometry and the Process of Islamic Design" in Chorbachi 1989a; for what it reveals about "Ornamental Patterns of Cubic Equations" in Özdural 1996; as a unique primary source in the formation of a design scroll containing comparable patterns without a text in Necipoğlu 1995; vis-à-vis a tiling pattern in the North Dome of the Friday Mosque of Isfahan in Hogendijk 2008; and

with reference to other patterns decorating monuments in Iran in Lu and Steinhardt 2007, as well as in Bier 2012.

6. The earlier, "critical" edition, Abū'l-Wafā' al-Būzjānī 1979 (edited by al-ʿAlī), is based on the Istanbul manuscript and two others in Cairo with variant titles (Appendix I, 1.1.i–iii), and has the slightly different title *Mā yaḥtāju ilayhi al-ṣāniʿ min ʿilm al-handasa* (On What Is Needed by the Craftsman from the Science of Geometry). The later facsimile edition to which I refer is Qurbānī and Shaykhān 1992 (on which see n. 1 above).

7. For the publication of the Persian texts of the *Anonymous Compendium*, and a translation of Abu'l-Wafa''s *Geometrical Constructions*, see Abū'l-Wafā' al-Būzjānī 1990–91 (1369). Edited and adapted to modern Persian by ʿAlī Rizā Jaẕbī, the full title of this volume is *Handasa-i Īrānī: Kitāb-i tijārat: Fī mā yaḥtāju ilayhi al-ʿummāl wa-al-ṣunnāʿ min al-ashkāl al-handasiyya* [sic], *yā, Kārburd-i handasa dar ʿamal*. As noted on the English cover, these Persian texts have been "Rewritten into modern Persian." Both the main title, *Handasa-i Īrānī* (Iranian Geometry), and the subtitle, *Al-ashkāl al-handasiyya* (Geometrical Propositions/Figures) are misleading: the term "*Īrānī*" does not convey what aspect of the texts is being described, including the "Persian" language in which they were written, while the term "*ashkāl*" in the subtitle does not reflect the title of Abu'l-Wafa''s work (*al-Aʿmāl al-handasiyya* [Geometrical Constructions]).

8. This "edition" is based on an anonymous manuscript in Tehran (see Part III. The Persian Tradition and Appendix I, 5.1.i), but does not mention another known manuscript or variant in Mashhad, which is discussed in the present chapter (see Part III. The Persian Tradition and Appendix I, 4.1.i): see Abū'l-Wafā' al-Būzjānī 2010 (ed. Āqāyānī-Chāvushī), 77. This publication, which includes a useful linguistic comparison with a late Persian translation of the same work in the Paris Codex (see Part III in this chapter, and Appendices I and III, 9.1.i), refers to "Kūbanānī's Persian translation" ("*tarjuma-i Fārsī … Abū Isḥāq Kūhbanānī*": Abū al-Wafā' al-Būzjānī 2010, 23 and 54), which is, in fact, a completion of that translation, co-translated by Kubanani. In his review of Abū al-Wafā' al-Būzjānī 2010, De Young 2010 makes no mention of the curious way in which the editor neglects to mention or refer to the anonymous Persian manuscript in Mashhad.

9. The title of the originally published volume is *Ibn al-Haytham, théorie des coniques, constructions géométriques et géométrie pratique*: see Rāshid 2000; for the English translation, *Ibn al-Haytham's Theory of Conics, Geometrical Constructions and Practical Geometry*, see Rāshid 2013. The volume does not contain references to Abu'l-Wafa''s *Geometrical Constructions* or any of the texts in its tradition; Abu'l-Wafa' is cited in the context of arithmetic and correspondence with a jurist (see Appendix I, 1.4), and as referred to by Abu Rayhan al-Biruni, on pp. 506, 653n12 and 471, and 577n16, respectively.

10. On Ibn al-Haytham's "Book of Optics," see Ibn al-Haytham 1983, Ibn al-Haytham 1989, and Ibn al-Haytham 2002, edited and translated by Sabra (more translations forthcoming). See also Kheirandish 2015 on the transmission of the "Book of Optics" and related works on optics.

11. For comparable translations in geometry and optics, see Kheirandish 1999. Optics and other subdivisions of geometry such as mechanics are discussed in Kheirandish 2013, 84–108.

12. This work is treated as part of the tradition of ornamental geometry, and also examined in relation to treatises on practical geometry and surveying in Necipoğlu 1995; see also Chorbachi 1989a and Chorbachi 1989b.

13. On Euclid's *Elements of Geometry*, see Heath 1956; on Euclid's *Optics*, see Heiberg and Menge 1895. The relevant treatments in Arabic are in Kheirandish 2008 and Kheirandish 1999.

14. This includes the identification of names and dates involving figures such as "al-Khalil al-Tajir," "Kubanani Yazdi," and others, which are directly related to discussions in the other chapters in the present volume.

15. The title and headings in this chapter are adopted from the project I worked on and a presentation I gave as an Aga Khan Fellow at Harvard University in preparation for this volume.

16. See Appendix II, 1.1.c–f. The full text of this important passage was brought to scholarly attention long ago in Chorbachi 1989b, 751–89, esp. 766; and Chorbachi 1989a. Both the Arabic text and the English translation appear in Özdural 2000, 173–83 (English translation), and 194–99 (Arabic text). The English translation is reproduced in Berggren 2007, 612–16. A revised Arabic transcription and English translation of some critical passages are included in the present chapter; for specific differences, see Part II, Passages 1–4, and n. 70 (English); Appendix II, 1.1c–1f (Arabic). The term *ṣunnāʿ* (sing. *ṣāniʿ*) has been translated throughout this chapter as "craftsmen," rather than "artisans," as in my other publications. I translate *ṣanāʿa*, *ṣanʿa*, or *ṣunʿ* (corresponding to the ancient Greek: *technē*) with the English term "craft," rather than "art," as also commonly translated. See Kheirandish 2006b; Kheirandish 2008; Kheirandish, Bier, and Yousefi 2008. An exception is made in the case of a late Persian translation of the term *ṣunnāʿ* as "*hunarvarān*," which I translate as "artisans." See Appendices II, 1.1, and III, 9.1, respectively.

17. General questions of the historical meaning of "science" and "arts/craft" (the *ʿilm* and *ṣināʿa* or *ṣanʿa* of Arabic and Persian) are addressed in my other publications with reference to specific sciences and arts/crafts, and their classifications. See Kheirandish 2006b; Kheirandish 2008; and, more generally, Kheirandish, Bier, and Yousefi 2008, which is a special issue of *Iranian Studies*, entitled "Sciences, Crafts, and the Production of Knowledge: Iran and Eastern Islamic Lands (ca. 184–1153 AH/800–1740 CE)," edited by myself, Carol Bier, and Najm al-Din Yousefi, for which Kheirandish 2008 was written. For the historical contexts of both the theoretical and practical aspects of geometry (*handasa*) in related fields, from subdivisions of geometry, such as optics

(*manāẓir*) and mechanics (*ḥiyal*), to subjects as close to them as surveying (*misāḥa*), itself called both a "science" and "craft," see Kheirandish 2003, 55–83, 59–64, and Kheirandish 2013, 84–108, 97–108. On *al-Misāḥāt al-manāẓiriyya* (Optical Surveying [on both optics and surveying]), by the tenth-century author "Sinān ibn al-Fatḥ," see Kheirandish 1999, vol. 1, xlvi, and lvi, nn. 25–26. The latter text is bound in the same manuscript codex with Abu'l-Wafa''s *Geometrical Constructions* in Cairo, Dar al-Kutub: see Appendix I, 1.1.ii.

18. On *Fī aṣnāf al-ṣināʿāt* (Classes of Crafts), ed. Daiber, see Qusṭā ibn Lūqā 1990, 93–125, 102–4. For a partial translation of the introduction, see Kheirandish 2006b, 149–50. On the author, see Kheirandish 2007, and Kheirandish 2006a.

19. On Farabi's *Enumeration of the Sciences*, see al-Fārābī 1968, 94–98. Farabi's name appears in the colophon of the manuscript variant of the *Geometrical Constructions* now in Uppsala (Ms. Tornberg 324: see Appendix I, 1.1.v). The name "al-Fārābī" follows the title given as *Kitāb al-ḥiyal al-ruḥāniyya wa al-asrār al-ṭabīʿiyya fī daqāʾiq al-ashkāl al-handasiyya* (fol. 1r), but is not in the incipit (fol. 1v); the longer form "Abū Naṣr ibn Muḥammad ibn Ūzlugh ibn Ṭarkhān al-Fārābī" is in the colophon (fol. 60r), along with the date—"the eleventh of Rajab, year three hundred and twenty one"—of the completion of its "composition" (*taʾlīf*). On the attribution to Farabi, see Kubesov and Rozenfel'd 1969; on its disputed authorship, see Hogendijk, 1993, 145; Necipoğlu, 1995, 133 and 176n9, 138 and 177n15; Özdural 2000, 193; Abū'l-Wafāʾ al-Būzjānī 2010 (ed. Āqāyānī-Chāvushī), 24–29, 75–81; and Raynaud 2012, 35n1. Among arguments against Farabi as the text's author, the earliest is that "only the first and last page belong to a text by al-Fārābī, and the rest is a large continuous fragment of the treatise by Abu-l-Wafāʾ, which begins abruptly on fol. 2a in the middle of a sentence": Hogendijk 1993, 145. The latest published discussion states that "it has been convincingly argued" that, other than the first and last pages, "the rest of the treatise [is] a simple a copy of Abū al-Wafā''s work": Raynaud 2012, 35n1. This notable manuscript still merits further textual and linguistic analysis: textual, because of the appearance of the name and date of composition on the same page as the last figure corresponding to Abū al-Wafā''s text, and linguistic, because of the distinct language of the incipit as compared to the known manuscript copies in Istanbul (1.1.i) and Cairo (1.1.ii). Besides the opening and closing folios of the Uppsala manuscript, I have consulted the section corresponding to Abu'l-Wafa''s geometrical constructions by craftsmen and a geometer, marked in red script inside the illustrations as "wrong" (*fāsid*) for that of craftsmen and "correct" (*ṣaḥīḥ*) for the geometer's version.

20. On al-Kātib al-Khwārizmī's *Miftāḥ al-ʿulūm* (Key to the Sciences), see Bosworth 1963, 97–111; for the suggested composition date, 100. The section on geometry (pp. 110–111) is not as detailed as other sections, such as the one on mechanics (p. 111). For the *Jāmi al-ʿulūm* (= *Kitāb Sittīnī*) by Fakhr al-Din al-Razi (d. ca. 1209), see Rāzī 2003; for the suggested composition date, see p. 336; for the expressions "*sākhtan*" (production) versus "*shinākhtan*" (knowledge), see pp. 375 and 380, respectively. See also Rāzī 1967–68 (facsimile reproduction), 157–60. Some "productions" under the heading "Science of War Instruments" (*ʿilm-i ālāt al-ḥurūb*) are included in Kheirandish 2008, 465–89.

21. Kheirandish 2006b refers to Farabi's *Enumeration of the Sciences*, along with later classifications of "geometry" in both Arabic and Persian by figures such as Ibn Sina, Tusi, and Shirazi. The subject of the Arabic classifications of the sciences, with reference to practical geometry, is generously covered in the chapter titled "Theory and Praxis" in Necipoğlu 1995, 132–41.

22. Al-Fārābī 1968, 95–96. This important passage is paraphrased in English and discussed in Necipoğlu 1995, 132, along with some Arabic classification works of a later period.

23. For further discussion, along with the linguistic distinctions between different parts of a geometrical proposition and methods of proof within geometry, from syllogistic reasoning (*ʿilla*) and demonstrations (*barāhīn*) to visible forms (*ṣūra*) or visual illustrations (*mithāl*), see Kheirandish 2008, 469–70. On the distinctions between geometrical proofs and physical set ups, see Kheirandish 2009, 79–104, 93–97. On "*mithāl*" as a "visual commentary," see Bier 2008, 491–511.

24. Such aspects of surveying are represented in works by al-Kindi, Sinan Ibn Fatḥ, and Ibn al-Haytham. The titles of those works are itemized in Kheirandish 1999, vol. 1, pp. xlv–lvii. The text by "Sinān ibn al-Fatḥ," titled *al-Misāḥāt al-manāẓiriyya* (Optical Surveying), which is bound in the same manuscript codex (Dār al-Kutub al-Miṣriyya, Ms. 260 (31024): see Appendix I. 1.1.ii.) with Abu'l-Wafa''s *Geometrical Constructions,* specifies methods "without measurements" (*ghayru misāḥa*).

25. Rāzī 2003, 359–65 (facsimile reproduction); Rāzī 1967–68, 148–53.

26. Ibn al-Akfānī 1978; Ibn al-Akfānī 2005, 187–275, esp. 205–10. The section on geometry is discussed in detail in Necipoğlu 1995, 140. Saliba 1995, 642–43, includes a discussion with English translation of the section titled "Erection of Structures."

27. On the *Elements* of Euclid (ca. 300 B.C.), see Heath 1956. On the *Optics* of Euclid, see Heiberg 1895. On the suggested expression of "assemblage" (*taʾlīf*) of areas (*suṭūḥ*) and its distinctions from the composition (*tarkīb*) or interlocking (*tadākhul*) of figures (*ashkāl*), and from the derivation (*istikhrāj*) of magnitudes (*qadr, miqdār*) applicable to geometrical and numerical magnitudes, and their ratios (*nisba*), see Appendix II, 1.1.

28. On this Arabic commentary, titled *Sharḥ al-Aʿmāl al-handasiyya li Abī al-Wafāʾ*, by Kamāl al-Dīn ibn Yūnus Mawṣilī, see Appendix I and II, 3.1; see also Part II in this chapter.

29. On the fifteenth-century Persian translation of Abu'l-Wafa''s *Geometrical Constructions,* titled *Tarjuma-i Aʿmāl-i handasiyya*, completed by Abū Isḥāq Kūbanānī Yazdī, see Appendices I and III, 9.1; also Part III in this chapter.

30. On the seventeenth-century Persian commentary on Abu'l-Wafa''s *Geometrical Constructions*, titled *Barāhīn al-Aʿmāl al-handasiyya*, by Muḥammad Bāqir Zayn al-ʿAbidīnī Yazdī (d. ca. 1638), and titled *Futūḥāt al-Ghaybiyya* by the author (*Futūḥāt al-Ghaybiyya fī al-ʿilm al-handasa* in the text's opening title), see Appendices I and III, 10.1; also Part III in this chapter.
31. The combined Arabic form "*al-ḥāsib al-muhandis*" appears after the name "Abī al-Wafā Muḥammad ibn Muḥammad" in a single manuscript folio of what is a related work, although I have not yet been able to gather any details concerning the length, location, and number of that manuscript (there are no folio numbers on the double page with the above reference). The Persian expression "*ʿālim-i ʿāmil*" is used by Kubanani Yazdi in reference to the deceased Persian translator of Abu'l-Wafa''s *Geometrical Constructions*, whose work Kubanani Yazdi reportedly completed. See Part III in this chapter, and Appendices I and III, 9.1.
32. The terms "connection" and "collaboration" appear in the respective subtitles of Özdural 2000 ("Mathematics and Arts: Connections between Theory and Practice in the Medieval Islamic World") and his unpublished book ("Interlocks of Similar or Complementary Figures: Collaboration of Mathematicians and Artisans in the Islamic World"). Such meetings were originally termed "conversations," as in Özdural 1995 ("Omar Khayyam, Mathematicians, and *Conversazioni* with Artisans").
33. The last two cases are included in Saliba 1999, 641–43.
34. The title of the text is *Mas'ala al-iqbāl va al-idbār ʿala mā ikhtiraʿuhu Ibrāhim ibn Naṣr ibn Sinān*. On this author, see also Berggren 1986, 85–89; on the manuscript folio of his area of a parabolic section, see 23–24.
35. Arabic edition: Bīrūnī 1964 (ed. Bulgakov and Aḥmad), 250; English translation: Bīrūnī 1967 (trans. ʿAlī), 214.
36. See Necipoğlu 2015.
37. The related compositions involving geometrical figures and areas are: *Fī jamʿ al-aḍlāʿ al-murabbaʿāt wa al-mukaʿabāt* (On the Addition of Sides of the Quadrangles and Cubes [Appendix I, 1.2]); *Fī Aʿmāl al-misāḥat* (Surveying Constructions [section of another text; Appendix I, 1.3: *Kitāb Fī mā yaḥtāju ilayhi al-kuttāb wa al-ʿummāl min ʿilm al-ḥisāb* (Saidan 1971)]); *Jawāb Abī al-Wafā' Muḥammad ibn Muḥammad al-Būzjānī ʿammā sa'alahu al-Faqīh Abū ʿAlī al-Ḥasan ibn al-Ḥārith al-Ḥubūbī, wa huwa al-burhān ʿala misāḥat al-muthalathāt* ... (Buzjani's Answer to Questions Put to Him by the Jurist al-Hububi, on the Demonstration of Measuring the Area of Triangles ... [Appendix I, 1.4]); and *Mas'alat Arshimīdis fī misāḥat al-muthallath* (Problem of Archimides on Measuring a Triangle [Appendix I, 1.5]). For an introduction and chronology, see Part II in this chapter.
38. The texts of *al-Misāḥa* by Baghdādī, and of *Kitāb al-Īḍāh*, long attributed to him and here re-evaluated, are known through manuscripts assumed to be unique and held, respectively, in Istanbul, Süleymaniye Library, Ms. Lâleli 2708/2, fols. 1–19, and Mashhad, Āstān-i Quds-i Rażavī Library, Ms. 5429. For details, see Part II in this chapter, and Appendix I, 2.1–2.2.
39. Mashhad, Āstān-i Quds-i Rażavī Library, Ms. 5357: Catalogue of Gulchīn Maʿānī 1971 (1350), vol. 8, 200. For details, see Appendix I, 3.1, and Part III in this chapter.
40. The name given to Abu'l-Wafa' in Ibn al-Nadīm 1871 (ed. Flügel), 283, and Ibn al-Nadīm 1970 (ed. and trans. Dodge), vol. 2, 667–68, is "Abū al-Wafā' Muḥammad ibn Muḥammad ibn Yaḥyā ibn Ismāʿīl ibn al-ʿAbbās." His place of birth is given as "Būzjān," from the region of Nishapur in northeastern Iran; the year is given as 328 (939–40), the day as Wednesday, the month as Ramadan, and the time of month as the crescent (moon) (*mustahall*). Ibn al-Nadim provides the names of Abu'l-Wafa''s earliest teachers (his paternal and maternal uncles) and the subjects he studied with them (numbers [*ʿadadiyyāt*] and arithmetic [*ḥisābiyyāt*]), and also states that his subject was geometry (*handasa*). Ibn al-Nadim describes his move to Iraq "during the year 348 [959–60]," and gives a list of his works in both arithmetic and astronomy, except for his *Book of Geometrical Constructions* (*Aʿmāl al-handasa*): Ibn al-Nadīm 1871, 283; Ibn al-Nadīm 1970, vol. 2, 667. Later on, practical geometry would appear before theoretical geometry in classification works such as Abū Naṣr Fārābī's *Iḥṣā al-ʿulūm*.
41. On Ibn al-Nadim, see Ibn al-Nadīm 1871, 283, 328; Ibn al-Nadīm 1970, vol. 2, 667–68; Bayhaqī 1935; Bayhaqī 2005, 76–77; Meyerhof 1948; Bayhaqī 2005, 154–55 (*ḥakīm*). On Ibn al-Qifṭī, see Ibn Qifṭī 1903 (ed. Lippert), vol. 1, 287–88, 353; Ibn Khallikān 1936, vol. 2; Ibn Khallikān 1961 (trans. de Slane), vol. 3, 320–21. On Ṣafadī, see Ṣafadī 1931, 209 (136) (*ḥāsib*); Kâtip Çelebi [Ḥājjī Khalīfa] 1850 (ed. Flügel), vol. 5, 172 (where *muhandis* Abu'l-Wafa''s *Geometrical Constructions* is cited).
42. The description "*al-ḥāsib al-muhandis*" appears after the name "Abū al-Wafā' Muḥammad ibn Muḥammad" in an unidentified manuscript folio: see above n. 31. The form "*al-muhandis*" after the name "*al-Būzjānī*" appears in the Arabic edition Abū'l-Wafā' al-Būzjānī 1979 (ed. al-ʿAlī), 23, without manuscript specification in the apparatus, or an indication of its absence in the Istanbul copy.
43. Bīrūnī 1954–56; *Taḥdīd nihāyāt al-amākin li-taṣḥīḥi masāfāt al-masākin*: Bīrūnī 1964, 250; Bīrūnī 1967, 214. On the latter, see also Kennedy 1973; on the author and his diverse patrons, see the entry on Biruni in the *Dictionary of Scientific Biography*. On passages in Biruni's works regarding direct interactions between "mathematicians" and "artisans," see Saliba 1999, 642. On a reference by "Bīrūnī" to "*Shaykhunā* [our Shaykh] *Abī'l-Wafā' Muḥammad ibn Muḥammad al-Būzjānī*," see Rāshid 2000, 653n12.
44. On the Buyid dynasty, see Donohue 2003; on the individual treatment of the rulers and patrons, see Madelung 1969, 84–108, 168–83; Madelung 1992. See also Mottahedeh 2012, 153–60.
45. The full title of this work by Abu'l-Wafa' is *Kitāb Fī mā yaḥtāju ilayhi al-kuttāb wa al-ʿummāl min ʿilm al-ḥisāb* (also known as *Kitāb al-Manāzil*), published in Saidan 1971, based on two manuscripts cited in Appendix I. See Part II in this chapter. An important passage from this text will be discussed below.

46. For this unpublished work, bearing the name of "Abu'l-Wafā' ... Būzjānī," see Appendix I, 1.2. Pingree 1983b, 329, conjectures that the patron of this text was "Mu'ayyad al-Dawla," the "regent of Isfahan" (the brother of ʿAdud al-Dawla and his uncle). This is, most likely, based on the title "Mu'ayyad" of Baha' al-Dawla, the actual patron of the text.

47. Four manuscript copies are cited with the quoted description in Kennedy and Ragep 1981, 85–108. Kennedy and Mawaldi 1979 includes a manuscript folio. Rāshid 2000, 653n12, includes a citation. See Part II in this chapter, and Appendix I, 1.4.

48. This unpublished work is cited in Rozenfel'd and İhsanoğlu 2003, 97. See Part II in this chapter and Appendix I, 1.5. The name "Archimides" is also cited in the text of Abu'l-Wafa''s *Geometrical Constructions*: Qurbānī and Shaykhān 1992; Bīrūnī 1967.

49. For the manuscript copies in Istanbul and Cairo (Appendix I, 1.1.i–ii), see, respectively, Krause 1936, 3, 4: 437–532, 466; King 1981, vol. 2, 844; and King 1986, 42. For the manuscript variant in Milan (Appendix I, 1.1.iv), see Löfgren-Traini 1975, 101–2. For the variant in Uppsala (1.1.v) naming Farabi, see Tornberg 2009, 221–22, with no mention of attribution problems; see also n. 19 above.

50. Sezgin 1974, 324, does not cite the two Cairo manuscripts (Appendix I, 1.1.ii–iii) or the variant in Uppsala (Appendix I, 1.1.v). Rozenfel'd and İhsanoğlu 2003, 97, makes no reference to the published texts (i.e., Qurbānī and Shaykhān 1992, and Abū'l-Wafā' al-Būzjānī 1979), or the catalogues (i.e., Krause 1936, King 1986, and Löfgren and Traini 1975) for the Istanbul, Cairo and Milan manuscripts, but does cite the Russian and German translations of the text in Krasnova 1966 and Suter 1928 (based on the Istanbul and Milan manuscripts). Sezgin 1974, 328, cites Krause, Suter, "Krasnowa," and "Qurbānī." Chorbachi 1989a describes the Istanbul and Cairo manuscripts without citing the Arabic edition (Abu'l-Wafā' al-Būzjānī 1979).

51. Besides the abovementioned erroneous reference to the Istanbul manuscript as a unique copy (Qurbānī and Shaykhān 1992, 13), this source refers to the incomplete Arabic manuscript of Abu'l-Wafa''s text in Milan (Appendix I, 1.1.iv) as being in Persian (*"nuskha-i khaṭṭī-yi nāqiṣ-i dīgarī az tarjuma-i Fārsī"*) (p. 15).

52. For details of manuscript copies, see Appendix, I, 1.1. Both "Cairo manuscripts are under the second title," according to Rozenfel'd and İhsanoğlu 2003, 97. See that variant title in Appendix I, 1.1.ii, Cairo manuscript dated 600 (1203). I have a copy of this Cairo manuscript and can thus verify its title, as well as its defective opening. I have not had access to the other Cairo manuscript, Appendix I, 1.1.iii, dated 831 (1427). The dates of the two manuscripts in Cairo are from David King's catalogue of the Dar al-Kutub: for the Arabic, see King 1981, vol. 2, 844; for the citation of the texts in the volume in English, see King 1986, 42.

53. The term *"Nijāra"* in this variant title is best regarded, in historical rather than modern terms, as describing a more general book of joinery, rather than carpentry, which has become more specific to woodwork over time. For a misreading of the term *"nijāra"* as *"tijāra/tijārat"* in the Persian title of Abu'l-Wafa''s text, see Abū'l-Wafā' al-Buzjānī 1990–91 (1369): *Handasa-i Īrānī: Kitāb-i tijārat*, where the first letter in *"nijāra"* (*nūn*) is read as *tā*, with two dots rather than one, thus rendering that term *"tijāra,"* as though the text's subject were trade and commerce.

54. My translation of this difficult passage is based on the subject of the second part of the sentence being *"ithbāt"* (proof), the only masculine singular term to which the verbs following it refer.

55. For literacy among some master practitioners, including master carpenters-masons and architects, see Necipoğlu's chapter.

56. The full citation of the patron's name and titles is: Our Master (*Mawlānā*), the Monarch (*al-Malik*), the King of Kings (*Shāhānshāh*), the Highest (*Ajall*), [Divinely]-assisted (*al-Mu'ayyad*), Victorious (*al-Manṣūr*), Noble (*Sayyid*) Baha' al-Dawla, Light of the Nation (*Ḍiyā' al-Milla*), Aid of the Community (*Ghiyāth al-ʿumma*), may God prolong and preserve his high standing, power and reign .... The title *"al-Mu'ayyad"* seems to be the cause of the mistaken conjecture that *"Mu'ayyad al-Dawla"* was the name of the patron of one of Abu'l-Wafa''s texts (Appendix I, 1.2): see Pingree 1983b, 392. In light of the article *"al-"* before the term *"Mu'ayyad,"* the word seems to refer to the title of someone such as Baha' al-Dawla rather than to part of a name such as "Mu'ayyad al-Dawla." Given the other titles of Baha' al-Dawla that are repeated along with it, the term is most likely intended as another title for that same patron. The additional expression *"Walī al-naʿīm"* (Guardians of Blessings, or a close customary title) appears along with *"al-Mu'ayyad"* in another text (Appendix I, 1.2). This is an expression also used in the dedication of Abu'l-Wafā''s related arithmetical composition naming the much earlier ʿAdud al-Dawla (Appendix I, 1.3). In Abu'l- Wafā''s geometrical text (Appendix I, 1.2), the expression *"Walī al-naʿīm"* occurs after the two titles *"al-Mu'ayyad"* and *"al-Manṣūr"* and before the name "Bahā' al-Dawla," followed by the two later titles *"Ḍiyā' al-Milla,"* and *"Ghiyāth al-ʿumma."*

57. On the title of *"shāhānshāh"* for Buyid rulers including Baha' al-Dawla, as well as some of his other titles, see Madelung 1969, esp. pt. 2. Özdural 2000, 172n2, who has *"Shāhinshāh"* instead of the documented form *"Shāhānshāh,"* believes that Baha' al-Dawla first assumed that title in 388 (998), and uses that date to support the argument that "the text was probably compiled by a student of Abu'l-Wafa' after the death of his master in 998." The term *al-Manṣūr*, which Özdural assumes to be the patron's first name, is also documented as one of his titles: it follows the title *al-Mu'ayyad* for Baha' al-Dawla in at least two of Abu'l- Wafa''s compositions (Appendix I, 1.1 and 1.2). The title *al-Manṣūr* is also found, without *al-Mu'ayyad* preceding it, as the title of the earlier ʿAdud al-Dawla, the patron of the related arithmetical composition by Abu'l-Wafā' (i.e., Appendix I, 1.3).

58. Raynaud 2012, 35, refers to Özdural 2000, 172, in the context of the note above, and follows Donohue, 2003, 98, in adopting the earlier date of 383 (993) as the year Baha' al-Dawla assumed this title. Raynaud 2012, 35 and n. 2, states that "from information contained in the preface it is possible to date it to between 993 and 1008," and that Abu'l-Wafa''s "treatise of geometry was written in the span of years 993–998." Raynaud's work contains much useful information, but does not refer to the dated coins discussed in this chapter. Raynaud also includes an Arabic transcription and English translation of Abu'l-Wafa''s dedication lines. I provide my own transcription and translation of that important passage (Appendix III, 1.1), as does Necipoğlu in her article (n. 71).
59. The details of the two coins are as follows. Silver dirham (British Museum Coins [BMC], no. 682, year 381), Mint of "*Madinat al-Salam*" [=Baghdad]: catalogues: Lane-Poole 1967, 217; Treadwell 2001, 154. Gold dinar (BMC no. 668, year 387/89), Mint of "*Suk al-Ahwāz*": Lane-Poole, 1967, 213; Treadwell, 2001, 118. The staff of the British Museum's coin division additionally communicated to me that Luke Treadwell "has no BM number" for the former coin (silver dirham: BMC, no. 682, year 381 [reg. no. OR 9404, side with titles and year]), and the "wrong registration number" for the latter coin (gold dinar coin: BMC, no. 668, 387/89 [reg. no.: 1850,0823.1, side with titles]).
60. The date 381 (991–92) coincides with the reported date of the assumption of the title "*Ghiyāth al-ʾumma*" by Baha' al-Dawla in Dhu'l-Qaʿda, according to Ibn al-Jawzī 1939–40 (1357–58), 163; see also Madelung 1969, 174–75, who gives "Jan-Febr. 992" as the date when the "third title" was conferred upon Baha' al-Dawla. But it is important to note that the coin dated 381 could be as early as 991, if it was minted before March 9, when 381 was still 991. Madelung's important work has not been used before in that connection. For references to that work in other matters related to the Buyids, see Mottahedeh 2012, 153–54, where an inscription with the same titles for Baha' al-Dawla is reported for the much later date of 392 (1001–2). This inscription features the title "*Shāhānshāh*" in its Sasanian form and also shows the titles granted to Baha' al-Dawla by the Abbasid caliph, including "*Ghiyāth al-ʾumma*," which is repeated twice.
61. A possible composition date as early as 381 (991–92) is consistent with Baha' al-Dawla's assumption of the title ending with "*Ghiyāth al-ʾumma*," which appears on both the dedication of Abu'l-Wafa''s text and the coin dated 381 (see n. 59 above). See also Raynaud 2012 and Woepcke 1855.
62. On the formal divisions of Euclidean propositions, see Heath 1926, vol. 1, 129–31. On the corresponding Arabic terms, including "*mithāl*," see Kheirandish 2008, 470–71.
63. On these lost works of Abu'l-Wafa', see his biographical entry in Pingree 1983b, 43. Surveyors are mentioned in Abu'l-Wafa''s *Geometrical Constructions* (1.1; see Appendix II, 1.1.c), as well as his earlier arithmetical work (1.3): see n. 68 below.
64. The word *muhandisīn* (sing. *muhandis*), from the root *h-n-d-s*, for *handasa* (geometry), as in Persian *andāza* (measure), is translated as "geometers"; *ṣunnāʿ* (sing. *ṣāniʿ*), from the root *ṣ-n-ʿ* is translated as "craftsmen," consistent with my translations of the terms *ṣanāʿa, ṣanʿa* or *ṣunʿ* (ancient Greek: *technē*) in English as "craft" elsewhere. A late Persian translation of the term *ṣunnāʿ* as "*hunarvarān*" can be accordingly rendered as "artisan." Also, the term *murabbaʿ*, from the root *r-b-ʿ*, is translated here according to the intended meaning as "quadrangle" or "square number," as well as "square," as the case may be. The term *taʾlīf* is consistently translated as "assemblage," as distinct from "composition" for *tarkīb*. The terms *barāhīn* and *aʿmāl* are consistently translated as "demonstrations" and "constructions."
65. The term "*awwal*" in the first line of this important and variously transmitted passage (see below) does not seem to have a temporal or causal sense as "first," as there is nothing corresponding to something coming next. The terms *ghalaṭ* and *khaṭāʾ* are consistently translated as "mistake" and "error." The contextual meaning of this part of the passage, which is critical for understanding any direct relation between the constructions of geometers and craftsmen mentioned by Abu'l-Wafa', has been further complicated by the particularly problematic transmission of the passage, which is different in both its Arabic manuscripts (see Appendix II, Variants), as well as a late Persian translation (see Appendices I and III, 9.1), and a variation as late as that in the Persian commentary by Muhammad Baqir Yazdi (Appendix I, 10.1).
66. Qurbānī and Shaykhān 1992, facs. 52–53.
67. Qurbānī and Shaykhān 1992, facs. 53.
68. The full section title given in Saidan's edition is: *Fī mā yaḥtāju ilayhi al-kuttāb wa al-ʿummāl wa ghayrahim min ʿilm al-ḥisāb wa hiya fi aʿmāl al-misāḥāt* (On What Is Needed by Secretaries and Administrators and Others from the Science of Arithmetic, and That Is on Surveying Constructions): Saidan 1971 (Arabic ed.), 202. A partial English translation of the passage under discussion is included in Saliba 1999, 643, where he refers to Abu'l-Wafa''s "book on arithmetic for the bureaucrats" and the latter's "grave errors" (*fāḥish al-khaṭāʾ*) in the context of Abu'l-Wafa''s similar remarks in his *Geometrical Constructions*. The English translation for the section title and the expressions "secretaries" (*kuttāb*), "administrators" (*ʿummāl*), and "surveyors" (*massāḥ*) from the full opening passage of the book are my additions. Note the slightly different form *māsiḥ* for surveyor in Abu'l-Wafa''s *Geometrical Constructions* (passage 1: 1.1c). Here, Abu'l-Wafa' extends his discussion by mentioning cases leading to loss/waste (*ḥayf*), due to area measures (*misāḥa*) impacting the ruler (*sulṭān*), agent (*muʿāmil*), and client (*mushtarī*): Saidan 1971, 203. For this important passage, see also Necipoğlu's chapter.
69. Euclid's *Elements*, Book II, e.g., proposition 4: Heath 1956, vol. 1, 379. A numerical case of that proposition is marked on the figures with *abjad* letters to represent the letter *alif* as 1, *bāʾ* as 2, *jīm* as 3, and *dāl* as 4, according to the historical labels found in the *Anonymous Compendium* itself (start-

ing with fol. 180v [1]). When the sides of the two enclosed squares are marked as 1 and 2 units, respectively, and that of the enclosing square as 3 units, the area of the larger enclosing square (9 units) equals the areas of the enclosed squares (1 and 2, total 5 units) plus the two rectangles with sides 1 and 2 each (2 units each, total 4 units) (fig. 3a).

70. Qurbānī and Shaykhān 1992, facs. 53–54. It is important to note that the expression "*niṣf sub*ʿ" (Qurbānī and Shaykhān 1992, facs. 55) here refers to half of one seventh, namely 1/14, and not 1/4, as found in the translations in Özdural 2000, 179 and 179n15, and Berggren 2007, 614. The value itself does not affect Abu'l-Wafa''s argument. But the following explanatory note, left out of the original text and the secondary sources referring to it, helps interpret the passage: in the case of the three smaller squares, if each has a side as 10 units (as given in the text), when they are assembled to be equal in area to a larger square to be constructed, the side of the encompassing square, once constructed would be 17 1/14, as opposed to 17 1/3, if mathematically approximated: namely, the total assembled areas add up to 300 units: 100 (area of first square with a side of 10 units) + 2 × 50 (areas of two equilateral triangles each with side 10 units) + 25 (area of 1 triangle of side 10 and height 5) + 2 × 37.5 (areas of 2 trapezas: each made up of a square and a triangle with sides 5) = total 300 units. Sum of 3 square areas, 300 = area of a larger square; side = square root of 300 = ~ 17 1/3. The values of the *abjad* letters in my labeled drawing (fig. 1e) are added here to mark area correspondences, in this case making the side of the larger square a whole number = 4. Area of the encompassing constructed square = 4 × 4 = Sum of the assembled areas: 1 (middle square) + 2 + 2 (side triangles) + 3 (bottom triangle) + 8 (2 trapeza) = 16.

71. Qurbānī and Shaykhān 1992, facs. 56–58. The division of the square by the "correct" method then follows (fig. 1f): of the three quadrilaterals, one is the single square (marked *alif* = 1), one is cut in two equal halves (*bā'* = 2), and the third, this time, is cut into two equal halves (*jīm* = 3). These are subsequently rearranged inside the encompassing square such that a central square sits within it, surrounded by four triangles. With the more exact method of the geometer specified, the two triangles cut out from each square (2, 2 and 3, 3) are arranged around the central square so that their vertices fall on each corner of the large square and their longest sides or hypotenuses form a smaller square at the center (1).

72. The citation and dating of the wooden door is in Chorbachi 1989b, 767. For the dating of part of the Isfahan tiling to 1080 and a printed photograph of it, see Seherr-Thoss 1968, 186; the original construction and later redecoration of the mosque are dated to 515 (1121–22) and 1112 (1700–1701) on p. 88; for a close-up of that tiling, see pl. 85 on p. 188. For a related discussion, see Chorbachi 1989a, 767.

73. عمل ابن محمد مؤمن محمد امين : note that "Ibn Muḥammad Mu'min Muḥammad Amīn" is read as "Muḥammad Ḥusayn ibn Muḥammad Qawāmī" in Farnūd 1997, 13; and as "Muhammad Ibin Mu'min Muḥammad Amīn...[sic]" in Chorbachi 1989b, 767.

74. چون نامه جرم ما بهم پیچیدند
بردند و به میزان عمل سنجیدند
بیش از همه کس گناه ما بود ولی
ما را به محبت علی بخشیدند

75. The name "Abu'l-Qāsim al-ʿAndijānī" is given in Chorbachi 1989a, 155, in reference to a marginal note on the Istanbul manuscript of Abu'l-Wafa''s *Geometrical Constructions*, where this name appears on folio 18 (see the facsimile edition of Qurbānī and Shaykhān 1992, 25 [on the margin], as well as pp. 59–60). Rashid refers to an "Aḥmad ibn Aḥmad ibn Jaʿfar al-Ghundijānī," as "probably [a] second half of the fifth or early sixth century hijri [late eleventh/early twelfth-century] Arabic commentator," who is mentioned in connection with the marginal commentary on Abu'l-Wafā''s *Geometrical Constructions*: see Rāshid 2005, 13n36. A mathematician named "Aḥmad ibn Aḥmad ibn Jaʿfar" (with no last name or date) is cited as the author of *Kitāb Fī qismat al-arāḍī* (Book on Division of Lands) in Rozenfel'd and İhsanoğlu 2003, 424: no. 032. For references to Chorbachi 1989a, 154–56, 270, dating ʿAndijani to the fifteenth century, and Rāshid 2005, 13n36, citing al-Ghundijani for a treatise on burning mirrors copied in 690 (1291) (p. 142), see the chapter in this volume by Necipoğlu, who brought the publication by Rashid to my attention.

76. For corresponding Babylonian, Greek, and Arabic problems, and issues with expressions such as "geometric algebra," see Unguru 1975, 67–114.

77. For a summary of this primary source, see summary in Berggren 1986, 104–8.

78. Kheirandish 2004, 197–213. On "*ṣūra*" as a visual form of proof, see also Kheirandish 2008.

79. An Arabic text by Thabit ibn Qurra under the title *Kitāb fī Misāḥat al-mujassamāt al-mukāfiyya* (Book on the Measurement of Parabolic Bodies) is cited in Necipoğlu 1995, 141–44, with an image of a manuscript folio from a "copy by al-Sijzī in 969," nos. 103–6. Surveying texts by Yaʿqub al-Kindi that contain the term "*Īḍāh*" in their titles are itemized with details in Kheirandish 1999, vol. 1, pp. xlv–lvi, lvn24, and lvi.

80. Mashhad, Āstān-i Quds-i Rażavī Library, Ms. 5429 (unique manuscript?). For the published editions and discussions, see the notes immediately below, as well as Appendix I, 2.1, and Part II in this chapter.

81. The name of the reported author may be read as "al-Yazdī" or "al-Mardī/Mur[ā]dī" in the absence of determining letters and dots on that name. Here it is kept as "Yazdī," in the form given in the text's manuscript catalogue: Gulchīn Maʿānī 1971 (1350), 40.

82. For a facsimile reproduction of the Arabic composition *Kitāb al-Īḍāḥ* (Appendix I, 2.1), and a reported Persian "translation" by Abū al-Futūḥ Muntajab al-Dīn ʿIjlī Iṣfahānī titled *Kitāb Dar ʿIlm-i misāḥat* (Appendix I, 6.1), which attributes the original Arabic to "Abū Manṣūr ʿAbd al-Qāhir ibn Baghdādī" without any specific title, see Ibn Ṭāhir al-Baghdādī 1969 (1347) (ed. Gulchīn Maʿānī), 1–85,

and 87–166, respectively. The volume includes an introduction to the texts' manuscript history by the modern editor Gulchin Ma'ani, who reports bringing the two manuscripts together in that same volume. For a more recent publication of both texts, and a few facsimile pages by Awjabi, as well as discussions of the problematic attribution of the Arabic text to al-Baghdadi, see Ibn Ṭāhir al-Baghdādī 2009 (1388) (ed. Awjabī): Arabic and Persian transcriptions, 50–69, 71–152; facsimile reproduction of opening pages, 156–57, and 158–59; introduction, 15–50 (Persian), and 3–5 (English). The modern editor, Awjabi, reprints related articles by Afshār 1969 (1348), 325–26, and Muḥīṭ Ṭabāṭabā'ī (see Muḥīṭ Ṭabāṭabā'ī 1982 [1361], 167–73), on the issue of the authorship of the Arabic text by Abu Mansur al-Baghdadi, and mentions another Arabic text titled *al-Misāḥat* by that author, but Awjabi adds that the latter text has not been seen by him: see Ibn Ṭāhir al-Baghdādī 2009 (1388), 23. On the problematic treatments of the main Arabic and Persian texts under discussion, namely the Arabic text *Kitāb al-Īḍāḥ*, and its attribution to Abu Mansur al-Baghdadi as author, and the Persian text *Misāḥat* bound with it in Ibn Ṭāhir al-Baghdādī 1969 (1347) (ed. Gulchīn Ma'ānī) as a Persian translation of that same Arabic text, see Parts II and III in this chapter.

83. The full name is given in the opening lines of the Persian text: "Abū al-Futūḥ As'ad ibn Abī al-Faḍā'il ibn Khalaf 'Ajalī," who may be identifiable as "Abū al-Fatḥ ...Maḥmūd ibn ...Muḥammad ...Iṣfahānī." On "Abu'l-Fath Esfahani," see Pingree 1983a, where the sources for his dates are discussed: "Abu'l-Fatḥ claims to have found the manuscript of an Arabic translation in the library of a king," who, according to Pingree, "may be identical with the Abū Kālījār Garšāsp b. 'Alī b. Farāmarz b. Moḥammad who ruled Yazd from 1095 till about 1119." The name "Abū al-Fatḥ ibn Maḥmūd [sic] al-Iṣfahānī" is reported in connection with "the Arabic version of Borelli's Latin translation" of parts of Apollonius's *Conics* in 1661 Florence, according to the Bodleian Quarterly Record 1921.

84. Among those who comment on the issue of the authorship of the Arabic text *Kitāb al-Īḍāḥ* by Abu Mansur al-Baghdadi, Afshar 1969 (1348), 325–26, raises questions about the author attribution with reference to the first part of the text's colophon, and reads the final parts of that name as "Mardī," and possibly "Yazdī." Muḥīṭ Ṭabāṭabā'ī 1982 (1361), 167–73, includes an additional phrase about the "owner" in the colophon immediately following the latter colophon, though he misdates the transcription as 718 rather than 728. In Ibn Ṭāhir al-Baghdādī 2009, the editor Awjabi cites references to the earlier facsimile editions in Arabic and Persian published in Ibn Ṭāhir al-Baghdādī 1969 (1347) (ed. Gulchīn Ma'ānī). Awjabi also reprints the same two Arabic and Persian texts and Afshar and Muhit Tabataba'i's aforementioned articles, and includes a preliminary comparison of the Arabic and Persian texts. On the related works under discussion, see Appendix I, 2.1 and 6.1.

85. For earlier problems and discussions of attribution, see n. 84 above.

86. Özdural refers to the name and date given for the transcription: see Özdural 2000, 172n4, and Özdural, "Preliminaries," where the name "'Alī ibn al-Khalīl al-Tājir" is associated with that of "Abū Bakr al-Khalīl al-Tājir" in the context of the latter's works within the Paris Codex, and as the "father of the copyist" of the Arabic text *Kitāb al-Īḍāḥ*, based on information provided by his colleagues in Tehran. One of these two colleagues has recently confirmed to me that the relation was based on a "conjecture" (see below). Necipoğlu 1995, 168–69, and 197n109, treats the name "Abū Bakr al-Khalīl al-Tājir" more fully in connection with specific constructions and citations in the *Anonymous Compendium*, and refers to Sezgin 1974, 386, regarding an "Abu Bakr" as a possible author of a *Risāla Fī misāḥat al-ashkāl* in a Latin work by "Abhabuchr." For the alternative transcription "Abū Bakr ibn al-Khalīl" al-Tājir, with the additional "ibn" indicating a generational gap, as a "copyist," in the transcription of the manuscript of a work by Ibn al-Haytham (cited in Rāshid 2013, 504), see the chapter by Necipoğlu, who kindly brought that manuscript and publication to my attention. The form "Abū Bakr *ibn* Khalīl," does appear in a different part of the Paris Codex as well, Ms. Persan 169, fol. 118v, and is referred to in Özdural 2000, 174n4, but not discussed by him with respect to the consequence of that form for the familial relationship he posits. See Part III in this chapter, as well as Appendix I, 6.1, 7.1, and 8.1.

87. See the introduction by Necipoğlu, as well as her chapter in this volume, and Özdural, "Preliminaries." See also Muḥīṭ Ṭabāṭabā'i 1982 (1361), 167–73.

88. On the unique manuscript of this Arabic commentary in Mashhad (Āstān-i Quds-i Rażavī Library, Ms. 5357), see Gulchīn Ma'ānī 1971 (1350), 200, as well as Appendix I, 3.1, and Part II in this chapter. Some discussion of the content and author of the text is included in Chorbachi 1989a, and Chorbachi 1989b, 753–55; the author is discussed in Raynaud 2012 in the context of his connection to Europe.

89. The author of the text and his patron have been the subject of a few biographical surveys in Persian: Qurbānī 1986–87 (1365), 398–400, and Karāmatī and Qalandarī 2011, 93–122.

90. Chorbachi's unpublished dissertation refers to the transcription date as "608 or 680 A.H" (Chorbachi 1989a, vol.1, 408). But the reading of the date in the colophon as 680 (*thamānīn*) (1280) seems preferable to 608 (*thāmin*) (1211), as the former date is also given in the catalogue of the Astan-i Quds-i Razavi Library in Mashhad, which holds the only manuscript.

91. On the Maragha Observatory, see Sayılı 1960; on the larger compound, see Varjavand 1987 (1366). For relevant astronomical activities, see Ragep 2000 and Ragep 2014; for activities relevant to the *Anonymous Compendium*, see Necipoğlu 1995, and Özdural 2003, 15.

92. For examples of some constructions, see Bier 2012, 251–73.

93. Details of the textual transcription of this early Arabic text (Mashhad, Ms. 5535), attributed in the catalogue to "'Abd al-Raḥmān ... Ṣūfī d. 376 [986]" and in the colophon to "Abū al-Ḥusayn al-Ṣūfī," where a so-called "Sun of Scholars" is

cited after the copy date (1289), can be found in Gulchīn Maʿānī 1971 (1350), 38: "*sana 688* [1289] *ʿalaquhā Shams al-Muḥaqiqqīn* (i.e., the Sun of Scholars, namely, al-Sijzi) … *bi Marāgha al-raṣad.*" No other information about this text is currently known to me other than that catalogue citation. The date and name included from the colophon above add new information to that in Ragep 2014, where the date 1283 is given as a "likely *terminus ad quem*" for known activities at the Maragha observatory (p. 233), along with a few name forms starting with "Shams" associated with a slightly later period. For another manuscript transcribed in Maragha in 1289, see n. 94 below.

94. Istanbul, Topkapı Sarayı, Ahmed III Library, Ms. 3457, fols. 1v–34v. On the possible relation of this text to the work of Abu'l- Wafa', see Jackson 1987–88, 369–90, 375–78, and Kheirandish 2006a, 207–22, esp. 214.

95. Kamāl al-Dīn Ibn Yūnus, *Sharḥ al-Aʿmāl al-handasiyya li Abī al-Wafā'*: Mashhad, Āstān-i Quds-i Rażavī Library, Ms. 5357, 680 (1281), Maragha: preface, folio 66r–66v. On this manuscript, see above n. 88.

96. The Arabic manuscript containing the colophon naming "al-Sijzī" is in Istanbul, Topkapı Sarayı, Ahmed III Library, Ms. 3457, folios 1v–34v, 34v. The possible relation between the copy transcribed by Sijzi and the *Geometrical Constructions* of Abu'l-Wafa' is discussed in Jackson 1987–88, 375–78, and Kheirandish 2006a, 214. In addition, the year of Sizji's copy, transcribed as "352" (963), is consistent with my dating of Abu'l-Wafa''s text as possibly being as early as 381 (991), though the suggested relation requires further documentation. On al-Sijzi and Abu'l-Wafa', who are both associated with problems within the Arabic tradition of Euclid's *Division of Figures*, see Hogendijk 1993, 144–45, where Sijzi is named for his "autograph" of a copy of the surviving Arabic text of Euclid included there (Text A), along with a related problem from Abu'l-Wafa''s *Geometrical Constructions* from the variant attributed to Farabi (Text C), and part of the anonymous compilation entitled *Various Geometrical Problems* (Text B).

97. The five "simple machines," which are at least as old as the *Mechanics* of Heron of Alexandria (second century), and were widely transmitted in Persian, include the lifting device known as a pulley (*bakra*). On the Greek mechanical tradition, see Schiefsky 2008, 15–49. I am preparing a publication on the Arabic and Persian traditions of Pappus of Alexandria's "Introduction to Mechanics" (*Madkhal fī al-Ḥiyal*), as part of a project supported by the National Science Foundation.

98. There are two statements on the title page: according to the one in Persian, the text "was among books that have been lost" (*min jumla kutubīst ki mafqūd shuda būd*); the other, in Arabic, states that it was "transferred to … Iṣfahān" (*waqaʿa intiqāluhu li-….Iṣfahān*), and names the "late" (*marḥūm*) Shaykh Bahā'ī as the endower, who owned other texts within both the Arabic and Persian traditions. See Part II in this chapter.

99. *Tarjuma-i Aʿmāl-i handasiyya*, Mashhad, Āstān-i Quds-i Rażavī Library, Ms. 5264: see Appendix I, 4.1: Library's oldest catalogue 1345, 311, no. 37; Gulchīn Maʿānī 1971 (1350), 80–81; Munzavī 1969 (1348), 147; Fikrat 1990 (1369), 61: "*naskh, bī tā*"; Qurbānī and Shaykhān 1992, 14, mentions the Mashhad manuscript as "another incomplete manuscript of the same [Persian] translation" (*nuskha-i khaṭṭī-yi nāqiṣ-i dīgarī az hamīn tarjama* [*Fārsī*]). In the comprehensive catalogue of Munzavī 1969 (1348), 147–51, the Mashhad manuscript is listed under the title *Tarjuma-i Aʿmāl-i handasiyya* (p. 147), and the Tehran manuscript under *Tarjuma-i Kitāb Tijārat* [=*Nijārat*] (p. 151). See Part II in this chapter.

100. *Kitāb-i Nijārat*, Tehran, Ms. 2876: for the full title, see Appendix I, 5.1: Dānish Pazhūh 1962 (1340), 1720; Munzavī 1969 (1348), 151: *Tarjuma-i Kitāb Tijārat* [= *Nijarat*] … *Tarjama-i Aʿmāl-i handasiyya.* Abū'l-Wafa' al-Būzjānī 2010 (ed. Āqāyānī-Chāvushī), 77, states that the edition of the Persian translation of Abu'l-Wafa''s *Geometrical Constructions* is based on the anonymous manuscript in Tehran, and refers to "two old Persian translations" ("*dū tarjama-i qadīm-i Fārsī*" and "deaux traductions Persanes": pp. 23, 37). This editor cites the manuscript known in Tehran, and the apparently later Persian translation now in Paris, as "Kūbanānī's Persian translation" (*tarjuma-i Fārsī … Abū Isḥāq Kūbanānī*), i.e., not a "completion" of that translation as it is. These two Persian translations are then used for comparison, without any mention of the translation in the Mashhad manuscript in Abū'l-Wafa' al-Būzjānī 2010 (ed. Āqāyānī-Chāvushī), and De Young 2010. See Part III in this chapter.

101. The transcription date of the Tehran manuscript is given in Dānish Pazhūh 1962 (1340), 1720: "*naskh ṣada-i 5 va 6*"; the transcription date of the Mashhad manuscript is given in Gulchīn Maʿānī 1971, 81: "*naskh, taḥrīr-i ḥudūd-i qarn-i shishum va haftum-i Hijrī.*" The Tehran Ms. colophon has the year 1017 (1608) next to the name one of its owners.

102. The titles of the two anonymous Persian translations are followed by the phrase "the start of the text of the translator from the Arabic is as follows …" (*āghāz-i matn-i mutarjam az zabān-i tāzī īn-ast*). The full titles are: *Dar Ānch* [*Ānchih* in Tehran Ms.] *ṣunnāʿ wa muḥtarifa bi ān muḥtāj bāshand az aʿmāl-i handasa, va īn kitābrā Kitāb-i Nijārat khwānand.* The Tehran manuscript has the following problematic title on the cover page of its holding library: "*Kitāb al-Tijāra* [=*Nijara*], *Fī mā yaḥtāju ilayhi al-ʿummāl wa al-ṣunnāʿ min al-ashkāl* [=*aʿmāl*] *al-handasiyya.*" Qurbānī and Shaykhān 1992, 15–17, includes a comparison based on chapter number and content: it identifies eleven out of thirteen chapters for both the Persian manuscript in Tehran and the Arabic manuscript in Istanbul that have different defects. Qurbani and Shaykhan also specify twelve chapters in the Persian manuscript in Paris, pointing out that it contains one chapter more than the versions represented by the Tehran and Istanbul manuscripts.

103. Ms. Persan 169, items 19, 25, and 24 (Appendix I, 7.1, 8.1, and 9.1, respectively): see Blochet 1912, 46–47; Richard 1989, 186–87.

104. Āqāyānī-Chāvushī's linguistic analysis in Abū'l-Wafā' al-Būzjānī 2010 is based only on the anonymous manu-

script in Tehran and the version in the Paris Codex, which he refers to as "Kubanani's translation," rather than co-translation, or completion of the translation, as reported by that author, without use of the Mashhad manuscript.

105. The opening titles in Persian contain the name of the patron of the earlier Arabic text as follows: "I followed (*imtithāl namūdam*) the order of (*farmān-i*) Malik Manṣūr Bahā' al-Dawla ..." The brief comparison here of the two anonymous manuscripts is based on their opening pages, the only part so far available to me from the manuscript in Mashhad: see Appendix I, 4.1 and 5.1, and Part III in this chapter.

106. In the Mashhad manuscript in the Āstān-i Quds-i Rażavī Library (Ms. 5264), the library's title page for the manuscript includes the note: "*khaṭṭ-i Shaykh dar ẓahr-i kitāb dīda mīshavad.*" The first few folios from that manuscript that were accessible to me do not include the reverse of that folio. The old catalogue of that library, Āstān-i Quds-i Rażavī Library 1926 (1345), no. 37, has "the stamp and signature" of Shaykh Bahā'ī "on the back of the first page" (*pusht-i ṣafha-i avval*). The catalogue of Munzavī 1969 (1348), 147, adds "*vaqf-i Shaykh Bahā' ī dar 1030* [=*1620*] ...." Related manuscripts in that library with notes about their endowment by "Shaykh Bahā'ī" include the Arabic commentary of Abu'l-Wafa"s *Geometric Constructions*: see Appendix I, 3.1, and Part II in this chapter.

107. This text is published as part of A. S. Saidan's *Al-Takmila fī al-ḥisāb* (The Completion of Arithmetic), with a tract on mensuration by "Abū Manṣūr 'Abd'l-Qāhir ibn Ṭāhir al- Baghdādī," edited and annotated by Saidan in Ibn Ṭāhir al-Baghdādī 1985, 333–75.

108. The *Risāle-i mi'māriyye* by Ca'fer Efendi (ca. 1614), on the "career of the Ottoman architect Mehmed," is discussed in Necipoğlu 1995, 152, along with the passage on the reported roles of "practical geometry (*handasa*) and mensuration (*misāḥa,* the science of measurement of lengths surfaces and volumes of geometric figures)." On the text, see Ca'fer Efendi 1987. The determination of "measures" (*misāḥa*) based on proportional relations and values is different from their numerical calculations based on physical "measurements" or "measurings," terms that are both represented by the same word (*misāḥa*).

109. On the scroll tradition, see Necipoğlu 1995.

110. For the manuscript details of these texts with the citations, see Appendix I, 7.1 and 8.1; for the folios included in this chapter, see figs. 7a–b, and 8a–b, where folios 7a and 8b contain name citations. For the earliest treatments of dating, see Necipoğlu 1995, 168–69, 197n109; Özdural 2000, 172, and n. 4; and Özdural, "Preliminaries." For various published and unpublished works on the *Anonymous Compendium*, see the introduction ("In Memory of Alpay Özdural") and the chapters by Necipoğlu, Hogendijk, and Özdural in this volume. Necipoğlu discusses the contents of the treatise titled "Of the Constructions by Abu Bakr al-Khalil al-Tajir al-Rasadi," in addition to the *Anonymous Compendium*, which is the main subject of her chapter, as well as Hogendijk's, and Özdural's. On treatments of dating, see note below.

111. Özdural 2000, 172, first stated that "that work can be dated to around 1300 because the mathematician Abu Bakr al-Khalil al-Tajir al-Rasadi (ca. 1300s) is cited twice as one of the participants of the discussions," and that "he is cited as the father of the copyist of a manuscript in Mashhad," according to information provided by colleagues in Tehran (172n4); he does not, however, provide the primary source for that information. In his chapter in this volume ("Preliminaries"), Özdural adds the "newly acquired information" of the Mashhad Ms. 5429, this time stating that "the father of the copyist, Khalil al-Tajir [sic], must have been Abu Bakr al-Khalil al-Tajir, who should have been active in the late thirteenth and early fourteenth centuries. Our document can thus be dated to circa 1300." I have recently confirmed with the colleague named in the acknowledgment in Özdural's "Preliminaries" that the father-and-son relation was a "conjecture," not a citation that may be documented. Even if the composition's date is regarded as close to the year 1300, the determination of the site of its production as Tabriz, based on a brief historical account of Ilkhanid developments, according to which that city suggests itself as "the most likely place of origin for 'Interlocks of Figures'" (Özdural "Preliminaries"), must await more concrete forms of evidence. The name "Abū Bakr al-Khalīl al-Tājir al-Raṣadī" and its shorter variants in the Paris Codex are cited in Necipoğlu 1995, 168–69, and 197n109, and in her chapter in this volume: she had noted in her earlier discussion (Necipoğlu 1995, 168–69) that "Abū Bakr al-Khalīl al-Tājir al-Raṣadī" specialized in observational astronomy. In her chapter in the present volume she repeats this observation and adds "ibn" to his name, with the alternative transcription "Abū Bakr [b.] al-Khalīl al-Tājir al-Raṣadī," with "ibn" after Abu Bakr, to suggest that Abu Bakr and 'Ali were possibly brothers. This implies a slightly later date than assumed for the former, and through it, for the *Anonymous Compendium*, which she tentatively dates in her chapter to ca. 1300s, that is, the fourteenth century, thereby further extending Özdural's dating of ca. 1300 (late thirteenth to early fourteenth century). I maintain that any family relation between the figures with the first names Abu Bakr, in their variant forms, and 'Ali, as found in a different codex (fig. 6a), remains to be established beyond the mere overlap of their last names as "al-Khalil al-Tajir." Regardless of a family relation, the part "ibn" appears in another manuscript folio of the same Paris Codex (Ms. Persan 169, 118v), in the form of "Abū Bakr ibn al-Khalīl," as well as in a codex in Saint Petersburg (Rāshid 2000, 531; Rāshid 2013, 504), the latter copied in the hand of "Abū Bakr ibn al-Khalīl," both noted by Necipoğlu. The "ibn" may have been part of the intended name elsewhere in the Paris Codex (7.1 and 8.1) but was dropped in the process of transmission.

112. Ms. Persan 169, fol.118v, ends with the statement that "the derivation (*ikhrāj*) of this problem (*mas'ala*) is from the pride of the geometers (*mufakhkhar al-muhandisīn*), Abū

Bakr ibn al-Khalīl." Özdural 2000, 172n4, refers to the figure praised as the "pride of geometers" (which he had read as *"fakhr al-muhandisīn"*) in "the same Paris manuscript as *Interlocks of Figures*," without mentioning the form of the name and the "ibn."

113. The reference to a copyist named "Abū Bakr ibn al-Khalīl al-Tājir" is in a manuscript codex in Saint Petersburg, which contains the fragment of a surveying treatise by Ibn al-Haytham and other mathematical texts; the details of that manuscript codex are in Rāshid 2000, 531. The quotation is from the English translation of that same work: Rāshid 2013, 504. Both Rashid and Necipoğlu (in her chapter in this volume) include the details of a manuscript fragment in Saint Petersburg titled *Fī Uṣūl al-misāḥa* (On the Principles of Measurement): Saint Petersburg, Library of the Oriental Institute, no. B 2139, fols. 100r–139v. The copyist's name is given in reference to a work by Ibn al-Khawwam al-Baghdadi, who is cited in Rāshid 2000, 650, and Rāshid 2013, 573. Ibn al-Khawwam al-Baghdadi is cited in Rozenfel'd and İhsanoğlu 2003, 231, no. 657; in Qurbānī 1986–87, 23–24 (in Persian); and in Sezgin 1974, 155, no. 55. According to a statement by Rashid about the copyist Abu Bakr, the "number of mistakes suggests that Abū Bakr did not belong to the learned classes."

114. Here are some possibilities with respect to variations in the forms of the names under discussion. In the case of the two names Abu Bakr and ʿAli each ending with "al-Khalīl al-Tājir," it is not inconceivable that they might refer to individuals who were not related as immediate members of the same family, or were unrelated beyond the part of their names that they had in common. But if the name as it appears in the manuscript of the two texts in the Paris Codex under discussion (7.1, fig. 7a; 8.1, fig 8b) represents his son through the part "ibn," that would suggest its own generation gap and date variation. For an interpretation of these name variations, see Necipoğlu's chapter, where both the transcription differences and the generation gap are noted. Finding it highly unlikely that the two individuals were unrelated, she discusses the form of the name that includes the "ibn" in relation to the signed copyist of texts documented in the Saint Petersburg Codex, and proposes that the two individuals may have been brothers rather than father and son.

115. In her chapter in the present volume, Necipoğlu discusses these in detail along with the other two treatises in the Paris Codex that mention Abu Bakr.

116. The term *"iʿtibār"* in the sense of examination is found in both the Arabic and Persian versions of Abu'l-Wafa''s *Geometrical Constructions*. On the Arabic term, see the facsimile edition of Qurbānī 1992, 7; on the anonymous Persian translation, see Abū'l-Wafā' al-Būzjānī 2010 (ed. Āqāyānī-Chāvushī), 11, 13, where the term *"imtiḥān"* occurs (12). On a more advanced use of the same term in reference to Ibn al-Haytham, see Sabra 1971 and Rāshid 2005. On methodological terminology, see Kheirandish 2009 and Kheirandish 2013.

117. The reading "Sayyid al-Sharif al-Jurjani (1339–1413)" appears in Necipoğlu's chapter, with reference to Richard 1989, 1:187, and also in Özdural's chapter ("Preliminaries"). The date given in Rozenfel'd and İhsanoğlu 2003, 265, is "(1340–1413)." The information I have added about the acknowledged benefits of glosses by Jurjani is from the following statement in Dhanani 2007, 603–604: "The 16th-century historian Khwāndamīr noted that Jurjānī has glosses on most books by the ancients and moderns in the curriculum. Indeed, from his own time until the present, no lesson is given without the benefit of his glosses and studies" (p. 603).

118. The upper limit of 1413 is in keeping with the date circa 1300 or 1300s that Necipoğlu proposes in her chapter for the initial composition of these works, prior to their being copied in a modified compilation in the fifteenth century, perhaps by or for Kubanani Yazdi.

119. The term "postface," suggested by Gülru Necipoğlu based on its use in Blochet's catalogue, is a more suitable expression than "epilogue" or "postscript," which may wrongly evoke a sense of afterthought. That important postface, called by Alpay Özdural an "epilogue," is rightly described by him as a "difficult passage caus[ing] several erroneous translations." The details of his own treatment and acknowledgement of Maria Subtelny's help with that "difficult postscript" are summarized in Necipoğlu's introduction to this volume. Moreover, her own chapter cites Subtelny's translation of the "poetry," which Özdural excluded.

120. See, in this volume, Özdural, "Preliminaries," and the introduction by Necipoğlu ("In Memory of Alpay Özdural"), as well as her chapter.

121. See Özdural, "Preliminaries," for expressions that refer differently to the author as "explaining (*ḥall*) and executing (*ʿabr*) the figures (*ashkāl*)" of the work. These are presented without consideration of the intended rhyming pairs "facilitating (*taysīr*) the passages (*ʿibar*)" and "solving the problems (*ishkāl*) of the propositions/figures (*askhāl*)" of this "wondrous book" (*kitāb-i ʿujāb*). The guiding rhymes of the challenging wording, starting with the postface's opening lines, have not saved native Persian readers from publishing the transcription of the whole passage with some problematic readings throughout: see Abu'l-Wafa' 1990–91 (ed. Jaẕbī).

122. In this first passage from the postface of the Persian translation of Abu'l-Wafa''s *Geometrical Constructions* (Appendix III, 9.1c), starting from Ms. Persan 169, fol. 178v (fig. 9a), the words *"taysīr"* (root: *y-s-r*) meaning "occasioned" and *"ʿibar,"* the plural of *ʿibāra* (root: *ʿ-b-r*), can also be read as *"tasyīr"* (root: *s-y-r*), to mean "set in motion," and *"ʿabar"* (same root), in the sense of a comparative examination. Sabra 1971; Rāshid 2005, 1039–45; and Kheirandish 2009, 79–104, contain analyses of the historical meaning of *iʿtibār* in the context of early methods of experimentation.

123. See Appendix III, 9.1d. For the title *ustād-i bashar* (teacher of mankind) used in reference to Nasir al-Din al-Tusi (d. ca. 1274), see Kheirandish 2004.

124. For the "mention of bricks (*khist*) and workshop (*kārkhāna*)" in the so-called "poetical passages," see Özdural, "Preliminaries," where the work's requester is wrongly referred to as a "mathematician/master craftsman" (*ustādī yā muhandis*) (the correct reading being "master geometer" [*ustād-ī muhandis*]), and the "workings of the earth" (*kārhā-yi zamīn*) are equated with "geometry."

125. In Passage 2, Kubanani provides a chain of names for Shams al-Din, the individual who asked Kubanani to complete the translation of a certain Najm al-Din. The latter's untimely "passing away in the peak of youth" (*ʿunfuvān-i javānī*) is mentioned in Passage 3. This chain of names suggests a direct familial relation between the two (fig. 9b). The combined sequence of names—"Shams al-Dīn Abū Bakr Shāh" before "*ibn ustād-i kāmil ʿālim-i ʿāmil-i dānā* Najm al-Dīn al-Maḥmūd Shāh," itself before "ibn ... Ḥājī Tāj al-Dīn Kadūk"—all on one and the same line suggests the following in light of the positions of the word *ibn* to indicate "son [of]": namely, that the requester, Shams al-Din Abu Bakr Shah, called "unique in the arts," was the son of the original translator, Najm al-Din Mahmud Shah, called a "complete master" (*ustād-i kāmil*) and wise (*dānā*) "scholar-practitioner" (*ʿālim-i ʿāmil*). The second *ibn* following the latter's name indicates that he himself was the son of Zamh al-Hajj wa al-Muʿtamirin Taj al-Din Kaduk. Thus, from the text we may infer that Shams al-Din was the son of Najm al-Din Mahmud, the deceased individual who first started this translation in Persian of Abu'l-Wafa''s *Geometrical Constructions*, and who had previously (*sābiqan*) translated (*tarjuma*) the book, some of which was "lost" (*ḍāyiʿ*) in the stormy seas of time. Shams al-Din wanted this line (*saṭr*) from his works [to] be preserved (*masṭūr*).

126. See Appendix III, 9.1e.

127. See Appendix III, 9.1f

128. Regarding the phrase "dearest brother" (*barādar-i aʿiz*), see the passage that Subtelny translated for Özdural, where she interprets "brother" as "colleague": cited in Özdural's "Preliminaries," and in Necipoğlu, "In Memory of Alpay Özdural," where Subtelny's interpretation is accepted. My own examination in light of other works by Kubanani (9.3), and others (10.1), leaves this as an open question. See note below. On the "Intermediate [Book]s" (*mutawassiṭāt*), see Kheirandish 2006b, 39 and 39n9.

129. In Passage 3, Kubanani refers to Najm al-Din Mahmud as his "dearest brother" (*barādar-i aʿīz*), as if he were Kubanani's literal "brother." If so, then the figure "causing the pressing request" so that the effort (*saʿy*) of the original translator not remain unknown (*nā-mashhūr*) could be considered the author's nephew. Whether they were actually related to Kubanani, however, must remain an open question for a number of reasons: Kubanani's letters use the expression "my brother" (*barādaram*) in regard to "Shaykh Najm al-Din Mahmud," as well as a certain "Burhān" (9.3), in ways that could indicate both the literal and non-literal ways such expressions could be employed. The names "Burhān al-Dīn Maḥmūd" and "Ṣadr al-Din Kūbanānī" are also in the letters (9.3), without the use of the term "*barādar*," which if the case, may explain the expression "*aʿīz*" (dearest) instead of "*ʿazīz*" (dear) next to it for Najm al-Din Mahmud in the post-face citation (9.1). We can also find close associates referred to as "brothers" elsewhere, as in the late Persian commentary of Muhammad Baqir al-Yazdi, *Barāhīn al-Aʿmāl al-handasiyya/Futūḥāt al-ghaybiyya fī al-ʿilm al-handasa* (Demonstrations of Geometrical Constructions / Hidden Openings in the Science of Geometry) (10.1).

130. My research for this chapter has extended the years in which Kubanani lived to at least 887–88 (1482–83), as he is cited as being alive in 887 (1482), according to a "eulogy ... by the famous al-Dawānī" in a manuscript codex in London, and in 888 (1483), according to his own signed work in that same codex: see Appendix, 9.4, and Rieu 1894, 509–10, no. 751: *Commentary on Ṭūsī's Recension of Euclid's Elements.* In another codex in London, Ms. Add 23570, Kubanani's *Risāla taḍʿīfiyya*, which is in a different hand from the latter, assigns it to "Mawlānā Shaykh Abū Isḥāq al-Kūbanānī" (fol. 105). The name "Kūbanānī" is found along with "Sharaf al-Dīn Yazdī" (fl. 1454; fol. 5r), and "Maḥmūd Haravī" (fl. 1435; fol. 118r). All three are named as authors in different treatises in the Paris Codex.

131. In their respective catalogues of the Paris Codex, Blochet and Richard each suggest that the script of the entire codex is in the same hand; Blochet surmises that it was copied in the seventeenth century, and Richard in the sixteenth. In the introduction to this volume ("In Memory of Alpay Özdural"), Necipoğlu explains how Özdural adopted Priscilla Soucek's "redating of the calligraphy of the Paris codex on stylistic grounds to the mid-fifteenth century." See also the conclusion of the present chapter.

132. See Necipoğlu's chapter concerning a citation in Rāshid 2013, 572–73, of an annotation on fol. 8r of a manuscript in Paris, discussed in my chapter and below in n. 161. The word "*khaṭṭ*" in this annotation means "script."

133. Other than manuscript catalogues that list his writings, there are few modern works on Kubanani, and these primarily include references to his translation of Abu'l-Wafā''s *Geometrical Constructions* (Appendix I, 9.1): Woepcke 1855, 18–256 309–59. Bio-bibliographical works include: Qurbānī 1986–87 (1365), 60–61; Rūh al-Amīnī 1990 (1369), 42–49; Karāmatī and Nigahbān 2009; and Qanbarīnanīz and Qanbarīnanīz 2009 (1388). Of these, the first two specify his "completion" (with Najm al-Din Mahmud) of the Persian translation: Woepcke 1855, 359; Qurbānī 1986–87 (1365), 61. Rūh al-Amīnī 1990 (1369), 43–44, 48, refers to the collection of his works, and letters. The third biographical work, Karāmatī and Nigahbān 2009, 157, names the original translator, but continues to cite the Paris manuscript as "Kūbunānī"'s translation, as does Āqāyānī-Chāvushī in Abū'l-Wafā al-Būzjānī 2010, 77. Qanbarīnanīz and Qanbarīnanīz 2009 (1388), 70–71, refer to a work on an astrolabe by Kubanani, with an apparently unique manuscript in Qum. None of the modern works include references to Kubanani's works related to area measures and

geometry (see Appendix I, 9.4 and 9.5, respectively), or his dates beyond 886 (1481). His life is extendable to after ca. 888 (1483) according to this research, based on a recent find that refers to him as a living author on that date (see Appendix I, 9.4).

134. The name "Kūbanānī" is mistakenly cited as "Kubnawi" in Rozenfel'd and İhsanoğlu 2003, no. 833, 282, where manuscript information about an important but relatively unknown work by him, *Ilḥāq-i Abī Isḥāq* (Attachment by Abū Isḥāq, London, 1514: Rieu 1894, no. 751 [see Appendix I: 9.5]), is given under the inaccurate title "Finding by Abū Isḥāq," with an incorrect reference: i.e., "London Sup 151" rather than "Supp. 751 (Ms. 1514)"; so is the alternative name citation "Kuyanani" (*sic*, reading *yā* for *bā*) in Özdural, "Preliminaries." The form "Kūhbanānī," on the other hand, as found in Karāmatī and Nigahbān 2009, and Qanbarīnanīz and Qanbarīnanīz (1388) 2009, is based on a more recent form of that historical name, i.e., "Kūhbanān" rather "Kūbanān" (located between Yazd and Kirman).

135. The misattributions include the composition, not "completion," by Kubanani of a late Persian translation of Abu'l-Wafā''s *A'māl al-handasa* (see Appendix I, 9.1). There are also references to "Kūbanānī's translation," in works such as Karāmatī and Nigahbān 2009,157, and Abū'l-Wafā' al-Būzjānī 2010 (ed. Aqāyānī-Chāvushī), 77. The attribution of the authorship of the *Anonymous Compendium* following that Persian translation in the Paris Codex (see Appendix I, 8.1) to Kubanani, without any form of documentation or discussion, is found in Abu'l-Wafā' al-Būzjānī 1990–91 (1369) (ed. Jaẕbī).

136. Chorbachi 1989a, 155, hypothesizes that Qushji may have carried Abu'l-Wafā''s manuscript to Istanbul.

137. On some mathematical breakthroughs of al-Kashi, see Berggren 2007; on his published letters to his father, see Kennedy 1960, 29:191–213; and Bāqirī 1997, 24:241–56.

138. Most of these versified phrases can be verified or rejected by consulting Kubanani's works and letters. The *Munsha'āt* (Appendix I, 9.3) includes passages about Kubanani's training in both the rational and transmitted sciences in the "Sa'diyya" School as well as references to his mathematical works. Kubanani's scientific compositions are known through a collection (Appendix I, 9.2), as well as a largely uncited Euclidean work (Appendix I, 9.5). There is a reference to "Shaykh Najm al-Din Mahmud" as "my brother" (*barādaram*)—a reference also applied to a certain "Burhan"—in the *Munsha'āt*, but a certain "Sadr al-Din Kubanani" is cited there without such a specification (see Appendix I, 9.3).

139. This little-known work by Kubanani (9.5), which he describes in the same closing lines as a much-needed (*muḥtāj*) commentary (*sharḥ*) for the explanation (*tawḍīḥ*) of parts (*mawāḍi'*) of Tusi's *Recension* of Euclid's *Elements*, is much more than an "Attachment" (*Ilḥāq*), even if it contains only a few of the books of that multi-volume work. It is in the title that the dictates of verses and rhymes seem to have had an effect: the text is called *Ilḥāq-i Abī Isḥāq*, where the first name of Kubanani rhymes with the second part of a verse that follows that title as *"Rāz-i gardūn dar ū khaṭṭ-i Ilḥāq."* The full verse is included in the *Munsha'āt* (9.3) of Kubanani: see Appendix III, Verse 3. It has been identified as a Persian poem of the twelfth-century Persian poet Hakim Anvari by Dr. Vahid Qanbarinaniz, who facilitated my access to the manuscripts of the *Munsha'āt*. The verse is: *"Fikratash nuskhah-i vujūd amad ... Rāz-i gardūn dar ū khaṭṭ-i Ilḥāq."* The name "Naṣīr al-Dīn al-Ṭūsī," which is given at the bottom of the same folio (in 9.5), includes his title as *"ustād al-bashar"* (teacher of mankind), a title that Kubanani uses in his postface (9.1) for the one who requested the completion of the Persian *Geometrical Constructions*, namely, "Shams al-Dīn Abu Bakr Shāh."

140. See Appendices I and III, 9.1, and Appendix I, 9.5. Although some of them are not uncommon expressions, the identifiable phrase contained in the text's closing lines is the Arabic *in'ām al-naẓar bi 'ayn al-im'ān* (bestowing a glance with a scrutinizing eye), referring to the "sages of the age" (*afāzḍil-i 'aṣr*) and the "grandees of the times" (*akābir-i dahr*). This is comparable to the Persian "scrutinizing sight" (*im'ān-i baṣar*) and "giving views" (*in'ām-i naẓar*) in reference to the "sages of the time" (*fuẓalā'-i rūzgār*) in the postface. For a verse (in 9.5) identical to one appearing in Kubanani's *Munsha'āt* (9.3), see the note above.

141. I am grateful to Adam Gacek of McGill University, who kindly seconded my efforts in reviewing some relevant scripts. He also considered an item in the autograph codex of Kubanani dated 846 (1442–43) to be an authentic script by the figure who signed with the name "Kūbanānī," though not the "script" reported to have been in his hand in a manuscript codex in Paris (BnF, Ms. 4821, fol. 8r), all based on images only, leaving the case of the script in Kubanani's Persian *Geometrical Constructions* open. This requires a full examination of the physical aspects of all the relevant manuscripts, including paper, for more conclusive considerations.

142. The two positions are represented by Özdural 1996, and Hogendijk's chapter in this volume. Özdural dates the initial compilation, prior to its fifteenth-century copying in the Paris Codex, to ca. 1300, and Chorbachi 1989a to ca. 1200.

143. Özdural, "Preliminaries," describes the text's "disorganized nature" and states that "in light" of other cases, it appears to have been the "compiled notes of a scribe—like Ca'fer Efendi—who recorded the discussions between mathematicians and artisans at a series of their joint meetings." According to Özdural, Ca'fer Efendi "appears to have been a professional scribe" who composed a text "from the notes ... taken at a series of meetings." Özdural 2000, 172, states that the *Anonymous Compendium* "may be similar to the lost work on geometry by Ca'fer Efendi ... which seems to have been compiled from notes taken by a scribe at a series of meetings between geometers and artisans." He cites his earlier publication, Özdural 1996, where he refers to the

text of the *Anonymous Compendium* as "notes taken over a period of time"; in Özdural 2000, he asserts that the text "gives the impression of a collection of notes, in some cases hastily taken," and that "Ca'fer's treatise, which cannot be traced, was a compilation of notes taken during the course of twenty years" (p. 192).

144. Abū'l-Wafā' al-Būzjānī 1990–91 (1369) (ed. Jazbī); and, in this volume, the introduction by Necipoğlu and her chapter.

145. Pingree 1983b, 392–94, 394, refers to this figure and his Persian commentary as "Moḥammad Bāqer Zayn al-ʿĀbedīnī" (without the "ibn" (son [of]), before "Zayn ...").

146. His name appears as "Muḥammad Bāqir ibn Zayn al-ʿAbidīn-ī Yazdī" in King 1999, 131, according to whom he was apparently still alive in 1047 (1637–38). A list of his works includes those related to Euclid's *Elements* (in n. 13); according to a note later in the book, "it is not clear whether he compiled his major work in 1047 [=1637/–38] or whether he died in that year" (p. 256).

147. Geometrical shapes of increasing sides, with physical and spiritual correspondences, include the following: the triangle of natural realms (mineral, vegetable and animal), the quadrangle of elements (air, water, earth and fire), the pentagon of senses (vision, hearing, touch, smell, and taste), the hexagon of the (six) days (of creation), the heptagon of deep perception (*darakāt*), the octagonal degrees of the bliss (*naʿīm*) of Paradise (*jannat*), the nonagon of heavenly bodies (*aflāk*), and the decagon of pure (*pāk*) minds (*ʿuqūl*). The comparable surveying text in the Paris Codex, Ms. Persan 169, fol. 65v, opens with the heptagon of (seven) heavenly bodies (*aflāk*), the hexagon of (six) directions, the pentagon of (five) senses, the quadrangle of elements, and the triangle of (three) natural realms, and also refers to tools such as the compass (*pargār*) in the same opening folio. In that text polygons are listed in the decreasing order of the number of their sides (from seven to three), but this may also be a mere reflection of common descriptions found in many other texts.

148. Al-Fārābī 1968, 94–98. On the appearance of Farabi's name in the colophon of one of the manuscript variants of Abu'l-Wafa''s *Geometrical Constructions* (the one in Uppsala), see n. 19 above, and Appendix I, 1.1.v, for the citation of the manuscript.

149. See Appendix II, 1.1. This is a complex passage in Abu'l-Wafa''s *Geometrical Constructions*, in which he states that a craftsman "sees what was [already] worked by the geometer and established (*iqāma*) its correctness (*ṣiḥḥa*)," and that both craftsmen and surveyors take the "choice" part (*zubda*) of such works without the ways (*wajh*) in which their correctness is proved (*thubita*). For these, and the practices he associates specifically with geometers in his reported "gathering" in the next sentence, see Appendix II, 1.1.d, as well as notes and textual variances below the translation, especially the part labelled "*qabl*," which I translated as "already" rather than "before." The passage seems to mean that the geometer has *already* established the correctness of a construction involving a craftsman, and not that the geometer established its correctness *before* the craftsman involved in that construction did so.

150. Jazarī 1979 (ed. al-Ḥasan), translated from the Arabic and annotated by Hill in Jazarī 1974 as *The Book of Knowledge of Ingenious Mechanical Devices*. Necipoğlu 1995: door: 150–52, figure: 151, no. 115; see also Necipoğlu's chapter in this volume.

151. An image of the door in Mosul is published in Chorbachi 1989b, 766, fig. 8, "Wooden door of the Mosque of Imam Ibrahim in Mosul dated 498 A.H./1104 A.D," and reproduced in this volume among Özdural's drawings (pl. 2.6). The geometrical construction in the passage by Abu'l-Wafa' (Qurbānī and Shaykhān 1992, facs. 59) is reconstructed for the door's repeat pattern in fig. 4a of this chapter.

152. The commentary of Kamal al-Din ibn Yunus al-Mawsili has been discussed in some detail in Chorbachi 1989a, and Chorbachi 1989b.

153. See Hogendijk 2008, 121, and his chapter in the present volume.

154. On the historical meaning of "*al-Jabr wa al-muqābala*," see Saliba 1972–73.

155. For a published transcription of the Arabic text of propositions from Book II of Euclid's *Elements*, see Brentjes 1994, 53–92, 56–69. For propositions from that same book used in reference to bricks, see Brentjes 2006, 169–97, fol. 14b, 12. More recently, she added the Syriac and Hebrew cognates for the term in a private communication.

156. A form (*ṣūra*) is a visual form of proof. It is sometimes represented by a figure without a logical demonstration, as in the expression "*ṣūratuhu* ..." (its [visible] proof ...). An illustration (*mithāl*) is an example, often representing the second part of a Euclidean proposition, as in the expression "*mithāluhu* ..." (its example ...). On the formal divisions of a Euclidean proposition, see Heath 1956, 1, 129–31, and Kheirandish 2008, 470–71.

157. Of these two drawings in the "Constructions by Abu Bakr," the one on the left is interpreted as Abu'l-Wafa''s solution to the problem of constructing a square from three squares, while the one on the right is attributed to Abu Bakr al-Khalil in Özdural's unpublished book manuscript (see his plates, published in the present volume). See also Necipoğlu's interpretation of both drawings in her chapter, which comes close to that of Özdural, yet differs in certain respects, particularly in the connection she establishes with the illustration of the accompanying measurement of a mountain, based on a comparable triangulation technique used in surveying. See also figs. 9e–f above.

158. A line of verse reportedly written by Kubanani, but apparently not in his hand, appears at the end of a work not by him, in a codex that does not contain any of his writings (other than that verse): see Appendix III.

159. For a discussion of the surveying problem and Abu'l-Wafa''s constructions in Persian on a single folio (fol. 136v [fig. 3a]), see Part II in this chapter; see also the chapter by Necipoğlu in this volume, where she considers all of the drawings on

that same folio (fol. 136v) as original parts of the "Constructions by Abu Bakr," without any later additions to that page by the copyist.

160 On the author Ibn al-Khawwam al-Baghdadi (ca. 1245–1325), see Qurbānī 1986–87, 23–24 (in Persian); Sezgin 1974, 155, no. 55; and Rozenfel'd and İhsanoğlu 2003, 231, no. 657. For references to the copyist of the work, "Abū Bakr ibn al-Khalīl al-Tājir," see Rāshid 2000, vol. 3, 650; and Rāshid 2013, 572–73. The full title of his text as copied is *al-Fawā'id al-bahā'iyya fī al-qawā'id al-ḥisābiyya*: see Rozenfel'd and İhsanoğlu 2003, 231, no. 657, M1: "Profitable Uses in the Rules of Arithmetic" (alternatively titled by Rozenfel'd and İhsanoğlu as "Bahā' al-Dīn's Uses," apparently after Baha' al-Din al-Juwayni, to whom the work is dedicated).

161. Kubanani is named as a commentator (*shāriḥ*) on Ibn Khawwam's *al-Fawā'id al-bahā'iyya* in the colophon of some verses that are claimed to be in his own hand. The manuscript of the transcription claimed to be in Kubanani's hand is in Paris, Bibliothèque nationale de France, Ms. 4821. This is cited by Rāshid 2000, vol. 3, 650, and Rāshid 2013, 572–73, and also referred to in Necipoğlu's chapter. I have examined the manuscript folio (fol. 8r), where, according to a second hand in the colophon, the verses are said to be in the "script" (*khaṭṭ*) of Kubanani. A third and more recent hand dated 1936 cites "Kūbanānī" as a commentator on "Khawwām Baghdādī'[s]" "*Fawā'id-i bahā'iyya.*" But here another work by that author may have been intended. For a transcription and English translation of the verses in question, see Appendix III, Verse 3, in this chapter.

162. For the relevant citations, see the note above.

163. In her chapter Necipoğlu makes the important point that "[t]he latest datable treatises of the Paris Codex were authored around the same time in the Timurid-Turkmen period, by the scholars Mahmud Haravi (ca. 1436) and the aforementioned Sharaf al-Din ʿAli Yazdi. Interestingly, the latter's *nisba* (component of a name indicating relation or origin) is connected with the city of Yazd, like that of Kubanani Yazdi." To this may be added other important links: the names of both figures, namely Mahmud Haravi and Sharaf al-Din ʿAli Yazdi, appear next to "Abū Isḥāq Kūbanānī" in a London manuscript codex (British Library, Ms. Add 23570); furthermore, the Arabic text *Īḍāḥ* (Appendix I, 2.1), which was transcribed by "ʿAlī ibn al-Khalīl al-Tājir" and has links to the Paris Codex through the name of that copyist, may have been composed by "Abū al-Fatḥ al-Yazdī," an apparent contemporary of that transcriber, who also carries the *nisba* "Yazdī" (see Part III in this chapter). It is not presently possible to confirm whether Kubanani composed any part of the Paris Codex: for details on a few modern authors who made such assumptions, including about his role in making additions to the *Anonymous Compendium*, see Necipoğlu's introduction to this volume ("In Memory of Alpay Özdural").

BIBLIOGRAPHY

Manuscripts

Cairo

Abū'l-Wafā' al-Būzjānī. *Kitāb al-Manāzil al-al-sab'* (also known as *Kitāb Fī mā yaḥtāju ilayhi al-kuttāb wa al-'ummāl min 'ilm al-ḥisāb*). (Arabic text.)
Dār al-Kutub al-Miṣriyya, Ms. Riyāḍa 42/1, dated 487 (1094).

Abū'l-Wafā' al-Būzjānī. *Kitāb al-Nijāra fī 'amal al-masṭara wa al-birkār wa al-kūniyā taṣnīf Abī al-Wafā' Muḥammad ibn Muḥammad al Būzjānī.* (Arabic text.)
Dār al-Kutub al-Miṣriyya, Ms. 260 (31024), 1, folios 1r–58v, dated 600 (1203).
Dār al-Kutub al-Miṣriyya, Ms. 366 (44795), 46 folios, dated 831 (1427) (Ms. inaccessible).

Damascus

Abū'l-Wafā' al-Būzjānī. *Jawāb Abī al-Wafā' Muḥammad ibn Muḥammad al-Būzjānī 'ammā sa'alahu al-Faqīh Abū 'Alī al-Ḥasan ibn al-Ḥārith al-Ḥubūbī.* (Arabic text.)
Ẓāhiriyya Library, Ms. 4871, 20, folios 82r–82v, dated after ca. 555 (1160).

Istanbul

Abū'l-Wafā' al-Būzjānī. *Kitāb Abī al-Wafā' Fī mā yaḥtāju ilayhi al-ṣāni' min a'māl al-handasa.* (Arabic text.)
Süleymaniye Library, Ms. Ayasofya 2753, dated ca. 851–53 (1447–49).

Ibn Ṭāhir al-Baghdādī, Abū Manṣūr 'Abd al-Qāhir. *Al-Misāḥa.* (Arabic text.)
Süleymaniye Library, Ms. Lâleli 2708, 2, folios 1–19.

Pappus of Alexandria. *Madkhal Babūs fī al-ḥiyal.* (Arabic text.)
Topkapı Sarayı, Ahmed III Library, Ms. 3457, folios 1v–34v, dated 688 (1289).

Leiden

Abū'l-Wafā' al-Būzjānī. *Kitāb Fī mā yaḥtāju ilayhi al-kuttāb wa al-'ummāl min 'ilm al-ḥisāb* (also known as *Kitāb al-Manāzil*): first three stations (*manāzil*). (Arabic text.)
Leiden University, Bibliotheca Universitatis Or. 103, no date.

London

Kūbanānī Yazdī, Abū Isḥāq. *Ilḥāq-i Abī Isḥāq.* (Arabic text.)
British Library, Ms. Or. 1514, dated 888 (1483).

Mashhad

'Abd al-Raḥmān al-Ṣūfī. *Ashkāl mutasāwī al-aḍlā'.* (Arabic text.)
Āstān-i Quds-i Rażavī Library, Ms. 5535, dated 688 (1289).

Abū'l-Wafā' al-Būzjānī. *Dar Ānch ṣunnā' va muḥtarifa bi ān muḥtāj bāshand az a'māl-i handasa va īn kitābrā Kitāb-i Nijārat khwānand.* (Anonymous Persian translation; referred to as *Kitāb-i Nijārat.*)
Āstān-i Quds-i Rażavī Library, Ms. 5264, dated in reference works as 6th–7th centuries (12th–13th centuries).

Abū'l-Wafā' al-Būzjānī. *Fī jam' al-aḍlā' al-murabba'āt wa al-muka'abāt.* (Arabic text.)
Āstān-i Quds-i Rażavī Library, Ms. 5521/1, dated 867 (1462).

Ibn Ṭāhir al-Baghdādī, Abū Manṣūr 'Abd al-Qāhir. *Kitāb al-Īḍāḥ 'an 'uṣūl-i ṣinā'at al-massāḥ* (attributed). (Arabic text.) Colophon name: "Abī al-Fatḥ al-Yazdī" (or "Mardī/Murādī").
Āstān-i Quds-i Rażavī Library, Ms. 5429, dated 728 (1327–28), based on the Arabic text in the same codex.

Ibn Yūnus (Mūsā Shāfi'ī), Kamāl al-Dīn. *Sharḥ al-A'māl al-handasiyya li Abī al-Wafā'.* (Arabic commentary.)
Āstān-i Quds-i Rażavī Library, Ms. 5357, dated 680 (1281).

Iṣfahānī, Abū'l-Fatḥ. *Dar 'Ilm-i misāḥat.* (Persian translation; referred to as *Dar Misāḥat.*)
Āstān-i Quds-i Rażavī Library, Ms. 5462, dated 728 (1327–28), based on the Arabic text in the same codex.

Milan

Abū'l-Wafā' al-Būzjānī. *A'māl-i handasiyya.* Arabic manuscript variant (defective).
Biblioteca Ambrosiana, Ms. Arab 68.

Oxford

Abū'l-Wafā' al-Būzjānī. *Jawāb Abī al-Wafā' Muḥammad ibn Muḥammad al-Būzjānī 'ammā sa'alahu al-Faqīh Abū 'Alī al-Ḥasan ibn al-Ḥārith al-Ḥubūbī, wa huwa al-burhān 'ala misāḥat al-muthalathāt min istikhrāj al-'amūd wa masqaṭ al-ḥajar.* (Arabic text.)
Bodleian Library, Ms. Thurston 3, folios 53r–53v.

Abū'l-Wafā' al-Būzjānī. *Mas'alat Arshimīdis fī misāḥat al-muthallath.* (Arabic text.)
Bodleian Library, Ms. Thurston 3, folios 55v–58r.

Paris
Anonymous. *Faṣl fī mas'alat al-Uqlidisiyya.* (Arabic text.)
Bibliothèque nationale de France, Ms. Persan 169, 8, folios 102v–104v.

Anonymous. *Fī tadākhul al-ashkāl al-mutashābiha aw al-mutawāfiqa.* (Persian text.)
Bibliothèque nationale de France, Ms. Persan 169, 24, folios 180r–199r.

Anonymous. *Mukhtaṣar dar ʿilm-i misāḥat.* (Persian text, referred to as *Mukhtaṣar dar misāḥat.*)
Bibliothèque nationale de France, Ms. Persan 169, 9, folios 65v–69v.

Abū Bakr al-Khalīl al-Tājir al-Raṣadī. *Lil-Aʿmāl Abī Bakr al-Khalīl al-Tājir al-Raṣadī.* (Persian text.)
Bibliothèque nationale de France, Ms. Persan 169, 19, folios 124v–137v.

Abū Sahl al-Qūhī. *Risāla Fī ʿamal ḍilʿ al-musabbaʿ al-mutasāwī al-aḍlāʿ fī al-dā'ira.* (Arabic text.)
Bibliothèque nationale de France, Ms. 4821, folios 1r– 8r.

Abū'l-Wafā' al-Būzjānī. *Tarjuma-i Kitāb-i Abī al-Wafā' Muḥammad ibn Muḥammad al-Būzjānī Dar Aʿmāl-i Handasiyya* (Persian translation of *Geometrical Constructions*, referred to as *Tarjuma-i Aʿmāl-i handasiyya*). Completed by Abū Isḥāq Kūbanānī Yazdī. Bibliothèque nationale de France, Ms. Persan 169, 23, folios 141v–179v.

Qum
Yazdī, Muḥammad Bāqir Zayn al-ʿAbidīnī. *Barāhīn al-Aʿmāl al-handasiyya / Futūḥāt al-ghaybiyya fī al-ʿilm al-handasa.* (Persian commentary.)
Marʿashī Najafī Library, Ms. 12936, 1, dated 1031 (1621).

Saint Petersburg
Ibn al-Khawwām al-Baghdādī. *Al-Fawā'id al-bahā'iyya.* (Arabic text.)
Library of the Oriental Institute, Ms. B 2139.

Tehran
Abū'l-Wafā' al-Būzjānī. *Dar Ānchih ṣunnāʿ va muḥtarifa bi ān muḥtāj bāshand az aʿmāl-i handasa va īn kitābrā Kitāb-i Nijārat khwānand.* (Anonymous Persian translation, referred to as *Kitāb-i Nijārat.*)
Tehran University, Central Library, Ms. 2876, folios 1–70 (dated in reference works 5th–6th centuries [11th–12th centuries]), dated 1017 (1608) for ownership.

Kūbanānī Yazdī, Abū Isḥāq. *Majmūʿa-i rasā'il va kutub.* (Persian and Arabic texts.)
Tehran University, Central Library, Ms. 2417.

Maybudī Yazdī, Qāḍī Ḥusayn ibn Muʿīn, compiler. *Munsha'āt.* (Writings [and Letters of Abū Isḥāq Kubanānī]). (Persian text.)
Malik Library, Ms. 1605; Majlis Library, Ms. 15330 (dated 979 [1571–72]); Hashimi Private Library, uncatalogued (dated 900 [1494–95]).

Uppsala
Al-Fārābī (attributed). *Kitāb al-ḥiyal al-ruḥāniyya wa al-asrār al-ṭabīʿiyya fī daqā'iq al-ashkāl al-handasiyya.* (Arabic text.)
Universitetbibliotek, Ms. Tornberg 324, Vet 27, folios 1r–60r, dated 321 (933).

Publications

Primary Sources

Abū'l-Wafā' al-Būzjānī. 2010. *Ketâb al-Nejârat: Sur ce qui est indispensable aux artisans dans les constructions géométriques.* Persian edition with French translation and commentary by Jaʿfar Āqāyānī-Chāvushī and introduction by Bernard Vitrac. Bibliothèque iranienne 71. Tehran: Institut français de recherche en Iran (IFRI).
———. 1990–91 (1369). *Handasa-i Īrānī: Kitāb-i Tijārat: Fī mā yaḥtāju ilayhi al-ʿummāl wa-al-ṣunnāʿ min al-ashkāl al-handasiyya, yā, Kārburd-i handasa dar ʿamal.* Edited and adapted (to modern Persian) by Sayyid ʿAlī Riżā Jazbī. Tehran: Surūsh.
———. 1979. *Mā yaḥtāju ilayhi al-ṣāniʿ min ʿilm al-handasa.* Edited by Ṣāliḥ Aḥmad al-ʿAlī. [Baghdād]: Jāmiʿat Baghdād, Markaz Iḥyā' al-Turāth al-ʿIlmī al-ʿArabī.
Bayhaqī, ʿAlī ibn Zayd. 2005. *Tatimma ṣiwān al-ḥikma: Texts and Studies.* Collected and reprinted by Fuat Sezgin, in collaboration with Carl Ehrig-Eggert and Eckhard Neubauer. Frankfurt am Main: Institute for the History of Arabic-Islamic Science at the Johann Wolfgang Goethe University.
———. 1935. *Kitāb Tatimmat ṣiwān al-ḥikma.* Edited by Moḥammad Shafīʿ. Lahore: L. Ishwar Das, Registrar, University of the Panjab.
Bīrūnī, Muḥammad ibn Aḥmad. 1967. *The Determination of the Coordinates of Positions for the Correction of Distances between Cities: A Translation from the Arabic of* Kitāb taḥdīd nihāyāt al-amākin litaṣḥīh masāfāt al-masākin. Translated by Jamīl ʿAlī. Beirut: American University of Beirut.
———. 1964. *Kitāb Taḥdīd nihāyāt al-amākin li-taṣḥīḥi masāfāt al-masākin.* Edited by P. G. Bulgakov and Imām Ibrāhīm Aḥmad. Cairo: Maṭbaʿat Lajnat al-Ta'līf wa-al-Tarjama wa-al-Nashr.
———. 1954–56. *Kitāb al-Qānūn al-Masʿūdī.* 3 vols. Hyderabad: Maṭbaʿat Majlis Dā'irat al-Maʿārif al-ʿUthmāniyya.
Caʿfer Efendi. 1987. *Risāle-i Miʿmāriyye: An Early-Seventeenth-Century Ottoman Treatise on Architecture.* Translated and edited by Howard Crane. Leiden: E. J. Brill.
Al-Fārābī, Abū Naṣr. 1968. *Iḥṣā' al-ʿulūm.* Edited by ʿUthmān Amīn. Cairo: Maktabat al-Anjalū al-Miṣriyya.
———. 1949. *Iḥṣā' al-ʿulūm.* Edited by ʿUthmān Amīn. Cairo: Dār al-Fikr al-ʿArabī.

Ibn Abī Uṣaybiʿa, Aḥmad ibn al-Qāsim. 1882–84. *ʿUyūn al-anbāʾ fī tabaqāt al-aṭibbāʾ*. Edited by August Müller. 2 vols. Königsberg i. Pr.: Selbstverlag.

Ibn al-Akfānī, Muḥammad ibn Ibrāhīm. 2005. *Irshād al-qāṣid ilā asnāʿ al-maqāṣid: Texts and Studies*. Edited by Fuat Sezgin, Carl Ehrig-Eggert, and Eckhard Neubauer. Historiography and Classification of Science in Islam 9, 187–275. Frankfurt am Main: Institute for the History of Arabic-Islamic Science at the Johann Wolfgang Goethe University.

———. 1978. *Irshād al-qāṣid ilā asnā al-maqāṣid*. Edited by ʿAbd al-Laṭīf Muḥammad al-ʿAbd. Cairo: Maktabat al-Anjlū al-Miṣriyya.

Ibn al-Haytham, Abū ʿAlī al-Ḥasan ibn al-Ḥasan. 2002. *Kitāb al-Manāzir, Books IV–V: On Reflection, and Images Seen by Reflection*. Edited by Abdelhamid I. Sabra. (Translation forthcoming.) 2 vols. Kuwait: The National Council for Culture, Arts and Letters.

———. 1989. *The Optics of Ibn al-Haytham, Books I–III: On Direct Vision = Kitāb al-Manāzir*. Translation and commentary by Abdelhamid I. Sabra. 2 vols. London: Warburg Institute, University of London.

———. 1983. *The Optics of Ibn al-Haytham, Books I–III: On Direct Vision = Kitāb al-Manāzir*. Edited by Abdelhamid I. Sabra. Kuwait: The National Council for Culture, Arts and Letters.

Ibn al-Jawzī, Abū al-Faraj ʿAbd al-Raḥmān ibn ʿAlī. 1939–1940 (1357–58). Vol. 7 of *Al-Muntaẓam fī tārīkh al-mulūk wa-al-umam*, 10 vols. Hyderabad: Maṭbaʿat Dāʾirat al-Maʿārif al-ʿUthmāniyya.

Ibn al-Nadīm, Muḥammad ibn Isḥāq. 1970. *The* Fihrist *of al-Nadīm: A Tenth-Century Survey of Muslim Culture*. Edited and translated by Bayard Dodge. 2 vols. New York: Columbia University Press.

———. 1871. *Kitāb al-Fihrist*. Edited by Gustav Flügel, Leipzig: Verlag von F. C. W. Vogel.

Ibn Khallikān. 1961. *Ibn Khallikān's Biographical Dictionary*. Translated by William MacGuckin de Slane. 4 vols. Paris: Printed for the Oriental Translation Fund of Great Britain and Ireland, 1842–71. Reprint, New York: Johnson Reprint Corp.

———. 1936. *Wafayāt al-aʿyān wa-anbāʾ al-zamān*. Cairo: Maṭbūʿāt Dār al-Maʾmūn.

Ibn Qifṭī, ʿAlī ibn Yūsuf. 1903. *Tārīkh al-ḥukamāʾ*. Edited by Julius Lippert. Leipzig: Dieterich'sche Verlagsbuchhandlung.

Ibn Ṭāhir al-Baghdādī, Abū Manṣūr ʿAbd al-Qāhir. 2009 (1388). *Al-Īḍāḥ ʿan uṣūl ṣināʿat al-massāḥ va Risāla dar ʿilm-i misāḥat*. Edited by ʿAlī Awjabī. Tehran: Kitābkhāna, Mūzih va Markaz-i Asnād, Majlis-i Shūrā-yi Islāmī.

———. 1985. *Al-Takmila fīʾl-ḥisāb; Maʿa risāla la-hu fī al-misāḥa*. Edited and annotated by A. S. Saidan. Kuwait: Maʿhad al-Makhṭūṭāt al-ʿArabiyya.

———. 1969 (1347). *Kitāb al-Īḍāḥ, ʿan uṣūl-i ṣināʿat al-massāḥ*. Facsimile edition by Aḥmad Gulchīn Maʿānī. Tehran: Bunyād-i Farhang-i Īrān.

Jazarī, Ismāʿīl ibn al-Razzāz. 1979. *Al-Jāmiʿ bayna al-ʿilm wa-al-ʿamal al-nāfiʿ fī ṣināʿat al-ḥiyal*. Edited by Aḥmad Yūsuf al-Ḥasan. Aleppo: Maʿhad al-Turāth al-ʿIlmī al-ʿArabī, Jāmiʿat Ḥalab.

———. 1974. *The Book of Knowledge of Ingenious Mechanical Devices = Fī maʿrifat al-ḥiyal al-handasiyya*. Translated by Donald R. Hill. Dordrecht: Reidel.

Kâtip Çelebi [Ḥājjī Khalīfa]. 1850. *Kashf al-ẓunūn*. Vol. 5 of *Kashf al-ẓunūn ʿan asāmī al-kutub wa-al-funūn*, edited by Gustav Flügel, 7 vols. London: R. Bentley for the Oriental Translation Fund of Great Britain and Ireland.

Qusṭā ibn Lūqā. 1990. *Fī Aṣnāf al-ṣināʿāt, Min Kalām Qusṭā ibn Lūqā*. Edited and translated (into German) by Hans Daiber. In "Qosṭā ibn Lūqā (9. Jh.) über die Einteilung der Wissenschaften." *Zeitschrift für Geschichte der Arabisch-Islamischen Wissenschaften* 6: 93–129.

Rāzī, Fakhr al-Dīn. 2003 (1382). *Jāmiʿ al-ʿulūm: "Sittīnī."* Edited by ʿAlī Āl Dāvūd. Tehran: Bunyād-i Mawqūfāt-i Duktur Maḥmūd Afshār Yazdī.

———. 1967–68 (1346). *Jāmi al-ʿulūm, yā, Ḥadāyiq al-anwār fī ḥaqāyiq al-asrār: Maʿrūf bih Kitāb-i Sittīnī*. Facsmile reproduction edited by Muḥammad Ḥusayn Tasbīḥī. Tehran: Kitābkhāna-i Asadī.

Ṣafadī, Khalīl ibn Aybak. 1962. *Kitāb al-Wāfī bi-al-wafayāt*. Vol. 1. Edited by Helmut Ritter. Leipzig: Deutsche Morgenländische Gesellschaft, in Kommission bei F. A. Brockhaus, 1931. Reprint, Wiesbaden: in Kommission bei Franz Steiner Verlag.

Secondary Sources

Afshār, Īraj. 1969 (1348). "Yād-dāsht." *Rāh-namā-yi Kitāb* 12: 325–26.

Āstān-i Quds-i Rażavī Library. 1926 (1345). *Fihrist-i kutub-i Kitābkhāna-i Āstān-i Quds-i Rażavī: Az jild-i sivvum*. Mashhad: Dār al-Ṭibāʿa "Ṭūs."

Bacharach, Jere L. 1974. *A Near East Studies Handbook, 570–1974*. Seattle: University of Washington Press.

Bāqirī, Muḥammad. 1997. "A Newly Found Letter of Al-Kāshī on Scientific Life in Samarqand." *Historia Mathematica* 24: 241–56.

Berggren, J. L. 2007. "Mathematics in Medieval Islam." In *The Mathematics of Egypt, Mesopotamia, China, India, and Islam: A Sourcebook*, edited by Victor J. Katz et al., 515–675. Princeton, N.J.: Princeton University Press.

———. 1986. *Episodes in the Mathematics of Medieval Islam*. New York: Springer-Verlag.

Bier, Carol. 2012. "The Decagonal Tomb Tower at Maragha and Its Architectural Context: Lines of Mathematical Thought." *Nexus Network Journal* 14, no. 2: 251–73.

———. 2008. "Art and Mithāl: Reading Geometry as Visual Commentary." In "Sciences, Crafts, and the Production of Knowledge: Iran and Eastern Islamic Lands (ca. 184–1153 AH/800–1740 CE)," edited by Elaheh Kheirandish, Carol Bier, and Najm al-Din Yousefi, special issue, *Iranian Studies* 41, no. 4: 491–509.

Blochet, Edgar. 1912. *Catalogue des manuscrits persans de la Bibliothèque nationale*. Vol. 2. Paris: Imprimerie nationale.

Bodleian Quarterly Record. 1921. Vol. 31: 152–53. Oxford: Bodleian Library.

Bosworth, C. E. 1963. "A Pioneer Arabic Encyclopedia of the Sciences: al Khwārizmī's Key of the Sciences." *ISIS* 54, no. 1: 97–111.

Brentjes, Sonja. 2006. "An Exciting New Arabic Version of Euclid's Elements: MS Mumbai, Mulla Firuz R.I.6." *Revue d'histoire des mathématiques* 12:169–197.

———. 1994. "Textzeugen und Hypothesen zum arabischen Euklid in der Überlieferung von al- Ḥaǧǧāǧ b. Yūsuf b. Maṭar (zwischen 786 und 833)." *Archive for History of Exact Sciences* 47, no. 1: 53–92.

Bulatov, Mitkhat S. 1988. *Geometricheskaia garmonizatsiia v arkhitekture Srednei Azii IX–XV vv.* Rev. ed. Moscow: "Nauka." Original edition Moscow, 1978.

Chorbachi, Wasma'a K. 1989a. "Beyond the Symmetries of Islamic Geometric Patterns: The Science of Practical Geometry and the Process of Islamic Design." PhD diss., Harvard University.

———. 1989b. "In the Tower of Babel: Beyond Symmetry in Islamic Design." *Computers and Mathematics, with Applications* 17:751–89.

Dānish-Pazhūh, M. T. 1961 (1340), 1962 (1340), 1966 (1345). *Fihrist-i Kitābkhāna-i Markazī-i Dānishgāh-i Tihrān*. Vol. 9 (1961 [1340]), manuscript nos. 2121–2629; vol. 10 (1962 [1340]), manuscript nos. 2630–3062; vol. 15 (1966 [1345]), manuscript nos. 4945–5400. Tehran: Dānishgāh-i Tihrān.

De Young, Gregg. 2010. Review of *Ketâb al-Nejârat: Sur ce qui est indispensable aux artisans dans les constructions géométriques*, edited by Ja'far Āqāyānī Chāvushī. *Suhayl: International Journal for the History of the Exact and Natural Sciences in Islamic Civilisation* 10:171–77.

Dhanani, Alnoor. 2007. "Jurjānī: 'Alī ibn Muḥammad ibn 'Ali al-Ḥusaynī al-Jurjānī (al-Sayyid al–Sharīf)." In *The Biographical Encyclopedia of Astronomers*, edited by Thomas Hockey et al., 603–4. New York: Springer.

Donohue, John J. 2003. *The Buwayhid Dynasty in Iraq 334 H./945 to 403 H./1012: Shaping Institutions for the Future*. Islamic History and Civilization 44. Leiden: Brill.

Farnūd, Riżā. 1997. *Farnūd*. Special issue, Teachers Society of Isfahan. Isfahan.

Fikrat, Muḥammad Āṣif, ed. 1990 (1369). *Fihrist-i alifbā'ī-i kutub-i khaṭṭī-i Kitābkhāna-i Markazī-i Āstān-i Quds-i Rażavī*. Supplemented by Muḥammad Vafādār Murādī. Mashhad: Kitābkhāna.

Gulchīn Ma'ānī, Aḥmad. 1971 (1350). *Fihrist-i kutub-i khaṭṭī-i Kitābkhāna-i Āstān-i Quds-i Rażavī*. Vol. 8. Mashhad: Kitābkhāna.

Heath, Thomas L., ed. and trans. 1956. *The Thirteen Books of Euclid's Elements*. Translated from the text of Heiberg. 2nd ed. Cambridge: Cambridge University Press, 1956.

Heiberg, J. L., and H. Menge, eds. 1895. *Euclidis Opera Omnia*. Vol. 7, *Optica; Opticorum recensio Theonis; Catoptrica, cum scholiis antiquis*. Leipzig: Teubner.

Hogendijk, Jan P. 2008. "Ancient and Modern Secrets of Isfahan." *Nieuw Archief voor Wiskunde*, 5th ser., 9, no. 2: 121.

———. 1993. "The Arabic Version of Euclid's 'On Divisions.'" In *Vestigia Mathematica: Studies in Medieval and Early Modern Mathematics in Honour of H. L. L. Busard*, edited by Menso Folkerts and Jan P. Hogendijk, 143–62. Amsterdam: Rodopi.

Jackson, David E. P. 1987–88. "Scholarship in Abbasid Baghdad with Special Reference to Greek Mechanics in Arabic." *Quaderni di Studi Arabi* 5–6: 369–90.

Karāmatī, Yūnus, and Farzīn Nigahbān. 2009. "Abū Isḥāq al-Kūbunānī." In *Encyclopaedia Islamica*, edited by Wilferd Madelung and Farhad Daftary. London: Brill, 2009, in association with the Institute of Islamic Studies.

Karāmatī, Yūnus, and Hanīf Qalandarī. 2011. "Zindagīnāma va kārnāma-i 'ilmī-i Kamāl al-Dīn ibn Yūnus al-Mawṣilī." In *Pizhūhishhā'ī dar tārīkh-i 'ilm*, edited by Ja'far Āqāyānī Chāvushī, 93–122. Tehran: Markaz-i Pizhūhishī-i Mīrāṯ-i Maktūb.

Kennedy, E. S. 1973. *A Commentary upon Bīrūnī's* Kitāb Taḥdīd al-amākin; *An 11th Century Treatise on Mathematical Geography*. Beirut: American University of Beirut.

———. 1960. "A Letter of Jamshīd al-Kāshī to His Father: Scientific Research and Personalities at a Fifteenth Century Court." *Orientalia* 29: 191–213.

Kennedy, E. S., and Mustafa Mawaldi. 1979. "Abū al-Wafā' and the Heron Theorems." *Journal for the History of Arabic Science* 3:19–30.

Kennedy, E. S., and Jamil Ragep. 1981. "A Description of Ẓāhiriyya (Damascus) MS. 4871: A Philosophical and Scientific Collection." *Journal for the History of Arabic Science* 5:85–108.

Kheirandish, Elaheh. 2015. "Light and Dark: The 'Checkered History' of Early Optics." In *God Is the Light of the Heavens and the Earth: Light in Islamic Art and Culture*, edited by Jonathan Bloom and Sheila Blair, 61–85. New Haven: Yale University Press.

———. 2013. "The Mixed Mathematical Sciences: Optics and Mechanics in the Islamic Middle Ages." In *The Cambridge History of Science*, vol. 2, *Medieval Science*, edited by David C. Lindberg and Michael H. Shank, 84–108. Cambridge: Cambridge University Press.

———. 2009. "Footprints of 'Experiment' in Early Arabic Optics." In *Evidence and Interpretation in Studies on Early Science and Medicine: Essays in Honor of John E. Murdoch*, edited by Edith Dudley Sylla and William R. Newman, special issue, *Early Science and Medicine* 14:79–104.

———. 2008. "Science and Mithāl: Demonstrations in Arabic and Persian Scientific Traditions." In "Sciences, Crafts, and the Production of Knowledge: Iran and Eastern Islamic Lands (ca. 184–1153 AH/800–1740 CE)," edited by Elaheh Kheirandish, Carol Bier, and Najm al-Din Yousefi, special issue, *Iranian Studies* 41, no. 4: 465–489.

———. 2007. "Qusṭā Ibn Lūqā al-Ba'labakkī." In *The Biographical Encyclopedia of Astronomers*, edited by Thomas Hockey et al. New York: Springer.

———. 2006a. "'The 'Fluctuating Fortunes of Scholarship': A Very Late Review Occasioned by a Fallen Book." *Early Science and Medicine* 11, no. 2: 207–222.

———. 2006b. "Organizing Scientific Knowledge: The 'Mixed' Sciences in Early Classifications." In *Organizing Knowledge: Encyclopaedic Activities in the Pre-Eighteenth Century Islamic World*, edited by Gerhard Endress, 135–54. Leiden: Brill.

———. 2004. "Mathematical Sciences through Persian Sources: The Puzzle of Ṭūsī's Optical Works." In *Sciences, techniques et instruments dans le monde iranien, Xe–XIXe siècle*, edited by

Naṣr Allāh Pūrjavādi and Živa Vesel, 197–213. Tehran: Institut français de recherche en Iran.

———. 2003. "The Many Aspects of 'Appearances': Arabic Optics to 950 AD." In *The Enterprise of Science in Islam: New Perspectives*, edited by Jan P. Hogendijk and Abdelhamid Sabra, 55–83. Cambridge: MIT Press.

———, ed. and trans. 1999. *The Arabic Version of Euclid's Optics: Kitāb Uqlīdis fī Ikhtilāf al-manāẓir*. 2 vols. Sources in the History of Mathematics and Physical Sciences 16. New York: Springer.

Kheirandish, Elaheh, Carol Bier, and Najm al-Din Yousefi, eds. 2008. "Sciences, Crafts, and the Production of Knowledge: Iran and Eastern Islamic Lands (ca. 184–1153 AH/800–1740 CE)." Special issue, *Iranian Studies* 41, no. 4: 433–600.

King, David A. 1999. *World-Maps for Finding the Direction and Distance to Mecca: Innovation and Tradition in Islamic Science*. Leiden, Brill.

———. 1986. *A Survey of the Scientific Manuscripts in the Egyptian National Library*. Winona Lake, Ind.: Eisenbrauns.

———. 1981–86. *Fihris al-makhṭūṭāt al-ʿilmiyya al-maḥfūẓa bi-Dār al-Kutub al-Miṣriyya*. 2 vols. Cairo: al-Hay'a al-Miṣriyya al-ʿĀmma lil-Kitāb: Bi-al-taʿāwun maʿa Markaz al-Buḥūth al-Amirīkī bi-Miṣr wa-Mu'assasat Samīthsūnyān.

Krasnova, S. A. 1966. "Abu-l-Vafa al-Buzdzhani, Kniga o tom, chto neobkhodimo remeslenniku iz geometricheskikh postroenii." *Fiziko-matematicheskie nauki v stranakh Vostoka* 1:42–140.

Krause, Max. 1936. "Stambuler Handschriften islamischer Mathematiker." *Quellen und Studien zur Geschichte der Mathematik, Astronomie und Physik* 3 (Abt. B): 437–532.

Kubesov, A., and Boris A. Rozenfel'd. 1969. "On the Geometrical Treatise of al- Fārābī." *Archives Internationales d'Histoire des Sciences* 22:50.

Lane-Poole, Stanley. 1967. *Catalogue of Oriental Coins in the British Museum*. Vol. 2, *The Coins of the Muhammedan Dynasties*. Edited by Reginald Stuart Poole, 10 vols. London: Printed by order of the Trustees [British Museum], 1875–90. Reprint, Bologna: Forni.

Löfgren, Oscar, and Renato Traini. 1975. *Catalogue of the Arabic Manuscripts in the Biblioteca Ambrosiana*. Vol. 1, *Antico Fondo and Medio Fondo*. Vicenza: Neri Pozza Editore.

Lu, Peter J., and Paul J. Steinhardt. 2007. "Decagonal and Quasi-Crystalline Tilings in Medieval Islamic Architecture." *Science* 315:1106–10.

Madelung, Wilferd. 1992. *Religious and Ethnic Movements in Medieval Islam*. Hampshire, Great Britain; Brookfield, Vt.: Variorum.

———. 1969. "The Assumption of the Title Shāhānshāh by the Būyids and 'The Reign of the Daylam (Dawlat al-Daylam)'." *Journal of Near Eastern Studies* 28: 84–108 and 168–83.

Meyerhof, Max. 1948. "'Alī al-Bayhaqī's *Tatimmat Ṣiwān al-Ḥikma*: A Biographical Work on Learned Men of Islam." *Osiris* 8:122–216.

Mottahedeh, Roy Parviz. 2012. "The Idea of Iran in the Buyid Dominions." In *Early Islamic Iran*, edited by Edmund Herzig and Sara Stewart, 153–60. London: I. B. Tauris.

Muḥīṭ Ṭabāṭabā'ī, M. 1982 (1361). "Naẓarī kūtāh bi du matn-i chāp shuda ʿArabī va Fārsī." In *Ārāmnāma: Majmūʿa-i maqālāt-i ʿilmī va adabī taqdīm shuda bi ustād Aḥmad Ārām*, edited by Mahdī Muḥaqqiq, 31–50. Tehran: Intishārāt-i Anjuman-i Ustādān-i Zabān va Adabiyyāt-i Fārsī.

Munzavī, Aḥmad. 1969 (1348). *Fihrist-i nuskhahā-yi khaṭṭī-i Fārsī*. Vol. 1, *Mathematics*. Tehran: Muʿassasa-i Farhangī-i Minṭaqaʿī.

Necipoğlu, Gülru. 1995. *The Topkapı Scroll: Geometry and Ornament in Islamic Architecture; Topkapı Palace Museum Library MS H. 1956*. Santa Monica, Calif.: Getty Center for the History of Art and the Humanities.

Özdural, Alpay. 2000. "Mathematics and Arts: Connections between Theory and Practice in the Medieval Islamic World." *Historia Mathematica* 27, no. 2: 171–201.

———. 1998. "A Mathematical Sonata for Architecture: Omar Khayyam and the Friday Mosque of Isfahan." *Technology and Culture* 39, no. 4: 699–715.

———. 1996. "On Interlocking Similar or Corresponding Figures and Ornamental Patterns of Cubic Equations." *Muqarnas* 13:191–211.

———. 1995. "Omar Khayyam, Mathematicians and *Conversazioni* with Artisans." *Journal of the Society of Architectural Historians* 54, no. 1: 54–71.

Pingree, David. 1983a. "Abu'l-Fatḥ Eṣfahānī." *Encyclopedia Iranica* (1983): 284–85.

———. 1983b. "Abu'l-Wafā Būzjānī." *Encyclopedia Iranica* (1983): 392–94.

Qanbarīnanīz, Vahīd, and Khorshīd Qanbarīnanīz. 2009 (1388). "Abū Isḥāq Kūhbanānī Kirmānī, riyāżīdān va munajjim-i buzurg-i qarn-i nuhum-i Hijrī." *ʿUlūm va funūn, kitāb-i māh* no. 122: 68–71.

Qurbānī, Abū al-Qāsim. 1986–87 (1365). *Zindagīnāma-i rīyāżīdānān-i dawra-i Islāmī az sada-i sivvum tā sada-i yāzdahum-i hijrī*. Tehran: Markaz-i Nashr-i Dānishgāhī.

Qurbānī, Abū al-Qāsim, and Muḥammad ʿAlī Shaykhān. 1992 (1371). *Būzjānīnāma: Sharḥ-i aḥvāl va āṯār-i riyāż-i Abū al-Vafā' Būzjānī*. Tehran: Intishārāt va Āmūzish-i Inqilāb-i Islāmī. Includes facsimile reproduction of the manuscript in Istanbul.

Ragep, F. Jamil. 2014. "New Light on Shams: The Islamic Side of Σάμψ Πουχάρης." In *Politics, Patronage, and the Transmission of Knowledge in 13th–15th Century Tabriz*, edited by Judith Pfeiffer, 231–47. Leiden: Brill, 2014.

———. 2000. "The Persian Context of the Ṭūsī Couple." In *Dānishmand-i Ṭūs: Majmūʿa-i maqālāt-i girdhamāyī-i ʿilm va falsafa dar āṯār-i Khwāja Naṣīr al-Dīn Ṭūsī = Naṣīr al-Dīn al-Ṭūsī: Philosophe et savant du XIIIe siécle*, edited by Naṣr Allāh Pūrjavādi and Živa Vesel, 113–30. Tehran: Institut français de recherche en Iran.

Rāshid, Rushdī. 2013. *Ibn al-Haytham's Theory of Conics, Geometrical Constructions and Practical Geometry*. Translated by J. V. Field. Vol. 3 of *A History of Arabic Sciences and Mathematics*, edited by Nader El-Bizri, 4 vols. London: Routledge.

———. 2005. *Geometry and Dioptrics in Classical Islam*. London: Al-Furqān Islamic Heritage Foundation.

———. 2000 (1421). *Ibn al-Haytham, théorie des coniques, constructions géométriques et géométrie pratique*. Vol. 3 of *Les mathématiques infinitésimales du IXe au XIe siècle*, 5 vols. London: Al-Furqān Islamic Heritage Foundation.

Raynaud, Dominique. 2012. "Abū al-Wafā' Latinus? A Study of Method." *Historia Mathematica* 39:34–83.

Richard, Francis. 1989. *Catalogue des manuscrits persans*. Vol. 1, *Ancien fonds*. Paris: Bibliothèque nationale de France.

Rieu, Charles. 1894. *Supplement of the Catalogue of the Arabic Manuscripts in the British Museum*. London: Longmans & Co.

Rozenfel'd, Boris. A., and Ekmeleddin İhsanoğlu. 2003. *Mathematicians, Astronomers, and Other Scholars of Islamic Civilization and Their Works (7th–19th c.)*. Istanbul: Research Center for Islamic History, Art and Culture (IRCICA).

Rūh al-Amīnī, Maḥmūd. 1990 (1369). "Abū Isḥāq Kūbanānī, riāżīdān va adīb-i qarn-i nuhum-i Hijrī." *Nashriyya-i Dānishkada-i Adabiyāt va 'Ulūm-i Insānī. Dānishgāh-i Shahīd Bāhunar-i Kirmān* 1:42–49.

Sabra, Abdelhamid I. 1971. "The Astronomical Origin of Ibn al-Haytham's Concept of Experiment." In *Actes XIIe Congrès International d'Histoire des Sciences*, vol. 3A, 133–36. Paris: Albert Blanchard.

Saidan, A. S., ed. 1971. *Arabic Arithmetic: The Arithmetic of Abū al-Wafā' al-Būzjānī, 10th Century Mss Or. 103 Leiden and 42 M Cairo*. In *Tārīkh 'ilm al-ḥisāb al-'Arabī*. Amman: Jam'iyyat A'māl al-Maṭābi' al-Ta'āwuniyya.

Saliba, George. 1999. "Artisans and Mathematicians in Medieval Islam." Review of *The Topkapı Scroll: Geometry and Ornament in Islamic Architecture* by Gülru Necipoğlu. *Journal of the American Oriental Society* 119, no. 4: 637–45.

———. 1972–73. "The Meaning of Al-Jabr wa al-Muqābalah." *Centaurus* 17:189–204.

Sayılı, Aydın. 1960. *The Observatory in Islam and Its Place in the General History of the Observatory*. Ankara: Türk Tarih Kurumu Basımevi.

Schiefsky, Mark. 2008. "Theory and Practice in Heron's *Mechanics*." In *Mechanics and Natural Philosophy before the Scientific Revolution*, edited by Sophie Roux and Walter Roy Laird, 15–49. New York: Springer.

Seherr-Thoss, Sonia P. 1968. *Design and Color in Islamic Architecture: Afghanistan, Iran, Turkey*. Washington D.C.: Smithsonian Institution Press, 1968.

Sezgin, Fuat. 1979. *Geschichte des arabischen Schrifttums. Band 7: Astrologie, Meteorologie, und Verwandtes bis ca. 430 H.* Leiden: E.J. Brill.

———. 1978. *Geschichte des arabischen Schrifttums. Band 6: Astronomie bis ca. 430 H.* Leiden: E.J. Brill.

———. 1974. *Geschichte des arabischen Schrifttums. Band 5: Mathematik bis ca. 430 H.* Leiden: E.J. Brill.

Storey, Charles A. 1958. *Persian Literature: A Bio-Bibliographical Survey*. Vol. 2, pt. 1, A. Mathematics, B. Weights and Measures, C. Astronomy and Astrology, D. Geography. London: Luzac.

Suter, Heinrich. 1922. "Das Buch der geometrischen Konstruktionen des Abu'l Wefā'." *Abhandlungen zur Geschichte der Naturwissenschaften und Medizin* 4:94–109.

Tornberg, C. J. 2009. *Catalogue of the Arabic, Persian, and Turkish Manuscripts of the Uppsala University Library = Codices arabici, persici et turcici Bibliothecae Regiae Universitatis Upsaliensis*. Osnabrück: Biblio, 1988. Reprint, Piscataway, N.J.: Gorgias Press.

Treadwell, Luke. 2001. *Buyid Coinage: A Die Corpus (322–445 A.H.)*. Oxford: Ashmolean Museum.

Unguru, Sabatai. 1975. "On the Need to Rewrite the History of Greek Mathematics." *Archive for the History of Exact Sciences* 15: 67–114.

Varjāvand, Parvīz. 1987 (1366). *Kāvish-i Raṣad-khāna-i Marāgha*. Tehran: Amīr Kabīr.

Woepcke, Franz. 1855. "Recherches sur l'histoire des sciences mathématiques chez les Orientaux d'après des traités inédits arabes et persans. Deuxième article. Analyse et extrait d'un recueil de constructions géométriques par Aboūl Wefā." *Journal Asiatique*, ser. 5, 5: 218–56, 309–59.

BRILL

JAN P. HOGENDIJK

A MATHEMATICAL CLASSIFICATION OF THE CONTENTS OF AN ANONYMOUS PERSIAN COMPENDIUM ON DECORATIVE PATTERNS

The purpose of this chapter is to provide a mathematical classification of the contents of the *Anonymous Compendium* comprising folios 180r–199r of Ms. Persan 169, located in the Bibliothèque nationale de France and referred to in this volume as the Paris Codex.[1] Ms. Persan 169 was written in a single hand, probably in the fifteenth to sixteenth centuries, and contains other mathematical texts as well.[2] Because the sixty-one patterns in the *Anonymous Compendium* occur in rather random order, it is likely that they were taken from different sources, probably dating back to different centuries.

The *Anonymous Compendium* is historically important because it differs in style and content from almost all other geometrical works that have come down to us from the Islamic tradition, including some of the other treatises in the Paris Codex. Most medieval Islamic authors who wrote on geometry were trained in the style of Euclid's *Elements* (ca. 300 B.C.), which was available in several good Arabic translations. These authors studied geometry in order to become astronomers and astrologers,[3] and in the majority of their writings no reference is made to decorative geometrical patterns—which may be surprising, as they must have often seen such patterns in palaces and mosques.

Most of the patterns appear to have been designed not by theoretically trained geometers but by craftsmen (Arabic: *ṣunnaᶜ*). The tenth-century geometer and astronomer Abu'l-Wafa' al-Buzjani wrote a short *Kitāb fī aᶜmāl al-handasa* (*Book of Geometrical Constructions*, henceforth *Geometrical Constructions*), which is the subject of the chapter by Elaheh Kheirandish in the present publication.[4] In the *Geometrical Constructions*, Abu'l-Wafa' comments on the practices of craftsmen and also reports on meetings at which craftsmen and theoretically trained geometers were present.[5] Abu'l-Wafa' says that the craftsmen did not distinguish between geometrically exact and approximate constructions, and were not concerned with geometric proofs. He also complains that they only wanted their decorative patterns to look right. The rather snobbish aim of his *Geometrical Constructions* was to make sure that the craftsmen henceforth used only mathematically exact constructions.

The *Anonymous Compendium* conforms to a large extent with Abu'l-Wafa''s description of the practices of medieval Islamic craftsmen. Mathematically exact and approximate constructions occur next to one another, and are presented without proofs. The approximate character of some constructions is sometimes indicated but not always. Moreover, the *Anonymous Compendium* also uses an instrument called the *gūnyā* (set square), which is mentioned in Abu'l-Wafa''s book, but not in Euclid's *Elements*. The *Anonymous Compendium* thus preserves textual evidence of how mathematical principles were applied in practice by craftsmen who designed decorative patterns—evidence, that is, of a mathematical tradition few written traces of which have otherwise come down to us.

The author(s) of the material in the *Anonymous Compendium* were not completely unfamiliar with the way in which theoretically trained geometers were doing mathematics. In many of the diagrams, points are labeled using the letters of the Arabic alphabet in increasing numerical value (*alif*=1, *bā'*=2, *jīm*=3, *dāl*=4, etc., hence the name *abjad* for this system of numbering). This labeling system was standard in Arabic translations of Euclid's *Elements* and in works by theoretically trained medieval Islamic geometers. Indeed, the letters of the entire Arabic alphabet are presented in the *abjad* order on folio 196v. These numbers may have been written as a personal note on a blank folio, but, alternatively, their presence may indicate that the audience for which the

 | DOI 10.1163/9789004315204_005

Anonymous Compendium was written was not wholly familiar with the *abjad* system. There are, however, slight differences between the *Anonymous Compendium* and the theoretically trained geometers' standard method for labeling points in cases where the number of points exceeded the number of letters in the alphabet (see fig. 11 and the commentary below).

Other differences can also be mentioned. The text in the *Anonymous Compendium* is terse, and sometimes unintelligible or defective; some of the diagrams are ambiguous; and, in one case, a single figure is used for two mutually inconsistent diagrams (fig. 11 below). By contrast, medieval Islamic geometrical works in the tradition of Euclid's *Elements* usually contain clear explanations.

Moreover, the author(s) of the geometrical constructions in the *Anonymous Compendium* often did not worry about issues that were important to theoretically trained geometers of the Islamic tradition in the tenth century and later. Many of the latter thought that all constructions should be made only by means that they considered legitimate, i.e., with an (unmarked) ruler, compass, and possibly conic sections. This restricted the range of possible constructions, as some—such as the trisection of an arbitrary arc, or the division of a circle in seven or nine equal arcs—could be made only by use of conic sections, and no legitimate solutions existed for problems such as the division of a circle into eleven equal arcs. In the *Anonymous Compendium*, by contrast, it is assumed, without any further explanation, that any arc can be trisected and that a circle can be divided into an arbitrary number of equal arcs. Works by theoretically trained medieval Islamic geometers rarely had approximate constructions, whereas the *Anonymous Compendium* contains many intricate ones. The quality of these approximations is often very good, and an approximate construction with an error of, say, one percent may produce a vastly better result than a theoretically "exact" construction that is difficult to execute with actual instruments. The *Anonymous Compendium* therefore preserves evidence of a type of practical geometrical knowledge that was perhaps not sufficiently appreciated by Abu'l-Wafa' and others trained in the theoretical geometry of the Greeks.

As for modern attempts to relate medieval Islamic patterns to mathematical concepts such as aperiodic tilings or group theory, it should be noted that the craftsmen who appear to have designed the decorative patterns were, with a only a few possible exceptions, not trained in the Euclidean deductive approach to mathematics, and so were not used to theoretical mathematical concepts and arguments. More generally, reading modern mathematical concepts into medieval texts and buildings is problematic.[6] Still, Islamic decorative patterns can be useful tools for teaching modern mathematics, even if no historical relationship between the two can be established.

The diagrams in the *Anonymous Compendium* do not represent a cross-section of the decorations on the remaining architectural monuments in Iran, and the decorative applications of some of them are uncertain at best. For example, a number of diagrams in the manuscript involve heptagons or heptagonal stars, which are found only infrequently in architectural remains in Iran, given the complexity of their design. Indeed, only one of the heptagonal diagrams in the *Anonymous Compendium* has hitherto been found on a building, namely in the North Dome of the Friday Mosque in Isfahan (see figs. 8–10 below). In one or two groups of diagrams to be discussed in Part VI, a connection between the work of the craftsmen and that of contemporaneous theoretically trained mathematicians can be demonstrated. One may well ask whether the *Anonymous Compendium*, although it contains material by craftsmen, piqued the interest of some theoretically trained geometers who were more open-minded than Abu'l-Wafa', and whether this is the reason it was copied and has come down to us. No answer can be given at this time because of the lack of written (and published) original documents about the mathematical traditions of craftsmen. Thus it is difficult to put some of the material in the *Anonymous Compendium* in its proper historical context. Libraries in Iran and other countries may well contain unexplored documents on the issue, but it may also be that parts of the tradition were transmitted orally and never put into writing.

In the following sections, the sixty-one patterns in the *Anonymous Compendium* will be classified into seven groups on mathematical grounds: (1) cut-and-paste geometrical constructions; (2) preliminary geometrical constructions and one theorem; (3) geometrical constructions using a single compass opening; (4) constructions of rectangular patterns; (5) divisions of a square; (6) difficult patterns involving kites (Persian: *turunj*, a quadrilateral with two equal opposite angles and two pairs of equal sides arranged in such a way that the sides containing each of the equal angles are unequal); and (7) *muqarnas*. Each section begins with a list of all the patterns in that group, and a brief description of each pattern with references; this is followed by a detailed discussion of one or two examples that were selected for the light they shed on some characteristics of the *Anonymous Compendium* and/or the mathematical practices of its author(s). I have refrained from giving a full mathematical commentary on all the patterns, not only because of lack of space, but also because the mathematical analysis of some of these patterns provides delightful research projects for students. My survey should be read in conjunction with Gülru Necipoğlu's survey in the present volume, which focuses on the cultural and art-historical aspects of the *Anonymous Compendium* in addition to analyzing its contents. It would also profit the reader to consult the drawings Alpay Özdural created based on the *Anonymous Compendium*, all of which have been reproduced here as well (see plates).

In this chapter (as throughout the present volume), the original geometrical constructions are cited by folio and number, as in "fol. 180v [1]" for Construction 1. Thus, in the section in this volume titled "Translation, Transcription, and Drawings," the reader can quickly locate the Persian text and English translation, as well as the commentaries summarized by Necipoğlu, which are based on Özdural's unpublished book manuscript. Using the further references provided in that section, the reader can find Özdural's extensive discussions on the constructions in his unpublished manuscript, which the editors have decided to make available to interested readers (online at http://dx.doi.org/10.1163/9789004XXXXXX). In the endnotes, I have added systematic references to the two previously published translations of the *Anonymous Compendium*, namely A.B. Vildanova's Russian translation, which is appended to Mitkhat S. Bulatov's *Geometricheskaia garmonizatsiia v arkhitekture Srednei Azii IX–XV vv.* (references are to the figures in the translation), and the modern Persian adaptation by Sayyid Ali Riza Jazbi, *Handasa-i Īrānī: Kitāb-i Tijārat: Fī mā yaḥtāju ilayhi al-ʿummāl wa-al-ṣunnāʾ min al-ashkāl al-handasiyya, yā, Kārburd-i handasa dar ʿamal.*[7]

I. CUT-AND-PASTE CONSTRUCTIONS

For each of these constructions the *Anonymous Compendium* presents a diagram that consists of two or three figures, divided into the same set of pieces. The purpose is to show how each of the two or three figures can be put together from the pieces.

The following constructions consist of figures only, without accompanying text:

Fol. 181r [3]:[8] an equilateral triangle and a hexagonal star, formed from six pieces.

Fol. 182v [10]:[9] a square and an octagonal star, formed from eight pieces.

Fol. 186bis(r) [22]:[10] a square and a regular octagon, formed from nine pieces.

Fol. 197r [58]:[11] a regular heptagon, an isosceles triangle, and a rectangle, formed from twenty pieces.

Fol. 197r [59]:[12] a regular hexagon, an equilateral triangle, and a rectangle, formed from ten pieces (this construction will be discussed below).

Fol. 197r [60]:[13] a regular octagon and an octagonal star, formed from twenty-four pieces.

The following constructions consist of a diagram together with instructions in the text:

Fol. 180r [1]:[14] two small decagons are cut into three pieces each, and these pieces plus one pentagonal star are put together to form one larger decagon.

Fol. 181r [4]:[15] the figure features a regular octagon cut into five pieces, which can be reassembled to form a square, as well as a geometrical construction. The text explains how to use this geometrical construction in order to derive the side of the square from the regular octagon.

Fol. 182r [8]:[16] the figure displays a square cut into eight pieces and a hexagonal star assembled from the same eight pieces, as well as part of a geometrical construction based on the regular hexagon. The text explains how the side of the square can be derived from the hexagonal star.

Fol. 182v [9]:[17] a geometrical construction in which an arbitrary rectangle is cut into four pieces that are reassembled to form a perfect square. The construction is based on the same idea as Euclid's *Elements* II:14, though it differs in the details.[18]

Fol. 183r [11]:[19] how a regular hexagon can be cut into twelve pieces that can be reassembled to form an equilateral triangle. Then, for an equilateral triangle, the text shows how the side of a regular hexagon with the same area can be found by an (approximate) geometrical construction.

Our example will be fol. 197r [59], a construction without any accompanying text (figs. 1–3).

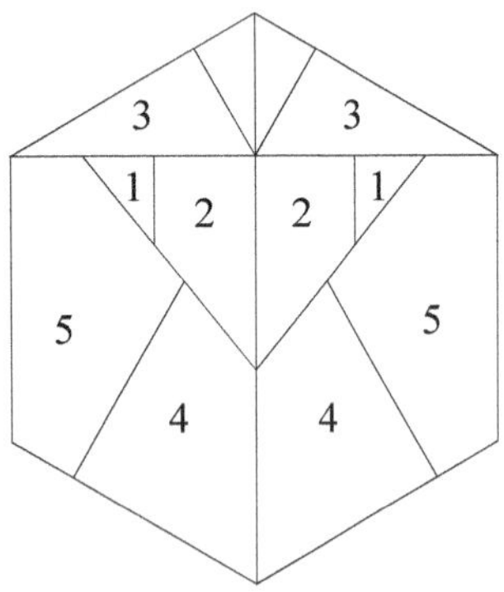

Fig. 1.

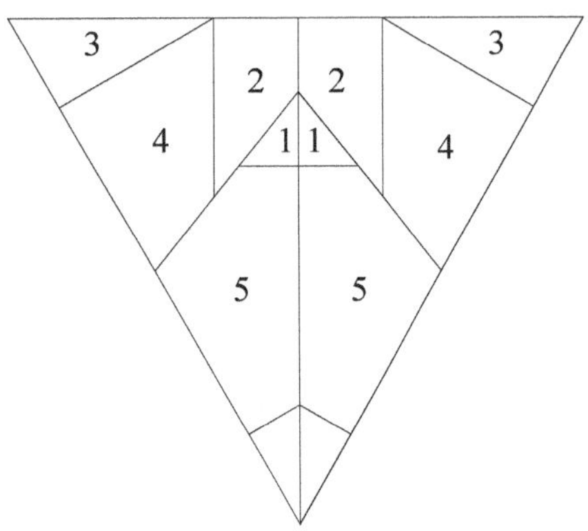

Fig. 2.

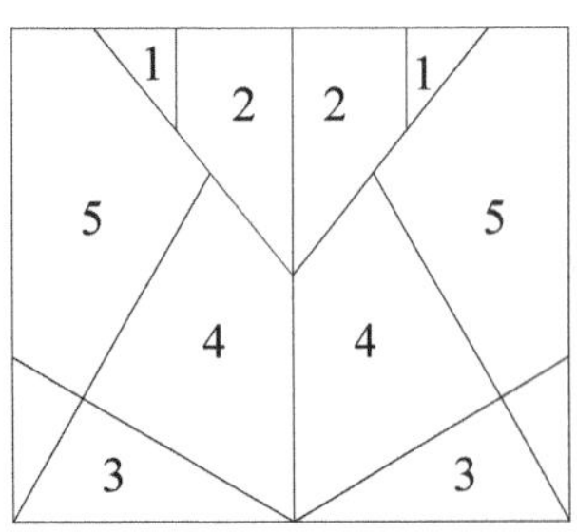

Fig. 3.

Figures 1–3 feature a regular hexagon, an isosceles triangle, and a rectangle, dissected into pieces from which all three figures can be composed. These figures have been derived from the manuscript with one difference: I have arbitrarily assumed the isosceles triangle to be equilateral, and have drawn the figures in a mathematically correct way. In the manuscript, the pieces are indicated by numbers (transcribed in the figures) so that the correspondence between the pieces is clear. There is no text in the manuscript to explain the exact way in which the pieces have to be cut; the reader is left to work out the details for himself/herself. After this exercise, she or he will probably be convinced that the *Anonymous Compendium* was meant to be studied under the guidance of a competent teacher who could supplement the visual information provided in the document.

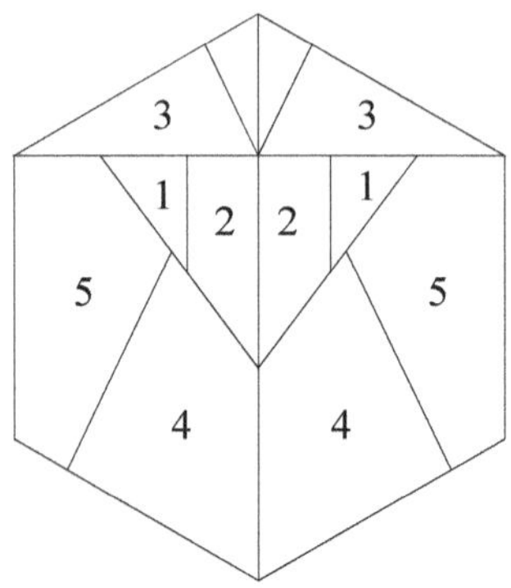

Fig. 4.

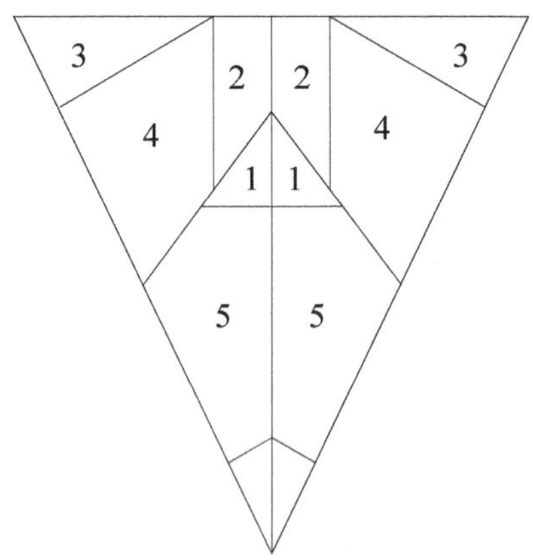

Fig. 5.

It should be noted that in the *Anonymous Compendium* the pieces I have reproduced in figures 1 and 2 are drawn in such a way that the piece numbered 1 is wider than the piece numbered 2. This does not happen in my reconstructed figure, but it may if the vertex angle of the isosceles triangle is less than 54°; figures 4 and 5 have been drawn for a vertex angle of (360/7)°. It is tempting to assume that the craftsmen had in mind the general dissection of an isosceles (rather than an equilateral) triangle, but because there is no accompanying text, one cannot be sure.

It is not necessary to assume that the fancy cut-and-paste constructions in this section were actually used in practice. Just like European arithmetic teachers in later centuries, medieval Islamic craftsmen may have challenged one another with gamelike problems that surpassed the requirements of their routine work.

II. PRELIMINARY GEOMETRICAL CONSTRUCTIONS AND ONE THEOREM

The *Anonymous Compendium* contains a few preliminary geometrical constructions and one theorem:

Fol. 181v [5]:[20] finding the center of a circle from an arc of the circle. The construction is unlike that in Euclid, *Elements* III:25.[21]

Fol. 181v [6]:[22] the completion of a full circle from the arc of a circle. The construction is not based on construction no. 5 and assumes that any arc of a circle can be trisected.

Fol. 182r [7]:[23] the division of a triangle into four pieces by means of three straight lines through the vertices, such that the small triangle is similar to the original triangle. The construction in the manuscript is only correct for an equilateral triangle.

Fol. 183v [12][24] contains the following theorem: for every triangle inscribed in a circle, the ratio of the angles is equal to the ratio of the arcs of the circumscribed circle subtended by the sides of the triangle. This theorem is stated without proof, and is illustrated by a triangle whose angles are in the proportion 1:2:4 (related to a regular heptagon). The theorem is a preliminary to the following construction, i.e., no. 13.

Fol. 184r [13]:[25] the construction of a triangle whose angles are in a given ratio (of whole numbers). The construction is illustrated by the ratios 1:2:6 (related to a regular nonagon) and 3:4:5. It is assumed that the circumference of the circle can be divided into any whole number of equal arcs. An accompanying table lists all seven triangles whose angles are integer multiples of one-ninth of a right angle (in modern terms, multiples of 20°). The text also explains how to construct a triangle with one given side whose angles are in a given ratio of whole numbers.

Fol. 186r [20]:[26] the construction of a regular pentagon with a given side (discussed below).

Fol. 192r [38]:[27] the construction of a circle through three given points (only a figure, without further explanation). The center is found as the point of intersection of the perpendicular bisectors of two of the line segments joining the three given points.

As our example we will discuss the construction of a regular pentagon with a given side in fol. 186r [20]: see figure 6, to which the words "side," "altitude," and "chord" were added in the Persian manuscript.

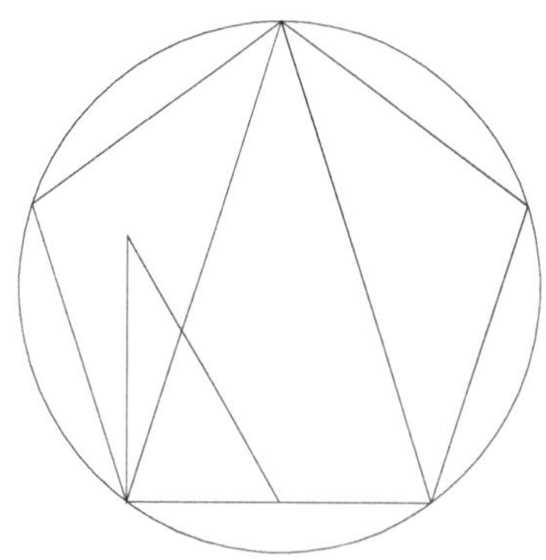

Fig. 6.

The construction is not indicated but must be inferred from figure 6 alone. In the following explanation, assume that the horizontal segment has length 1. First we construct the right-angled triangle in the figure, such that the sides containing the right angle are $\frac{1}{2}$ (horizontal side) and 1 (vertical side). Then the hypotenuse will have length $\frac{1}{2}\sqrt{5}$. Now construct the isosceles triangle in the figure with (horizontal) base of length 1 and sides of length $\frac{1}{2} + \frac{1}{2}\sqrt{5}$, that is, the sum of two sides of the right-angled triangle that was constructed before. We now will have found three vertices of the regular pentagon. The two others can be found by constructing two more isosceles triangles whose sides have length 1.

III. GEOMETRICAL CONSTRUCTIONS USING A SINGLE COMPASS OPENING

This category consists of constructions using a single compass opening. Such constructions eliminate errors that are caused by successive adjustments of the instrument. The following constructions are all approximate but the errors are small and often negligible in practice. Several scholars have proposed that these constructions were meant to serve as templates for triangles corresponding to different set squares.[28]

Fol. 184v [14]:[29] The purpose of this construction is, in modern terms, to find an angle that is 1/5 times two right angles (i.e., 36°), with a fixed compass opening (discussed below).

Fol. 186r [19]:[30] construction of an equilateral and equiangular pentagon with a [fixed compass] opening [equal] to the altitude.

Fol. 186v and 186bis(r) [21]:[31] another construction of a regular pentagon, with a compass opening equal to the altitude of the pentagon.

Fol. 186bis(v) [23]:[32] construction of an equilateral and equiangular pentagon with a [fixed compass] opening equal to the side.

Fol. 186bis(v) [24]:[33] a pentagonal star is inscribed in a semicircle. No instructions are given in the text, but the figure is titled, perhaps incorrectly, "on the construction of an equilateral and equiangular pentagon, with compass opening [equal to] the side."

Fol. 187r [25]:[34] the construction of a pentagon with a [fixed compass] opening equal to the chord. Near the figure is written "by Abu Bakr al-Khalil al-Tajir."

The construction fol. 184v [14] will serve as an example (fig. 7). Here points are labeled with letters[35] that have been transcribed in the figure. The *Anonymous Compendium* also explains the construction, which is as follows:

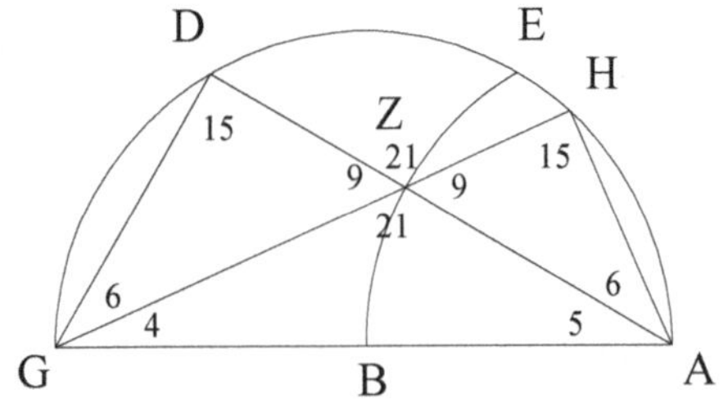

Fig. 7.

First draw a semicircle with center *B* and diameter *ABG*. The compass opening will always be equal to *r*, the radius of this circle. Draw an arc with center *A*, which will pass through *B* and meet the semicircle at point *E*. Find point *D* as the point of intersection of the semicircle and the circular arc with center *G* and radius *r*, and draw *GD*. Draw line *AD* and let *Z* be the point of intersection with arc *EB*. Draw line *GZ* to meet the circle at *H*. Draw line *AH*. Then ∠*DGZ* and ∠*HAZ* are the required angles. The text in the *Anonymous Compendium* does not inform the reader that we also obtain *DH*, the side of the regular pentagon inscribed in the semicircle, and that the construction is a good approximation but not exact.[36] Abu'l-Wafa' explains an exact construction of the regular pentagon by a single compass opening equal to the radius of the pentagon.[37] Abu'l-Wafa''s exact construction is complicated, so the easier approximation may well have given better results in practice.

The text states the purpose of this diagram as the construction of "set square (i.e., *gūnyā*) [number] 5 with a compass opening equal to the radius [of the circumscribing semicircle] from set square [number] 6." A "*gūnyā* n" is a right-angled triangle, with one angle equal to the n-th part of two right angles, that is, in modern terms, (180/n)°. Triangle *ADG* is *gūnyā-6*, and the two triangles *AZH* and *GZD* are *gūnyā-5*. The term *gūnyā* may ultimately be of Greek origin, from the word *gonia*, meaning angle, as defined in Definition 8 of Book I of Euclid's *Elements*.[38]

In the figure, the magnitude of the angles is indicated by numbers that have been transcribed in figure 7, where an angle of "1" would correspond in modern terms to 6°. The author(s) of the *Anonymous Compendium* were probably unfamiliar with the modern concept of "degrees" to measure angles (such that a right angle is 90°). Degrees with their sexagesimal subdivisions were introduced in late Babylonian astronomy to measure arcs of the ecliptic and are not found in Euclid's *Elements*. In later Greek and Islamic astronomy, degrees were used for measuring arcs on a sphere and on a circle, but not often for measuring angles.

IV. CONSTRUCTIONS OF RECTANGULAR PATTERNS TO BE USED IN DECORATIVE TILINGS

In this category, the constructions consist of a basic rectangle filled with a decorative pattern. In some cases, the whole plane can be tiled with the pattern by translating and mirroring the basic rectangle. In others, the decorative applications are less obvious. In most of the constructions, only one side of the basic rectangle is given; the other sides are determined in the construction process and usually the length and width have an irrational ratio.

Ten constructions in the manuscript belong to this category, the first seven of which have textual instructions.

Fol. 190v [35]:[39] a pattern containing a regular heptagon, and parts of a hexagonal and octagonal star.

Fol. 192r [37]:[40] another pattern containing a regular heptagon (discussed below).

Fol. 192v [42]:[41] a pattern containing two kites, as in figure 12 below, subdivided into three pieces, but with two equal angles of 75° rather than 90°.

Fol. 193r [43]:[42] a nice pattern with hexagonal symmetry; the pattern can be extended to tile the plane in a beautiful way with hexagonal figures.

Fol. 193v [44]:[43] a pattern with pentagonal stars and equilateral, but non-equiangular, pentagons in between.

Fol. 195v [54]:[44] a pattern to tile the plane with large dodecagonal and decagonal stars, with small pentagonal stars in between.

Fol. 196v [57]:[45] a division of a rectangle whose length and width are expressed as "three perpendiculars [i.e., altitudes] plus one and a half sides" and "4 and a half sides minus one perpendicular [altitude]." By "side" and "perpendicular [altitude]" the text means the hypotenuse and the intermediate side of a right-angled triangle with angles of 30°, 60°, and 90°, that is, a "set square 6" triangle. In modern terms the length and width of the basic rectangle are in the ratio $(3\sqrt{3}+3) : (9-\sqrt{3}) = 1.127\ldots$

The last three constructions in this category lack texts:

Fol. 189r [32][46] displays a rectangle whose sides are specified as "two parts" and "square root of 5," divided into eight pieces. Six pieces form two large subdivided kites as in figure 12 below, plus two smaller kites, similar to the larger ones. The pattern resembles the "variant pattern" in Part VI below, but can be constructed by ruler and compass, by dividing a line segment in extreme and mean ratio, as in Euclid's *Elements* II:11.[47]

Fol. 196r [56]:[48] a figure only, without text, displaying a pattern for tiling the plane with dodecagonal stars and somewhat smaller decagonal stars, but without pentagonal stars.

Finally, fol. 199r [61]:[49] an intricate division of a rectangle, filled with small pentagonal stars, intermediate figures, and parts of decagonal stars; copying the rectangles and their mirror image produces a beautiful tiling of the plane. This is the last pattern in the *Anonymous Compendium*; it is separated from the rest of the compendium by several blank pages in the manuscript. The points are not labeled with letters and there is no accompanying text explaining the construction of the pattern.

My example is fol. 192r [37], the resulting pattern for which is found in the North Dome of the Friday Mosque in Isfahan and can be related to *girih* (from Persian, meaning knot) tiles, as will be discussed below.[50] Figure 8 is a transcription of the construction in the manuscript, which has points labeled with letters. It also has some broken lines that are used in the construction but do not form part of the final decorative pattern.

The procedure is as follows. The length AB is assumed and at points A and B two perpendiculars, AF and BC, of unspecified length, are drawn. The text prescribes that line AG should be drawn such that angle $\angle BAG$ is equal to "three-sevenths of the right angle." In the figure in the manuscript, AG is drawn as a broken line, apparently indicating that it is part of an auxiliary construction but does not belong to the final pattern. Broken lines do not often occur in works by medieval Islamic geometers trained in the style of Euclid's *Elements*. The designers do not give further details about the construction of line AG, and were not worried by the fact that an angle of (in modern terms) $(3 \times 90)/7°$ cannot be constructed by ruler and compass. Then point E is chosen on BG such that $BE = AG/2$, and the broken line EZ is drawn parallel to AG. According to the text, point L should be constructed on EZ in such a way that if LK is drawn parallel to EB to meet AB at K, we have $LK = EL/2$. Point K is found through the following auxiliary construction: through an arbitrary point T on EZ draw a line parallel to BE, and choose I on this parallel such that $TI = TE/2$ as in the figure, and draw EI, which will intersect AB at the required point K. Again, some of the lines appear as broken lines in the figure in the manuscript, as well as in figure 8.

Then a circular arc is drawn with center Z and radius ZK to meet EZ at M, and point N is chosen on the arc such that $MN = KM$. Note that point N is the mirror image of K in line ZE. Finally, LN is extended to meet the horizontal side of the rectangle through A at point S. The text says that S is "the center of the heptagon" and laconically concludes: "the construction should be easy, God, may He be exalted, willing."

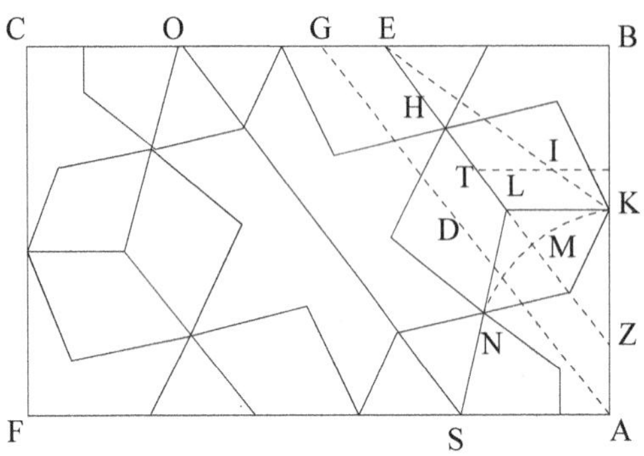

Fig. 8.

Then a few alternative methods are presented in the manuscript to find point S. If point O is on BE, which is extended such that $EL = EO$, then point O is also the center of a heptagon and OS is parallel to AG. The text does not give more information.

To analyze this pattern I use the notation R for a right angle, and $\alpha = 2R/7$; in this notation $\angle EZB = \angle GAB = 3\alpha/2$. In the pattern, rectangle $ABCF$ is divided by line SO into two congruent trapezia, and each trapezium is divided (in the way of $ABOS$) into three quadrangles: $BKLE$ (angles R, R, 5α, 2α), $AKLS$ (angles R, R, 4α, 3α), and $OELS$ (angles 2α, 5α, 5α, 2α). Note that $OE = EL = LS = 2LK$. To complete the pattern, we note that N is the midpoint of LS, and we use point K, point N, and the midpoints of segments LE and EO. Points S and O are centers of regular heptagons, and the other lines in the pattern can be defined on the basis of arguments of symmetry, but the manuscript does not give any details, nor does it mention that the ratio between the length and width of the rectangle is $(2 + \cos 2\alpha + \cos 3\alpha)/(\sin 2\alpha + \sin 3\alpha) = 1.62003\ldots$, although it would be useful to know this number if one wanted to draw copies of the rectangle.

Bulatov observed that copies of the rectangle and of its mirror image produce a nice tiling of the plane, as shown by the thick continuous lines in figure 9.[51] As mentioned above, this tiling appears in the North Dome in the Friday Mosque in Isfahan; figure 10 shows two rectangles together with two mirror images superimposed on the pattern in that mosque.

Using figure 9, we can relate the pattern to a procedure in which a decorative pattern is based on a set of auxiliary figures called "*girih* tiles" (mentioned above). The actual pattern was drawn through the midpoints of these *girih* tiles. The auxiliary figures themselves could then be erased. We note that the thin lines in figure 9 do not show up in the final pattern in the Friday Mosque in figure 10. Thus, the thin lines in figure 9 define a set of two hexagonal *girih* tiles (with angles 5α, 5α, 4α, 5α, 5α, 4α and 6α, 4α, 4α, 6α, 4α, 4α) on which the actual pattern in figure 10 is based. One complete *girih* tile is visible in figure 8 as the trapezium $OELS$ together with its mirror image in OS.

In conclusion, the concise and laconic instructions in the text are insufficient to explain the design and application of the pattern. The *Anonymous Compendium* was clearly part of an oral tradition of instruction.

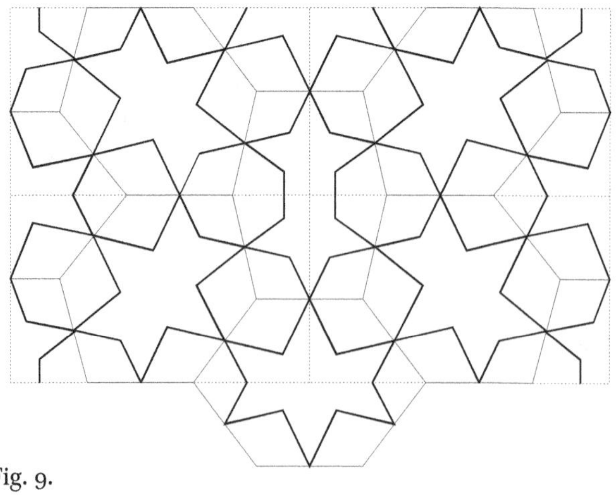

Fig. 9.

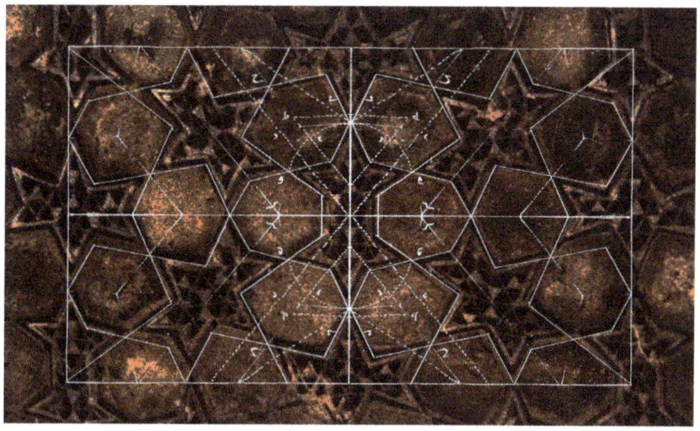

Fig. 10.

V. DIVISIONS OF SQUARES

Many constructions in the manuscript start from a square that is then divided into parts. These constructions will be listed here in subcategories according to the probable purposes for which these divisions were designed. Some of these constructions are of considerable mathematical difficulty and will be deferred to the next section, which describes kite patterns.

a. Fol. 180v [2][52] features a big square divided into 97 pieces, which can be reassembled into 2 decagons, 2 nonagons, 2 octagons, and 2 small squares. Rather than showing how to assemble the two decagons, nonagons, or octagons, the text explains how to assemble a decagonal, nonagonal, or octagonal star. The construction was probably designed in order to show how a square (for example, a colored square tile) could be cut up and reassembled to form other figures.

b. The following two patterns were drawn in a square that could then be repeated to produce a tiling for a larger area.

Fol. 194r [45]:[53] an octagonal star is constructed in the interior of a square, and parts of hexagonal stars are drawn in such a way that the centers of these stars are the vertices of the square. Using copies of this square diagram, the plane can be tiled with hexagonal and octagonal stars, as well as intermediate figures.

Fol. 194v [48]:[54] inside a square the halves of four heptagonal stars and intermediate figures are constructed. The explanation is laconic, and the diagram shows that the construction is inexact, in the sense that one of the angles of every heptagonal star protrudes beyond the square.

c. Divisions of a single square that may have been used as an independent decorative element and also to tile a larger area.

Fol. 188v [30]:[55] division of a large square into a border area and a concentric intermediate square, which is in turn subdivided into four kites, four non-convex hexagons, and a small square in the middle. The division is drawn in such a way that many line segments and distances are equal (see the explanation below).

Fol. 189r [31]:[56] another division of a larger square in a way somewhat similar to fol. 188v [30]. In fol. 189r [31] many of the angles should be, in modern terms, 60° and 120° (see the explanation below).

Fol. 192v [41]:[57] another division of a big square, somewhat similar to fol. 188v [30] and fol. 189r [31], with an intermediate square, four kites, and four quadrilaterals, but without a small square in the middle.

Fol. 196r [55]:[58] decorative division of a square by means of lines that are parallel to the sides and diagonals of the square. The text presents what is probably an approximate construction.

d. Eight divisions of a square are drawn with either no accompanying text or so little that it is not clear what exactly was intended by the diagrams. In many cases, Bulatov has a reconstruction (for references, see the endnotes below).

Fol. 188r [29]:[59] a complicated division of a square. The diagram also reveals some intermediate steps in the construction by means of two dotted lines and two dotted semicircles, but the precise position of many of the points and lines is unclear.

Fol. 192v [39]:[60] a division of a square into 17 pieces, including two kites, as subdivided in figure 12, without text.

Fol. 192v [40]:[61] a division of a square into 16 pieces, including four subdivided kites, as in figure 12, without text.

The following five divisions of a square are found on the same page of the manuscript: fol. 195r [49],[62] fol. 195r [50],[63] fol. 195r [51],[64] fol. 195r [52] (a construction without text),[65] and fol. 195r [53].[66]

Our examples will be fol. 188v [30] and fol. 189r [31], which are illustrated by the same figure in the manuscript (i.e., fol. 188v [30], fig. 11), although the patterns are mathematically different. Figure 11 includes transcriptions of all the letters that are used to label points in the figure in the manuscript. We note that the letters F and O are used as labels for more than one point, in contrast to the procedures of theoretically trained mathematicians, who, if there were more points than could be accommodated by the letters of the alphabet (the *abjad* system), used two-letter combinations beginning with the letter *lām*, such as *lām-alif*, *lām-bā'*, etc., to designate the extra points.

In fol. 188v [30], the large square with the broken lines as sides is given. One has to construct a concentric intermediate square *KIML*, and divide it by combinations of line segments *FCQRJ*, *XhtJ*, etc., in such a way that the intermediate square is divided into four kites, that is to say, a quadrilateral with two pairs of equal sides and two equal angles, which are right angles in this case; a small square in the middle (with side *RJ*) and four non-convex hexagons *RJthK*, etc. The lines *Xht*, *FCQ*, etc., should be perpendicular to the sides *KhL*, *KCI*, etc., of the intermediate square. All of this should be done in such a way that $FC = CQ = QR = RJ = Jt = th = hX$. In other words, the thickness of the "band" between the big and the intermediate square ($CF = Xh$) should be equal to the side of the small square (*JR*) and also equal to the short sides (*QR*, *CQ*, etc.) of the kites.

The text presents the solution as follows: the sides of the intermediate square (*KL*) and of the big square (*OF*) should be "two units plus the square root of 5" and "four units plus the square root of 5," so if one of them is known, the other can be found. It is also specified that line *Qt* (broken line in the figure) should be "the square root of 5." The text assumes (but does not specify) that $FC = CQ = QR = RJ = Jt = th = Xh$ are equal to 1. Then triangle *JTQ* is a right-angled triangle with sides $Jt = 1$ and $QJ = 2$, so the hypotenuse *Qt* is $\sqrt{5}$ by the Pythagorean theorem. The triangle should be drawn in such a way that its hypotenuse *Qt* is parallel to the side *KI* of the intermediate square.

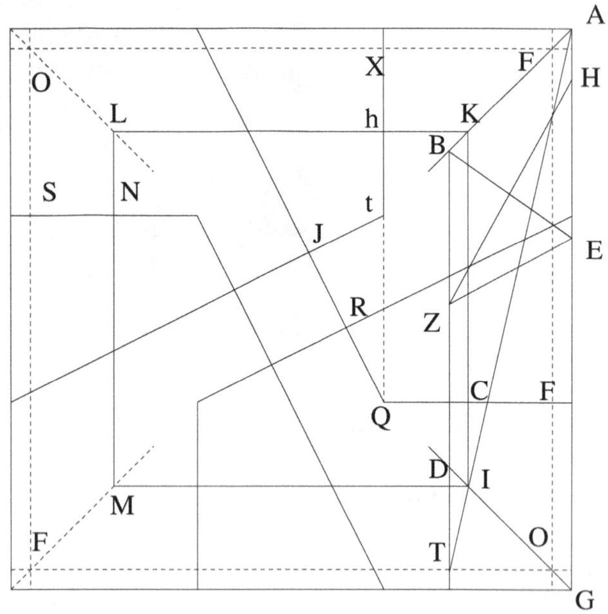

Fig. 11.

The same figure 11 is then used in the manuscript for the pattern textually described in fol. 189r [31]. Through points *A* and *G* we draw bisectors *AB*, *GD*, and a parallel *BD*. Line *AB* is called a "line of four," meaning that one of the angles with the sides of the square is 1/4 times two right angles. Then the text constructs an equilateral triangle *BZE* such that the base *BZ* is part of segment *BD* and point *E* is on *AG*. One then constructs *H* such that *EH* is equal to the side of the equilateral triangle (so *BEH* is another equilateral triangle [fig. 11 is not drawn to scale]). Then the text constructs *T* on *BZ*, extended such that $TZ = HZ$. Line *AT* is joined to intersect *GD* at point *I*, and line *IK* is drawn parallel to *AG* to meet *AB* in *K*. Then *IK* is taken as the side of an intermediate square. Finally, the text constructs in the intermediate square *IKLM* four "kites 6" (Persian: *turunj-6*). Apparently, a "kite 6" is a combination of two congruent *gūnyā-6* triangles, which produce a kite with angles of 60°, 90°, 90°, and 120°. The text does not say how to find these kites: this can be done by dividing the sides of the intermediate square in the ratio 1:$\sqrt{3}$. Apparently the lines *FO*, *Qt*, etc., are no longer needed.

The manuscript uses the same figure for two different patterns only in the case of fol. 188v [30] and fol. 189r

[31]. One wonders whether these two constructions and the figure originated from a different source than the rest of the *Anonymous Compendium*.

VI. DIFFICULT KITES AND THEIR PATTERNS

The two mathematically most interesting patterns in the manuscript are related to two special kites (*turunj*). See *ABCD* in figure 12, where *AD* = *AB*, *CD* = *CB*, angles *D* and *B* are equal, and *AB* ≠ *BC*; almost always, angles *B* and *D* are right angles.

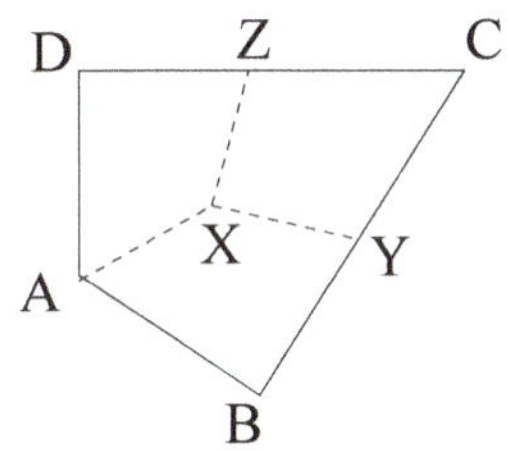

Fig. 12.

In Isfahan, one often sees decorations such as the pattern in the Hakim Mosque (fig. 13), where a big square is divided into four congruent kites and a small square; such an ornament is called a "four kite" (Persian: *chahār-turunj*).[67] The small squares and the four kites are sometimes inscribed with calligraphy. The size of the square in the middle of the *chahār-turunj* depends on the shape of the kite, which in turn is completely determined by its smallest angle.

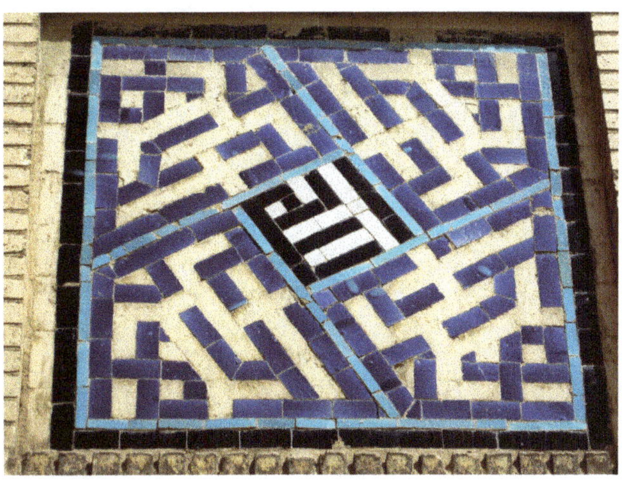

Fig. 13.

In some of the diagrams in the manuscript, the kites are subdivided into three quadrilateral pieces as in figure 12 by choosing points *Y* and *Z* on the longer sides *BC* and *DC* and point *X* on the diagonal *AC* in such a way that *BY* = *BA* = *DA* = *DZ* and *AX* = *XY* = *XZ*. The division is indicated by dotted lines in figure 12 but will mostly be omitted in the following figures to minimize the clutter. Kites subdivided as in figure 12 occur in fol. 189r [32], fol. 192v [39], fol. 192v [40], and fol. 192v [42], which have been mentioned in the previous sections.

Two special types of kites in the *Anonymous Compendium* are more difficult than the others because they cannot be constructed by ruler and compass in the manner of Euclid's *Elements*. Mathematically, the construction of these two types of kites is equivalent to the solution of a cubic equation, as we will see below. The two special kites occur in patterns that I call the "twelve kite-pattern" and the "variant pattern" (my terminology). The twelve-kite pattern was also studied by the famous mathematicians Ibn al-Haytham (965–ca. 1040) and ʿUmar Khayyam (1048–1131).[68] Before explaining the patterns in detail I will list the relevant constructions in the *Anonymous Compendium*:

a. *Twelve-kite pattern*

Fol. 188r [28]:[69] three approximate constructions of the twelve-kite pattern, called "circular," "linear," and "another type." It turns out that the three are not equivalent and that the "linear" one is the best; the error in the angle is only a few minutes (if angles are measured in degrees); the others have errors in the order of half a degree.

Fol. 189v [33]:[70] another approximate construction of the twelve-kite pattern, mathematically equivalent to the "linear" construction in fol. 188r [28].

Fols. 191r–191v [36]:[71] a trial-and-error "construction" of the twelve-kite pattern by means of a T-shaped instrument, which may have been the invention of someone called "Kātib."[72] The diagram does not display the complete twelve-kite pattern but only the part corresponding to the intermediate square *EFGH* in figure 14 below, which is not found in the manuscript. The manuscript also tells us that the pattern can only be found by means of conic sections, and that finding the pattern is equivalent to the construction of a right-angled triangle, such that the length of the shortest side plus the length of the perpendicular drawn on the hypotenuse are equal to the length of the hypotenuse itself (see below). According to the manuscript, Ibn al-Haytham wrote a treatise on this triangle in which he constructed it by means of two conic sections, namely, a parabola and hyperbola. His treatise and construction are now lost. The second part of the construction is a description of the T-shaped instrument, which is compared to the alidade of the "boat" astrolabe (*usṭurlāb-i zawraqī*). [73]

b. Variant pattern

The variant pattern occurs in the following five constructions:

Fol. 185r [16]:[74] a construction by trial and error of a certain right-angled triangle, which can be used to construct the variant pattern. The triangle is defined below, at the end of this section. The construction is executed by means of a moving ruler and a compass.

Fol. 185v [18]:[75] an approximate construction of the variant kite pattern. It has an error of approximately a quarter of a degree.

Fol. 187v [26]:[76] an approximate construction of the variant pattern, based on a regular octagon. In the middle of the page is an annotation stating that the ratio between the length and the width of the pattern is 7:6; this is another good approximation. Then the construction based on the octagon is described again in general terms.

Fol. 187v [27]:[77] This seems to be an approximate construction of the variant kite pattern, which is based on the same idea as the approximate construction of fol. 187v [26], but the two constructions differ in some details. The diagram is labeled with letters but there is no corresponding text in the manuscript. The text of fol. 187v [27] begins with another very easy approximation, to the effect that the angles in triangle *ABK* in figure 17 below are in the ratio 4:5:9. The error in this approximation of angles *A* and *B* is approximately half a degree.

Fol. 190r [34]:[78] another construction by trial and error of the variant kite pattern, using a T-shaped ruler.

It is likely that the following construction relates to the variant pattern:

Fol. 185v [17]:[79] a figure with a right-angled triangle in which, according to the text, a kite has to be inscribed. It is stated that "[o]btaining a triangle of this sort is difficult. It falls outside Euclid's *Elements* and concerns the science of conics. The construction is achieved by the motion of a ruler, if the magnitude of the perpendicular is assumed." Bulatov does not translate the text but his interpretation of the figure is plausible. In Bulatov's interpretation (followed by Özdural), the ratio of the hypotenuse of the triangle to its intermediate side is 3:2. It follows that the angles of the triangle differ only a little more than one degree from the corresponding ones in the variant pattern. The placement of this construction on fol. 185v between nos. 16 and 18 suggests a relationship to the variant kite, although the altitude of the right-angled triangle (drawn as a broken line in the manuscript) only occurs in the triangle related to the twelve-kite pattern, which triangle was constructed by Ibn al-Haytham and ʿUmar Khayyam by means of conic sections.

We now turn to the details. The pattern that I call the "twelve-kite pattern" consists of a big square with a small, rotated square in the middle, surrounded by a number of kites. A precise description now follows.

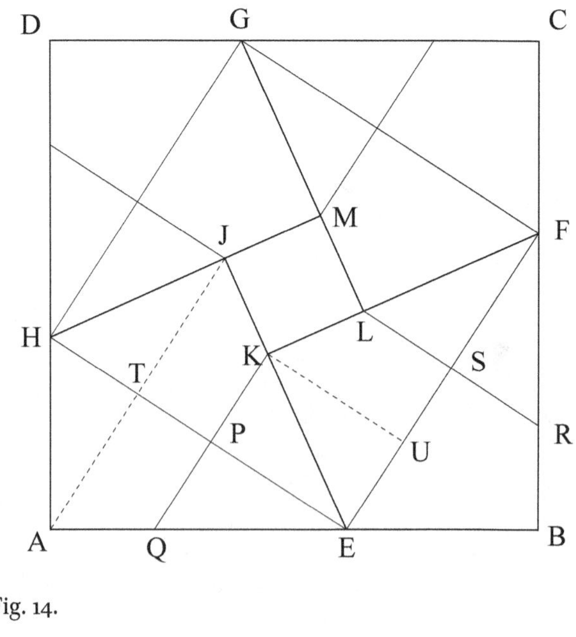

Fig. 14.

In figure 14 we consider a (large) square *ABCD* and choose a point *E* on side *AB* such that *AE* > *EB*. We also choose points *F*, *G*, and *H* on the three other sides of the square such that *AE* = *BF* = *CG* = *DH*, so also *EB* = *FC* = *GD* = *HA*. Then figure *EFGH* will also be an (intermediate) square. Let points *J*, *K*, *L*, and *M* be the mirror images of points *A*, *B*, *C*, and *D* in sides *HE*, *EF*, *FG*, and *GH*. Then figures *AHJE*, *BEKF*, *CFLG*, and *DGMH* are four (big) kites and *JKLM* is a (small) square. Draw perpendicular *KPQ* to *HE* to meet *HE* at *P* and *AE* at *Q*, and draw perpendicular *LSR* to *EF* to meet *EF* at *S* and *BF* at *R*.

In the general case, *JK* and *KP* need not be equal but there is one particular choice of point *E* for which *JK* = *KP*. Since also *HK* = *HK* and the two angles *J* and *P* are right angles, the two triangles *JKH* and *PKH* are congruent; so *JH* = *HP* and *JKPH* is also a (small) kite. Similarly, *AQPH* must also be a kite because it is the mirror image of *JKPH* in side *HE*. In the same way, it can be shown that there are six more (small) kites *KLSE*, *BRSE*, etc.; hence there is a total of twelve kites in the big square. Therefore I call the resulting pattern the "twelve-kite pattern."

How should point *E* be chosen so as to produce this nice arrangement? This question can be resolved if we know the ratio *AE* : *EB*; since *AH* = *EB*, the ratio determines (and is determined by) the shape of the right triangle *HAE*. For the next argument we draw two dotted lines in figure 14: diagonal *ATJ*, which is perpendicular to *HE*, and perpendicular *KU* onto *EF*.

We now concentrate on this triangle *HAE* and note that segment *AT* is the altitude of this triangle. By symmetry, *AT* = *KU* = *PE* and also *AH* = *PH* in the kite *AQPH*; so *AT* + *AH* = *PE* + *PH* = *EH*. Thus triangle *AEH* is a right-angled triangle such that the shortest side (*AH*) plus the altitude (*AT*) is equal to the hypotenuse (*EH*). On fol. 191r–191v [36], we learn that Ibn al-Haytham wrote a treatise in which he constructed this triangle by means of a parabola and a hyperbola. ʿUmar Khayyam knew that constructing the triangle is equivalent to solving a cubic equation.[80]

The manuscript contains the following intriguing approximate construction of the twelve-kite pattern in fol. 189v [33].

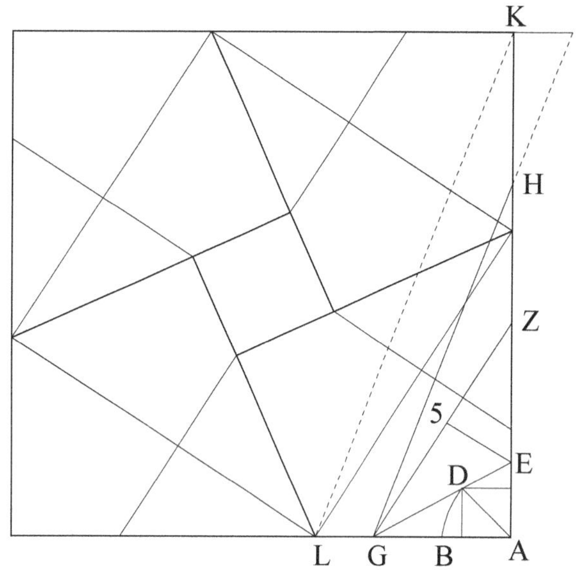

Fig. 15.

On vertex *A* of the big square we construct a little square of arbitrary size, with diagonal *AD*. Find point *G* on one of the sides through point *A* of the big square such that *AG* = 2*AD*; draw *GD* and extend it to meet the other side *AK* at point *E*. Then find point *H* on *AE* extended such that *EH* = 2*AG* and draw *GH*. Through vertex *K* draw *KL* parallel to *GH*. Point L is one of the vertices of the intermediate square. Now the rest of the pattern can easily be found. The construction is approximate but the error is slight; if side *AK* is one meter, the error in the position of L is only a few millimeters.[81]

In order to introduce the "variant pattern," we reconsider a kite from the standard pattern: figure 16 displays a right-angled triangle such that the altitude (not drawn in figure 16) plus the shorter side *BK* is equal to the hypotenuse *AB*. If we add the dotted lines of the "decorative" division of the smaller kite as in figure 12 above, we have *DG* = *GE* = *EK* = *KL*; it is also possible to show that *AE*, *BK*, and *BG* are equal.[82]

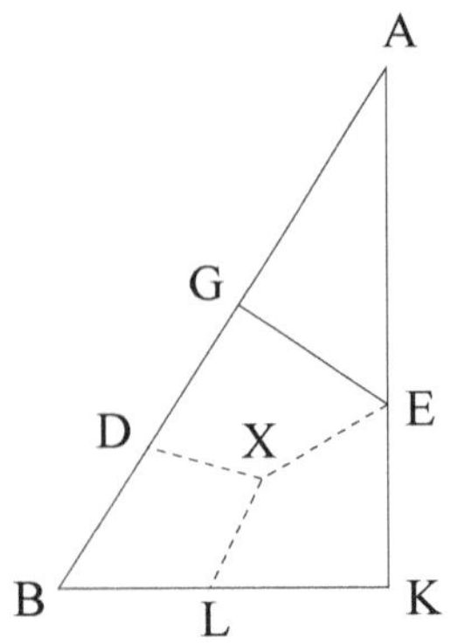

Fig. 16.

By playing with figure 16, the designers may have considered the variant triangle of figure 17. There, we have $BG = BK \neq AE$, but $BD = AG$. In this way, two triangles with congruent kites *BGEK* and *ADPQ* can be placed in a rectangle *BKAQ*; the result is what I call the variant pattern. The construction of this pattern is also equivalent to a cubic equation.[83] Again, this kite cannot be constructed by ruler and compass.

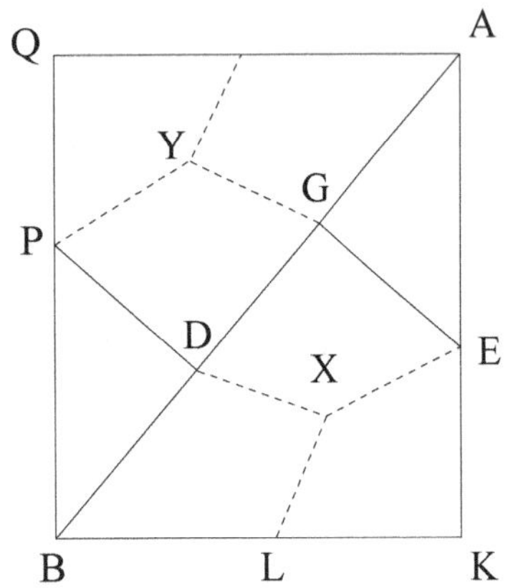

Fig. 17.

The text in fol. 185v [17] suggests that at least some scholars in the medieval Islamic tradition knew that an exact construction of the variant pattern is possible by means of conic sections but not by means of a ruler and a compass. It seems to me that the pattern was discovered as a variant of the twelve-kite pattern, in the way suggested above. Another indication is the fact that just like the twelve-kite pattern, the "variant pattern" was related to the problem of constructing a right-angled triangle, which is mentioned in fol. 185r [16]. In this case, the triangle is less natural and more complicated; the text in fol. 185r [16] is corrupted, but it should have been "a right-angled triangle in which the ratio of the difference between the shortest side and the hypotenuse to the difference between the [former] difference and the shortest side [is the same as the ratio of the intermediate side to the shortest side]." In figure 17 this property corresponds to $AG : GE = AK : KB$; we note that $AG = AB - BG = AB - BK$ and $GE = KL = BK - BL = BK - AG$. Although the "variant pattern" may never have received the same mathematical prestige as the "twelve-kite pattern," the craftsman or craftsmen who were the author(s) of the *Anonymous Compendium* nevertheless came up with a number of intriguing approximate constructions.

VII. *MUQARNAS*

Three diagrams in the manuscript, fol. 184v [15],[84] fol. 194r [46],[85] and fol. 194r [47],[86] are in all probability horizontal projections of decorative stalactite-like construction in Islamic vaults, i.e., a *muqarnas*. These simple diagrams are furnished with numbers but not accompanied by instructions in the text. The numbers refer to the successive altitudes of layers of the *muqarnas*-structure.[87]

CONCLUSION

For a historian of mathematics, the *Anonymous Compendium* is interesting because it is different from the standard literature on mathematics in the medieval Islamic tradition. The constructions of the *Anonymous Compendium* are witness to types of mathematical expertise and training few traces of which have come down to us thus far. The material in the *Anonymous Compendium* is heterogeneous and in rather random order, and may well date back to different authors who may have lived centuries apart. Because the *Anonymous Compendium* is one of a kind it is difficult to draw general conclusions about it. It may well be that more manuscripts of the same type existed in the Middle Ages, but were used up until they were worn out and discarded. Nevertheless, some more manuscripts of this type may well have survived the ravages of time in Iran and neighboring countries and await discovery by scholars. If such manuscripts are found, we will be able to say more with certainty about the contents and context of the mathematical material in the *Anonymous Compendium*.

Department of Mathematics
University of Utrecht

NOTES

Author's note: I wish to thank Karen A. Leal, managing editor of *Muqarnas*, for her very careful editing of this paper.

1. For a discussion of the possible title of the work, "On Similar and Complementary Interlocking Figures," see, in the present volume, the introduction ("In Memory of Alpay Özdural and His Unrealized Book Project") by Gülru Necipoğlu, and chapter 2, "An Early Tradition in Practical Geometry: The Telling Lines of Unique Arabic and Persian Sources" by Elaheh Kheirandish.
2. See Richard 1989, 183–87, where the manuscript is tentatively dated to the sixteenth century; cf. also the older catalogue Blochet 1912, 41–47, who dated the Paris Codex to the seventeenth century. See Alpay Özdural's "Preliminaries" (from his unpublished book manuscript), published as chapter 4 in this volume, for the fifteenth-century date proposed by Priscilla Soucek. For further information on the manuscript codex and the other treatises in it, with references to the catalogues, see the chapter by Necipoğlu ("Ornamental Geometries: A Persian Compendium at the Intersection of the Visual Arts and Mathematical Sciences") and the one by Kheirandish in the present volume, and Brentjes 1994, 55–56.
3. Geometry was considered a prerequisite for studying astronomy, as can be seen from the numerous extant manuscripts of the "Middle Books" (Arabic: *mutawassiṭāt*), which were meant to be studied between the *Elements* of Euclid and the *Almagest* of Ptolemy: see Steinschneider 1865, 456–98; see also Kheirandish 2006, 135–54.
4. Complete references to editions and translations of Abu'l-Wafa' al-Buzjani's book are found in Kheirandish's chapter in this volume. See also Necipoğlu's chapter for a discussion of Abu'l-Wafa"s *Geometrical Constructions* in relation to the *Anonymous Compendium*'s contents.
5. These discussions were published in Özdural 2000, 171–201.
6. On this issue, see Cromwell 2015: http://link.springer.com/article/10.1007/s00283-015-9538-9.
7. See Bulatov 1988, which contains an abbreviated Russian translation of the *Anonymous Compendium* by A. B. Vildanova; and Abū'l-Wafā' al-Būzjānī 1990–91 (1369), ed. Jazbī, appendix 1, pp. 73–95. On some of the problems with the latter work, see Kheirandish's chapter in the present volume.
8. Bulatov 1988, 318, fig. 8; Abū'l-Wafā' al-Būzjānī 1990–91 (1369), ed. Jazbī, 74, fig. between no. 2 and no. 3.
9. Bulatov 1988, 319, fig. 13; Abū'l-Wafā' al-Būzjānī 1990–91 (1369), ed. Jazbī, 76, no. 10.
10. Bulatov 1988, 323, fig. 25; Abū'l-Wafā' al-Būzjānī 1990–91 (1369), ed. Jazbī, 80, no. 24.
11. Bulatov 1988, 339, fig. 60, where the isosceles triangle is omitted; Abū'l-Wafā' al-Būzjānī 1990–91 (1369), ed. Jazbī, 92, no. 63.
12. Bulatov 1988, 339, fig. 58; Abū'l-Wafā' al-Būzjānī 1990–91 (1369), ed. Jazbī, 93, no. 64.
13. Bulatov 1988, 339, fig. 59; Abū'l-Wafā' al-Būzjānī 1990–91 (1369), ed. Jazbī, 92, no. 62.
14. Bulatov 1988, 316, fig. 1; Abū'l-Wafā' al-Būzjānī 1990–91 (1369), ed. Jazbī, 73, no. 1. The construction is mentioned in Chorbachi 1989b, 751–89, at 774. See also Özdural 2000, 171–201, esp. 187–90.
15. Bulatov 1988, 317, fig. 7; Abū'l-Wafā' al-Būzjānī 1990–91 (1369), ed. Jazbī, 74, no. 3.
16. Bulatov 1988, 319, fig. 11ab; Abū'l-Wafā' al-Būzjānī 1990–91 (1369), ed. Jazbī, 75, no. 8.
17. Bulatov 1988, 319, fig. 12; Abū'l-Wafā' al-Būzjānī 1990–91 (1369), ed. Jazbī, 76, no. 9. See also Özdural 2000, 185–86.
18. See Heath 1956, vol. 1, p. 205; and Özdural 2000, 185–86.
19. Bulatov 1988, 320, fig. 14; Abū'l-Wafā' al-Būzjānī 1990–91 (1369), ed. Jazbī, 76–77, nos. 11–12.
20. Bulatov 1988, 318, fig. 9ab; Abū'l-Wafā' al-Būzjānī 1990–91 (1369), ed. Jazbī, 74, nos. 4–5.
21. Heath 1956, vol. 2, 54–56.
22. Bulatov 1988, 318, fig. 9cd; Abū'l-Wafā' al-Būzjānī 1990–91 (1369), ed. Jazbī, 75, no. 6.
23. Bulatov 1988, 319, fig. 10; Abū'l-Wafā' al-Būzjānī 1990–91 (1369), ed. Jazbī, 75, no. 7.
24. Bulatov 1988, 320, fig. 15; Abū'l-Wafā' al-Būzjānī 1990–91 (1369), ed. Jazbī, 77, no. 13.
25. Bulatov 1988, 321, fig. 16; Abū'l-Wafā' al-Būzjānī 1990–91 (1369), ed. Jazbī, 77–78, nos. 14–17. See Necipoğlu's chapter in the present volume for the relationship to a work by Ibn al-Haytham on the regular heptagon.
26. This construction, which consists of a figure only, is omitted in Bulatov 1988, and also in Abū'l-Wafā' al-Būzjānī 1990–91 (1369), ed. Jazbī.
27. Abū'l-Wafā' al-Būzjānī 1990–91 (1369), ed. Jazbī, 87, no. 44. The construction is omitted in Bulatov 1988.
28. See Chorbachi 1989a; see also Bulatov 1988, 322–23, and Özdural's drawing in pl. 14.1 in the present volume.
29. Bulatov 1988, 321, fig. 17; Abū'l-Wafā' al-Būzjānī 1990–91 (1369), ed. Jazbī, 78, no. 18.
30. Bulatov 1988, 321, fig. 21; Abū'l-Wafā' al-Būzjānī 1990–91 (1369), ed. Jazbī, 79, no. 21.
31. Bulatov 1988, 323, fig. 24; Abū'l-Wafā' al-Būzjānī 1990–91 (1369), ed. Jazbī, 80, nos. 22–23.
32. Bulatov 1988, 322, fig. 22; Abū'l-Wafā' al-Būzjānī 1990–91 (1369), ed. Jazbī, 81, at the beginning of no. 25.

33. Bulatov 1988, 323, fig. 23; Abū'l-Wafā' al-Būzjānī 1990–91 (1369), ed. Jazbī, 81, at the end of no. 25.
34. Bulatov 1988, 324, fig. 26; Abū'l-Wafā' al-Būzjānī 1990–91 (1369), ed. Jazbī, 81, no. 26. On Abū Bakr [b.] al-Khalīl al-Tājir al-Raṣadī, see the extensive discussions in Necipoğlu's chapter as well as in Kheirandish's chapter.
35. The letter *B* does not occur in the figure but has been supplied following Bulatov 1988, 321, fig. 17; Abū'l-Wafā' al-Būzjānī 1990–91 (1369), ed. Jazbī, 78, no. 18; and Özdural, on the basis of the information in the text.
36. Suppose that the radius of the circle is 1, and drop a perpendicular ZP onto AG. Then we have $ZA = 1$; $\angle ZAP = 30°$; $ZP = 1/2$; $AP = (\sqrt{3})/2$; $GP = 2 - (\sqrt{3})/2$; $\angle ZGP = \arctan(ZP/GP) \approx 23.8°$. Because $\angle DGP = 60°$, $\angle ZAH = \angle ZGD \approx 36.2°$.
37. Suter 1922, 102–4, and Woepcke 1855, 328.
38. Heath 1956, vol. 1, p. 176.
39. Bulatov 1988, 327, fig. 35; Abū'l-Wafā' al-Būzjānī 1990–91 (1369), ed. Jazbī, 85, nos. 38–39.
40. Bulatov 1988, 329, figs. 38–39; Abū'l-Wafā' al-Būzjānī 1990–91 (1369), ed. Jazbī, 86–87, nos. 41–43.
41. Bulatov 1988, 330, fig. 41; Abū'l-Wafā' al-Būzjānī 1990–91 (1369), ed. Jazbī, 88, no. 47.
42. Bulatov 1988, 332, figs. 42–44; Abū'l-Wafā' al-Būzjānī 1990–91 (1369), ed. Jazbī, 88, no. 48.
43. Bulatov 1988, 332, fig. 45; Abū'l-Wafā' al-Būzjānī 1990–91 (1369), ed. Jazbī, 88-89, nos. 49–51.
44. Bulatov 1988, 337, fig. 53; Abū'l-Wafā' al-Būzjānī 1990–91 (1369), ed. Jazbī, 91, part of no. 59.
45. Bulatov 1988, 338, figs. 56–57; Abū'l-Wafā' al-Būzjānī 1990–91 (1369), ed. Jazbī, 92, no. 61.
46. Bulatov 1988, 327, fig. 32; Abū'l-Wafā' al-Būzjānī 1990–91 (1369), ed. Jazbī, 84, no. 34.
47. Heath 1956, vol. 1, 402–3.
48. Bulatov 1988, 338, fig. 54; Abū'l-Wafā' al-Būzjānī 1990–91 (1369), ed. Jazbī, 91, part of no. 59.
49. The figure is in Bulatov 1988, 339, fig. 61, and in Abū'l-Wafā' al-Būzjānī 1990–91 (1369), ed. Jazbī, 95, and also on the front page.
50. See Hogendijk 2008, 121, for the discovery of this pattern on the North Dome. For a further discussion of this pattern and the related *girih* tiles, see also Necipoğlu's chapter in the present volume.
51. Bulatov 1988, 329, 332–33.
52. Bulatov 1988, 316–17, figs. 2–6; Abū'l-Wafā' al-Būzjānī 1990–91 (1369), ed. Jazbī, 73, no. 2. See also Özdural 2000, 190–92.
53. Bulatov 1988, 333–34, figs. 46 and 47; Abū'l-Wafā' al-Būzjānī 1990–91 (1369), ed. Jazbī, 89, no. 52.
54. Bulatov 1988, 335, figs. 50 and 51; Abū'l-Wafā' al-Būzjānī 1990–91 (1369), ed. Jazbī, 90, nos. 53 and 54.
55. Bulatov 1988, 325, fig. 30; Abū'l-Wafā' al-Būzjānī 1990–91 (1369), ed. Jazbī, 83, between no. 32 and no. 33.
56. Bulatov 1988, 326, fig. 31; Abū'l-Wafā' al-Būzjānī 1990–91 (1369), ed. Jazbī, 83, no. 33.
57. Bulatov 1988, 330, fig. 40; Abū'l-Wafā' al-Būzjānī 1990–91 (1369), ed. Jazbī, 87, no. 46. See also Chorbachi 1989b, 777–85, and Kappraff 1991, 206–7.
58. Bulatov 1988, 338, fig. 55; Abū'l-Wafā' al-Būzjānī 1990–91 (1369), ed. Jazbī, 91, no. 60.
59. Bulatov 1988, 325, fig. 29; omitted in Abū'l-Wafā' al-Būzjānī 1990–91 (1369), ed. Jazbī.
60. Both Bulatov 1988 and Abū'l-Wafā' al-Būzjānī 1990–91 (1369), ed. Jazbī, omit this construction, but it is mentioned in Chorbachi 1989b, 777.
61. Abū'l-Wafā' al-Būzjānī 1990–91 (1369), ed. Jazbī, 87, no. 45; see also Chorbachi 1989b, 777. The construction is omitted in Bulatov 1988.
62. Bulatov 1988, 336 fig. 52, upper left; Abū'l-Wafā' al-Būzjānī 1990–91 (1369), ed. Jazbī, 90, no. 55.
63. Bulatov 1988, 336, fig. 52, upper right; Abū'l-Wafā' al-Būzjānī 1990–91 (1369), ed. Jazbī, 90, no. 56.
64. Bulatov 1988, 336, fig. 52, lower left; Abū'l-Wafā' al-Būzjānī 1990–91 (1369), ed. Jazbī, 90, no. 57.
65. Abū'l-Wafā' al-Būzjānī 1990–91 (1369), ed. Jazbī, 91, part of no. 58. The construction is omitted in Bulatov 1988.
66. Bulatov 1988, 336, fig. 52, lower right; Abū'l-Wafā' al-Būzjānī 1990–91 (1369), ed. Jazbī, 91, part of no. 58.
67. See Necipoğlu's chapter in the present volume, and also Cromwell and Beltrami 2011, 84–93.
68. See Özdural 1996, 191–211, esp. 197, and Özdural 1995, 54–71, esp. 56–64, where attention is drawn to an extant treatise on algebra by the mathematician and poet ʿUmar Khayyam. In this treatise, a fundamental triangle in the twelve-kite pattern is constructed by means of conic sections, but Khayyam does not inform the reader that the triangle is related to decorative patterns.
69. Bulatov 1988, 325, fig. 28; Abū'l-Wafā' al-Būzjānī 1990–91 (1369), ed. Jazbī, 82–83, nos. 29–32. See Özdural 1995, 67, fig. 12.
70. Bulatov 1988, 327, fig. 33; Abū'l-Wafā' al-Būzjānī 1990–91 (1369), ed. Jazbī, 84, no. 35. See Özdural 1995, 66, fig. 11.
71. Bulatov 1988, 328, figs. 36 and 37; Abū'l-Wafā' al-Būzjānī 1990–91 (1369), ed. Jazbī, 85–86, no. 40. On this construction, see Özdural 1996, 197, and Özdural 1995, 64–65, but note that fol. 191r–191v [36] is not a "verging construction" in the sense of ancient Greek geometry. For a further discussion of the relationship with Ibn al-Haytham and the T-shaped instrument, see Necipoğlu's chapter in the present volume.
72. I interpret the Persian *istinbāṭ* as "discovery," and not as "deduction," as in the English translation in this volume. In my opinion, the discovery in question is the mathematical insight that many constructions by conic sections can be realized by trial-and-error constructions by the T-shaped instrument. Therefore, I find it implausible to assume that the following word "*kātib*" refers to the scribe of the present manuscript, who was not trained in mathematics. In Özdural 1996, 196, Özdural (who based his interpretation on that of his Persian translator Zaka Siddiqi) interprets *istinbāṭ* as "inference," and concludes that "*kātib*" is "more likely the compiler himself, who could hardly be a mathematician, than the copyist or the translator." In her chapter, Necipoglu interprets "*kātib*" to mean "scribe," and thus has

the word been rendered in the English translation in this volume.

73. On the second part of fol. 191r–191v [36], see Özdural 1996, 196; on the "boat astrolabe" (*usṭurlāb-i zawraqī*), see Frank 1920, 18–21.

74. Abū'l-Wafā' al-Būzjānī 1990–91 (1369), ed. Jaẕbī, 78, no. 19. Bulatov 1988, 323, states that the "text of this construction is not understood," and the corresponding figure in Bulatov 1988, 321, fig. 18, is incorrect.

75. Bulatov 1988, 321 fig. 20; Abū'l-Wafā' al-Būzjānī 1990–91 (1369), ed. Jaẕbī, 79, second part of no. 20. On this construction, see Özdural 1996, 201, fig. 6

76. Abū'l-Wafā' al-Būzjānī 1990–91 (1369), ed. Jaẕbī, 81, no. 27; omitted in Bulatov 1988. On this construction, see Özdural 1996, 200, fig. 4.

77. Bulatov 1988, 324, fig. 27; Abū'l-Wafā' al-Būzjānī 1990–91 (1369), ed. Jaẕbī, 82, no. 28. See Özdural 1996, 200, fig. 5.

78. Abū'l-Wafā' al-Būzjānī 1990–91 (1369), ed. Jaẕbī, 84, no. 36-37. The construction was misinterpreted in Bulatov 1988, 329–30, and the corresponding figure in Bulatov 1988, 327, fig. 34, is incorrect. On this construction, see Özdural 1996, 198–99.

79. Bulatov 1988, 321, fig. 19; Abū'l-Wafā' al-Būzjānī 1990–91 (1369), ed. Jaẕbī, 79, beginning of no. 20.

80. See Özdural 1996, 197; Özdural 1995, 56–64; and Hogendijk 2012, 41, for more information and references. Here is an easy method to find a cubic equation in a simple form. In figure 14 we no longer assume that the big square is known, but we choose HA as our point of departure. Put $HA = BE = 1$, $AE = BF = x$, then $HE = EF = \sqrt{1+x^2}$ by the Pythagorean theorem. By similar triangles $HA : AT = HE : EA$, so $1 : AT = \sqrt{1+x^2} : x$ whence $AT = x/\sqrt{1+x^2}$. Because $AH + AT = EH$ we have $1+x/\sqrt{1+x^2} = \sqrt{1+x^2}$, which boils down to $x^3 - 2x^2+2x - 2 = 0$ with positive real root $x = 1.54369$ .... Hence, the larger acute angle in the right-angled triangle is $\angle AHE = \angle BEF = \arctan(x) = 57.0649$ ... degrees. Özdural 1995, 64, argues that the scribe of the manuscript was confused, and that his reference to Ibn al-Haytham was incorrect and should have been a reference to ʿUmar Khayyam. I disagree with Özdural's argument on this point. Ibn al-Haytham was one of the foremost experts in conic sections in the Islamic tradition, much more so than ʿUmar Khayyam; the fact that a treatise on the relevant triangle did not occur in any of Ibn al-Haytham's lists of his own works is unimportant because other extant works by Ibn al-Haytham do not occur in the same lists either. See also Özdural 1996, 205n45.

81. If in figure 15 we make the side of the little square equal to 1, we have $AG = 2\sqrt{2}$, and by similar triangles $AE/AG = 1/(2\sqrt{2} - 1)$. Since $EZ = AG$, it follows that $AZ/AG = 1 + 1/(2\sqrt{2}-1) = 2\sqrt{2}/(2\sqrt{2}-1) = (8 + 2\sqrt{2})/7$; therefore $\angle ZGA = \arctan(8 + 2\sqrt{2})/7 = 57.1195...^\circ$, which is only 0.056 degrees more than the exact value.

82. In figure 14, by symmetry $EQ = EK$, and since the triangles EKF and AHE are congruent, we have $EK = AH$, and so $EQ = AH$ in figure 14; thus $BK = AE$ in fig. 16.

83. If we put in fig. 17 $BK = 1$ and $AK = x$, and also $DG = GE = EK = KL = y$, we have $BD = AG = 1 - y$. By similar triangles $AG : GE = AK : KB$ or $(1 - y) : y = x : 1$ or $y = 1/(1 + x)$ and $AB = 2 - y = (2x + 1)/(x + 1)$. Then, by the Pythagorean theorem $(2x+1)^2/(x+1)^2 = 1+x^2$, which equation boils down to $x^3+2x^2-2x-2 = 0$. This equation has a positive real root $x = 1.170086$ .... which produces $\angle ABK = \arctan(x) = 49.481 ...^\circ$ and $\angle BAK = 40.518 ...^\circ$. In fol. 185v [17], one of the angles of the right-angled triangle is $\arcsin(2/3) = 41.81 ...^\circ$, which may be interpreted as an approximation to $\angle BAK$ in fig. 14.

84. This figure is omitted in Bulatov 1988, and also in Abū'l-Wafā' al-Būzjānī 1990–91 (1369), ed. Jaẕbī.

85. Bulatov 1988, 334, fig. 48; Abū'l-Wafā' al-Būzjānī 1990–91 (1369), ed. Jaẕbī, 89, figure between nos. 52 and 53.

86. See Bulatov 1988, 334, fig. 49; Abū'l-Wafā' al-Būzjānī 1990–91 (1369), ed. Jaẕbī, 89, figure between nos. 52 and 53.

87. These three patterns were first identified as *muqarnas* projections by Özdural: see his chapter ("Preliminaries") in this volume, and his plates reconstructing the *muqarnas* tiers (pls. 15, 46, and 47). On *muqarnas*, see Necipoğlu 1995, 4–27, 44–50, 159–60, 179, 350, 353, 359, and 173, as well as al-Asad's appendix on "The Muqarnas: A Geometric Analysis." I thank Professor Necipoğlu for providing me with these detailed references. See also Dold-Samplonius 1992, 193–242.

BIBLIOGRAPHY

Abū'l-Wafā' al-Būzjānī. 1990–91 (1369). *Handasa-i Īrānī: Kitāb-i Tijārat: Fī mā yaḥtāju ilayhi al-ʿummāl wa-al-ṣunnāʿ min al-ashkāl al-handasiyya, yā, Kārburd-i handasa dar ʿamal.* Edited and adapted (to modern Persian) by Sayyid ʿAlī Riżā Jaẕbī. Tehran: Surūsh.

Al-Asad, Mohammad. 1995. "The Muqarnas: A Geometric Analysis." In *The Topkapı Scroll: Geometry and Ornament in Islamic Architecture; Topkapı Palace Museum Library MS H. 1956* by Gülru Necipoğlu, 349–59. Santa Monica, Calif.: Getty Center for the History of Art and the Humanities.

Blochet, Edgar. 1905–34. *Catalogue des manuscripts persans.* 4 vols. Paris: Imprimerie nationale.

Brentjes, Sonja. 1994. "Textzeugen und Hypothesen zum arabischen Euklid in der Überlieferung von al-Ḥaǧǧāǧ b. Yūsuf b. Maṭar (zwischen 786 und 833)." *Archive for History of Exact Sciences* 47, no. 1: 53–92.

Bulatov, Mitkhat S. 1988. *Geometricheskaia garmonizatsiia v arkhitekture Srednei Azii IX–XV vv.* Rev. ed. Moscow: "Nauka." Original edition Moscow, 1978.

Chorbachi, Wasma'a K. 1989a. "Beyond the Symmetries of Islamic Geometric Patterns: The Science of Practical Geometry and the Process of Islamic Design." PhD diss., Harvard University.

———. 1989b. "In the Tower of Babel: Beyond Symmetry in Islamic Design." *Computers and Mathematics, with Applications* 17:751–89.

Cromwell, Peter R. 2015. "Cognitive Bias and Claims of Quasiperiodicity in Traditional Islamic Patterns." *The Mathematical Intelligencer* (June 20, 2015). doi: 10.1007/s00283-015-9538-9.

Cromwell, Peter R., and Elisabetta Beltrami. 2011. "The Whirling Kites of Isfahan: Geometric Variations on a Theme." *The Mathematical Intelligencer* 33, no. 3: 84–93.

Dold-Samplonius, Yvonne. 1992. "Practical Arabic Mathematics: Measuring the Muqarnas by al-Kāshī." *Centaurus* 35:193–242.

Frank, Josef. 1920. *Zur Geschichte des Astrolabs.* Erlangen: Universitäts-Buchdruckerei von Junge & Sohn.

Heath, Thomas L., ed. and trans. 1956. *The Thirteen Books of Euclid's Elements.* Translated from the text of Heiberg. 2nd ed. Cambridge: Cambridge University Press, 1956.

Hogendijk, Jan P. 2012. "Mathematics and Geometric Ornamentation in the Medieval Islamic World." *Newsletter of the European Mathematical Society* 86:37–43.

———. 2008. "Ancient and Modern Secrets of Isfahan." *Nieuw Archief voor Wiskunde,* 5th ser., 9, no. 2: 121.

Kappraff, Jay. 1991. *Connections: The Geometric Bridge between Art and Science.* New York: McGraw-Hill Pub. Co.

Kheirandish, Elaheh. 2006. "Organizing Scientific Knowledge: The 'Mixed' Sciences in Early Classifications." In *Organizing Knowledge: Encyclopaedic Activities in the Pre-Eighteenth Century Islamic World,* edited by Gerhard Endress, 135–54. Leiden: Brill.

Necipoğlu, Gülru. 1995. *The Topkapı Scroll: Geometry and Ornament in Islamic Architecture; Topkapı Palace Museum Library MS H. 1956.* Santa Monica, Calif.: Getty Center for the History of Art and the Humanities.

Özdural, Alpay. 2000. "Mathematics and Arts: Connections between Theory and Practice in the Medieval Islamic World." *Historia Mathematica* 27, no. 2: 171–201.

———. 1996. "On Interlocking Similar or Corresponding Figures and Ornamental Patterns of Cubic Equations." *Muqarnas* 13:191–211.

———. 1995. "Omar Khayyam, Mathematicians and *Conversazioni* with Artisans." *Journal of the Society of Architectural Historians* 54, no. 1: 54–71.

Richard, Francis. 1989. *Catalogue des manuscrits persans.* Vol. 1, *Ancien fonds.* Paris: Bibliothèque nationale de France.

Steinschneider, Moritz. 1865. "Die 'mittleren' Bücher der Araber und ihre Bearbeiter." *Zeitschrift für Mathematik und Physik* 10:456–98.

Suter, Heinrich. 1922. "Das Buch der geometrischen Konstruktionen des Abu'l Wefā'." *Abhandlungen zur Geschichte der Naturwissenschaften und Medizin* 4:94–109.

Woepcke, Franz. 1855. "Recherches sur l'histoire des sciences mathématiques chez les Orientaux d'après des traités inédits arabes et persans. Deuxième article. Analyse et extrait d'un recueil de constructions géométriques par Aboūl Wefā." *Journal Asiatique,* ser. 5, no. 5: 218–56, 309–59.

BRILL

ALPAY ÖZDURAL

PRELIMINARIES

The subject of this study is a Persian work on ornamental geometry. Discovered in the 1970s, the only extant manuscript copy known so far comprises folios 180r–199r of Ms. Persan 169, a volume of collected works in the Bibliothèque nationale de France, Paris. This volume, referred to henceforth as the Paris Codex, consists of twenty-five treatises on the mathematical sciences, covering a variety of topics ranging from the astrolabe to finger counting and a period from the late tenth to the mid-fifteenth century. Altogether they give the impression of having been collected and bound for somebody who was interested in the practical applications of mathematics in various fields.[1]

The document to be considered here is listed as the twenty-fourth treatise in the Paris Codex. This work (i.e., the *Anonymous Compendium*) was mentioned only briefly in the catalogues of Edgar Blochet and Francis Richard, who described it as containing "solutions for different geometric problems accompanied by figures."[2] After its discovery by historians of Islamic art and architecture, however, its significance for both fields began to be appreciated. It is a unique document in which both verbal and pictorial descriptions of geometric constructions for scores of ornamental patterns are provided. It thus combines the information that can be supplied by mathematical sources and gathered from architectural drawings preserved in the form of scrolls. This dual character sheds new light on the creative process of the architectural arts.

The main hypothesis underlying this study is that mathematicians played an active role in Islamic art and architecture. This unique document provides us with ample evidence in support of this argument. Even at first glance, it appears to embody the informal link between theory and praxis in the Islamic world. To demonstrate the collaboration of mathematicians and artisans that it represents, in Chapter 2, titled "Analyses," each construction will be analyzed with respect to the history of mathematics on the one hand, and the history of art and architecture on the other. Then, in Chapter 3, "Synthesis," the treatise as a whole will be assessed in terms of topics around which the constructions are loosely classified, in order to illustrate its role as the documentation of a series of meetings between theoreticians and practitioners of geometry.

A SURVEY OF THE LITERATURE

In the last thirty years, our document began to draw scholarly attention, particularly from the fields of Islamic art and architecture. The information it provides sheds light on how those geometric ornaments, which still intrigue us (today), were actually conceived and designed. We thus gain a sharp insight into what the intimate link between theory and praxis really meant in terms of Islamic art. It may signify why geometry gained predominance in Islamic culture. Despite its far-reaching implications for the history of art and architecture, however, it was until recent years considered an insignificant work with respect to the history of mathematics.

Franz Woepcke, who in the mid-nineteenth century published an abridged French translation of the *Geometrical Constructions* by the Iranian mathematician Abu'l-Wafa' al-Buzjani (d. 998), acknowledged that some of his constructions could have been utilized in the execution of architectural monuments and particularly in their decoration.[3] Woepcke's translation was

 | DOI 10.1163/9789004315204_006

based on the abridged Persian translation of Abu'l-Wafa''s book on practical geometry, which immediately precedes our document in the Paris Codex. (That is, it is the twenty-third treatise in Ms. Persan 169.) As both works seem to share the same objective, some constructions in each are treated similarly; our anonymous document (i.e., the *Anonymous Compendium*) seems to have been appended to the *Geometrical Constructions* at a later date. This situation could not have escaped Woepcke's attention, yet he made no mention of that document. His silence suggests that he did not regard it as worthy of study. It is not in fact surprising that it did not appeal to nineteenth-century historians of mathematics, since they were interested in the advances of Arab mathematics rather than in the applied geometry of the ornamental arts. A similar interpretation can be made for Charles Storey's lack of interest in our document. He, too, made no reference to it, although he cited most of the treatises in the Paris Codex.[4]

Wasma'a K. Chorbachi tells us that it was she who first identified the document under examination here, in 1971, during the early stages of her doctoral studies, and that she introduced it to the scholarly world. Her doctoral dissertation, completed at Harvard University in 1989, was mainly focused on this document.[5] To the regret of interested scholars, however, it was not available for public use until 1997. The historical information she gathered is valuable for understanding the milieu of the *Geometrical Constructions* and, by extension, our document. She is of the opinion that these were two completely independent texts and dated the latter to a few centuries after the time of Abu'l-Wafa'. To give an idea of the contents of the document, she analyzed ten patterns as "example problems" in her dissertation.[6] In an attempt to find "a scientific language and methodology with which to understand and systemically categorize Islamic geometric patterns" and to "search for the proper scientific languages and tools to generate new forms and expressions," she based her analyses on group theory and crystallography. These modern notions and the complex set of notations that they yield may be acceptable from the viewpoint of a present-day artist, but they are of little help in understanding how medieval artisans conceptualized those patterns.

In 1989 Chorbachi also published a summary of her views on this subject.[7] After explaining the story of her research in a personal manner, she commented on previous studies on the application of geometry in Islamic art and architecture, and analyzed the underlying geometric principles of two constructions in our document (fols. 191r and 192v). She regarded the complex geometric shapes in this anonymous treatise as representing "a much higher and later stage of development" than the simple shapes and polygons of Abu'l-Wafa''s work.[8] In their joint publication on the same subject, she and Arthur L. Loeb analyzed the first construction in our document (fol. 180r) and dated it to around the late twelfth or early thirteenth century.[9]

A Russian translation of the anonymous treatise (i.e., the *Anonymous Compendium*) was published by Mitkhat Bulatov as an appendix to his book on the architecture of Central Asia.[10] The translation itself, undertaken by A. B. Vildanova, is not completely accurate, and some passages were omitted altogether because they were deemed undecipherable. The mathematical commentary is deficient at times because Bulatov did not distinguish between accurate constructions and approximations. Also, the neatly drawn figures are unreliable in certain instances because they contain many additions by Bulatov. He regarded the anonymous treatise as a textbook that explained the basic principles of geometric decorations and reflected the state of applied geometry at the time (of its composition). A Persian edition of our document together with Abu'l-Wafa''s *Geometrical Constructions* was published by ʿAli Riza Jazbi in 1990–91.[11]

With Bulatov's publication, our document began to attract the attention of leading Western scholars. Renata Holod, as well as Lisa Golombek and Donald Wilber, treated it as a scarce bit of evidence for establishing design theories in Islamic art and architecture.[12] In a comprehensive study on the role of geometry and ornament in Islamic architecture, Gülru Necipoğlu emphasized that the importance of the anonymous treatise lies in documenting the intimate link between the practical geometry that was popularized in late tenth-century Baghdad by such mathematicians as Abu'l-Wafa' and the later scroll tradition of architectural drawings for which we have examples starting from the late fifteenth century.[13] She is of the opinion that the anonymous author of our document was a *muhandis* (geometer) with practical rather than theoretical training in geometry, and that the document also records the interaction of professional mathematicians and practitioner artisans. She drew attention to the content of the treatise, composed of both scientifically correct geometric constructions and practical approximations, and made the following observation:

> Although a detailed analysis of the *Aʿmāl wa ashkāl* [the title Necipoğlu adopted from Richard for the anonymous treatise, after citing its original heading] lies beyond the scope of this book, its relationship to later *girih* [the Persian term, meaning "knot," for interlocking star-and-polygon patterns] scrolls preserved from the Timurid period onward needs to be assessed. This twenty-page "how-to" manual, providing instructions for the construction of sixty-one repeat units for planar geometric patterns, is of particular interest because it shows how the general principles outlined in the practical geometries of [Abu'l-Wafa'] al-Buzjani and others were used in developing interlocking star-and-polygon compositions.[14]

I wrote a series of articles on what Necipoğlu described as lying beyond the scope of her book. In the first one, I concentrated on an ornamental pattern based on a cubic equation the solution for which was discovered by means of a triangle, as discussed in a treatise by the eleventh-century poet-mathematician ʿUmar Khayyam.[15] This pattern demonstrated the contribution of a celebrated mathematician to the ornamental arts in the Islamic world. For that article I used the Russian translation of Bulatov's publication and assessed the mathematical content of the treatise—particularly in comparison with the *Geometrical Constructions*. In this piece I formed the opinion that the anonymous author was an artisan who had only a shallow acquaintance with geometry, rather than a geometer immersed in the ornamental arts. I also started to develop the idea that at major urban centers in the Islamic world, where the sciences and arts had flourished together, meetings (sing. *majlis*) were held between mathematicians and artisans for purposes of collaboration and the exchange of architectural and geometric knowledge.

When I was able to study the original manuscript itself, rather than relying on the defective translations in Bulatov's publication, I reached the conclusion that our anonymous treatise actually documented the collaboration between theoreticians and practitioners. There was apparently no single author who composed the treatise. It was rather a compilation of notes taken by a professional scribe during a series of meetings attended by mathematicians and artisans who produced geometric constructions in response to different needs, ranging from simple procedures to delicate patterns involving conic sections. Having discovered another construction based on a cubic equation, which was solved again by "moving geometry," I wrote my second article on the anonymous treatise in 1996 and rectified my views.[16]

Lately, our document has begun to attract the attention of historians of mathematics as well. In addition to the ornamental patterns of cubic equations, Jan Hogendijk commented upon the constructions utilizing a practical method of dissecting and reassembling.[17] I also wrote an article on this cut-and-paste method, which was frequently used in Abu'l-Wafa''s *Geometrical Constructions* and the anonymous treatise in order to visually demonstrate certain geometric theorems to practical-minded artisans, and showed how theory and practice were intimately connected in the medieval Islamic world.[18]

According to the idea that I have been developing, some of the aesthetic, structural, and spatial innovations that we observe in the architectural products of major urban centers may be explained as the contributions of certain talented mathematicians. I wrote another article arguing that ʿUmar Khayyam (d. 1131), one of the greatest intellects that the Islamic world has produced, was behind the conceptual design of the North Dome in the Friday Mosque of Isfahan, which is generally considered one of the masterpieces of Islamic architecture.[19] The main point of my argument was that the proportions of the North Dome were generated wholly from the triangle ʿUmar Khayyam had discovered for formulating the cubic equation that later served as the basis of an ornamental pattern in our document. I summarized my findings concerning the use of cubic equations in Islamic art and architecture in another article in 2002.[20]

THE MANUSCRIPT

Ms. Persan 169 is written on rag paper, probably from Samarqand, entirely in the *nastaʿlīq* script. It came to the Bibliothèque nationale from the library of Jean de Thévenot (d. 1667). The bookplate of a certain Sinan Çavuş, which appears on fol. 2, indicated that before being acquired by Thévenot the manuscript belonged to a low-ranking Ottoman officer. It is not known how or from where Sinan Çavuş obtained the manuscript. According to Blochet, it was copied in the early seventeenth century; Richard assumed an earlier date in the sixteenth century.

Both Blochet and Richard are of the opinion that the same hand copied all twenty-five of the treatises that are bound together in Ms. Persan 169. Priscilla Soucek dates the calligraphy of the entire manuscript to the mid-fifteenth century on stylistic grounds.[21] The figures in various treatises also seem to have been drawn by the same hand. It appears that the copyist, who rewrote all the treatises compiled in the manuscript volume along with their newly drawn figures, was familiar with geometric constructions—that is, was a mathematician himself. Although the name of the copyist is not specifically mentioned, we can establish his identity fairly confidently through the information provided in the twenty-third treatise in the Paris Codex.

In the epilogue to the Persian translation of Abu'l-Wafa''s *Geometrical Constructions* (section XXIII, fols. 179r–179v), we are told that the person who explained (or "solved," *hall*) and executed (*ʿabr*) the figures (or "propositions," *ashkāl*) of this treatise is Abu Ishaq ibn ʿAbdallah Kubanani (or "Kuyanani") Yazdi. Referring to himself as "I, the miserable mortal," Abu Ishaq Kubanani finished his task in the course of two months at the request (*bar hasb-i iltimās*) of a mathematician/master craftsman (*ustādī yā muhandis*) [*sic*, actually, master geometer (*ustād-i muhandis*)] named Ustad Shams al-Din Abu Bakr Shah ibn Ustad Najm al-Din Mahmud Shah ibn Hajji Taj al-Din, who was described as unique in his art (*yagān-i bar hunar*). The poetical passages related to him mention bricks (*khisht*) and a workshop (*kārkhāna*). Abu Ishaq Kubanani also says that he wrote during a distressing time, following the death of his colleague (literally "brother"), Shaykh Najm al-Din Mahmud, who was the father of the *muhandis* who had requested that Abu Ishaq Kubanani do the book. Shaykh Najm al-Din Mahmud was a fixer of astronomical tables (*rasad-band*) and the author of a commentary (*sharḥ*) on the *Almagest* (Ptolemy's famous book on astronomy) and of the marginal notes (*ḥāshiyya*) to the Amendment of the Book of Menelaus on the Spherical Figures (*Iṣlāḥ kitāb Manālfī al-ashkāl al-kuriyya*) by Amir Mansur Nasr ibn ʿIraq. Najm al-Din Mahmud had also done a translation of the present book but had died young, and part of it (*baʿżī az ān*) had been lost to the vicissitudes of time. His son, Shams al-Din Abu Bakr, had not wanted his work to have been in vain, and desired that it be completed. Abu Ishaq Kubanani, who then completed the translation, asked learned people (*fużalā*) of his time to examine the intricate problems discussed in the book so that they might appreciate the extent to which human ability was able to grasp "the works of the earth" (*kārhā-yi zamīn*, i.e., geometry).[22]

According to the information gathered from Ahmad Munzavi's comprehensive catalogue of Persian manuscripts, Abu Ishaq Kubanani, who was also known as Shaykhzada Hadim Burhani, was a mathematician and astronomer active in 1442–43 at Sari and then in 1459 at Kirman and Hormuz.[23] We can infer from his words that he, Shams al-Din Abu Bakr, and Najm al-Din Mahmud (the three mathematicians involved in that particular Persian translation of *Geometrical Constructions*) lived in the mid-fifteenth century. Their dates agree perfectly with Soucek's dating of the calligraphy of the manuscript. Therefore, it would be reasonable to assume that in addition to translating Abu'l-Wafa''s work, Abu Ishaq Kubanani copied and redrew the figures of all the treatises that are bound together in Ms. Persan 169.

Although all the treatises show similar calligraphic features, our document differs from the rest with respect to the disorganized nature of its folios. Some constructions lack explanatory texts, and are sometimes not even properly laid out. Instead they are inclined in various directions—vertically, following a curve, or upside down. The accompanying texts often exhibit the same type of disorganization. In some cases they are squeezed in between constructions and stop at mid-sentence. All these irregularities give the impression that these were notes spontaneously jotted down at a meeting. It appears that Abu Ishaq Kubanani copied the original document without changing its layout.

THE AUTHOR

Our document is truly an anonymous work, seemingly without an individual author who was intellectually responsible for the contents as a whole. We cannot even call it a treatise since it lacks a predetermined structure, an argument, or even a logical sequence. At first glance, it appears to be a collection of geometric constructions loosely grouped around certain topics. Upon closer inspection, it reveals an instructive pattern, which will be discussed in the last chapter, indicating that it was the result of the collaborative work of a group of artisans and mathematicians.

The person who actually wrote the document referred to the craftsmen as *ṣunnā* (artisans) or *ustādān* (masters), while for the mathematicians he used the phrase "experts in geometry" (*aṣḥāb-i handasa*). The copyist likely belonged to neither of these groups but was probably a professional scribe charged with recording the discussions held at gatherings of artisans and mathematicians.

In mathematical and literary sources there are quite a number of direct or indirect references to meetings between mathematicians and artisans in the Islamic world. The meeting that Abu'l-Wafa' described in great detail will be discussed later, as it provides very useful information for understanding our document. In an untitled treatise ʿUmar Khayyam remarked on the meeting that prompted him to write that treatise, which appears to have been addressed to both mathematicians and artisans in eleventh-century Isfahan.[24] In a letter to his father, the Timurid mathematician Giyath al-Din al-Kashi hinted at a problem he solved at a gathering of dignitaries, mathematicians, and artisans near the construction site of the Samarqand observatory in the fifteenth century.[25] The Ottoman author Caʿfer Efendi, who appears to have been a professional scribe, tells us in the *Risāle-i miʿmāriyye* that he had earlier composed a treatise on geometry from the notes he had taken at a series of meetings, over a period of twenty years, between architects and mathematicians in late-sixteenth- and early-seventeenth-century Istanbul.[26]

In light of these references, our anonymous document appears to consist of the compiled notes of a scribe such as Caʿfer Efendi, who recorded the discussions held between mathematicians and artisans at a series of meetings (over a period that was likely much less than twenty years). The similarities between the anonymous treatise and Caʿfer Efendi's now-lost work on geometry suggest that there may have been other proceedings of such technical meetings. Our document seems to be the only extant example of its kind.

THE TITLE

A note written vertically on the right-hand side of the first folio, 180r, reads: "*Fī tadākhul al-ashkāl al-mutashābiha aw al-mutawāfiqa.*" Both Blochet and Richard regarded this as a mere marginal note and in their respective catalogues treated the document as untitled. Richard assigned it the title *Aʿmāl va ashkāl* (Geometrical Constructions and Figures), which describes its contents succinctly but appears nowhere in the text. Necipoğlu, too, adopted this invented title, while at the same time accepting and citing the original heading. In other studies, the marginal note was regarded as the title of the work with slightly different wordings. In Bulatov's publication Vildanova misread the second word as *madākhil* (introduction). Chorbachi read it correctly as *tadākhul* and translated it as the adjective "interlocking." In my earlier articles I accepted Chorbachi's translation. After analyzing the document closely, I now think that the word conveys the main idea for creating the document and so should be translated as a noun: "interlock." Another problematic word is "*al-mutawāfiqa.*" Chorbachi translated this term as "congruent," but in medieval Islamic mathematical literature it was never used in that sense. Steingass defines it as: "agreed, leagued, confederate; (in arithmetic) commensurable."[27] In ancient and medieval mathematics the term "commensurable" is normally used to qualify the magnitudes that can be expressed by numbers. This word is not suitable for our case because most of the figures in our document are comprised of incommensurable magnitudes produced by geometric means. I had earlier used the word "corresponding," in the sense of "agreeing" or "fitting together," but to convey the idea more directly, I now translate the word as "complementary."

Thus, I translate the entire marginal note as "On interlocks of similar or complementary figures."

In the course of compiling the notes he had taken at the meetings, the scribe seems to have added the marginal note to describe succinctly the peculiar nature of the topics that had been discussed. Since it is authentic and a fairly accurate description of the contents, it would be more appropriate to regard this as the title, rather than using Richard's invented title. I will thus refer to our document in abbreviated fashion as "Interlocks of Figures."

THE DATE

No specific information is given in "Interlocks of Figures" about the date it was compiled. Various scholars have proposed different dates. Bulatov, assuming that it was created as an appendix to the Persian translation of *Geometrical Constructions* in the Paris Codex, dated both works to the early eleventh century.[28] He correctly identified Mansur Nasr ibn ʿIraq, who was mentioned in the epilogue, as Abu Nasr Mansur ibn ʿAli ibn ʿIraq, who was active in the early eleventh century, but misinterpreted the relation between him and Najm al-Din Mahmud.[29] Saying that the latter made the translation at the request of the former, Bulatov dated the Persian translation of *Geometrical Constructions* to the early eleventh century. But the fact that Abu Ishaq Kubanani and Najm al-Din were contemporaries disproves Bulatov's argument. Golombek and Wilber accepted the eleventh-century date but suggested that some of the constructions that appear at the end may have been added to the original set during the Timurid period, when it was copied.[30] On the basis of the likeness of a pentagonal star figure in "Interlocks of Figures" to one on the decorations of the so-called Abbasid Palace in Baghdad, Chorbachi and Loeb attributed the approximate date of the building, between 1180 and 1230, to our document as well.[31] Necipoğlu cautiously proposed a wider range of dates, between the early eleventh and early thirteenth centuries, adding that the more exact date of the anonymous treatise could be firmly established once the identity of Abu Bakr al-Khalil, whose name appears twice in "Interlocks of Figures" as a contemporary mathematician, is ascertained.[32]

Fortunately, some newly acquired information about Abu Bakr al-Khalil enables us to establish a secure date for our document. Ms. 5429, an Arabic manuscript in the Library of the Holy Shrine in Mashhad, gives the name of the copyist and the date of copying: ʿAli b. Khalil al-Tajir in 728 (1327–28).[33] The father of the copyist, Khalil al-Tajir, must have been Abu Bakr al-Khalil al-Tajir, who should have been active in the late thirteenth and early fourteenth centuries. Our document can thus be dated to circa 1300.

PROVENANCE

The established date of our document, which falls in the Ilkhanid period of the Persian-speaking world, plays a pivotal role in identifying its provenance. For almost one hundred and fifty years of Seljukid rule, until the beginning of the thirteenth century, Iran enjoyed material prosperity and architectural richness. Following a lull in artistic activity with the advent of the Mongols in 1220, Central Asia, Iran, Iraq, and part of Anatolia were all conquered by the invading armies in a matter of a couple of decades, causing catastrophic destruction in scores of urban centers. Artisans were sometimes spared, but conditions were certainly not favorable for artistic activities during that period: very few monuments have survived that can be dated to between 1220 and 1270. Some artisans escaped the destruction by fleeing to the unoccupied parts of Anatolia and working on Seljukid construction there. During these years of inactivity in Iran, it is likely that almost two generations of artisans preserved their hereditary training within their own family circles. The revival of architecture, which began under Hülegü (r. 1245–65), was a slow and laborious process. A notable example was the construction of the observatory at Maragha (ca. 1258) under the supervision of the famous mathematician-astronomer Nasir al-Din Tusi (d. 1273). During this period Tabriz became the capital of the Ilkhanids, and the wealth and culture of the country shifted to Azerbaijan.

During the reign of Abaqa (1265–82) artistic and architectural activity was still scarce, and only a single structure at Takht-i Sulayman bears witness to his construction work. Tabriz, however, was on its way to becoming a rich commercial and cultural center. Arghun

(r. 1284–91) was the first ruler to express real interest in large-scale architecture. Before he died, a good deal of construction work was undertaken in Arghuniya, a splendid suburb he had established in the neighborhood of Tabriz. When the three Polo brothers passed through Tabriz several years later, they described it as "a great noble city and a very prosperous commercial center."[34]

During the disastrous years of the Mongol invasion, southern Iran had been spared and local Iranian dynasties continued to rule by paying tribute to the Ilkhanids. According to literary sources, Shiraz in particular was embellished with many fine buildings during those years. It is logical to expect that architectural activity continued uninterruptedly in this region, albeit on a modest scale. It may be that this area, together with the secure parts of Anatolia, furnished the stock of artisans necessary for the large-scale architectural campaigns that took place in Azerbaijan at the turn of the fourteenth century.

The reign of Ghazan Khan (1295–1304) started the golden age of the Ilkhanids. A great statesman who showed a personal and active interest in architecture, he also had practical knowledge and skill in various other fields, including astronomy. In a very short time he completed the construction of the suburb established by Arghun and started a new one, Ghazaniya, which soon rivaled Tabriz in size. As evidenced by the construction campaigns he undertook, he was a man of great vision and equipped with tremendous organizational skills. A great number of scholars, scientists, and artisans were gathered in Tabriz, which had by then become a very cosmopolitan city. His famous vizier, Rashid al-Din (d. 1318), tells us that many master artisans (sing. *ustād*) and geometers (sing. *muhandis*) were involved in the construction of his tomb.[35] Rashid al-Din himself established the suburb known as the "Quarter of Rashid" outside Tabriz and dedicated it to the encouragement of the arts and sciences. Crowds of scholars and men of science were drawn to the quarter, which contained twenty-four caravanserais, 1,500 shops, 30,000 houses, and many other public buildings. This busy pace of architectural activity continued during Öljeytü's reign (1304–16). He started and completed the construction of Sultaniya to replace Tabriz as the capital, and his mausoleum still stands there as testament to the great achievements of Ilkhanid architecture.[36]

From this brief historical account, Tabriz seems to have been the most likely place of origin for "Interlocks of Figures." Owing to the scarcity of architectural activity in the thirteenth century, the number of artisans had dwindled. The unprecedented scale of construction undertaken at the turn of the fourteenth century must have lured builders and craftsmen of all trades to the city. Those who were especially skilled must have been highly regarded, along with other technical people and mathematicians. As suggested by Rashid al-Din's account of Ghazan Khan's tomb and other evidence, the outburst of creative architectural energy in Tabriz at that time led to an intimate collaboration between artisans and mathematicians. It was to this collaboration, apparently, that we owe the creation of our unique document.

ABU'L-WAFA''S *GEOMETRICAL CONSTRUCTIONS*

What we observe in our document is a unique type of geometry that would be used exclusively by artisans—that is to say, a work on ornamental geometry.[37] It is not a textbook but a collection of geometric procedures and patterns that would be useful for artisans. Nor is it the work of a single author but the joint product of a number of mathematicians and artisans. The mathematicians who contributed to "Interlocks of Figures" (i.e., the *Anonymous Compendium*) appear to have been well aware of Abu'l-Wafa''s *Geometrical Constructions*, which was probably intended as a textbook on geometry for artisans. But, as we observe in "Interlocks of Figures," produced three centuries later, it was mathematicians rather than artisans who in the meantime had become familiar with Abu'l-Wafa''s work. Because of its implicit connection with our document, we need to have more information about this book before analyzing the constructions of "Interlocks of Figures" in the following chapter.

Unlike our document, *Geometrical Constructions* is a well-organized, well-structured work that deals with geometry in a systematic manner. Abu'l-Wafa''s treatise, which is divided into thirteen chapters, starts with the instruments and basic procedures of geometry and then gradually proceeds to more complex issues such as regular polygons, superimposed figures, divisions of triangles and squares, and, finally, polyhedrons.[38] Certain advanced subjects such as trisecting the angle, duplicating the cube, and "burning mirrors" (i.e., parabolic curves) are regarded as basic procedures. For these constructions, which involve cubic equations, Abu'l-Wafa' offered practical methods of "moving geometry" (namely, verging procedures), rather than solutions based on conic sections, which were popular among mathematicians at that time. All the constructions he proposed were deliberately designed in accordance with the needs and intellectual capacity of artisans. For instance, for the constructions of the heptagon—which preoccupied prominent mathematicians in the tenth and eleventh centuries who produced solutions involving conic sections but not moving geometry—his proposals were based on approximation methods. All his other constructions were theoretically correct, but he did not offer axiomatic proof for any of them. His aim was not to make geometers out of artisans, who were mostly illiterate, but to instruct them in the correct methods of geometry so that they would attain perfection in their works.

Abu'l-Wafa' was certainly aware of the tendency among artisans to learn through observation rather than deductive reasoning. To teach artisans the basic elements of geometry, he preferred to utilize a method that they were already familiar with in demonstrating the proofs. According to this method, which he called "dissecting and assembling" (*qismat wa tarqīb*), a figure is dissected into pieces and then reassembled into another figure, illustrating visually that their areas are equal. This method resembles the "cutting-and-pasting" today most commonly associated with word processing and will be referred to hereafter as the "cut-and-paste" method. In the hands of resourceful and artistically imaginative mathematicians, it also had the potential of serving for ornamental purposes. Some very elegant solutions later became favorite ornamental motifs throughout the Islamic world.

In the introduction to chapter 10 of his *Geometrical Constructions*, Abu'l-Wafa' noted that artisans widely used the technique of dissecting figures. He stated that this chapter aimed to describe rules that artisans should follow to avoid making mistakes in dissecting and assembling squares. Later in the chapter he elaborated on the differing attitudes of mathematicians and artisans toward geometry:

> A number of geometers (sing. *muhandis*) and artisans have erred in the matter of these squares and their assembling. The geometers [have erred] because they have little practice in constructing, and the artisans [have erred] because they lack knowledge of the proofs. The reason is that since the geometers do not have experience in construction, it is difficult for them to approximate [i.e., construct in practice]—in the way required by the artisan—what is known to be correct by proofs by means of lines [i.e., diagrams]. The aim of an artisan is what is approximated by the construction, and [for him] correctness is apparent by what he perceives through his senses and by inspection. He is not concerned with the proofs by means of lines. If, for a geometer, the proof of something is established by way of imagination, he is not concerned with the [apparent] correctness or incorrectness of something by inspection. But we do not doubt that everything that the artisan sees is taken from what the geometer had worked out previously and what had been proved to be correct. Therefore, the artisan and the surveyor take the choice parts [literally, the cream] of the thing, and they do not think about the methods by which correctness is established. Thus occur errors and mistakes.
>
> The geometer knows the correctness of what he wants by means of proofs, since he is the one who has derived the notions on which the artisan and the surveyor base their work. However, it is difficult for him to transform what he has proved into a [practical] construction, since he has no experience with the practical work of the artisan and the surveyor. If the skillful among these geometers are asked about something in dividing the figures or multiplying the lines, they are confused and need a long time to think. Sometimes they are successful, and it is easy for them; but sometimes it is difficult for them and they do not find its construction.[39]

We infer from Abu'l-Wafa''s words that in his time surveyors and artisans were considered appliers of practical geometry and that only those who dealt with theoretical geometry were qualified for the title *muhandis* (geometer). In later centuries, however, surveyors, too, were accorded the title *muhandis*.

Abu'l-Wafa' continued:

> I was present at some meetings (*majālis*) in which a group of geometers and artisans participated. They were asked about the construction of a square from three [given] squares. A geometer easily constructed a line such that the square on it is equal to three [given] squares, but none of the artisans were satisfied with what he had done. The artisan wants to divide those squares into pieces from which one square can be assembled, as we have described for two squares and five squares.[40]

This passage is crucial for our argument. Abu'l-Wafa' explicitly stated that he took part in a number of meetings between mathematicians and artisans. As a senior member of the group, he seemed to have had authority over both parties owing to his expertise in geometry and to his familiarity with the work of artisans. The problem discussed at that meeting was essentially about finding the square root of three. The artisans themselves, who were not satisfied with the algebraic solution offered by a mathematician, proposed two constructions based on the cut-and-paste method. Abu'l-Wafa' patiently proved that their methods were wrongly conceived, before proposing an elegant solution. At the same meeting he also produced a novel proof for the famous Pythagorean theorem by dissecting a square into four congruent triangles rotating around a central square and then assembling them into another square. The four-fold rotational symmetry of the dissection apparently appealed to the artisans so much that it later became a popular motif of the Islamic ornamental arts.

Abu'l-Wafa' was well aware of the needs and problems of artisans, and he tried to accommodate them, and encouraged other mathematicians to do the same. At the meeting he recorded he seems to have enjoyed being involved with architecture and the architectural arts—two fields that require the application of geometry. It appears that such meetings motivated him to write the *Geometrical Constructions* in order to contribute to the advancement of Islamic art. In his book, he displayed knowledge of pure geometry, familiarity with practical applications, and skill in teaching theoretical subjects to practical-minded artisans. His book, which is a unique combination of practical and theoretical geometry, was judged "the best work of pure geometry ever written by a Muslim mathematician."[41] Several extant manuscript copies bear witness to its popularity throughout the Middle Ages. At the turn of the fourteenth century, the mathematicians who gathered in Tabriz were in all likelihood familiar with the *Geometrical Constructions*, and it may have motivated them to collaborate with skilled artisans in the town in order to design fresh ornamental patterns by utilizing advanced methods of geometry. This collaboration probably led to the creation of our document.

NOTES

Editors' note: As mentioned in the introduction, "Preliminaries" was the first chapter of Alpay Özdural's unpublished book manuscript titled "Interlocks of Similar or Complementary Figures: Collaboration of Mathematicians and Artisans in the Islamic World." This chapter was edited for publication in this volume.

1. I. Fols. 2v–17. *Bīst bāb dar ma'rifat-i usṭurlāb* (Twenty chapters on the science of the astrolabe) by Naṣīr al-Dīn Muḥammad b. Muḥammad al-Ḥasan al-Ṭūsī (d. 1274).
   II. Fols. 17–18. Four short anonymous excerpts on the astrolabe: *'Amal bar jadval al-inḥirāf* (Construction with the Declination Table); *Dar ma'rifat-i samt-i Qibla* (On the Knowledge of the Direction of the Qibla); *Dar ma'rifat-i samt-i Qibla* (another version of the same subject); *Dar ma'rifat-i qaws-i inḥirāf* (On the Knowledge of the Curve of Declination).
   III. Fols. 17v–34v. *Risāla*. A treatise on the science of computation by Maḥmūd b. Muḥammad b. Qawwām Qāżī Wālishtānī (active ca. 1436).
   IV. Fols. 35–36. Anonymous and untitled sine and cosine tables.
   V. Fols. 34v–41. Some anonymous and untitled chapters extracted from a treatise on algebra.
   VI. Fols. 42–48. Anonymous and untitled treatise on the direction of the qibla, with diagrams showing how to determine the qibla at Shiraz, Abarquh, and Yazd. It is preceded on fol. 42 by a schematic table that allows the calculation of the qibla in relation to the position of Isfahan.
   VII. Fols. 49v–63. *Durj al-jawāhir va burj al-ẓawāhir* (The Casket of Substances and the Tower of Appearances). An abridged treatise on surveying by an author whose name cannot be deciphered.
   VIII. Fols. 63–64. *Dar shinākhtan-i chigūnagī sanjīdan-i zamīnhā wa makānhā* (On Knowing the Manner of Preparation of Lands and Construction Sites), anonymous.
   IX. Fols. 65v–69v. *Mukhtaṣar dar 'ilm-i masāḥat* (Compendium on the Science of Surveying), anonymous.
   X. Fols. 70v–89v. *Kitāb-i Uṣūl* (Book of Elements). An anonymous treatise on surveying, divided into three parts: *ālāt-i masāḥat* (surveying tools); *ashkāl* (geometric shapes); and *ṭuruq-i masāḥat* (surveying methods).
   XI. Fols. 90v–100. *Risāla dar iṣlāḥ-i ahl-i masāḥat* (Treatise on the Improvement of Surveyors), anonymous.

XII. Fols. 100v–102. Various applications of algebraic problems, such as dividing an inheritance among four sons, anonymous and untitled.
XIII. Fols. 102v–104v. *Faṣl fī masā'il al-lqlīdisiyya min maqāla al-thāniyya* (A Section on the Problems of the Second Chapter of Euclid's *Elements*), anonymous.
XIV. Fols. 105–105v. *Burhān ḥisāb al-ḫaṭā'īn* (Proof for Computing Errors), anonymous.
XV. Fols. 106–107. *Faṣl fī al-muḍli'āt* (A Section on Algebraic Notations), which includes a table showing the algebraic notations of the powers of the unknown; anonymous.
XVI. Fols. 108–118v. *Fī qismat al-musallaṣāt* (On the Division of Triangles), anonymous. The last construction on fol. 118v is attributed to Abū Bakr b. al-Khalīl (see below, no. XIX).
XVII. Fols. 119–199v. *Masā'il* (Problems). Three arithmetic problems, anonymous.
XVIII. Fols. 120v–123v. Three sets of constructions showing a series of progressively larger right-angled triangles inscribed in semicircles, explanations for these, and twelve sets of relations between the sides of right-angled triangles; anonymous and untitled.
XIX. Fols. 124v–138v. *A'māl-i Abū Bakr al-Khalīl al-Tājir al-Raṣadī* (Of the Constructions by Abu Bakr al-Khalil al-Tajir al-Rasadi). A work on dividing right-angled triangles, a treatise on surveying concerning pulleys, the laws of reflection, measuring the heights of inaccessible objects, etc. The name of the author is repeated on fol. 126, and a marginal note on fol. 135v refers to a correction made by Sayyid Sharif (Jurjani) (d. 1413).
XX. Fol. 139. *Dawā'ir al-khamsa fī istikhrāj al-ḥurūf muẓmīrāt* (Five Circles on Deriving the Cabalistic Letters), anonymous.
XXI. Fol. 139v. An astrological table of the twelve signs of the zodiac.
XXII. Fols. 140–140v. *Ikhtiyārāt-i musayyar al-Qamar* (Selections of the Mansion of the Moon). The *maṣnavī* poem, in accordance with twelve constellations, is attributed to Nāṣir al-Dīn al-Ṭūsī.
XXIII. Fols. 141v–179v. *Tarjama-i Kitāb A'māl-i handasiyya* (Translation of the Book of Geometrical Constructions). An abridged translation of Abu'l-Wafa' al-Buzjani's *Fī mā yaḥtāju ilayhi al-ṣāni' min a'māl al-handasa* (On the Geometric Construction Necessary for the Artisan [which will be referred to as *Geometrical Constructions* hereafter]) from Arabic into Persian by Abū Isḥaq ibn 'Abdallah Kubananī Yazdī, a mathematician and astronomer who flourished in the mid-fifteenth century.
XXIV. Fols. 180–197. An anonymous and untitled work on ornamental geometry with constructions and accompanying explanations. Abu Bakr al-Khalil al-Tajir's name is cited twice (on fols. 187 and 189), as he was the mathematician who solved two of the problems.
XXV. Fols. 200–201. An untitled treatise on finger reckoning (*angusht-shumārī*) by Sharaf al-Dīn 'Ali Yazdī (d. 1454).

2. Blochet 1905–1934, 2: no. 772, 41–47; Richard 1989, 183–87.
3. Woepcke 1855, 225–56.
4. Storey 1972, 2:3, 9, 25, 27–30, 52, 54.
5. Chorbachi 1989a.
6. Her examples are from fols. 196v, 192v, 193r, 184v, 193v, 190v, 192r, 195r, 196r, and 180r. They correspond respectively to geometrical constructions 64, 48, 49, 14, 50, 39, 43, 59, 61, and 1 in the present study.
7. Chorbachi 1989b.
8. Chorbachi 1989b, 755.
9. Chorbachi and Loeb 1992. On the basis of the pentagonal seal (star) seen both in the first construction of our document and on the decorations of the so-called Abbasid Palace in Baghdad, Chorbachi and Loeb suggested that the treatise might have been compiled before the palace's construction between 1180 and 1230. For a more detailed discussion of this point, see n. 31 below.
10. Bulatov 1988, 315–40 (in 1978 edition, 325–54).
11. Abū'l-Wafā' 1990–91.
12. Holod 1986; Holod 1988; Golombek and Wilber 1988, 137–73.
13. Necipoğlu 1995, 138.
14. Necipoğlu 1995, 169.
15. Özdural 1995.
16. Özdural 1996.
17. Hogendijk 1996.
18. Özdural 2000.
19. Özdural 1998a.
20. Özdural 2002.
21. Soucek kindly offered her opinion on this matter in a private conversation.
22. This complicated passage caused several erroneous translations. According to Woepcke's French translation, four students of Abu Ishaq Kubanani assisted him in translating Abu'l-Wafa's work into Persian: Woepcke 1855, 359. According to Bulatov's Russian publication, Amir Nasr initiated the Persian translation of the *Geometrical Constructions*: Bulatov 1988, 89. Necipoğlu adopted Bulatov's translation: Necipoğlu 1995, 168. I am grateful to Eva Maria Subtelny, who translated this difficult passage for me after studying it scrupulously.
23. Munzavī 1969, 1:143, 152, 318, 368.
24. Özdural 1995, 59–64.
25. Özdural 1990, 34–35; al-Kāshī 1960b, 101–2.
26. Özdural 1996, 192; Ca'fer 1987, 22–23.
27. Steingass 1970, 1170.
28. Bulatov 1988, 88–89.
29. For Abū Naṣir ibn 'Iraq and his works, see Sezgin 1974, 338.
30. Golombek and Wilber 1988, 169.
31. They wrote: "Since this architectural decoration dates from the period 1180–1230 A.D., we may conclude that knowledge about the Pentagonal Seal had by that time filtered down to artisans, so that the material presented in folio 180a may be dated to before 1200 A.D., circa 1180–1230 A.D.": Chorbachi and Loeb 1992, 302. This argument, however, is not convincing. The pentagonal seal that fits in the space surrounded by five regular pentagons is a common geometric figure to which a certain date cannot possibly be assigned. This and other kinds of pentagonal stars were

widely known in history. It can scarcely be used as evidence to date the document. Moreover, the pre-thirteenth-century examples of architectural decorations in and around Baghdad have completely disappeared. We cannot know, therefore, when this type of pentagonal star first appeared in architectural decorations.

32. Necipoğlu 1995, 168–69. Abū Bakr al-Khalīl is not cited, unfortunately, in the following biographical works: Sezgin 1974; Sezgin 1978; Suter 1972; Storey 1972; Brockelmann 1943–49.
33. For this information I am indebted to Muhammad Baqiri in Tehran. He informed me in a private correspondence that a friend of his, Mihran Akhbarifar, had located this manuscript in the library of Mashhad. He also sent a photocopy of the modern edition of this manuscript, published in 1968 by Ahmad Gulchin Ma'ani. The title of this Arabic treatise is *al-Īżāḥ*. The published edition mainly consists of facsimiles of *al-Īżāḥ* and a Persian manuscript on the same subject, which was written by the same hand. The editor erroneously attributed the former to Abū Manṣūr ʿAbd-al-Ṭāhir b. Ṭāhir b. Muḥammad b. ʿAbdallāh Tamīnī, and supposed the latter to be the translation of the former. [Editor's note: On this manuscript, see Elaheh Kheirandish's chapter in this volume.]
34. Wilber 1969, 15.
35. Rashīd al-Dīn 1940, 117, 208.
36. For the architectural activities of Ghazan Khan, Rashid al-Din, and Öljeytü, see Rashīd al-Dīn 1940; Browne 1920, 80–87; Morton 1999; Wilber and Minovi 1938; Boyle 1971; Blair 1984; Blair 1986c; and Sims 1988.
37. The citations in this study are from Ms. Ayasofya 2753, located in the Süleymaniye Kütüphanesi, Istanbul. The manuscript was copied for Ulugh Beg's library in Samarqand and brought to Istanbul in 1471, probably by ʿAli Kuşçu, when he moved to Istanbul and started teaching in the Ayasofya Madrasa, where the manuscript was located. Abu'l-Wafa' dedicated his book to the Buyid ruler Baha' al-Dawla (r. 987–1012), who was referred to by the title *Shāhinshāh*. This title could only have been used for the rulers of Iran; therefore the manuscript was probably compiled by a student of Abu'l-Wafa' after the death of his master in 998. This argument can be supported by phrases that appear in the text such as "Abu'l-Wafa' said"; "the professor said"; and "the sage said." [Editor's note: In actuality, the manuscript was written before the author's death and has been dated to ca. 991–92 to 998 in Elaheh Kheirandish's chapter.] The following is a list of the extant copies of *Geometrical Constructions* that I was able to locate in addition to Ms. Ayasofya 2753: Mashhad, Ms. Riḍā 37 (Persian), end of the tenth or early eleventh century; Tehran, Ms. Dānishgāh 2876 (Persian), the eleventh or twelfth century; Mashhad, Ms. Riḍā 5337/139 (Arabic commentary), by Kamāl al-Dīn Mūsā b. Muḥammad b. Maʿnā (1175–1242), early thirteenth century; Cairo, Ms. Dār Riyāḍa 260 (Arabic), thirteenth century; Paris, Bibliothèque nationale de France, Ms. Persan 169 (Persian), translated by Abū Isḥaq Yazdī, mid-fifteenth century; Mashhad, Ms. Riḍā 144 (Persian commentary), by Muḥammad Bāqir Zayn al-ʿĀbidīn (active 1637–38), seventeenth century. For the modern translations and editions of *Geometrical Constructions*, see Woepcke 1855; Suter 1922; Krasnova 1966; Abū'l-Wafā' 1990–91; Qurbānī and Shaykhān 1992. For more information on the manuscript copies of the *Geometrical Constructions*, see Sezgin 1974, 5:321–25; Sezgin 1978, 7:408; Storey 1972, 2:2–3; Özdural 1995, 67; Necipoğlu 1995, 175.
38. The titles of the chapters of the *Geometrical Constructions* are as follows: On the straightedge, compass, and setsquare; On the basic procedures that should precede the explanations; On constructing regular polygons; On constructing polygons inscribed in circles; On constructing circles that circumscribe polygons; On constructing circles inscribed in polygons; On constructing superimposed polygons; On the division of triangles; On the division of squares; On composing and dividing a square; On the division of irregular quadrangles; On tangential circles; On the division of spheres by polygons (i.e., polyhedrons). A complete copy of Abu'l-Wafa''s book could not be found in any of the manuscripts. Chapters 11 and 12 are missing in every copy. Some other chapters are also missing in various copies.
39. Abū'l-Wafā', Ms. Ayasofya 2753, fols. 52–53.
40. Abū'l-Wafā', Ms. Ayasofya 2753, fol. 53.
41. M. Soussi, "ʿIlm al-handasa," in *The Encyclopedia of Islam, New Edition* (Leiden, 1954–2005).

BIBLIOGRAPHY

Abū'l-Wafā' al-Būzjānī. 1990–91 (1369). *Handasa-i Īrānī: Kitāb-i Tijārat: Fī mā yaḥtāju ilayhi al-'ummāl wa-al-ṣunnā' min al-ashkāl al-handasiyya, yā, Kārburd-i handasa dar 'amal.* Edited and adapted (to modern Persian) by Sayyid 'Alī Rịżā Jazbī. Tehran: Surūsh.

Afshar, Iraj. 1979. "Architectural Information through the Persian Classical Texts." In *Akten des VII. Internationalen Kongresses für Iranische Kunst und Archäologie, München 7.–10. September 1976*, 612–16. Berlin: D. Reimer.

Altan, Kemal. 1936. "Eski Mimârların Plânları." *Arkitekt* 6:199.

Archimedes. 1964. *Archimedes in the Middle Ages.* Edited by Marshall Clagett. Vol. 1, *The Arabo-Latin Tradition.* Madison, Wis.: University of Wisconsin Press.

———. 1912. *The Works of Archimedes.* Translated and edited by Thomas Heath. Cambridge: Cambridge University Press.

Al-Asad, Mohammad. 1995. "The Muqarnas: A Geometric Analysis." In *The Topkapı Scroll: Geometry and Ornament in Islamic Architecture; Topkapı Palace Museum Library MS H. 1956* by Gülru Necipoğlu, 349–59. Santa Monica, Calif.: Getty Center for the History of Art and the Humanities.

Baklanov, Nikolai B. 1947. "Gerikh: Geometricheskii ornament Srednei Azii i metody ego postroeniia." *Sovetskaia Arkheologiia* 9:101–20.

———. 1944. "Arkhitekturnye chertezhi uzbekskogo mastera XVI veka." *Soobshcheniia Instituta istorii i teorii arkhitektury, Akademii arkhitektury SSSR* 4:1–21.

Bayhaqī, Abū al-Fażl Muḥammad ibn Ḥusayn. 1862. *The Tá'ríkh-i Baihaki: Containing the Life of Masaúd, Son of Sultán Mahmúd of Ghaznín.* Edited by William H. Morley. Calcutta: Asiatic Society of Bengal.

Begley, Wayne E., and Ziyaud-Din A. Desai, eds. 1989. *Taj Mahal: The Illumined Tomb; An Anthology of Seventeenth-Century Mughal and European Documentary Sources.* Cambridge, Mass.: Aga Khan Program for Islamic Architecture.

Berggren, J. L. 1986. *Episodes in the Mathematics of Medieval Islam.* New York: Springer-Verlag.

———. 1981. "An Anonymous Treatise on the Regular Nonagon." *Journal for the History of Arabic Science* 5:37–41.

Blair, Sheila. 1996. "Abū Tāhir." In the *Dictionary of Art*, edited by Jane Turner, vol. 1, 97–98. New York: Grove's Dictionaries.

———. 1987. "The Epigraphic Program of the Tomb of Uljaytu at Sultaniyya: Meaning in Mongol Architecture." *Islamic Art* 2:43–96.

———. 1986a. *The Ilkhanid Shrine Complex at Natanz, Iran.* Cambridge, Mass: Center for Middle Eastern Studies, Harvard University.

———. 1986b. "A Medieval Persian Builder." *Journal of the Society of Architectural Historians* 45, no. 4: 389–95.

———. 1986c. "The Mongol Capital of Sultāniyya, 'the Imperial.'" *Iran* 24:139–51.

———. 1984. "Ilkhanid Architecture and Society: An Analysis of the Endowment Deed of the Rab'-i Rashīdī." *Iran* 22:67–90.

Blochet, Edgar. 1905–34. *Catalogue des manuscripts persans.* 4 vols. Paris: Imprimerie nationale.

Bloom, Jonathan M. 1993. "On the Transmission of Designs in Early Islamic Architecture." *Muqarnas* 10:21–28.

Bosworth, Clifford Edmund. 1963. A Pioneer Arabic Encyclopedia of the Sciences: Al-Khwārizmī's 'Keys of the Science.'" *Isis* 54:97–111.

Boyle, John Andrew. 1971. "Rashīd al-Dīn: The First World Historian." *Iran* 9:19–26.

Brockelmann, Carl. 1943–49. *Geschichte der arabischen Litteratur.* Leiden: E.J. Brill.

Browne, Edward G. 1920. *A History of Persian Literature under Tartar Dominion, A.D. 1265–1502.* Vol. 3 of *A Literary History of Persia.* Cambridge: Cambridge University Press.

Bulatov, Mitkhat S. 1988. *Geometricheskaia garmonizatsiia v arkhitekture Srednei Azii IX–XV vv.* Rev. ed. Moscow: "Nauka." Original edition Moscow: "Nauka," 1978.

Bulmer-Thomas, Ivor. 1967–68. *Selections Illustrating the History of Greek Mathematics.* 2 vols. London: W. Heinemann, 1939–41. Reprint, London: W. Heinemann.

Ca'fer Efendi. 1987. *Risāle-i Mi'māriyye: An Early-Seventeenth-Century Ottoman Treatise on Architecture.* Translated and edited by Howard Crane. Leiden: E.J. Brill.

Cezar, Mustafa. 1971. *Sanatta Batı'ya Açılış ve Osman Hamdi.* Istanbul: Türkiye İş Bankası.

Chorbachi, Wasma'a K. 1989a. "Beyond the Symmetries of Islamic Geometric Patterns: The Science of Practical Geometry and the Process of Islamic Design." PhD diss., Harvard University.

———. 1989b. "In the Tower of Babel: Beyond Symmetry in Islamic Design." *Computers and Mathematics, with Applications* 17:751–89.

Chorbachi, Wasma'a K., and Arthur L. Loeb. 1992. "An Islamic Pentagonal Seal (from Scientific Manuscripts of the Geometry of Design)." In *Fivefold Symmetry*, edited by István Hargittai, 283–305. Singapore: World Scientific.

Christie, Archibald H. 1929. *Traditional Methods of Pattern Designing: An Introduction to the Study of Formal Ornament*. 2nd ed. Oxford: Clarendon Press. Original edition Oxford: Clarendon Press, 1910.

Clarke, Caspar Purdon. 1893. "The Tracing Board in Modern Oriental and Medieval Operative Masonry." *Ars Quatuor Coronatorum, Being the Transactions of the Quatuor Coronati Lodge No. 2076* 6:99–107.

Copernicus. 1875. *Reliquiae Copernicanae*. Edited by Maximilian Curtze. Leipzig: B.G. Teubner.

Dimand, M. S. 1930. *A Handbook of Mohammedan Decorative Arts*. New York: Metropolitan Museum of Art.

Diocles. 1976. *Diocles, On Burning Mirrors: The Arabic Translation of the Lost Greek Original = Kitāb fī al-marāyā al-muḥriqa*. Translated and edited by G. J. Toomer. Berlin: Springer-Verlag.

Dold-Samplonius, Yvonne. 1996. "How al-Kashi Measures the *Muqarnas*: A Second Look." In *Mathematische Probleme im Mittelalter: Der lateinische und arabische Sprachbereich*, edited by Menso Folkerts, 56–90. Wiesbaden: Harrassowitz Verlag.

———. 1992. "Practical Arabic Mathematics: Measuring the Muqarnas by al-Kāshī." *Centaurus* 35:193–242.

Dürer, Albrecht. 1977. *The Painter's Manual: A Manual of Measurement of Lines, Areas, and Solids by means of compass and ruler, assembled by Albrecht Dürer for the use of all lovers of art with appropriate illustrations arranged to be printed in the year MDXXV*. Translated and edited by Walter L. Strauss. New York: Abaris Books.

Erdenen, Orhan. 1965. "Baki Yapılarımızda Plân Meselesi." *Mimarlık* 26:19–23.

Ettinghausen, Richard. 1936. "Evidence for the Identification of Kāshān Pottery." *Ars Islamica* 3:44–75.

Euclid. 1956. *The Thirteen Books of Euclid's Elements*. Translated by Thomas L. Heath. 2nd ed. 3 vols. New York: Dover Publications.

Al-Fārābī, Abū Naṣr Muḥammad ibn Muḥammad. 1961. *Fuṣūl al-madanī: Aphorisms of the Statesman*. Translated and edited by D. M. Dunlop. Cambridge: Cambridge University Press.

Fleurent, Gilbert M. 1992. "Pentagon and Decagon Designs in Islamic Art." In *Fivefold Symmetry*, edited by István Hargittai, 263–81. Singapore: World Scientific.

Frederickson, Greg N. 1997. *Dissections: Plane and Fancy*. Cambridge: Cambridge University Press.

Gaganov, G. A. 1958. "Geometricheskii ornament Srednei Azii." *Arkhitekturnoe nasledstvo* 11:181–208.

Al-Ghazālī, Muḥammad. 1970. *Ninety-Nine Names of God in Islam: A Translation of the Major Portion of al-Ghazālī's al-Maqṣad al-asnā*. Translated by Robert Charles Stade. Ibadan, Nigeria: Daystar Press.

Golombek, Lisa. 1992. "Discourses of an Imaginary Arts Council in Fifteenth-Century Iran." In *Timurid Art and Culture*, edited by Lisa Golombek and Maria Subtelny, 1–17. Leiden: E.J. Brill.

———. 1988. "The Function of Decoration in Islamic Architecture." In *Theories and Principles of Design in the Architecture of Islamic Societies*, edited by Margaret Bentley Ševčenko, 35–45. Cambridge, Mass.: Aga Khan Program for Islamic Architecture.

Golombek, Lisa, and Donald Wilber. 1988. *The Timurid Architecture of Iran and Turan*. 2 vols. Princeton, N.J.: Princeton University Press.

Grabar, Oleg. 1988. "The Iconography of Islamic Architecture." In *Content and Context of Visual Arts in the Islamic World: Papers from a Colloquium in Memory of Richard Ettinghausen, Institute of Fine Arts, New York University, 2–4 April 1980*, edited by Priscilla P. Soucek, 51–65. University Park, Pa.: Published for the College Art Association of America by the Pennsylvania State University Press.

Grabar, Oleg, and Renata Holod. 1979–80. "A Tenth-Century Source for Architecture." In "Eucharisterion: Essays Presented to Omeljan Pritsak," edited by Ihor Ševčenko and Frank E. Sysyn. Special issue, *Harvard Ukrainian Studies* 3–4, part 1: 310–19. Cambridge, Mass: Harvard Ukrainian Research Institute.

Gunther, Robert T. 1932. *The Astrolabes of the World: Based upon the Series of Instruments in the Lewis Evans Collection in the Old Ashmolean Museum at Oxford*. 2 vols. Oxford: Oxford University Press.

Harb, Ulrich. 1978. *Ilkhanidische Stalaktitengewölbe: Beiträge zu Entwurf und Bautechnik*. Archäologische Mitteilungen aus Iran: Ergänzungsband 4. Berlin: Reimer.

Hargittai, István. 1992. *Fivefold Symmetry*. Singapore: World Scientific

Hartner, Willy. 1938. "The Principle and the Use of the Astrolabe." In *A Survey of Persian Art from Prehistoric Times to the Present*, edited by Arthur Upham Pope and Phyllis Ackerman, 3:2530–54, 6:1397–1402. London: Oxford University Press.

Hautecoeur, Louis. 1931. "De la trompe aux «mukarnas»." *Gazette des Beaux-Arts* 6:27–51.

Heath, Thomas. 1965. *A History of Greek Mathematics*. 2 vols. Oxford: Clarendon Press.

Heiberg, J. L., and Eilhard Wiedemann. 1909–10. "Ibn al-Haiṯams Schrift über parabolische Hohlspiegel." *Bibliotheca Mathematica*, 3rd ser., 10:201–37.

Heron of Alexandria. 1964. *Codex Constantinopolitanus Palatii Veteris No. 1*. Translated and edited by Evert M. Bruins. 3 vols. Leiden: E.J. Brill.

Herzfeld, Ernst. 1942. "Damascus: Studies in Architecture—1." *Ars Islamica* 9:1–53.

Hogendijk, Jan P. 1996. "Een workshop over iraanse mozaïken." *Nieuwe Wiskrant* 16, no. 2: 38–42.

———. 1985. "Al-Kūhī's Construction of an Equilateral Pentagon in a Given Square." *Zeitschrift für Geschichte der arabisch-islamischen Wissenschaften* 1:100–144.

———. 1984. "Greek and Arabic Constructions of the Regular Heptagon." *Archive for History of Exact Sciences* 30:197–330.

———. 1981. "How Trisections of the Angle Were Transmitted from Greek to Islamic Geometry." *Historia Mathematica* 8:417–38.

———. 1979. "On the Trisection of an Angle and the Construction of a Regular Nonagon by Means of Conic Sections in Medieval Islamic Geometry." Utrecht University, Department of Mathematics. Presented at the Second International Symposium of the History of Arabic Science, Aleppo, 5–12 April 1979. Preprint 113. http://www.jphogendijk.nl/publ/tris.pdf.

Holod, Renata. 1988. "Text, Plan, and Building: On the Transmission of Architectural Knowledge." In *Theories and Principles of Design in the Architecture of Islamic Societies: A Symposium Held by the Aga Khan Program for Islamic Architecture at Harvard University and the Massachusetts Institute of Technology, Cambridge, Massachusetts, November 6–8, 1987*, edited by Margaret Bentley Ševčenko, 1–12. Cambridge, Mass.: Aga Khan Program for Islamic Architecture.

———. 1986. "Defining an Art of Architecture." In *Architecture Education in the Islamic World: Proceedings of Seminar Ten in the Series Architectural Transformations in the Islamic World, held in Granada, Spain, April 21–25, 1986*, edited by Ahmet Evin, 1–32. Singapore: Concept Media for the Aga Khan Award for Architecture.

Ibn al-Haytham, Abū ʿAlī al-Ḥasan ibn al-Ḥasan. 1989. *The Optics of Ibn al-Haytham: Books I–III: On Direct Vision = Kitāb al-Manāzir*. Translated and edited by Abdelhamid I. Sabra. 2 vols. London: Warburg Institute, University of London.

———. 1985. *Ibn al-Haytham's Completion of the Conics = Al-Maqāla fī tamām Kitāb al-makhrūṭāt*. Translated and edited by Jan P. Hogendijk. New York: Springer.

Ibn Bībī, Nāṣir al-Dīn Ḥusayn ibn Muḥammad. 1959. *Die Seltschukengeschichte des Ibn Bībī: Mukhtaṣar-i Saljūqnāma*. Translated by Herbert W. Duda. Copenhagen: Munksgaard.

Ibn Miskawayh, Aḥmad ibn Muḥammad. 1968. *The Refinement of Character = Tahdhīb al-akhlāq wa taṭhīr al-aʿrāq*. Translated by Constantine K. Zurayk. Beirut: American University of Beirut.

Ibn Sinān, Ibrāhīm. 2000. *Ibrāhīm ibn Sinān, Logique et Géométrie au Xe siècle*. Translated and edited by Rushdī Rāshid and Hélène Bellosta. Leiden: Brill.

Iushkevich, Adol'f Pavlovich. 1976. *Les mathématiques arabes: VIIIe–XVe siècles*. Translated by Michel Cazenave and Khalil Jaouiche. Paris: J. Vrin. (Translation of part 3 of *Istoriia matematiki v srednie veka*. Moscow: Gos. izd-vo fiziko-matematicheskoi lit-ry, 1961.)

———. 1970. "Abū'l-Wafāʾ al-Būzjānī, Muḥammad Ibn Muḥammad Ibn Yaḥyā Ibn Ismāʿīl Ibn Al-ʿAbbās." In the *Dictionary of Scientific Biography*, edited by Charles Coulston Gillispie, vol. 1, 39–43. New York: Scribner.

Jackson, David Edward Pritchett. 1972. "The Arabic Translation of a Greek Manual of Mechanics." *Islamic Quarterly* 16, nos. 1–2: 96–103.

Jaʿfarī, Jaʿfar ibn Muḥammad. 1960 (1338). *Taʾrīkh-i Yazd*. Edited by Īraj Afshār. Tehran: Bungāh-i Tarjuma va Nashr-i Kitāb.

Al-Jazarī, Ismāʿīl ibn al-Razzāz. 1974. *The Book of Knowledge of Ingenious Mechanical Devices = Kitāb fī maʿrifat al-ḥiyal al-handasiyya*. Translated by Donald R. Hill. Dordrecht: Reidel, 1974.

Jones, Dalu, and George Michell. 1972. "Squinches and Pendentives: Problems and Definitions." *Art and Archaeology Research Papers* 1:9–25.

Jones, Owen. 1982. *The Grammar of Ornament*. London: Day and Son, 1856. Reprint, New York: Van Nostrand Reinhold.

Al-Kāshī, Jamshīd ibn Masʿūd. 1977. *Miftāḥ al-ḥisāb*. Edited by Nādir Nabulsī. Damascus: al-Jamhūriyya al-ʿArabiyya al-Sūriyya, Wizārat al-Taʿlīm al-ʿĀlī.

———. 1969. *Miftāḥ al-ḥisāb*. Edited by Aḥmad Saʿīd al-Dimirdāsh and Muḥammad Ḥamdī al-Hifnī al-Shaykh. Cairo: Dār al-Kātib al-ʿArabī.

———. 1960a. *The Planetary Equatorium of Jamshīd Ghiyāth al-Dīn al-Kāshī*. Translated by Edward S. Kennedy. Princeton, N.J.: Princeton University Press.

———. 1960b. *Uluğ Bey ve Semerkand'deki İlim Faaliyeti hakkında Giyasüddin-i Kâşî'nin Mektubu = Ghiyâth al Dîn al Kâshî's Letter on Ulugh Bey and the Scientific Activity in Samarqand*. Edited by Aydın Sayılı. Ankara: Türk Tarih Kurumu Basımevi.

———. 1954. *Miftāḥ al-ḥisāb*. Translated into Russian by Boris A. Rozenfel'd. Edited by V. S. Segal and Adol'f Pavlovich Iushkevich. In "Matematicheskie trakaty," *Istoriko-matematicheskie issledovaniia* 7:13–326.

Kātib, Aḥmad ibn Ḥusayn ibn ʿAli. 1966 (1345). *Tārīkh-i jadīd-i Yazd*. Edited by Īraj Afshār. Tehran: Intishārāt-i Ibn Sīnā.

Kennedy, Edward S. 1968. "The Exact Sciences of Iran under the Saljuks and Mongols." In *Cambridge History of Iran*, vol. 5, *The Saljuq and Mongol Periods*, edited by J. A. Boyle, 659–79. Cambridge: Cambridge University Press.

Khānlarī, Parvīz Nātil, ed. 197–?. *Girih-sāzī dar hunar-i miʿmārī-yi Īrān*. With tile patterns by Muḥammad ʿAlī Fuṣhatī and ʿAbd al-ʿĀlī Fuṣhatī, photographed by Karāmat Raʿnā Ḥusaynī. Tehran: Bunyād-i Farhang-i Īrān.

Krachkovskaia, V. A. 1946. *Izrazts'i Mavzoleia Pir-Khuseina*. Tbilisi: Akad. Nauk Gruz. SSR.

Krasnova, S. A. 1966. "Abu-l-Vafa al-Buzdzhani, Kniga o tom, chto neobkhodimo remeslenniku iz geometricheskikh postroenii." *Fiziko-matematicheskie nauki v stranakh Vostoka* 1:42–140.

Krause, Max. 1936. "Stambuler Handschriften islamischer Mathematiker." *Quellen und Studien zur Geschichte der Mathematik, Astronomie und Physik* 3 (Abt. B): 437–532.

Kuran, Aptullah. 1968. *The Mosque in Early Ottoman Architecture*. Chicago: University of Chicago Press.

Lee, A. J. 1987. "Islamic Star Patterns." *Muqarnas* 4:182–97.

Lurzādeh, Ḥusayn. 1979. *Iḥyā-yi hunarhā-yi az yād rafta: Mabānī-i miʿmārī-i sunnatī dar Īrān*. Tehran: Muḥammad Khvājavī.

Al-Maqdīsī, Muṭahhar ibn Ṭāhir. 1903. *Le Livre de la création et de l'histoire de Moṭahhar ben Ṭāhir el-Maqdisī = Kitāb al-Baʿd wa'l-Taʾrīkh*. Translated and edited by Clément Huart. Paris: Ernest Leroux.

Morton, A. H. 1999. "The Letters of Rashīd al-Dīn: Īlkhānid Fact or Timurid Fiction?" In *The Mongol Empire and Its Legacy*, edited by Reuven Amitai-Preiss and David O. Morgan, 155–99. Leiden: Brill.

Mülayim, Selçuk. 1982. *Anadolu Türk Mimarisinde Geometrik Süslemeler: Selçuklu Çağı*. Ankara: Kültür ve Turizm Bakanlığı Yayınları.

Munzavī, Aḥmad. 1969 (1348). *Fihrist-i nuskhahā-yi khaṭṭī-i Fārsī*. 6 vols. Tehran: Muʿassasa-i Farhangī-i Minṭaqaʾī.

Necipoğlu, Gülru. 1995. *The Topkapı Scroll: Geometry and Ornament in Islamic Architecture; Topkapı Palace Museum Library MS H. 1956*. Santa Monica, Calif.: Getty Center for the History of Art and the Humanities.

———. 1992. "Geometric Design in Timurid/Turkmen Architectural Practice: Thoughts on a Recently Discovered Scroll and Its Late Gothic Parallels." In *Timurid Art and Culture*, edited by Lisa Golombek and Maria Subtelny, 48–66. Leiden: E.J. Brill.

———. 1986. "Plans and Models in Fifteenth- and Sixteenth-Century Ottoman Architectural Practice." *Journal of the Society of Architectural Historians* 45, no. 3: 224–43.

Neugebauer, Otto. 1969. *The Exact Sciences of Antiquity*. 2nd ed. New York: Dover Publications.

Notkin, Iosif Isakevich. 1995. "Decoding Sixteenth-Century Muqarnas Drawings." *Muqarnas* 12:148–71.

———. 1970. "Stalaktity XIV v. iz ansambli Shakhi-Zinda." *Zodchestvo Uzbekistana* 2:239–54.

———. 1961. *Bukharskaia rez' ba po ganchu v rabotakh Usto Shirina Muradova*. Tashkent: Goslitizdat UzSSR.

Ödekan, Ayla. 1977. *Osmanlı Öncesi Anadolu Türk Mimarisinde Mukarnaslı Portal Örtüleri*. Istanbul: İ.T.Ü. Mimarlık Fakültesi Baskı Atölyesi.

O'Kane, Bernard. 1992. "Poetry, Geometry and the Arabesque: Notes on Timurid Aesthetics." *Annales Islamologiques* 26:63–78.

———. 1987. *Timurid Architecture in Khurasan*. Costa Mesa, Calif.: Mazdâ Publishers.

Omar Khayyām. 2000. *Ḥakīm 'Umar Khayyām: Bi-'unvān-i 'ālim-i jabr: Mushtamil bar mutūn va tarjuma-i Fārsī-i ās̱ār-i Khayyām dar 'ilm-i jabr = Hakim Omare Khayyam as an Algebraist*. Translated (into Persian) and edited (in Arabic) by Ghulām Ḥusayn Muṣāḥib. Tehran: Anjuman-i Ās̱ār-i Millī, 1960. Reprint, Tehran: Anjuman-i Ās̱ār va Mafākhir-i Farhangī: Kumīsiyūn-i Millī-i Yūniskū dar Īrān.

———. 1981. *Rasā'il al-Khayyām al-jabriyya = L'oeuvre algébrique d'al-Khayyām*. Translated and edited by Rushdī Rāshid and Aḥmad Jabbār. Aleppo: Jāmi'at Ḥalab, Ma'had al-Turāth al-'Ilmī al-'Arabī.

———. 1963a. "Omar Khayyam: Pervy algebraicheskii trakta." Translated by S. A. Krasnova and B. A. Rozenfel'd. *Istoriko-matematicheskie issledovaniia* 15:445–72.

———. 1963b. "A Paper by Omar Khayyam." Translated by Ali R. Amir-Moéz. *Scripta Mathematica* 26:323–37.

Orgun, Zarif. 1938. "Hassa Mimarları." *Arkitekt* 12:333–42.

Öz, Tahsin. 1936. "Eski Câmi Plânları ve Mimâri Vesikâlar." *Yedigün* 172, nos. 3–4: 25.

Özdural, Alpay. 2002. "The Use of Cubic Equations in Islamic Art and Architecture." In *Nexus IV: Architecture and Mathematics*, edited by José Francisco Rodrigues and Kim Williams, 165–79. Fucecchio, Italy: Kim Williams Books.

———. 2000. "Mathematics and Arts: Connections between Theory and Practice in the Medieval Islamic World." *Historia Mathematica* 27, no. 2: 171–201.

———. 1998a. "A Mathematical Sonata for Architecture: Omar Khayyam and the Friday Mosque of Isfahan." *Technology and Culture* 39, no. 4: 699–715.

———. 1998b. "Sinan's *Arşin*: A Survey of Ottoman Architectural Metrology." *Muqarnas* 15:101–115.

———. 1996. "On Interlocking Similar or Corresponding Figures and Ornamental Patterns of Cubic Equations." *Muqarnas* 13:191–211.

———. 1995. "Omar Khayyam, Mathematicians and *Conversazioni* with Artisans." *Journal of the Society of Architectural Historians* 54, no. 1: 54–71.

———. 1991. "An Analysis of the Geometry of Stalactites: Buruciye Medrese in Sivas." *Middle East Technical University, Journal of the Faculty of Architecture* 11, nos. 1–2: 57–71.

———. 1990. "Giyaseddin Jemshid el-Kashi and Stalactites." *Middle East Technical University, Journal of the Faculty of Architecture* 10, nos. 1–2: 31–49.

Paccard, André. 1979. *Le Maroc et l'artisanat traditionnel islamique dans l'architecture*. 2 vols. Saint-Jorioz: Éditions Atelier 74.

Penrose, Roger. 1974. "The Role of Aesthetics in Pure and Applied Mathematical Research." *Bulletin of the Institute of Mathematical Applications* 10:266–71.

Pope, Arthur U., and Phyllis Ackerman, eds. 1938. *A Survey of Persian Art from Prehistoric Times to the Present*. 6 vols. London: Oxford University Press.

Ptolemy. 1984. *Almagest*. Translated by G. J. Toomer. London: Duckworth.

Qādī Aḥmad. 1959. *Calligraphers and Painters: A Treatise by Qādī Aḥmad*. Translated from the Persian by V. Minorsky; translated from the Russian by T. Minorsky. Smithsonian Institution Publication 4339, Occasional Papers (Freer Gallery of Art) 3, no. 2. Washington, D.C.: Smithsonian Institution.

Qaisar, Ahsan Jan. 1988. *Building Construction in Mughal India: The Evidence from Painting*. Delhi: Oxford University Press.

Qurbānī, Abū al-Qāsim and Muḥammad 'Alī Shaykhān, eds. 1992 (1371). *Būzjānīnāma: Sharḥ-i aḥvāl va ās̱ār-i riyāẓ-i Abū al-Vafā' Būzjānī*. Tehran: Intishārāt va Āmūzish-i Inqilāb-i Islāmī.

Rāshid, Rushdī. 1979. "La construction de l'heptagone régulier particular Ibn al-Haytham." *Journal of the History of Arabic Science* 3:309–87.

Rashīd al-Dīn Fażl Allāh Ṭabīb. 1978. *Vaqfnāma-i Rab'-i Rashīdī*. Edited by Īraj Afshār and Mujtabā Minuvī. Tehran: Anjuman-i As̱ār-i Millī.

———. 1940. *Geschichte Ġāzān-Ḫān's aus dem Ta'riḫ-i-mubārak-i-Ġāzānī des Rašīd al-Dīn Faḍlallāh b. 'Imād al-Daula Abūl-Ḫair*. Persian text edited by Karl Jahn. London: Printed for the Trustees of the "E.J.W. Gibb Memorial," and published by Luzac & Co.

Richard, Francis. 1989. *Catalogue des manuscrits persans*. Vol. 1, *Ancien fonds*. Paris: Bibliothèque nationale de France.

Rosintal, Josef. 1928. *Pendentifs: Trompes et stalactites de l'architecture orientale*. Paris: Librairie orientaliste Paul Geuthner.

Rozenfel'd, Boris A., and Iushkevich, Adol'f P. 1996. "Geometry." In *Encyclopedia of the History of Arabic Science*, edited by Rushdī Rāshid, vol. 2, 447–94. New York: Routledge.

Rypka, Jan. 1968. *History of Iranian Literature*. Dordrecht: D. Reidel.

Sabra, Abdelhamid I. 1982. "Ibn al-Haytham's Lemmas for Solving 'Alhazen's Problem.'" *Archive for History of Exact Sciences* 26:299–324.

———. 1972. "Ibn al-Haytham, Abū 'Alī al-Ḥasan Ibn al-Ḥasan." In the *Dictionary of Scientific Biography*, edited by Charles Coulston Gillispie, vol. 6, 189–210. New York: Scribner.

Said, Hakim Mohammed, ed. 1970(?). *Ibn al-Haitham: Proceedings of the Celebrations of the 1000th Anniversary Held under the Auspices of the Hamdard National Foundation, Pakistan.* Karachi: Hamdard Academy.

Saidan, A. S. 1979. "The Trigonometry of al-Biruni." In *Al-Biruni: Commemorative Volume; Proceedings of the International Congress Held in Pakistan on the Occasion of the Millenary of Abū Raihān Muḥammad ibn Aḥmad al-Bīrūnī, 973–ca. 1051 A.D., November 26, 1973– December 12, 1973*, edited by Hakim Mohammed Said, 675–90. Karachi: Hamdard National Foundation.

———, ed. 1971. *Arabic Arithmetic: The Arithmetic of Abū al-Wafā' al-Būzjānī, 10th Century Mss Or. 103 Leiden and 42 M Cairo.* In *Tārīkh 'ilm al-ḥisāb al-'Arabī.* Amman: Jam'iyyat A'māl al-Maṭābi' al-Ta'āwuniyya

Schoy, Carl. 1927. *Die trigonometrischen Lehren des persischen Astronomen Abū'l-Raihân Muh. Ibn Ahmad al-Bîrûnî: Dargestellt nach al-Qânûn al-Mas'ûdî.* Hanover: Orient-Buchhandlung Heinz Lafaire K.-G.

Sezgin, Fuat. 1978. *Geschichte des arabischen Schrifttums. Band 6: Astronomie bis ca. 430 H.* Leiden: E.J. Brill.

———. 1974. *Geschichte des arabischen Schrifttums. Band 5: Mathematik bis ca. 430 H.* Leiden: E.J. Brill.

Shelby, Lon R., ed. and trans. 1977. *Gothic Design Techniques: The Fifteenth-Century Design Booklets of Mathes Roriczer and Hanns Schmuttermayer.* Carbondale: Southern Illinois University Press.

Shi'rbāf, Aṣghar. 1982–83 (1361). *Girih va Kārbandī.* Vol. 1. Tehran: Sāzmān-i Millī-yi Ḥifāẓat-i Āṣār-i Bāstāni-yi Irān.

Sims, Eleanor. 1988. "The 'Iconography' of the Internal Decoration in the Mausoleum of Uljaytu at Sultaniyya." In *Content and Context of Visual Arts in the Islamic World: Papers from a Colloquium in Memory of Richard Ettinghausen, Institute of Fine Arts, New York University, 2–4 April 1980*, edited by Priscilla P. Soucek, 139–76. University Park, Pa.: Pennsylvania State University Press.

Al- Sizjī. 1996. *Al-Sijzī's Treatise on Geometrical Problem Solving = Risāla-i Sizjī dar ravishhā-yi ḥall-i Masā'il-i handasī.* Edited by Jan P. Hogendijk and Muḥammad Bāqirī. Tehran: Fatemi Publications.

Smart, Ellen S. 1974. "Graphic Evidence for Mughal Architectural Plans." *Art and Archaeology Research Papers* 6:22–23.

Smith, Myron B. 1937. "Material for a Corpus of Early Iranian Islamic Architecture—II: Manār and Masdjid, Barsīān (Iṣfahān)." *Ars Islamica* 4:6–41.

Sprenger, Aloys. 1849. *Two Works on Arabic Bibliography.* Calcutta: Asiatic Society of Bengal.

Steingass, Francis Joseph. 1970. *A Comprehensive Persian-English Dictionary.* Beirut: Librairie du Liban.

Storey, Charles A. 1972. *Persian Literature: A Bio-Bibliographical Survey.* 5 vols. London: Luzac, 1958–71. Reprint, London: Luzac.

Subtelny, Maria. 1986. "A Taste for the Intricate: The Persian Poetry of the Late Timurid Period." *Zeitschrift der Deutschen Morgenländischen Gesselschaft* 136, no. 1: 56–79.

———. 1983. "Art and Politics in Early 16th Century Central Asia." *Central Asiatic Journal* 27, nos. 1–2: 121–48.

Suter, Heinrich. 1922. "Das Buch der geometrischen Konstruktionen des Abu'l Wefā'." *Abhandlungen zur Geschichte der Naturwissenschaften und Medizin* 4:94–109.

———. 1972. *Die Mathematiker und Astronomen der Araber und ihre Werke.* Leipzig: B.G. Teubner, 1900. Reprint, New York: Johnson Reprint Corp.

Swelim, M. Tarek Nabil. 1994. "The Mosque of Ibn Tūlūn: A New Perspective." PhD diss., Harvard University.

Tabbaa, Yasser. 1985. "The Muqarnas Dome: Its Origin and Meaning." *Muqarnas* 3:61–74.

Tillotson, Giles Henry Rupert. 1987. *The Rajput Palaces: The Development of an Architectural Style, 1450–1750.* New Haven: Yale University Press.

Tott, Baron François de. 1785. *Memoirs of Baron de Tott: Containing the State of the Turkish Empire and the Crimea, during the Late War with Russia.* Translated from the French edition. 2 vols. London: G. G. J. and J. Robinson.

Al-Ṭūsī, Nāṣir al-Dīn Muḥammad ibn Muḥammad. 1964. *The Nasirean Ethics.* Translated by G. M. Wickens. London: Allen & Unwin.

Ünsal, Behçet. 1963. "Topkapı Sarayında Bulunan Mimarî Plânlar Üzerine." *Türk Sanat Tarihi Araştırma ve İncelemeleri* 1:168–97.

Watson, Oliver. 1985. *Persian Lustre Ware.* London and Boston: Faber and Faber.

Wiedemann, Eilhard. 1970. *Aufsätze zur arabischen Wissenschaftsgeschichte.* 2 vols. Hildesheim: G. Olms.

Wilber, Donald N. 1976. "Builders and Craftsmen of Islamic Iran: The Earlier Periods." *Art and Archaeology Research Papers* 10:31–39.

———. 1969. *The Architecture of Islamic Iran: The Il Khānid Period.* New York: Greenwood Press.

———. 1939. "The Development of Mosaic Faïence in Islamic Architecture in Iran." *Ars Islamica* 6:16–47.

Wilber, Donald N., and M. Minovi. 1938. "Notes on the Rab'-i Rashidi." *Bulletin of the American Institute for Iranian Art and Archaeology* 5:246–54.

Winter, H. J. J., and W. 'Arafāt. 1949. "Ibn al-Haitham on the Paraboloidal Focusing Mirror." *Journal of the Royal Asiatic Society of Bengal*, 3rd ser., *Science* 15:25–40.

Woepcke, Franz. 1874. *Trois traités arabes sur le compas parfait.* Notices et extraits des manuscrits de la Bibliothèque impériale et autres bibliothèques 22, 1–175. Paris: Imprimerie nationale.

———. 1855. "Recherches sur l'histoire des sciences mathématiques chez les Orientaux d'après des traités inédits arabes et persans. Deuxième article. Analyse et extrait d'un recueil de constructions géométriques par Aboûl Wefâ." *Journal Asiatique*, ser. 5, no. 5: 218–56, 309–59.

Wulff, Hans E. 1966. *The Traditional Crafts of Persia: Their Development, Technology, and Influence on Eastern and Western Civilizations.* Cambridge, Mass.: MIT Press.

TRANSLATION, TRANSCRIPTION, AND DRAWINGS

An English translation by Alpay Özdural of

Fī tadākhul al-ashkāl al-mutashābiha aw al-mutawāfiqa

(*On Similar and Complementary Interlocking Figures*),

edited and revised by Wheeler M. Thackston, with contributions by the other authors

Accompanied by Wheeler M. Thackston's transcription of the Persian text

and Alpay Özdural's drawings, with commentaries by Gülru Necipoğlu (based on "Analyses,"

the second chapter of Alpay Özdural's unpublished book)

 | DOI 10.1163/9789004315204_007

A NOTE TO THE READER

Every effort has been made to ensure the accuracy of the following translation and transcription of Ms. Persan 169, fols. 180r–199r, *Fī tadākhul al-ashkāl al-mutashābiha aw al-mutawāfiqa* (*On Similar and Complementary Interlocking Figures* [also referred to in this volume as the *Anonymous Compendium*]). The translation hews closely to the original text and is extremely literal (even numerals that are spelled out or indicated as numbers in the original document have been recorded as such). Although the resulting prose is sometimes awkward, the authors of the volume believe that this manner of rendering the text into English will enhance the reader's understanding of the original document.

The translation and transcription of each folio (or a portion thereof) are accompanied on the facing page by Alpay Özdural's redrawing of the corresponding geometrical figure (referred to as a "construction" and always labeled with the construction number followed by the numeral 1; for instance, Construction 1 on fol. 180r is labeled "Pl. 1.1"). The supplementary drawings that often follow the initial redrawing of the original figure in the document represent Özdural's efforts to provide additional commentary on the geometrical constructions. (These drawings are labeled with the construction number followed by consecutive numbers starting with 2; thus, those related to Construction 1 on fol. 180r are labeled as Pls. 1.2, 1.3, 1.4, 1.5, 1.6, and 1.7.) Throughout the chapters of the present volume the original geometrical figures and their accompanying texts are cited by folio and construction number (for instance, Construction 1 is referred to as "fol. 180v [1]").

The commentaries that accompany the drawings were merely summarized from "Analyses," the second chapter of Alpay Özdural's unpublished book manuscript; the authors of the present volume have not evaluated the contents of his analyses. Jan P. Hogendijk has checked the mathematical terminology and incorporated small additions to make the mathematical details clearer in Gülru Necipoğlu's textual summary. Although not all of the figures in Özdural's drawings are equally relevant to the constructions in the *Anonymous Compendium*, we have decided not to omit any and to include all of his drawings. Please note that in his redrawings, which record the uninked construction lines invisible in the facsimile, he has occasionally added letters of his own that are not included in the original geometrical constructions in order to facilitate the explanations provided in his detailed commentaries on each construction.

Readers should refer to Alpay Özdural's bibliography for the full citations of the works mentioned below. For the full text of his chapter titled "Analyses," on which these commentaries are based, interested readers should refer to the unedited manuscript of his book manuscript available online at https://doi.org/10.6084/m9.figshare.5255416.

DRAWING CONVENTIONS USED IN THE GEOMETRICAL CONSTRUCTIONS AND THEIR REDRAWINGS

Özdural's redrawings and supplementary drawings have been discussed in "A Note to the Reader." It is worth quoting in full his observations on the drawing methods employed in the *Anonymous Compendium*, as well as his explanation of the conventions used in his own drawings:

> All the figures in the original manuscript were drawn in two colors of ink: black and red. They were not used consistently though. In some cases, the whole construction was drawn in either red or black. In others, both colors were used together. In the latter cases, the ornamental pattern drawn in black ink is distinguished from the red auxiliary lines. In the figures of the analyses, which were drawn by me, the black-inked lines are shown as continuous lines, the red ones as long-broken lines. Some of the original figures contain dotted lines to indicate the auxiliary lines that were part of the construction procedure. Such lines are shown here as short-broken lines. The letters designating points of the construction or parts of the figures were mostly written in black, but in a certain case (namely, Construction 31) red ink was deliberately used to differentiate the ones that are on a special instrument; these are distinguished in our figure by the prime symbol.
>
> In some of the figures in the manuscript there are incised lines that can be detected only by close inspection. Apparently, the constructions were first executed with a sharp instrument like a stylus, including all the guiding lines, and then the lines relevant to the pattern or to the construction procedure explained in the accompanying text were inked. The auxiliary lines that were left uninked in the manuscript are invisible in photographs of the folios. The analyses here are based on the original figures. In the figures drawn by me, the incised lines underlying the patterns are shown as dotted lines. These guiding lines provide us with the necessary information, particularly when an accompanying text is missing, for understanding the construction procedure.[1]

---

1 Alpay Özdural, "Analyses," the second chapter in his unpublished book, "Interlocks of Similar or Complementary Figures: Collaboration of Mathematicians and Artisans in the Islamic World," 32–33.

THE *ABJAD* NUMERICAL SYSTEM FOR LABELING THE GEOMETRICAL CONSTRUCTIONS

In many constructions of the *Anonymous Compendium* points are labeled with the letters of the Arabic alphabet in increasing numerical value, in accordance with the *abjad* system of numbering (*alif* = 1, *bāʾ* = 2, *jīm* = 3, *dāl* = 4, etc.). This was the traditional system for labeling used by medieval Islamic geometers/astronomers. The letters of the alphabet are listed in the *abjad* order on folio 196v. Because not all of the twenty-eight Arabic letters have equivalents in the Latin alphabet, Professor Özdural used various upper- and lowercase Latin letters (sometimes arbitrarily selected) in the transcription system he created for designating points in his redrawings of the the geometrical constructions.

The notation "*lA*" stands for "*lām-alif*." In some cases, geometrical figures had more points that required labeling than the *abjad* system could accommodate (i.e., beyond the Arabic letter *ghayn* (= *gh*), which equals 1000 in the *abjad* order). It was then standard procedure for mathematicians to begin again with the letters of the alphabet in the same *abjad* order, but prefixed with the Arabic letter *lām* (= *l*): hence, *lām-alif* (*lA*), *lām-bāʾ* (*lB*), etc. These letters have no numerical value. Özdural had need of only one extra letter (*lA*), which, in his redrawings, he designated with the Greek letter *lambda* (λ).

Özdural's Redrawings	Ms. Persan 169, fols. 180r–199r	*Abjad* numerical values	Özdural's Redrawings	Ms. Persan 169, fols. 180r–199r	*Abjad* numerical values
A	(*a*)	1	O	(ʿ)	70
B	(*b*)	2	F	(*f*)	80
G	(*j*)	3	C	(*ṣ*)	90
D	(*d*)	4	Q	(*q*)	100
E	(*h*)	5	R	(*r*)	200
W	(*w*)	6	J	(*sh*)	300
Z	(*z*)	7	t	(*t*)	400
H	(*ḥ*)	8	h	(*th*)	500
T	(*ṭ*)	9	X	(*kh*)	600
I	(*y*)	10	V	(*ẕ*)	700
K	(*k*)	20	d	(*ḍ*)	800
L	(*l*)	30	z	(*ẓ*)	900
M	(*m*)	40	g	(*gh*)	1000
N	(*n*)	50	λ	(*lA*)	n/a
S	(*s*)	60			

GLOSSARY OF SELECTED TECHNICAL TERMS

ʿaqd (lit. knot): repeat-unit or pattern

gūnyā: set square, a right-angled triangle that approximated angle measurements in terms of six degrees; a tool used for designing geometric patterns with particular angles

gūnyā-5: set square [number] 5, corresponding to 36°-54°-90° (degrees in ratio 6:9:15), facilitated the construction of regular pentagons and decagons

gūnyā-6: set square [number] 6, corresponding to 30°-60°-90° (degrees in ratio 5:10:15), simplified drawing regular hexagons and dodecagons

gūnyā-15: set square [number 15], corresponding to 24°-66°-90° (degrees in ratio 4:11:15), could be used for drawing regular fifteen-sided polygons

gūnyā misṭar: a certain T-square (as explained in the text), used for rare proportions and angles that could not be drawn with standard set squares, or with ruler and compass

makhrūṭāt: conic sections

khāna (lit. compartment), or *khāna-i ʿaqd* (lit. compartment of the repeat-unit): refers to the square or rectangular outer frame of a repeat-unit

khātim (lit. seal): star polygon, that is, an equiangular and equilateral non-convex polygon

khātim-i mukhammas (lit. pentagonal seal): pentagonal star

khātim-i chahār (lit. tetragonal seal): tetragonal star

khaṭṭ (lit. line): in a more specific sense, *khaṭṭ* followed by a number is used for a line referring to angle measurements, with the number in question corresponding to certain angles formed by the radii of circular construction lines:
khaṭṭ-i chahār (line [number] four)
khaṭṭ-i panj (line [number] five)
khaṭṭ-i 6 (line [number] six)
khaṭṭ-i haft (line [number] seven)
khaṭṭ-i 8 (line [number] eight)

Regarding the lines of angles, Chorbachi writes:

To draw a "line [number] n" refers to an angle required for the construction of a polygon or star polygon, corresponding to the modern use of a protractor with…degrees. Line [number] 4 refers to the diagonal of a square, that is, a line at an angle of 45° in the right-angled corner of a repeat unit. In other cases the line numbers refer to the lines of polygons "that radiate from the center of a circle and cut its circumference at a point that generates half the length of the side of the polygon, or the polygonal star. This angle is usually equal to half the center angle of the polygon," as explained in the only axiom cited in the *Anonymous Compendium* (fol. 183v [12]). It can be deduced that "artisans must have had a mechanical means or an instrument to draw these angle lines without the need to reconstruct them each time they are required, and that this instrument is a proportionate right triangle set having the value of the desired angle." The following list provides the values of the lines mentioned in the textual commentaries and their relative polygons:[1]

Line [number] 4: Value 45° (half the central angle of a square)
Line [number] 5: Value 36° (half the central angle of a pentagon)
Line [number] 6: Value 30° (half the central angle of a hexagon)
Line [number] 7: Value 26°, exactly 25.51° (half the central angle of a heptagon)
Line [number] 8: Value 22.5° (half the central angle of an octagon)

misṭar or *misṭara:* ruler

nisbat: ratio or proportion

shabīh, mutashābiha (lit. similar): in a more specific sense, referring to geometrical figures or solids, also numbers

turunj (or *ṭurunj*): kite, i.e., a quadrilateral consisting of two congruent triangles but not a parallelogram. A kite is also called a "pinecone figure" (*shakl-i ṣanawbarī*) or "pinecone-like quadrilateral" (*murabbaʿ-i ṣanawbarī*) in the *Anonymous Compendium.*

ṭurunj-6: translated as "kite [number] 6," corresponds to two *gūnyā-6*, that is to say, to two right-angled triangles of 30°-, 60°-, and 90°-angles

1 Chorbachi 1989a, 208–10.

Translation and Transcription (fol. 180r [1])

Fol. 180r [1]*

On similar and complementary interlocking figures.[a]

When two equilateral and equiangular decagons with equal areas are drawn, and a pentagonal star (*khātim-i mukhammasī,* lit. pentagonal seal) [is also drawn] whose sides are equal to the sides of the decagon[s] and its outer diameter [i.e., the diameter of the circle circumscribing the pentagonal star] is equal to the outer diameter of the decagon[s], and half of its inner diameter [i.e., the radius of the circle inscribed within the pentagonal star] is equal to the side of the decagon[s], [then] these three figures [i.e., the two regular decagons and one pentagonal star] can be placed within another [larger] decagon. This is a decagon whose sides are equal to half of the outer diameter [i.e., radius] of the assumed decagon[s], while half of its outer diameter [i.e., radius] is equal to half of the outer diameter plus one side of the assumed decagon[s], as is illustrated.[b]

a Two vertical lines written in red ink, from bottom to top, along the right margin of the page.
b The text is composed of nine oblique lines on the left side of the figures.

فی تداخل الأشکال المتشابهة او المتوافقة

هرگاه که دو معشر متساوی الأضلاع والزوایا متساوی السطوح رسم کند و خاتم مخمسی که اضلاع او مثل اضلاع معشر باشد و قطر خارج او مثل قطر خارج معشر بود و نصف قطر داخل او مثل ضلع معشر باشد این هر سه شکل در یک معشر توان نهاد و آن معشری باشد که اضلاع او مثل نصف قطر خارج معشر مفروض باشد و نصف قطر خارج او مثل نصف قطر خارج معشر مفروض باشد. با ضلع او بهم منضم باشد چنانک مرسوم است.

* The editors would like to thank Farshid Emami, who proofread the Persian transcription and English translation.

CONSTRUCTION 1 (fol. 180r [1])[1]

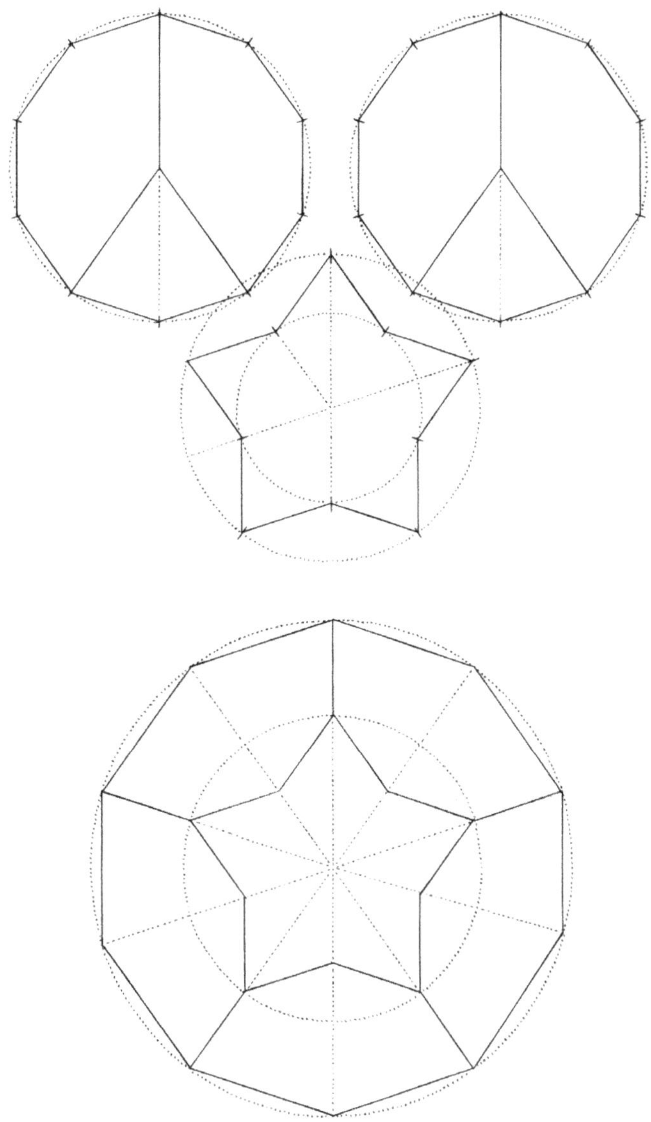

Pl. 1.1. Construction of a large decagon containing a central pentagonal star from two small decagons and a pentagonal star.

This cut-and-paste construction, depicted in two stages, shows how to transform a pentagonal star (*khātim-i mukhammas*, lit. pentagonal seal) and two congruent regular decagons (top) into a large decagon inscribed with a central pentagonal star (bottom). Both stages are based on a 36°-angle and probably required a set square with 36°-, 54°-, and 90°-angles (which is described later, in Construction 14; see pl. 14.1). Each regular decagon is divided into pieces that are reassembled into the large decagon, which comprises six pieces: 5 double-kites and the pentagonal star.

1 Özdural, "Analyses" (unpublished chapter), 33–41.

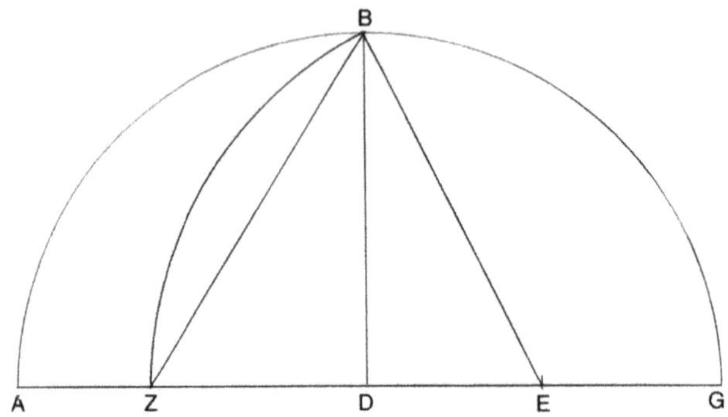

Pl. 1.2. Construction of the decagon and pentagon by using a line segment cut in extreme and mean ratio to find their respective sides, as described in Ptolemy's *Almagest.* (A straight line is said to have been cut in extreme and mean ratio when the ratio of the whole line to the large segment is the same as the ratio of the large segment to the small segment.) "Let there be a semicircle ABG about center D and on diameter ADG. Draw DB perpendicular to AG at D. Let DG be bisected at E, join EB, and let EZ be made equal to EB. Join ZB. I say that ZD is the side of the [regular] decagon and BZ the side of the [regular] pentagon."[2]

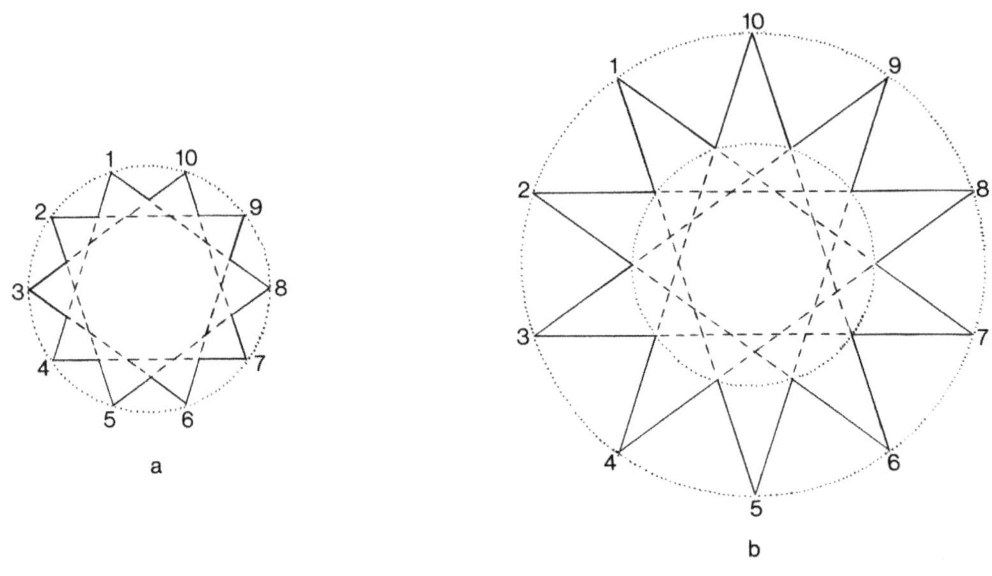

Pl. 1.3a. How to generate a decagonal star, inscribed in a circle, by connecting every fourth vertex.

Pl. 1.3b. Within the decagonal star, circumscribed by a circle, draw a concentric circle circumscribing another decagonal star, which is formed by the intersecting lines of the outer star. This star is generated by connecting each third vertex.

2 Ptolemy 1984, 48.

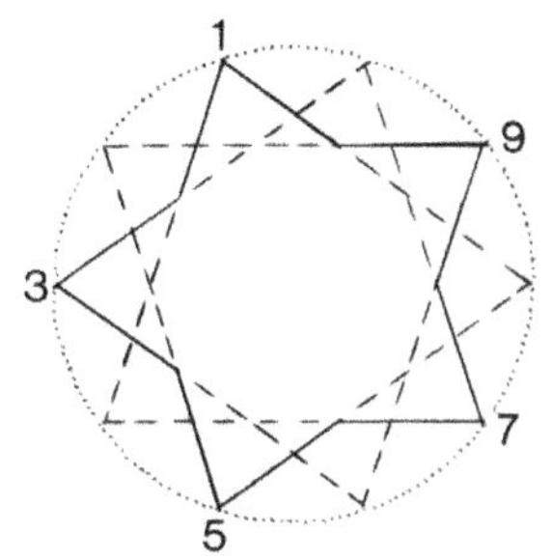

c

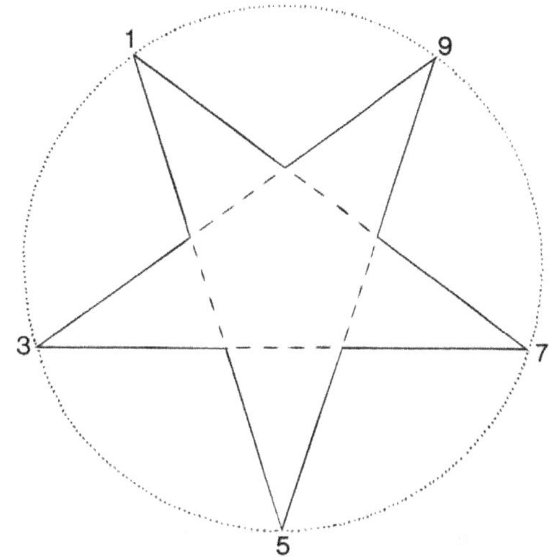

d

Pl. 1.3c. The pentagonal star can be defined in relation to the decagonal star in that the alternating vertices are missing.

Pl. 1.3d. This pentagonal star (corresponding to the *khātim-i mukhammas*, or pentagonal seal) is a version of the star depicted in pl. 1.3b, in which the alternating vertices are missing. Unlike the star in pl. 1.3c, it can only be constructed with the aid of the star depicted in pl. 1.3b. In the *khātim-i mukhammas* the opposite sides of neighboring vertices are parallel to each other. This is useful for creating ornamental patterns composed of rosettes.

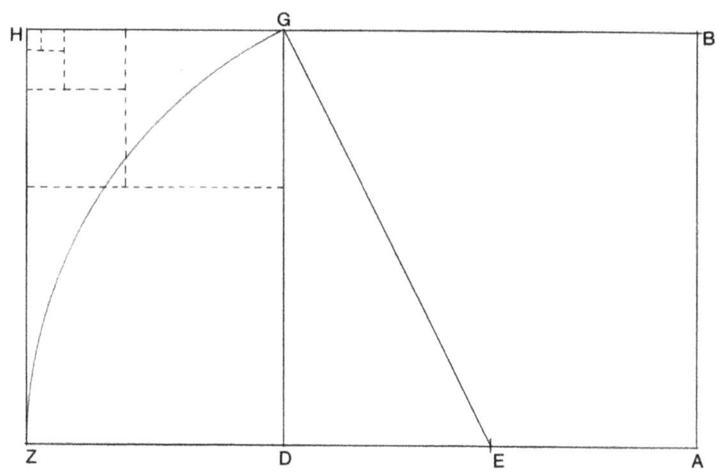

Pl. 1.4. Construction of a line segment cut in extreme and mean ratio ("golden mean") according to Euclid, *Elements*, 2.14 and 6.3 (definition).[3] The line AZ is cut in extreme and mean ratio: AZ:AD = AD:DZ. When perpendicular ZH = AD is erected, rectangle BAZH becomes a "golden rectangle." According to Özdural, this relation is the hidden objective behind Construction 1.

3 Euclid 1956, 1:402–3 and 2:188–89.

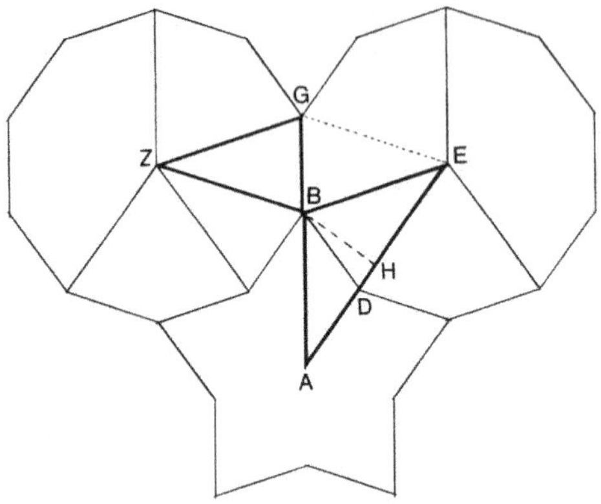

a

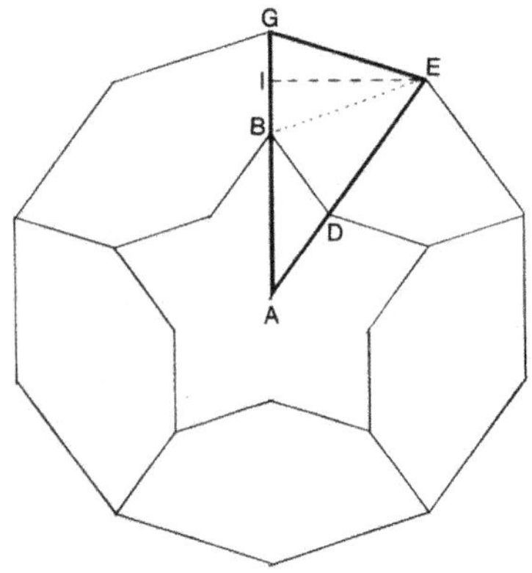

b

Pl. 1.5a. AE is the combined length of the side and radius of the assumed decagon in Construction 1. According to Euclid, *Elements*, 13.9, point D cuts it in extreme and mean ratio.[4] Adding triangle BGZ of the second assumed decagon to triangle ABE, so that point Z corresponds to point E, yields the isosceles triangle AGE of the large decagon depicted in Construction 1.

Pl. 1.5b. In triangle AGE point B cuts line AG into segments AB and BG in extreme and mean ratio, and line BE divides triangle AGE into parts ABE and BEG in the same ratio (all three triangles have the same altitude EI on the base AG). Hence, Construction 1 is, according to Özdural, a visual demonstration of the square of the extreme and mean ratio. Two congruent decagons plus a concave decagon (i.e., the pentagonal star) make up the large decagon through an elegant arrangement of dissected pieces.

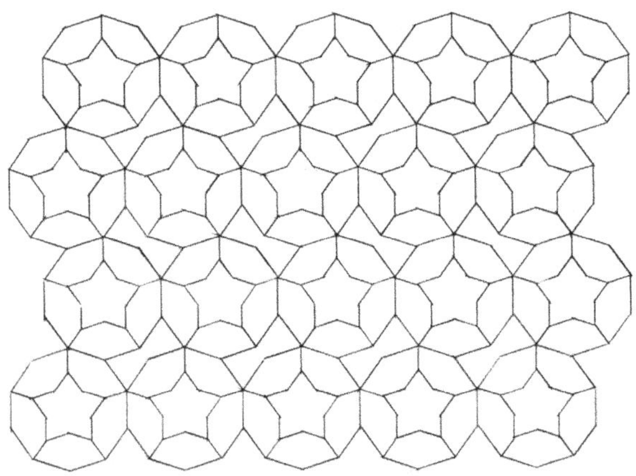

a

Pl. 1.6a. An ornamental possibility offered by Construction 1. It is obtained by the linear repetition of the large decagon that consists of the central pentagonal star surrounded by five pieces of the two congruent small decagons.

4 Euclid 1956, 3:455–57. "If the side of the hexagon and that of the decagon inscribed in the same circle be added together, the whole straight-line has been cut in extreme and mean ratio, and its greater segment is the side of the hexagon," since the side of the hexagon corresponds to the radius of the circle.

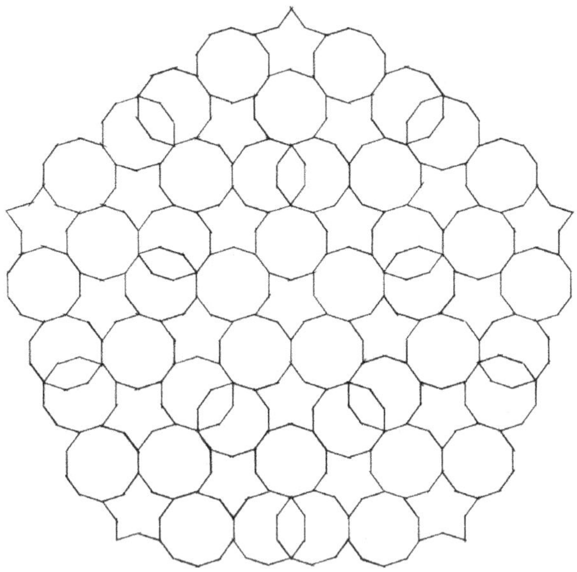

Pl. 1.6b. Another ornamental possibility offered by Construction 1. This decorative tiling can be generated from the pentagonal star and the surrounding decagons depicted in the first stage of Construction 1, through the growth of the central pattern in five directions.

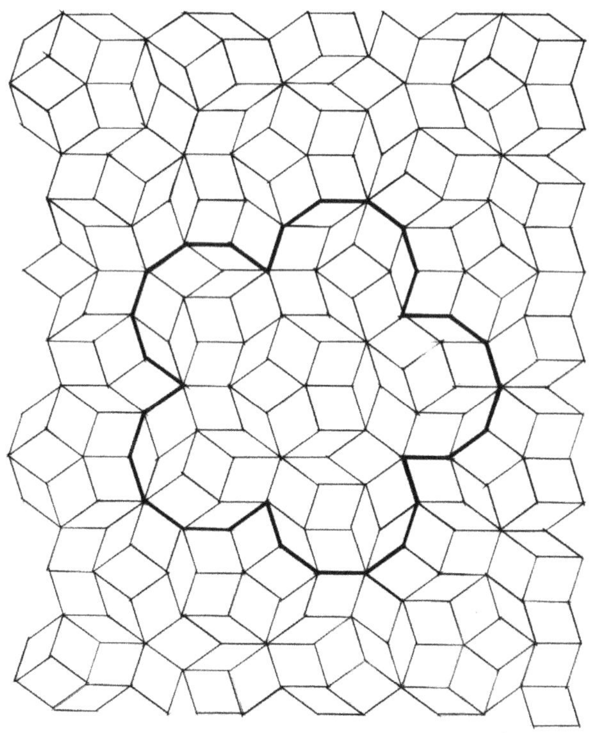

Pl. 1.7. A more complex tiling obtained by different arrangements of the same figures. Systematic studies by modern scientists on the pentagonal star and decagon have led to the discovery of what is known as a "Penrose tiling."[5]

5 Penrose 1974, 266–71.

Translation and Transcription (fol. 180v [2])

Fol. 180v [2]

From pieces W and Z altogether one can obtain one decagonal star, or, if one wants, one can make two decagons. Again from pieces E and D, one nonagonal star can be obtained, or else 2 nonagons. Again from the components G and B, one octagonal star can be obtained, or else 2 octagons. Similarly, from the components of A, one tetragonal star or two squares can be obtained, as in this example.[a]

It must be known that from the pieces of this [large] square, one tetragonal star, one octagonal star, one nonagonal star, and one decagonal star can be obtained. If one wants, 2 squares, 2 octagons, 2 nonagons, and 2 decagons can be composed. This [construction] is extremely elegant. God knows best.[b]

---

a The first part of the text, which is in two parts, is composed of five horizontal lines below the figure.

b The second part of the text is composed of two vertical lines from top to bottom along the right margin.

از مقطعات وٓ زٓ مجموع یک معشر مجنح حاصل شود و اگر خواهد دو معشر مضلع سازد و باز از مقطعات هٓ دٓ یک متسع مجنح حاصل شود و إلّا ۲ متسع مضلع و باز از اجزاء جٓ بٓ یک مثمن مجنح حاصل شود و إلّا ۲ مثمن مضلع. و هم چنین از اجزاء آ یک مربع مجنح یا دو مربع مضلع حاصل شود بدین مثال.

باید دانست که اجزاء این مربع یک مربع مجنح و یک مثمن مجنح و یک متسع مجنح و یک معشر مجنح حاصل توان کرد و اگر خواهد ۲ مربع مضلع و ۲ مثمن مضلع و ۲ متسع مضلع و ۲ معشر مضلع ترکیب توان کرد و بغایت لطیف است والله اعلم.

CONSTRUCTION 2 (fol. 180v [2])[1]

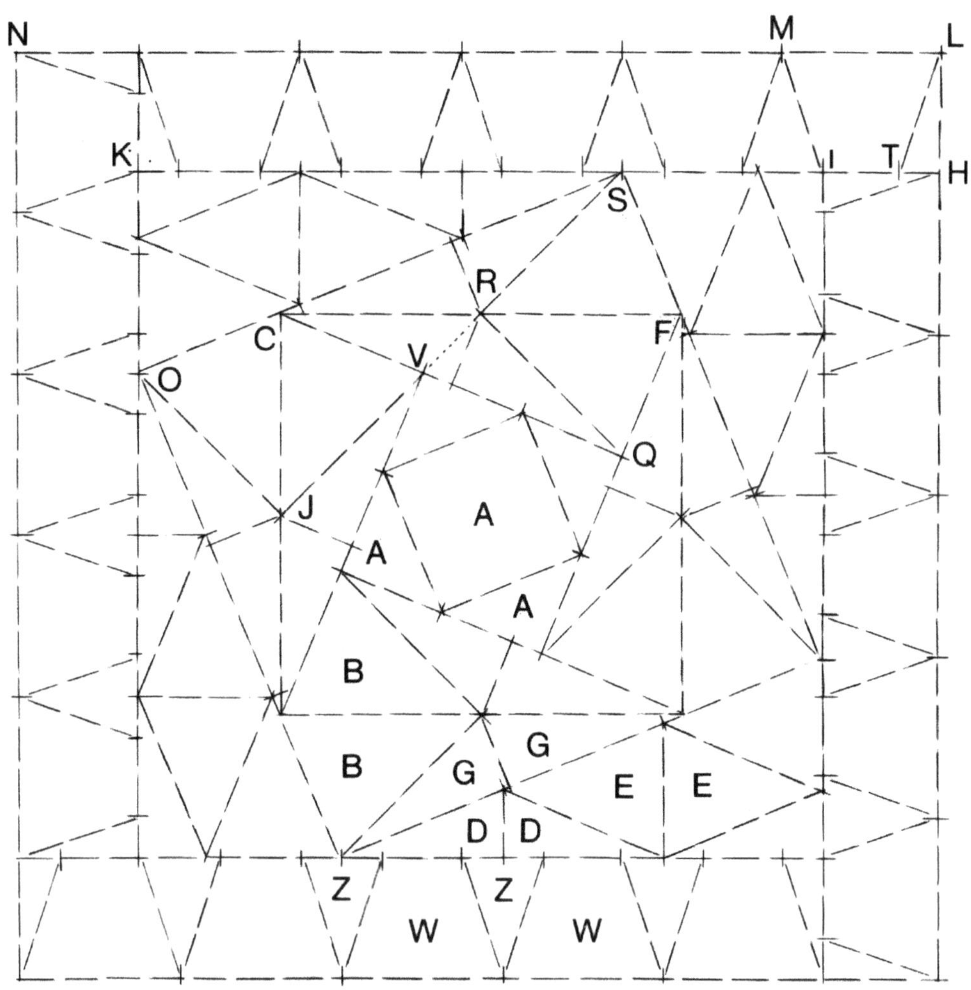

Pl. 2.1. Construction with parallel and rotating concentric squares, subdivided into triangular or trapezoidal pieces to be reassembled into polygons and star polygons.

The objective of this "cut-and-paste" construction is to unite, within a single square, smaller squares, and eight-, nine-, and ten-sided figures, which belong to three different geometrical categories. All the lines in the original drawing are in red ink, with letters in black ink. Judging by the lack of uninked guiding lines, Özdural suggests that the figure was not constructed with a ruler and compass. Instead, set squares were utilized for drawing 22.5°/67.5°- and 45°-angles (set squares are mentioned in Constructions 14, 34, and 36). According to the text, from the dissected components one tetragonal (square) star, one octagonal star, and one decagonal star can be obtained. Alternatively, two squares, two octagons, two nonagons, and two decagons can be composed.

1 Özdural, "Analyses" (unpublished chapter), 41–49.

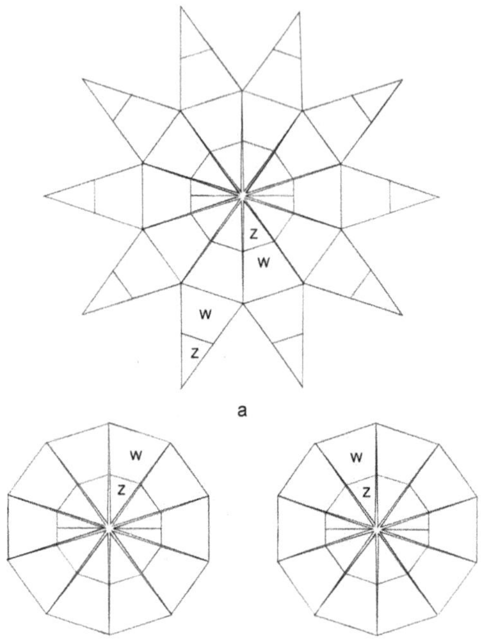

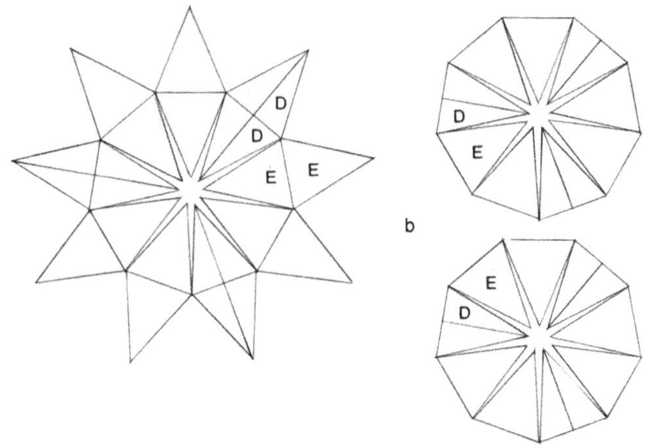

Pl. 2.2a. According to the text, pieces W and Z constitute the radial triangle of the decagon. Joining twenty such pieces makes up one decagonal star and two decagons by approximation (with only a small deviation). This deviation enables one to integrate the decagons and decagonal stars with other polygons and star polygons.

Pl. 2.2b. Construction of two congruent nonagons or one nonagonal star, by a crude approximation, from pieces D and E.

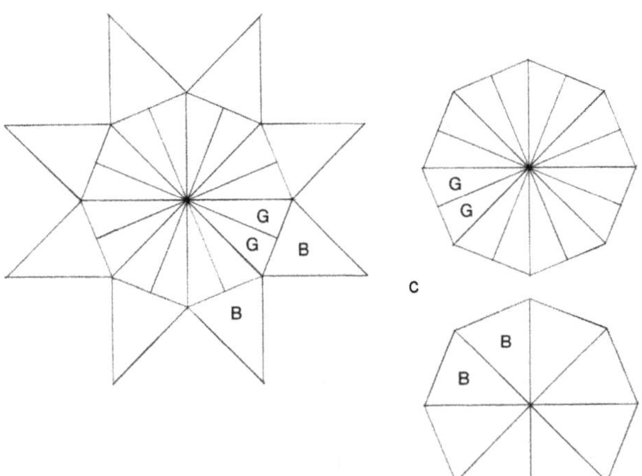

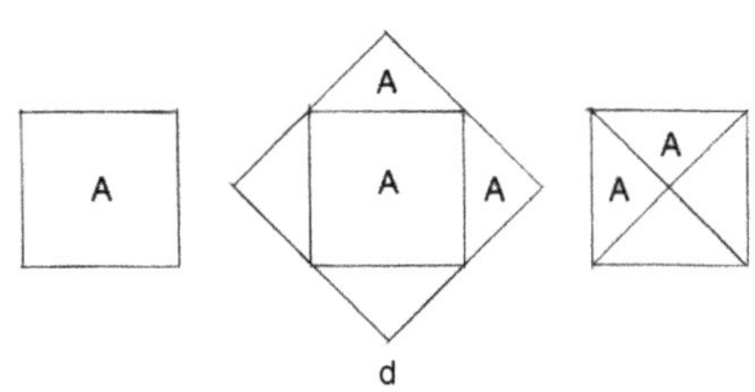

Pl. 2.2c. Construction of either one regular octagonal star or two regular octagons from pieces B and G.

Pl. 2.2d. The relationship between the two rotated squares at the middle of pl. 2.1. The side of the outer square is the $\sqrt{2}$ times the side of the innermost square; thus the area of the former is twice that of the latter.

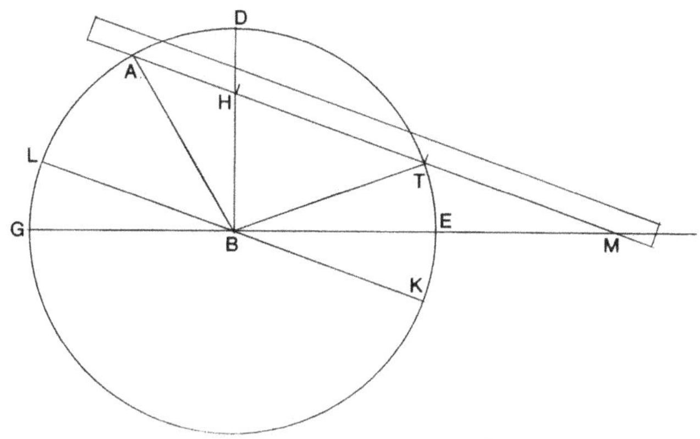

Pl. 2.3. Construction of the nonagon based on the trisection of an angle by means of a verging procedure, that is, with the use of a moving ruler, as described in Abu'l-Wafa' al-Buzjani's practical geometry treatise, titled *Geometrical Constructions.* (A verging construction, also called a Neusis construction, allows the classical rules of geometric construction to be relaxed in order to permit the insertion of a line segment of given length between two given lines [or circles] such that the extension of the segment passes through a given point. This procedure is called verging because the segment of given length "verges" to the given point. It is often visualized by means of a marked ruler rotating about the given point; the ruler is "marked" in the sense that distance measurement is supposed to be possible. Using a Neusis construction, cube duplication, angle trisection, and the construction of a regular heptagon are soluble). During this procedure, the moving ruler is never to be removed from point A, and the ruler is rotated until the distance between the intersection of points H and T is equal to the radius of the circle. Angle LBG becomes one-third of angle ABG. This theoretically correct but cumbersome procedure was not used in Construction 2 to construct the nonagon; instead, a crude approximation was preferred.

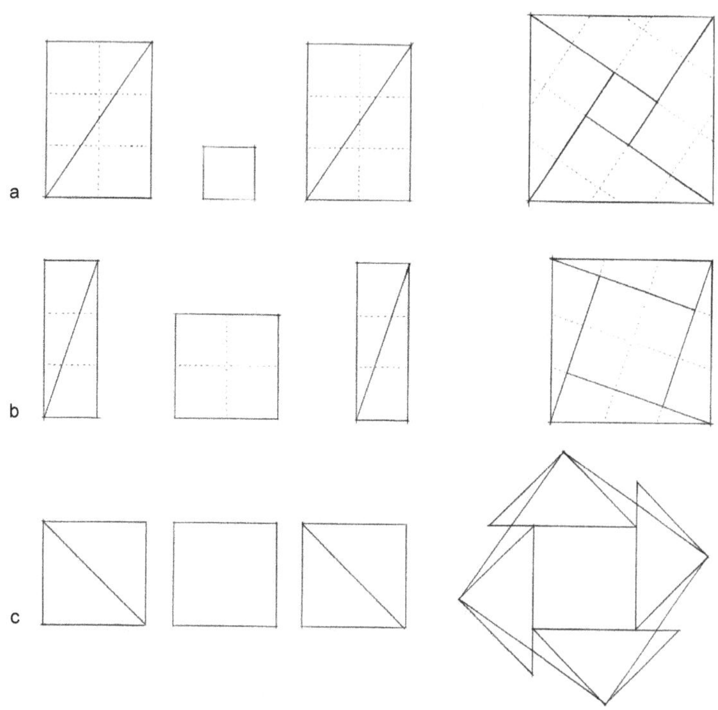

Pl. 2.4a–c.

d 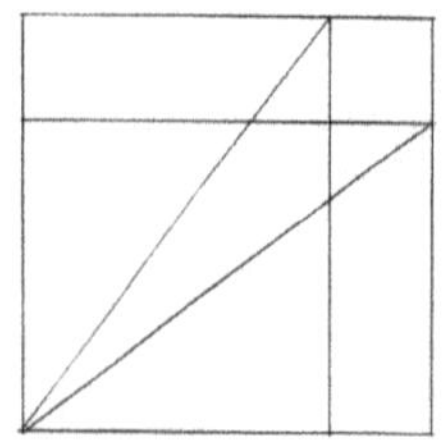 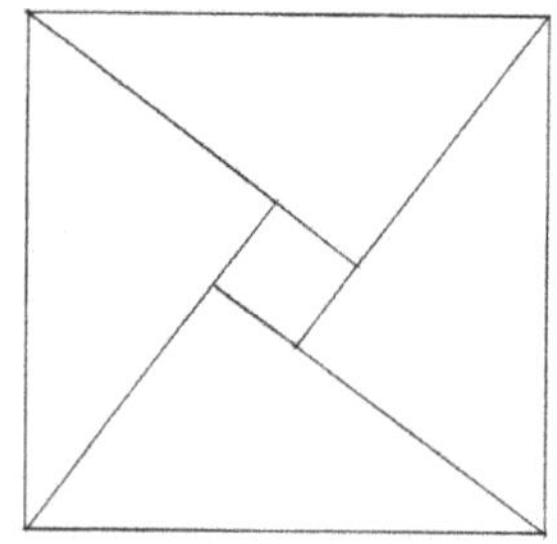

Pl. 2.4a–d. Constructions explicated in Abu'l-Wafa''s *Geometrical Constructions* concerning the problem of composing a square from several smaller squares, using the "cut-and-paste" method. Özdural connects the four-fold rotational symmetry of the diminishing squares in pl. 2.1 to Abu'l-Wafa''s account of a meeting between geometers and artisans in late tenth-century Baghdad, when this problem was discussed.[2]

(a): In composing a single square from 13 squares, Abu'l-Wafa' first separated the number into 9 and 4, the square roots of which are 3 and 2. Then he made two rectangles, with sides equal to 3 units and 2 units. As the difference between 3 and 2 is 1, he made another square of 1 unit. Dissecting the rectangles along their diagonals to obtain four triangles, he rotated these triangles (whose hypotenuses are each $\sqrt{13}$) around the 1-unit square. The sides of the desired square are thus $\sqrt{13}$ units. The exercise corresponds to the modern equation $a^2 + b^2 = (a - b)^2 + 2ab$ (if "a" is 3 and "b" is 2).[3]

(b) The case for 10 squares was based on the same equation, but this time "a" is 3 and "b" is 1, and the difference between "a" and "b" is 2. In this case, four rotating triangles having sides with 3 and 1 units were placed around the rotated small square whose sides are 2 units. In both cases, one obtains a figure as in 2.4a, composed of four triangles that rotate around a central rotated square, but the angle of rotation depends on the sides of the triangles and their difference.[4]

(c) Abu'l-Wafa''s solution for the problem of constructing a single square from 3 squares, again involving four triangles rotating around a central square.[5]

(d) Abu'l-Wafa''s construction composed of four rotating triangles, proposed under the general subtitle, "On constructing the square from two different squares with unknown dimensions of sides."[6]

2 Abū'l-Wafā' al-Būzjānī, Ms. Ayasofya 2753, analyzed in Özdural 2000, 174–76.
3 Abū'l-Wafā' al-Būzjānī, Ms. Ayasofya 2753, fols. 48–50 ([chapter] 10.6, 10.7).
4 Abū'l-Wafā' al-Būzjānī, Ms. Ayasofya 2753, fols. 50–51 (10.10, 10.11), on the cases of 10 and 20 squares in the subsection titled "On dividing squares if their number is the sum of two unequal squares."
5 Abū'l-Wafā' al-Būzjānī, Ms. Ayasofya 2753, fols. 56–57 (10.14).
6 Abū'l-Wafā' al-Būzjānī, Ms. Ayasofya 2753, fols. 58–59 (10.16).

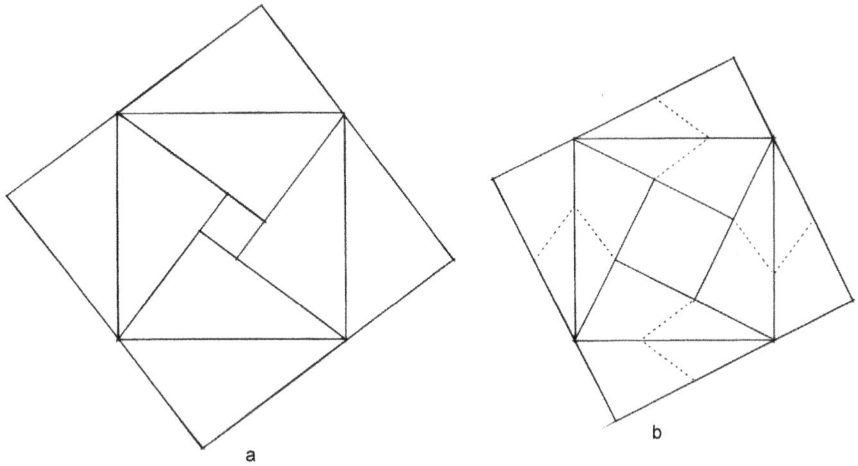

a b

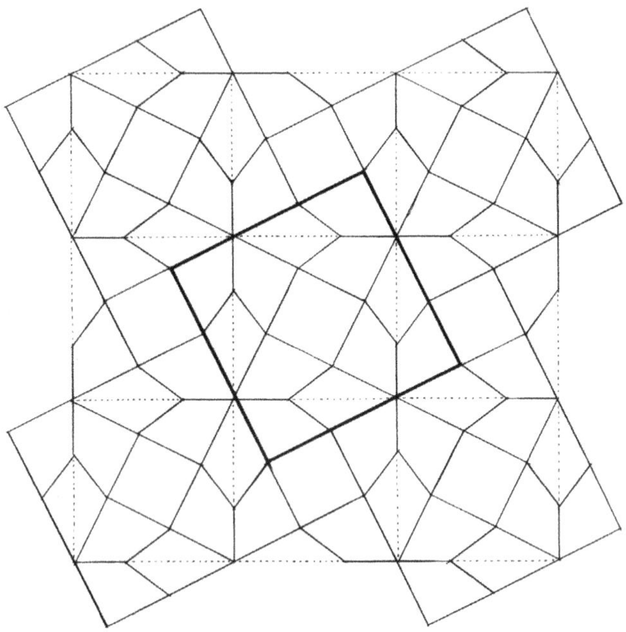

c

Pl. 2.5a. The transformation of the same type of rotational symmetry used by Abu'l-Wafa' into a favorite ornamental motif by artisans. Joining four more congruent triangles along their diagonals yields two rotating squares inscribed within one another. These two rotating squares visually express the following equation: $(a^2 + b^2) + 2ab = (a + b)^2$.

Pl. 2.5b. The same scheme used as an ornamental device, usually found on a single panel with the ratio of 1:2 between the perpendicular sides of the triangle.

Pl. 2.5c. With the ratio of 1:2 between the sides of the triangles, the rotating squares can be subdivided so as to create an interlocking composition.

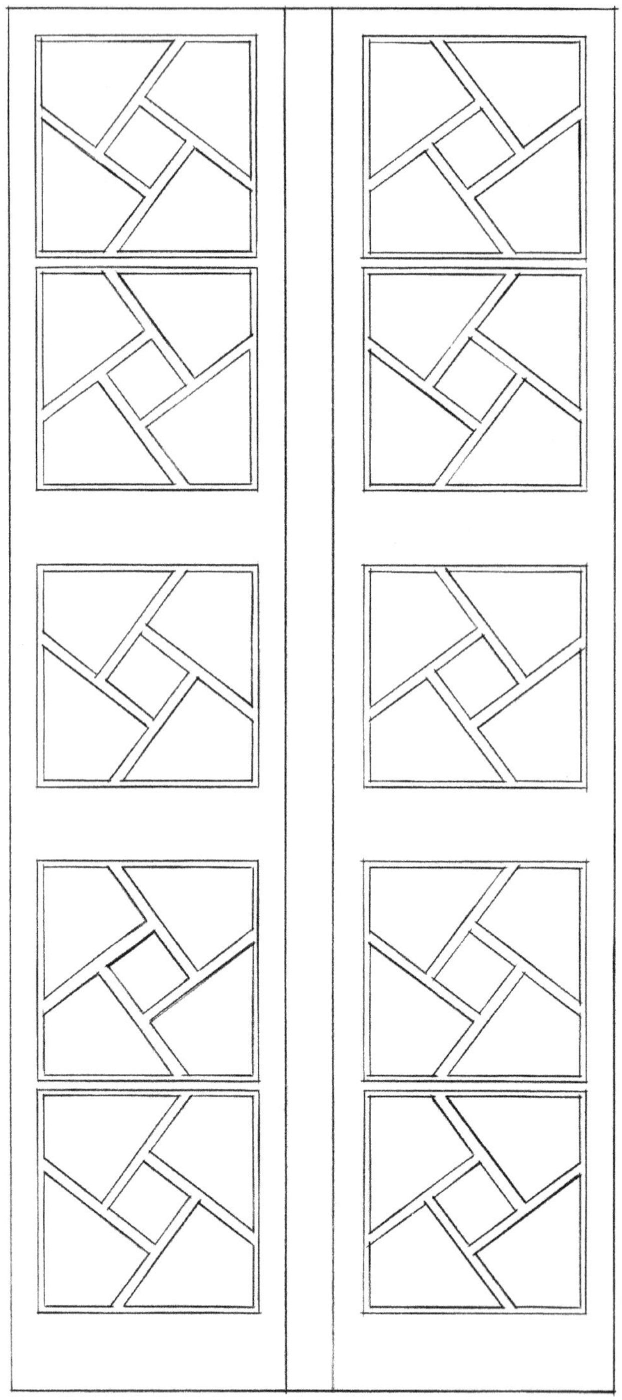

Pl. 2.6. The scheme of a rotating central square, surrounded by four rotating kites (each composed of two congruent right-angled triangles), within a larger outer square. This pattern was used as an ornamental device in the wooden door of the Imam Ibrahim Mosque in Mosul, dated 498 (1104).[7]

7 The picture of the door is published in Chorbachi 1989b, 766. It was made of ten square panels, each panel composed of triangles whose perpendicular sides are related in a 1:2 ratio.

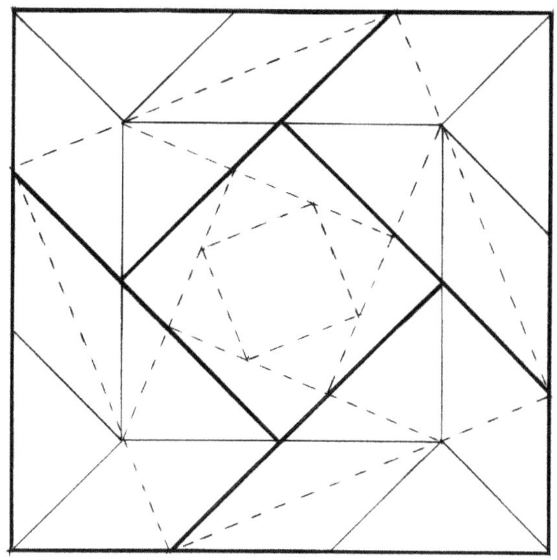

a

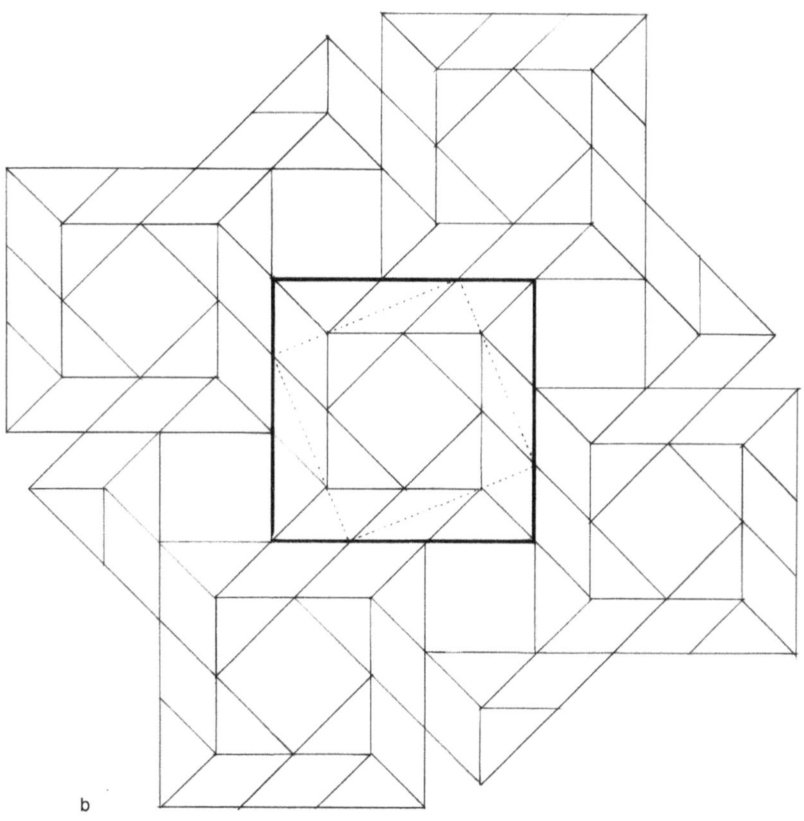

b

Pl. 2.7a. In Construction 2, the special angle of rotation (22.5°/67.5°) allows the squares to be subdivided into congruent rhombi and triangles.

Pl. 2.7b. The repetition of the pattern above generates an ornamental composition of rotating square belts, which was not the stated objective of Construction 2.

TRANSLATION AND TRANSCRIPTION (fol. 181r [3])

Fol. 181r [3]

No accompanying text.

Construction 3 (fol. 181r [3])[1]

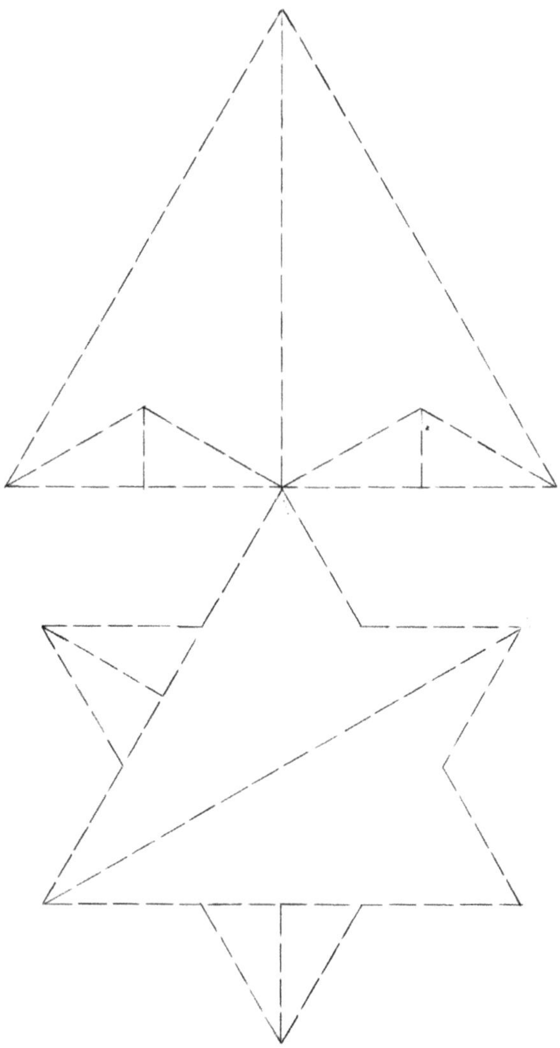

Pl. 3.1. Construction of a hexagonal star and equilateral triangle of equivalent areas.

The "cut-and-paste" construction, drawn in red ink, like the preceding one, lacks an explanatory text. The figure shows how to transform an equilateral triangle into a hexagonal star so that they have the same area. The side of the main equilateral triangle corresponds to the diameter (altitude) of the hexagonal star (i.e., the diameter of a circle circumscribing the hexagonal star). One of the sides of the equilateral triangle is divided into four equal parts by two perpendiculars. The four small triangles obtained are rearranged on the two sides of the inversely positioned equilateral triangles to complete the hexagonal star. This construction demonstrates the proportional relationship between the dimensions of the equilateral triangle and hexagonal star.

1 Özdural, "Analyses" (unpublished chapter), 49–50.

Translation and Transcription (fol. 181r [4])

Fol. 181r [4]

When one wants to determine the side of a square equal in area to the area of an octagon, [then] the square of the side of [a square inside] the octagon should be subtracted from the square of the diameter of the circle inscribed in the octagon. The square root of the remainder is the side of the square. God knows best.[a]

a The text is composed of three oblique lines from right to left.

هرگاه که خواهد تا ضلع مربعی که سطح او مساوی سطح مثمن باشد معلوم کند باید که مربع ضلع مثمن از مربع قطر دایره که داخل مثمن باشد نقصان کند مابقی حد روی ضلع مربع باشد والله اعلم.

Construction 4 (fol. 181r [4])[1]

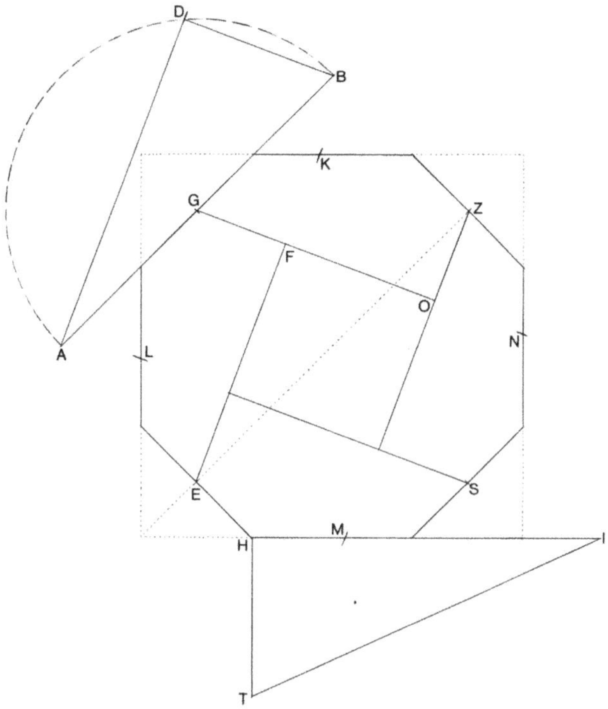

Pl. 4.1. Construction of a square and octagon with equivalent areas.

An octagon with two right-angled triangles attached to both of its sides; only the semicircle circumscribing one of the triangles is drawn in red ink. The text explains how to determine the side of a square whose area equals that of an octagon. It only gives the theoretical principles behind the construction, without specifying the procedure for dissecting the octagon into five pieces that can be reassembled into the required square (see pl. 4.2b).

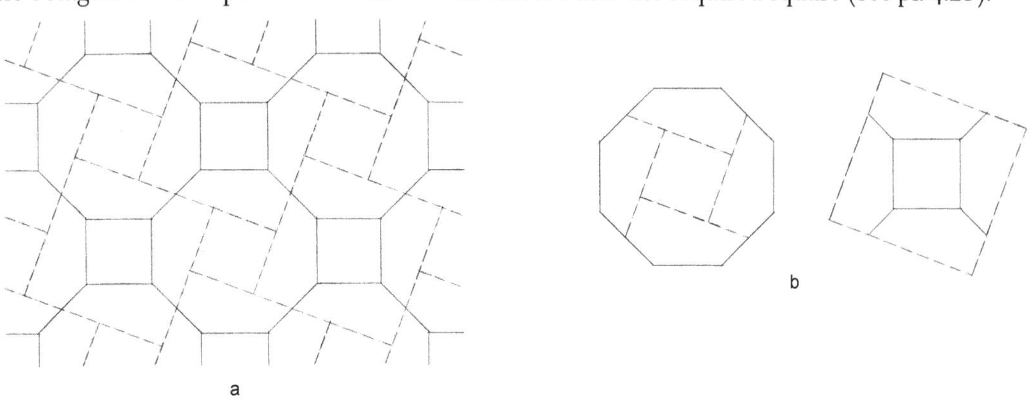

Pl. 4.2a–b. A tessellation formea ɒy the repetition of the octagon and the square having equal sides, superposed with another tessellation composed of two unequal squares, the larger one equal in area to the octagon, and the smaller one identical to the square in the first tessellation (pl. 4.2a, shown in broken lines). This superposition shows how the transformation of the octagon into a square can be achieved most economically, based on Geoffrey Thomas Bennett's novel technique called "completing the tessellation" (4.2b). In 1926 he demonsrated how to dissect an octagon into five pieces from which to compose a square.[2]

---

1 Özdural, "Analyses" (unpublished chapter), 50–53.

2 On this dissection by G.T. Bennett (d. 1943), see Frederickson 1997, 150.

Translation and Transcription (fol. 181v [5])

Fol. 181v [5]

Section (*faṣl*): If one wants to turn a segment, such as arc AGB, into a circle, but the center is not known, one should use points A and B [as centers] and inscribe with any compass opening greater than the radius of arcs ZD and HD so that they intersect at point D. From this point to the midpoint of the arc, erect perpendicular DG. Then from point G with the previous compass opening, that is AZ, mark point E[a] and join line AE. Again from point G[b] with compass opening AE, inscribe arc ZH so that it intersects [arcs ZD and HD] on both sides at points Z and H. Draw lines AZ and BH so that [point T[c]] is obtained. This is the center of the circle whose radius is the compass opening TG. God knows best.[d]

---

a Text: D.
b Text: ED.
c Text: DHA.
d The text is composed of six horizontal lines.

فصل اگر قطعه از دائره باشد مثل قوس آ ج̅ ب̅ و خواهد که آنرا دائره سازد و مرکز معلوم نبود باید که از نقطهٔ آ ب̅ بهر فتح که اعظم از نصف قطر باشد قوس ز̅ د̅ ح̅ د̅ بگرداند تا متلاقی شوند بنقطهٔ د̅، و ازین نقطه بوسط قوس عمود د̅ ج̅ اخراج کند و باز از نقطهٔ ج̅ بفتح اول اعنی آ ز̅ نقطهٔ ه̅[۱] نشان و خط آ ه̅ پیوندد و باز از نقطهٔ ج̅[۲] بفتح آ ه̅ قوس [ز̅] ح̅ بگرداند تا از طرفین متلاقی شود بنقطهٔ ز̅ ح̅ و خط آ ز̅ ب̅ ح̅ بکشد تا حاصل شود ط̅[۳]. این مرکز دائره است بفتح ط̅ ج̅ که نصف قطر دائره است والله اعلم.

---

۱ ه̅] د̅
۲ ج̅] ه̅ د̅
۳ ط̅] د̅ ح̅ آ

Construction 5 (fol. 181v [5])[1]

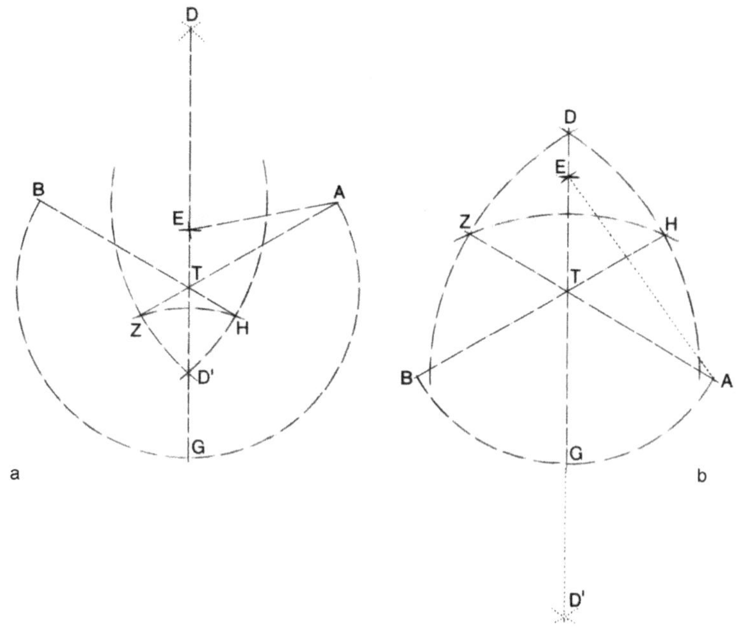

Pl. 5.1a–b. Construction of a circle from its parts.

The construction explicates two alternative cases for transforming a segment into a circle. All the lines except letters are drawn in red ink.

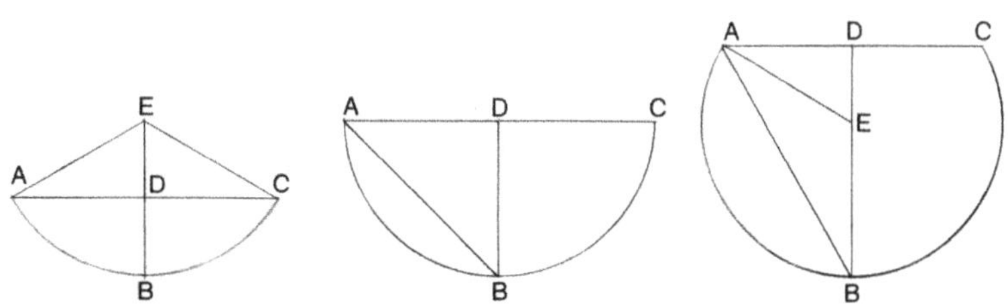

Pl. 5.2. The previous two procedures are a more practical version of Euclid, *Elements*, 3.25: "Given a segment of a circle, to describe the complete circle of which it is a segment."[2] In the easier solutions (pl. 5.1a–b), the points are determined by intersecting arcs, rather than constructing angle BAE equal to angle ABD, as in Euclid's proposition.

1 Özdural, "Analyses" (unpublished chapter), 53–54.
2 Euclid 1956, 2:54–56.

Translation and Transcription (fol. 181v [6])

Fol. 181v [6]

Section (*faṣl*): If one has a segment of a circle and wants to complete it, the procedure is to: divide arc AB into three equal parts at points G and D. From points A and D find intersection E with an arbitrary compass opening, and from points B and G reach intersection Z. From intersection E draw line EG,[a] and from intersection Z draw line ZD so that they intersect each other at point H. This point is the center of the original circle. With compass opening HD, which is the radius, complete the circle. God knows best.[b]

---

a Text: ED.
b The text is composed of four horizontal lines.

فصل اگر قطعه از دائره باشد و خواهد که آنرا[١] تمام گرداند طریق آن بود که قوس آ بَ را
بسه قسم کند بنقطه جَ دَ اقسام متساوی و از نقطه آ دَ تقاطع هَ برند هر فتح که خواهد
و از نقطه بَ جَ تقاطع زَ پیوندد و از تقاطع هَ خط هَ جَ[٢] بکشد و از تقاطع زَ خط زَ دَ
بکشد که یک دگررا قطع کند بنقطه حَ و این نقطه مرکز دائره اصل است بفتح حَ دَ که
نصف قطرست دائره تمام کند والله اعلم.

---

[١] آنرا] آنران
[٢] هَ جَ] هَ دَ

CONSTRUCTION 6 (fol. 181v [6])[1]

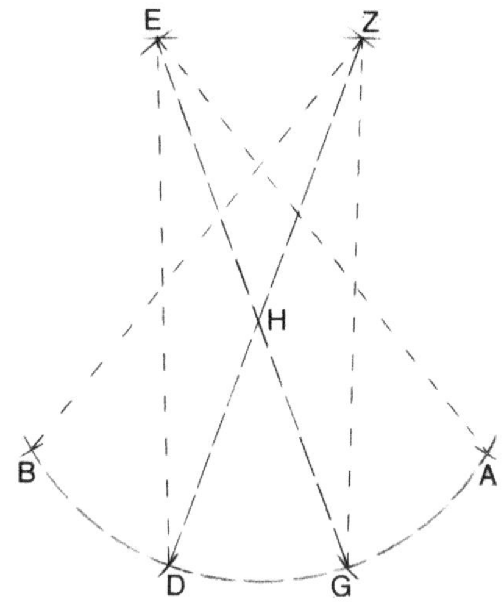

a

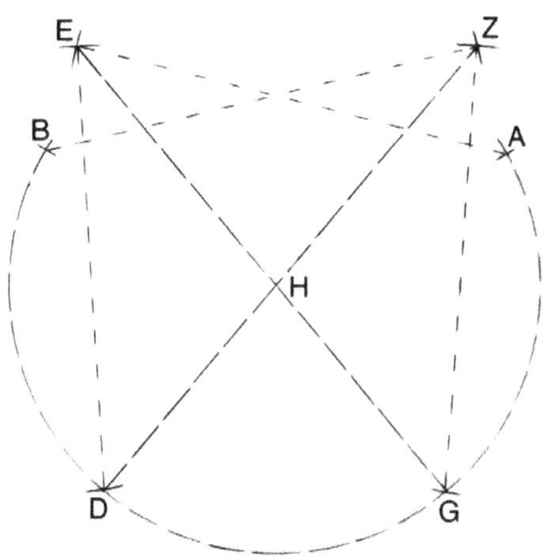

b

Pl. 6.1a–b. Construction for completing the segment of a circle.

An alternative version of the problem addressed in Construction 5, drawn in red ink. This is an even more practical version of Euclid's proposition (see pl. 5.2), involving the determination on the arc of three equal parts, by marking four equally spaced points with a compass.[2] The point of intersection of two lines (GE and DZ) passing through the two middle points defines the center of the circle.

1 Özdural, "Analyses" (unpublished chapter), 54–55.
2 Euclid 1956, 2:54–56.

Translation and Transcription (fol. 182r [7])

Fol. 182r [7]

Problem (*mas'ala*): If they say, "We want to divide a triangle into four equal parts such that one part is similar (*shabīh*) to the assumed triangle, and we draw a line that issues from each of the three angles, and the triangle is assumed to be divided into four triangles [by lines] from the angles of the assumed triangle."[a]

Solution (*javāb*): The procedure is as follows: Divide each side of the assumed triangle into two parts. In the triangle in the middle, make the greatest circle that is tangent to that triangle. From the interior [of the big triangle] to the circumference of this circle [lines] are joined, as in the example (*mis̱āl*).[b]

---

a This first part of the text, which is in two parts, is composed of five horizontal lines and two oblique lines, from upper right to lower left, on the right side of the figure.

b The introductory word, "solution" (*javāb*), is written above the two oblique lines in the first part of the text. The second part of the text is composed of three oblique lines, from lower right to upper left, on the left side of the same figure.

مسئله اگر گویند میخواهیم که مثلثی‌را بچهار قسم متساوی کنیم چنانک یک قسم مثلثی شبیه مثلث مفروض بود و از هر سه زاویه خط بیرون کرده باشیم و مثلث‌را بچهار مثلث منقسم فرض کند و از زوایا[ء] مثلث مفروض، جواب طریق آن باشد که هر ضلعی از اضلاع مثلث مفروض بدو قسم کند و در مثلث وسط اعظم دائره که مماس آن مثلث شود از داخل بمحیط آن دائره شوند و متصل بیکدیگر برین مثال.

CONSTRUCTION 7 (fol. 182r [7])[1]

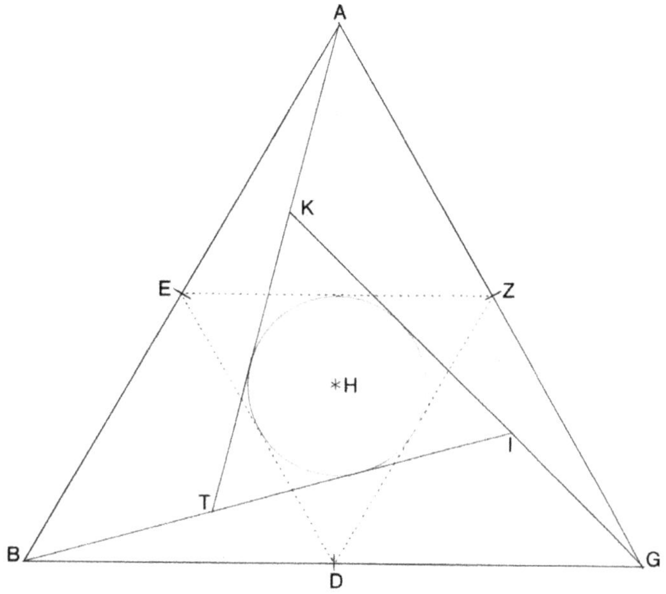

Pl. 7.1. Division of an equilateral triangle into four equal parts.

In this "cut-and-paste" construction, the triangle is dissected in such a way that one part is a rotated similar triangle inscribed with a circle, from each vertex of which a line is extended to join the vertex of the outer encompassing triangle. The dissected pieces display a three-fold symmetry.

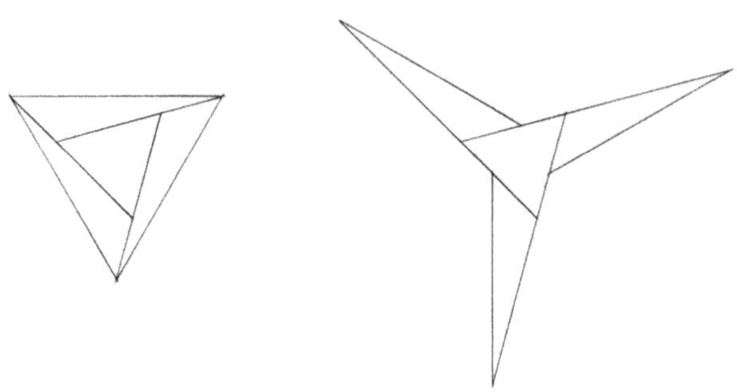

a

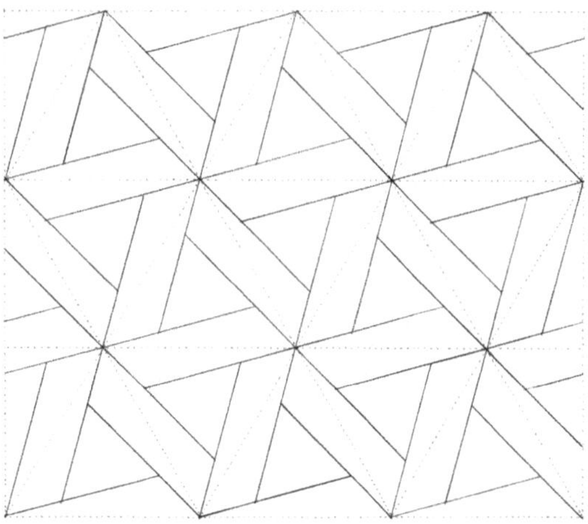

b

Pl. 7.2a. The dissected pieces may be reassembled into a triangular star.

Pl. 7.2b. A hexagonal ornamental composition generated by repeating Construction 7.

1 Özdural, "Analyses" (unpublished chapter), 55–57.

Translation and Transcription (fol. 182r [8])

Fol. 182r [8]

Problem (*mas'ala*): If they say, "We want to transform a hexagon into a square such that the parts of the hexagon appear in the square."[a]

Solution (*javāb*): I say the procedure is as follows: On the rectilinear extension of line AB we draw line AD equal to [line] AG. We bisect line DB at point E. With compass opening EB [and point E as the center], we draw arc DZB. Again on the rectilinear extension of AG[b] we draw AZ until it ends at the circumference of the arc. Thus line AZ is the side of the square that is equal to the hexagon[al star]. Then from line AB we take part AH, which is equal to AZ. When point H becomes known, the rest is easy. God knows best.[c]

---

a Two interrelated figures in the middle and at the bottom of the page, accompanied by a text in two parts. Beginning with the word "problem," the first part is composed of four horizontal lines above the bottom figure and to the left side of the middle figure.

b Text: AE.

c The second part of the text is written obliquely in twelve lines, from lower right to upper left, on the left side of the figure. The word "answer" is written vertically along the right margin of the text.

مسئله اگر گویند که میخواهیم تا مسدسی‌را مربع گردانیم چنانک اجزاء مسدس در مربع ظاهر بود، جواب گویم طریق آن بود که بر استقامت خط آ ب̅ خط آ د̅ مثل آ ج̅ وصل کنیم و خط د̅ ب̅ تنصیف کنیم بنقطهٔ ه̅، و بفتح ه̅ ب̅ قوس ب̅ ز̅ د̅ بگردانیم و باز بر استقامت آ ج̅[1] خط آ ز̅ بکشیم تا منتهی محیط قوس شود که خط آ ز̅ ضلع مربعی بود که آن مربع مثل این مسدس باشد آنگه از خط آ ب̅ قسم آ ح̅ مثل آ ز̅ وضع کنیم. چون نقطهٔ ح̅ معلوم شد مابقی سهل باشد والله اعلم.

---

[1] آ ج̅] آ ه̅

Construction 8 (fol. 182r [8])[1]

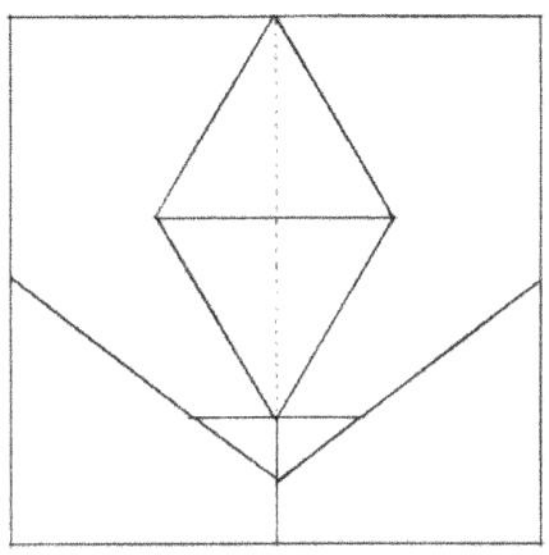

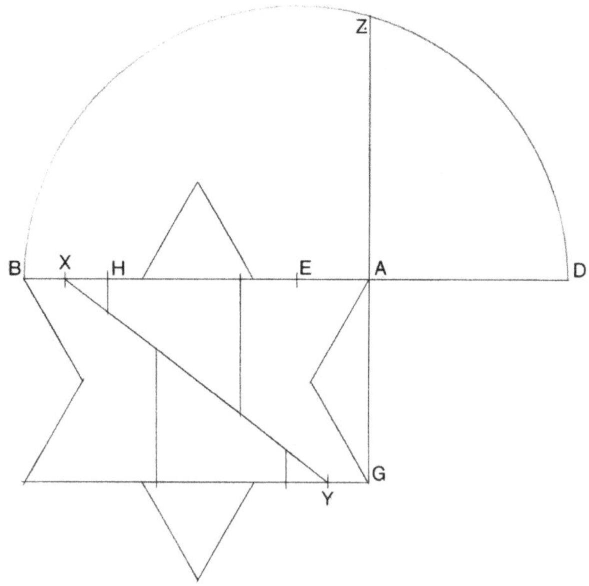

Pl. 8.1. Construction of a hexagonal star and a square with equivalent areas.

This “cut-and-paste” construction shows how to transform a hexagonal star into a square having the same area. The hexagonal star is drawn below and the square above it, with the latter assembled from the eight dissected pieces of the hexagonal star.

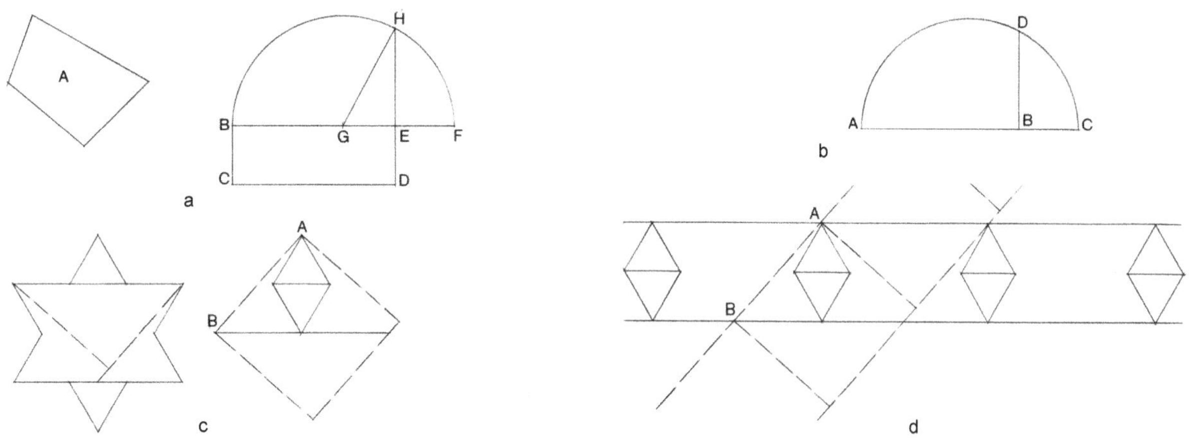

Pl. 8.2a–b. The problem above is first reduced to transforming a rectangle into a square of equal area, as provided in Euclid, *Elements*, 2.14 (pl. 8.2a): “To construct a square equal to a given rectilinear figure”; and 2.13 (pl. 8.2b): “To two given straight lines to find a mean proportional.” Euclid’s two propositions involve in modern terms the extraction of the square root, that is, the solution of the quadratic equation $x^2 = ab$, which is presented here in a visual demonstration.[2]

Pl. 8.2c–d. In 1921, Harry Bradley elegantly solved the same problem of recreational mathematics (pl. 8.2.c), which looks very similar. It involves dissecting the hexagonal star into five pieces (pl. 8.2d).[3] This more economical modern version of the dissection has three pieces less than that of the *Anonymous Compendium* (pl. 8.1).

1 Özdural, “Analyses” (unpublished chapter), 57–59.
2 Cf. Euclid 1956, 2:216, 409–10.
3 Frederickson 1997, 122–23.

Translation and Transcription (fol. 182v [9])

Fol. 182v [9]

Section (*faṣl*): If we want to produce an actual square from rectangle ABGD, it is first required that on the rectilinear extension of line AG we draw line GE equal to GD. We bisect line AE at point Z and we make that point the center. With compass opening ZA, we draw arc EHA. Then, on the rectilinear extension of line GD we draw line GH, until it reaches the circumference of the circle EHA. Thus, line GH is the potential side of the square. Then, of the side AG of the rectangle, from point G we cut off the length of GT equal to GH; from point B of side BD, we cut off the length of BI[a] equal to GH. We bisect the remainder from both sides at points L and K, and join line LK. Then, at points T and I, we draw perpendiculars TM and IN[b] until they meet diagonal LK. Four pieces are [thus] produced. We compose from these four pieces the actual square STHG. The construction is completed by means of confirmation (*taqrīr*, lit. discourse or proof). God knows best.[c]

---

a Text: BD.

b Text: ID.

c The text is written in fourteen slanting lines, following the curvature of the circle in this figure.

فصل اگر خواهیم که مربع آ ب ج د مستطیل‌را بالفعل مربعی متساوی الأضلاع گردانیم اول باید که بر استقامت خط آ ج خط ج ه بکشیم مثل ج د. خط آ ه تنصیف کنیم بنقطه ز و این نقطه‌را مرکز سازیم و بفتح ز آ قوس ه ح آ بگردانیم و باز باستقامت ج د خط ج ح بکشیم تا بمحیط دائره ه ج آ رسد که خط ج ح ضلع مربع متساوی الأضلاع باشد بالقوه. آنگه از ضلع آ ج مستطیل از نقطه ج مقدار ج ط مثل ج ح جدا کنیم و از نقطه ب از ضلع ب د و مقدار ب ی مثل ج ح جدا کنیم و مابقی از ضلع از طرفین تنصیف کنیم بنقطه ل ک و خط ل ک بپیوندیم و آنگه از نقطه ط ی عمود ط م ی ن[1] اخراج کنیم تا متصل قطر ل ک شود و چهار قطعه حاصل آید و این چهار قطعه مربع س ط ح ج بسازیم بالفعل که عمل اتمام بود بر طریق تقریر والله اعلم.

---

[1] ی ن] ی د

Construction 9 (fol. 182v [9])[1]

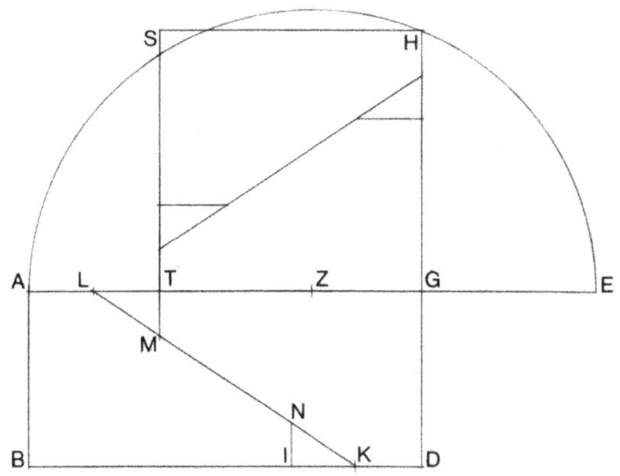

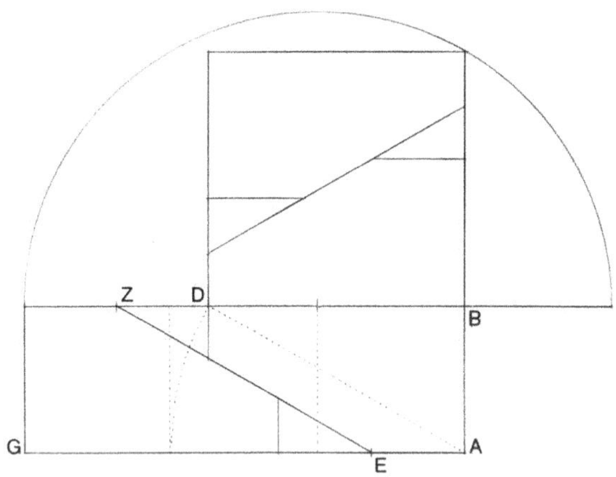

a

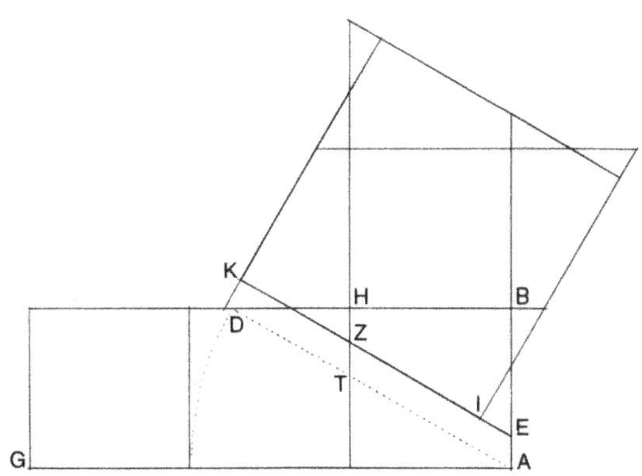

b

Pl. 9.1. Construction of a rectangle and square of equal areas.

According to Özdural, the preceding construction is related to this one and was probably produced by the same author. The text explains how to transform the rectangle into a square, a procedure used in Construction 8 with no further comment. The present construction, without the dissections, is based on Euclid, *Elements*, 2.14 (pl. 8.2a). Özdural suggests that the author referred to this theorem in the text, where it is stated that the construction was completed by means of confirmation (*taqrīr*, lit. discourse, proof).

Pl. 9.2a. If Euclid, *Elements*, 2.14 (pl. 8.2a), is applied to a rectangle where side BA corresponds to a unit length, and side AG is three times the unit length, then side BD of the square becomes √3. In this case, the hypotenuse of triangle ABD corresponds to two unit lengths and angle GAD is 30°.

Pl. 9.2b. Özdural speculates that if the author of Construction 9 had faced the problem of constructing a square composed of three squares (a problem solved in Abu'l-Wafa''s *Geometrical Constructions*, pl. 2.4c), he would have provided the same solution for pl. 9.2a. Instead of defining the hypotenuse AD, he would have bisected the rectangle through the parallel line ZE and then slid the pieces along this diagonal line. Another solution could have been found if line EZ bisects the unit square instead of the rectangle (pl. 9.2b). When two unit squares are bisected in this way and the four halves are placed rotating around the third unit square, the required square is obtained. This solution appeared in another treatise of Ms. Persan 169, titled *Lil-aʿmāl Abī Bakr al-Khalīl al-Tājir al-Raṣadī* (Of the Constructions, by Abu Bakr al-Khalil al-Tajir al-Rasadi). The name of this geometer (date of death unknown) is cited twice in the *Anonymous Compendium.*[2]

1 Özdural, "Analyses" (unpublished chapter), 59–62.
2 Abū Bakr al-Khalīl, Ms. Persan 169, item 19, fols. 124v–138r, on 136v.

Translation and Transcription (fol. 182v [10])

Fol. 182v [10]

No accompanying text.

Construction 10 (fol. 182v [10])[1]

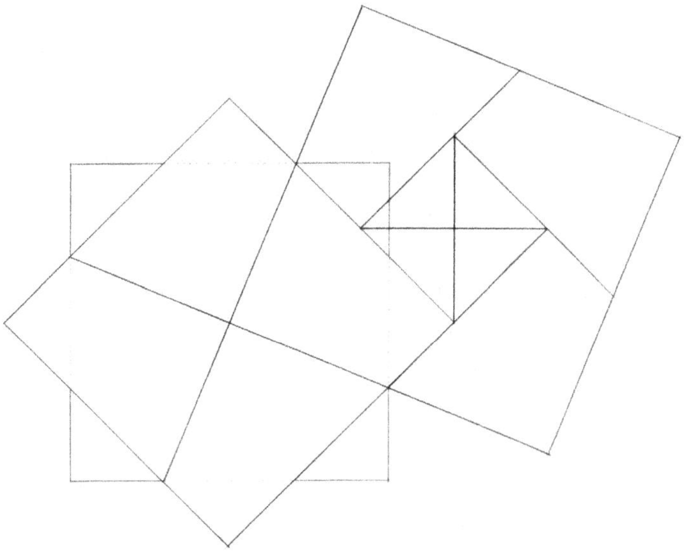

Pl. 10.1. Construction of an octagonal star and square of equal areas.

This "cut-and-paste" construction shows how to transform an octagonal star into an equivalent square. There is no accompanying text, but the procedure is apparent with the help of the uninked incised construction lines. The octagonal star constructed of two rotated squares at 45° is dissected into eight pieces that are assembled in the square. Based on the construction lines, Özdural suggests that the figure was drawn with the aid of two types of set squares, one with 45°-angles and the other with 22.5°-67.5°-angles. Its four-fold rotational symmetry is reminiscent of the elegant dissection of the octagon in Construction 4.

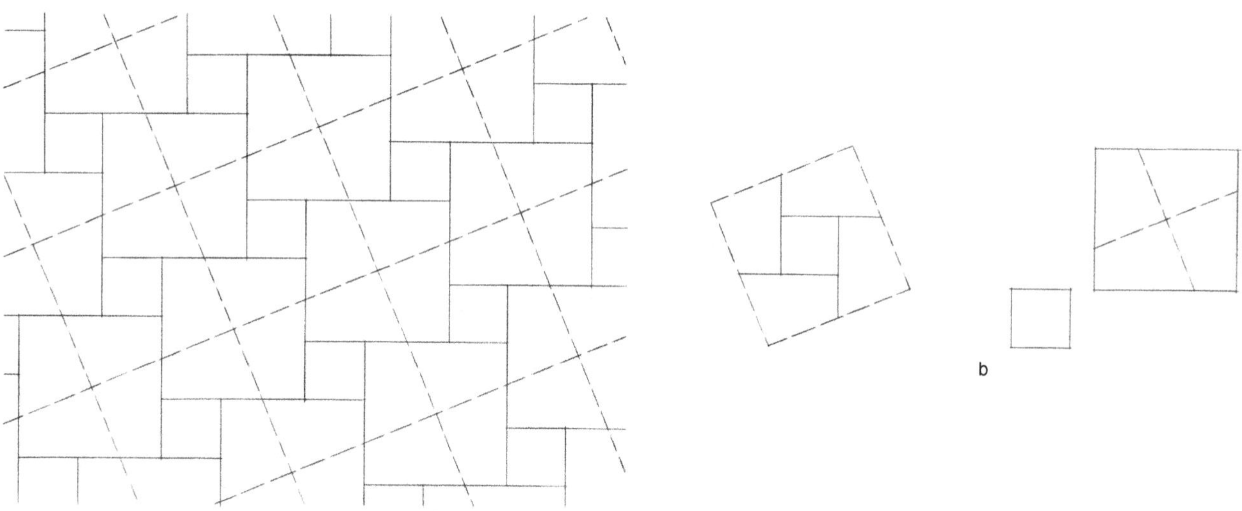

a

b

Pl. 10.2a–b. Henry Perigal discovered a modern and more economical version of Construction 10 in 1873, reducing the number of pieces to five. His dissection was derived by applying the technique of "superposing tessellations"; that is, overlaying two tessellations with the same repetition pattern (as in pl. 4.2a–b).[2]

1 Özdural, "Analyses" (unpublished chapter), 62–63.
2 Frederickson 1997, 30.

Translation and Transcription (fol. 183r [11])

Fol. 183r [11]

The procedure for dissecting a [regular] hexagon into triangles is as follows: Since triangle ADZ[a] is known, the triangle is divided into four parts by lines AT, DI, and ZH drawn from each angle, until they meet one another at points H, T, and I. Then triangle HTI emerges. This triangle is divided into six parts by perpendiculars IL, HN, and TM.[b] From all these pieces, one triangle is composed. God knows best.[c]

Problem (*mas'ala*): If they want to transform a triangle into a hexagon, or to transform a hexagon into a triangle, the division has to be in this manner so that it comes out correctly.

Solution (*javāb*): When one wants to determine magnitude AH, one must divide perpendicular AD into four parts equal to magnitudes AZ and AH. Magnitudes GT, GI, BL, and BK are then marked with the same magnitude. One draws the lines of triangle ZLT and finds the center E. [From it] one draws lines to the angles, and each of the triangles AZH, GIT, and BKL is divided into two parts. From all the pieces, if so wished, one hexagon can be composed, or, if so wished, two triangles can be composed. God knows best. [d]

---

a Text: AKZ.
b "M" is missing from the Persian figure.
c The text is composed of six vertical lines, from bottom to top, on the left side of the figure.
d The text is composed of twelve horizontal lines on the left side of the figure.

طریق قسمت مسدسی از جهت مثلثات آنست کچون مثلث آ د ز معلوم کند هر یک از زوایا مثلث بچهار قسم کند بخطوط آ ط د ی ز ح بکشد تا متلاقی یکدیگر شوند بنقطه ح ط ی و مثلث ح ط ی پیدا شود. این مثلث‌را بشش قسمت کند با عمده ی ل ح ن ط م و از مجموع مقطعات یک مثلث سازد والله اعلم.

مسئله اگر خواهد که مثلثی‌را مسدس گرداند ویا مسدسی‌را مثلث گرداند باید که قسمت وی بدین نسق کند تا راست آید.

جواب چون خواهد که مقدار خط آ ح معلوم کند باید که عمود آ د چهار قسم کند که آن مقدار آ ز آ ح بود. بهمین مقدار ج ط، ج ل، ب ل، ب ک نشان کند و خطوط مثلث ز ل ط بکشد و مرکز ه معلوم کند و خطوط به زوایا کشد و مجموع مثلثات آ ز ح، ج ی ط، ب ک ل را هر یک بدو قسم کند و از مجموع مقطعات اگر خواهد یک مسدس سازد و اگر خواهد دو مثلث سازد والله اعلم.

Construction 11 (fol. 183r [11])[1]

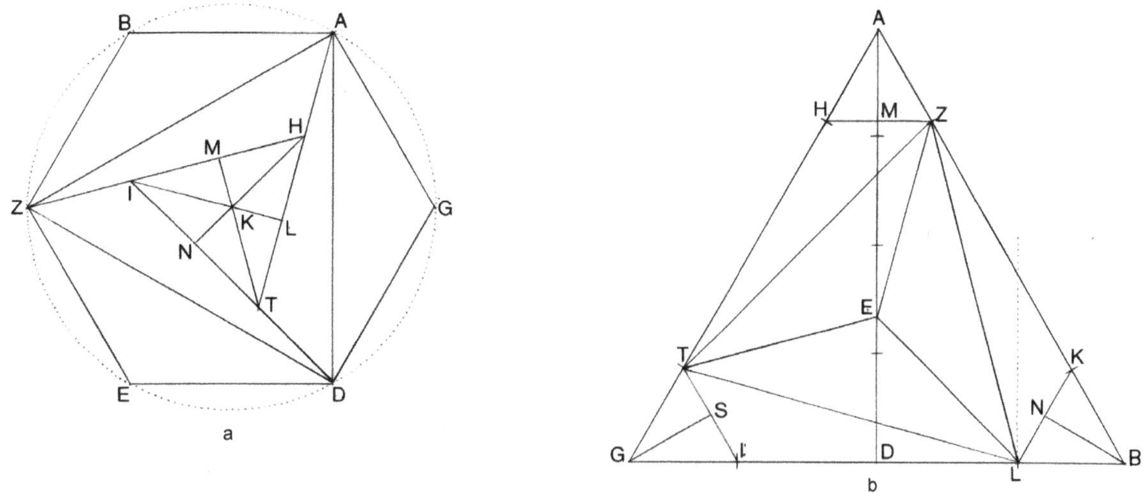

Pl. 11.1a–b. Construction of a hexagon and equilateral triangle of equal areas.

The text shows how to transform a triangle into a hexagon or a hexagon into a triangle. The purpose of the procedure of the first figure (a) is to demonstrate how to dissect the hexagon. Since triangle ADZ is known, one divides each angle of the triangle by lines AT, DI, and ZH, extending these lines until they meet to establish triangle HTI, which in turn is divided into six parts. All these pieces form one triangle. The aim of the second figure (b) is to establish the rotated triangle ZTL, which is equal to half of the given triangle ABG.

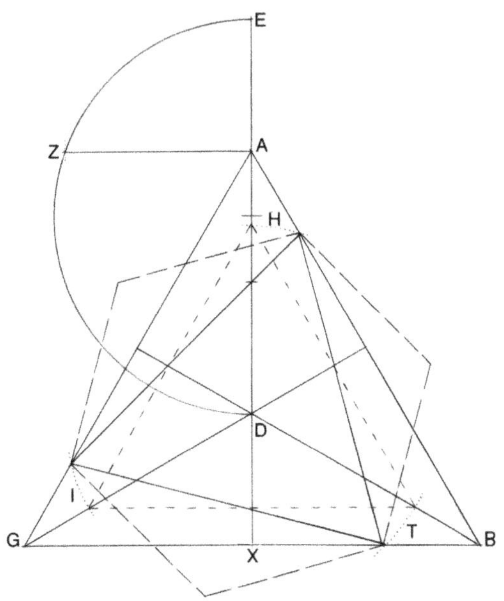

Pl. 11.2. Özdural speculates that the author of the "incorrect and cumbersome" Construction 11.1a, which is only an approximation, "was not, apparently well-versed in mathematical literature," since a more elegant solution to this problem is provided in Abu'l-Wafa''s *Geometrical Constructions.* It involves the problem of halving the triangle. The same construction, with the same letters designating the points as in Abu'l-Wafa''s construction, is found in an anonymous treatise in Ms. Persan 169. Titled *Fī qismat al-mus̲allas̲āt* (On the Division of Triangles), this work features a construction attributed to "*Abī Bakr ibn al-Khalīl*" (fol. 118v), whose name appears twice in the *Anonymous Compendium* and who was familiar with Abu'l-Wafa''s work (see Construction 9.2b). Both treatises describe the easy and more "correct" solution for "How to find the triangle whose area is equal to half of a given triangle."

1 Özdural, "Analyses" (unpublished chapter), 63–65.

Translation and Transcription (fol. 183v [12])

Fol. 183v [12]

Preliminary (*muqaddima*): For every triangle inscribed in a circle, the ratio (*nisbat*) of the angles of that triangle to each other is equal to the ratio of [the opposite] arcs of the chords of the angles. As in the [following] example (*miṯāl*): When angle A is half of angle B, arc BG is equal to half of arc AG. When angle B is half of angle G, necessarily angle G becomes four times angle A, and arc AG becomes half of arc ADB. Hence, arc BG is a quarter of arc ADB. The whole circle is seven times arc BG, and the line BG becomes the side of the heptagon. The sum of the all the angles of triangle ABG is seven times angle A. God knows best.[a]

---

a The text is composed of four horizontal lines above and five horizontal lines below the figure (the letters in the figure in the manuscript are wrong; thus A and G have been interchanged in Özdural's redrawing: see pl. 12.1).

مقدمه هر مثلثی که در دایره واقع شود نسبت زوایاء آن مثلث با یکدیگر مثل نسبت قوس
اوتار آن زوایا باشد با یکدیگر برین مثال. چون زاویهٔ آ نصف زاویه ب باشد قوس ب ح
نصف قوس ج آ بود، و چون زاویهٔ ب نصف زاویهٔ ج بود بضرورت زاویهٔ ج اربعه امثال
زاویهٔ آ باشد و قوس ج آ نصف قوس آ د ب بود پس قوس ب ج ربع قوس ب د آ باشد
و مجموع دایره سبعه امثال قوس ب ج باشد و خط ب ج ضلع مسبع بود و مجموع زوایا[ء]
مثلث آ ب ج و همچنین سبعه امثال زاویه آ باشد والله اعلم.

Construction 12 (fol. 183v [12])[1]

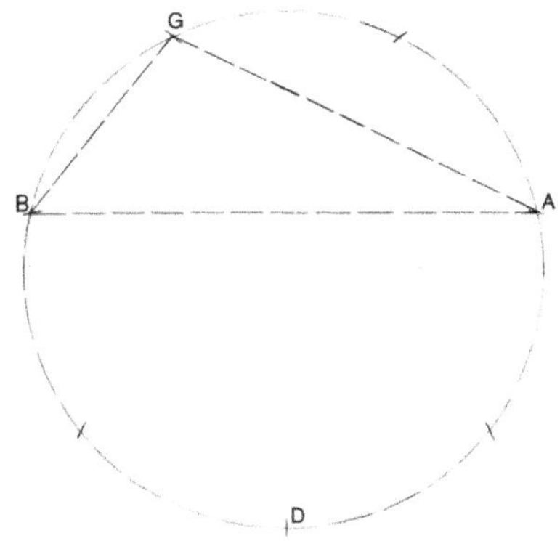

Pl. 12.1. Construction of a triangle inscribed in a circle.

The text states the theorem that, for every triangle inscribed in a circle, the ratio of its angles equals the ratio of the arcs of the chords subtended by the angles. It is explained that the depicted circle is seven times the arc BG, such that the line (chord) BG becomes the side of a heptagon, while the sum of the angles of triangle ABG is seven times angle A. This theorem is derived from Euclid, *Elements,* 6.33.[2] Thus, the construction of the heptagon is reduced to the problem of drawing the triangle in red ink, whose angles are in the ratio 1:2:4. As there are no uninked incised construction lines, Özdural proposes that this figure was probably drawn with the aid of a special set square for the heptagon. The set square must have been based on an approximation, rather than on an exact construction using intersecting conic sections.

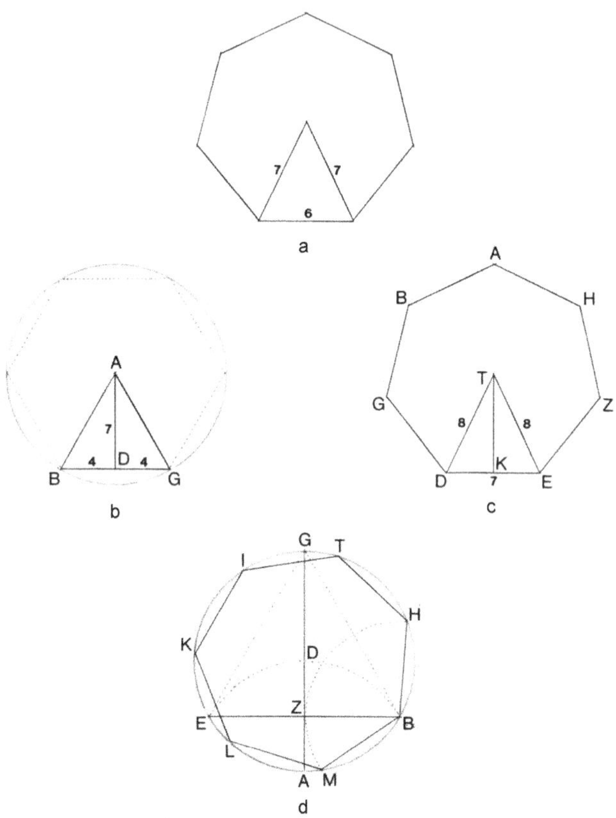

Pl. 12.2a. Old Babylonian approximation of the heptagon,[3] whose perimeter (i.e., seven times the side of the heptagon) is equal to six times the radius of a circumscribing circle.

Pl. 12.2b. A first approximation from the *Metrica* of Heron of Alexandria (first century A.D.)

Pl. 12.2c–d. An approximation offered in Abu'l-Wafa''s *Geometrical Constructions*,[4] equivalent to another approximation given by Heron of Alexandria in the *Metrica*.[5] Abu'l-Wafa''s approximation method was repeated in the construction of the heptagon in the anonymous treatise *Fī qismat al-muṯallaṯāt* (mentioned in pl. 11.2 above), which refers to the geometer Abu Bakr ibn al-Khalil, whose name appears twice in the *Anonymous Compendium*.[6]

---

1 Özdural, "Analyses" (unpublished chapter), 66–70.

2 Euclid 1956, 2:273–76.

3 The construction by the Babylonians is different from, and unrelated to, the more refined construction by Heron of Alexandria in pl. 12.2c–d.

4 Abū'l-Wafā' al-Būzjānī, Ms. Ayasofya 2753, 25 (4:12).

5 Heron of Alexandria 1964, 3:230–33.

6 Ms. Persan 169, item 16, fol. 114r (Construction 25).

Translation (fol. 184r [13])*

Fol. 184r [13]

Problem (*mas'ala*): If they say: "We want to construct a triangle for which angle A is half of angle B and angle B is one-third of angle G," the procedure is for us to assume that angle A is one, angle B is 2, and angle G is 6, and the sum of these three angles is 9. So we draw a circle and divide the circumference of that circle into nine parts. We join the chords of the arcs of one part, 2 parts, and 6 parts, until the required triangle ABG results. In it angle A is half of angle B, and angle B is one-third of angle G, as in the [following] example (*miṯāl*).[a]

If they assume a line and say: "We want to construct this triangle [i.e., similar to the one above] on this line," then with an arbitrary compass opening draw a circle. Construct triangle ABG in a position that might be asked according to the example that has been described [above]. Then check whether line AB is equal to the assumed line. If it is, then triangle ABG is the required one. If it is greater or less, then draw on the rectilinear extension of line AB either the addition BD, or the subtraction BE. Then let from points E or D line DH or line ET be parallel to BG. God knows best.[b]

These are all the triangles for which [the sum of the angles are] marked in 9 parts.[c]

A	B	G
1	2	6
1	3	5
1	4	4
2	3	4
1	1	7
2	2	5
3	3	3

Obtaining all the triangles, in which [the sum of] the angles are considered, will be according to this comparison, such that if they say: "We want a triangle in which angle A is three, angle B is 4, angle G is 5, and the sum of these three is the number 12," one must draw a circle and divide the circumference into 12 parts, and join the chords of 3, 4, and 5 parts so that the required triangle is obtained. As an example this much is sufficient.[d]

a The text is in four parts. This is the first part, composed of five horizontal lines above the figure.

b The second part of the text is written in six horizontal lines below the figure.

c The third part of the text concerns the table and is composed of two horizontal lines above it.

d The fourth part of the text is written inside the circular figure, in eight oblique lines.

* For the Persian transcription, turn to page 220.

Construction 13 (fol. 184r [13])[1]

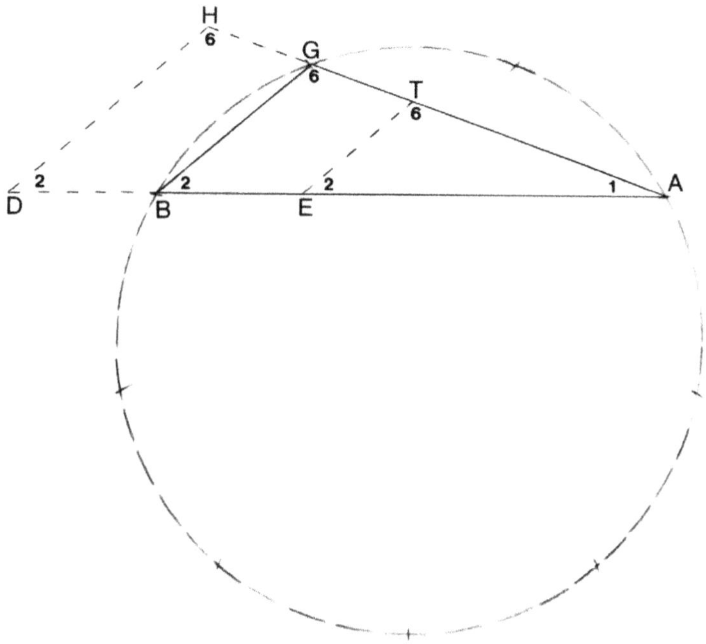

a

1	2	6
1	3	5
1	4	4
2	3	4
1	1	7
2	2	5
3	3	3

b

Pl. 13.1a–b. Application of the method in Construction 12.

This construction, drawn in red ink, is accompanied by a table composed of three columns. The text shows how to construct a triangle in which angle A is half of angle B, and angle B is one-third of angle G. The triangle of the nonagon is labeled "example" (*miṡāl*). As there are no uninked incised construction lines, Özdural believed that the figure was probably drawn with the aid of a special set square for the nonagon. The table lists all the triangles the sum of whose angles adds up to 9, which can be inscribed in a nonagon.

1 Özdural, "Analyses" (unpublished chapter), 70–73.

Transcription (fol. 184r [13])*

مسئله اگر گویند عمل مثلثی می‌خواهیم که زاویهٔ آ نصف زاویهٔ ب باشد و زاویهٔ ب ثلث زاویهٔ ج بود، طریق آن بود که زاویهٔ آ یکی فرض کنیم زاویهٔ ب ۲ باشد و زاویهٔ ج ۶ و مجموع هر سه زاویه ۹ باشد. دایره رسم کنیم و محیط آن دائره بنه قسم گردانیدیم و وتر قوس یک قسم و ۲ قسم و ۶ قسم بپیوندیم تا مثلث آ ب ج که مطلوبست حاصل آید و در وی زاویهٔ آ نصف زاویهٔ ب باشد و زاویهٔ ب ثلث زاویهٔ ج بود برین مثال:

و اگر خطی فرض کنند و گویند میخواهیم تا برین خط این مثلث عمل کنی بهر فتح که خواهد دائره رسم کند و مثلث آ ب ج بهر وضع که سوال کرده باشد عمل کند برین[1] مثال که گفته شد. آنگه بنگرد اگر خط آ ب مثل خط مفروض بود. مثلث آ ب ج مطلوب باشد و اگر زیاده یا نقصان باشد بر استقامت آ ب ب د زیادتی برافزاید یا ب ه ناقص بکاهد. آنگه از نقطهٔ ه یا د خط د ح یا ه ط موازی ب ج باشد والله اعلم.

مجموع مثلثات که از بهر آن] نشان به ۹ باید کرد اینست

و حصول جمله مثلثات که زوایا اعتبار کرده باشند درو برین قیاس باشد چنانک اگر گویند مثلثی میخواهیم که زاویهٔ آ سه باشد و زاویه ب ۴ و زاویه ج ۵ و مجموع این هر سه عدد ۱۲ بود دایرهٔ رسم کند و محیط به ۱۲ قسم گرداند و وتر ۳ و ۴ و ۵ بپیوندد تا مثلث مطلوب حاصل آید و در مثال این قدر کفایت است.

---

[1] برین] بدبرین

---

* For the English translation, see page 218.

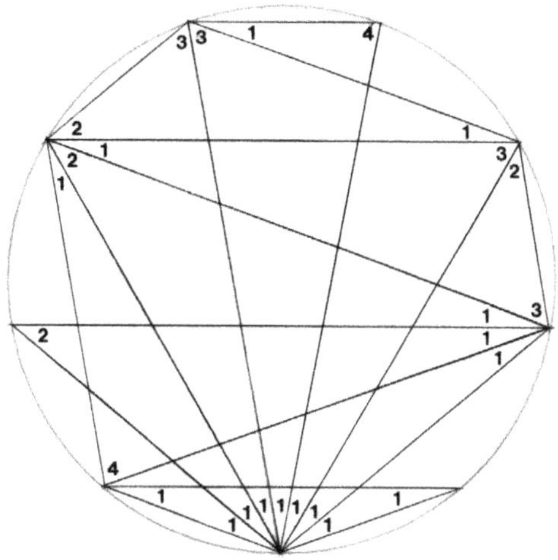

a

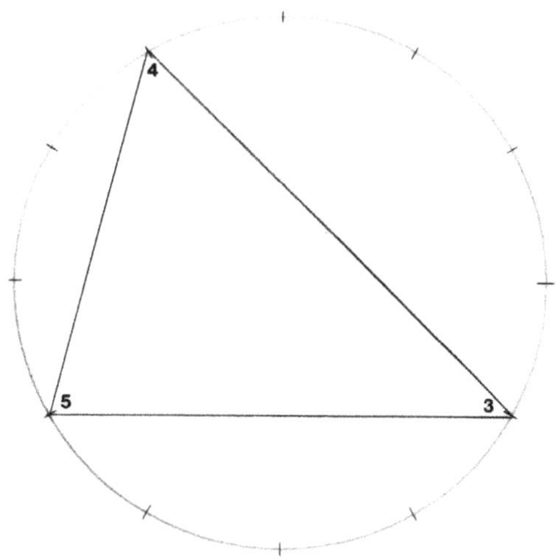

b

Pl. 13.2a. Özdural suggests that Ibn al-Haytham's treatise on all possible combinations of angles for the construction of the heptagon must have inspired the author of Construction 13. This drawing shows all the triangles that can be inscribed in a nonagon (listed in a table in pl. 13.1b).[2]

Pl. 13.2b. The text written inside the circle on folio 184r (depicted here in pl. 13.1a) concerns another application of the same theorem to the case of the dodecagon. The procedure involves dividing the circumference into 12 parts and joining the chords of 3, 4, and 5 parts to obtain the desired triangle. According to Euclid, *Elements*, 3.20,[3] the angle module corresponds to half of the angle at the center that the side of the polygon subtends. In the case of the heptagon, the module is 25.7°, in the nonagon 20°, and in the dodecagon 15°.

2 Özdural, "Analyses" (unpublished chapter), 72, 209n49, 210n57.

3 Euclid 1956, 2:46–49. "In a circle the angle at the center is double the angle at the circumference, when the angles have the same circumference as base."

Translation and Transcription (fol. 184v [14])

Fol. 184v [14]

To construct set square [number] 5 with a compass opening equal to the radius [of the circumscribing semicircle] from set square [number] 6, it is necessary that: On line AG, [with the compass opening] AB, inscribe semicircle ADG with center B. Then make point A the center and inscribe arc EB. Similarly, make point G the center and find point D on the circumference of the arc and join line AD so that it intersects arc EB at point Z. Then extend line GZ until it meets the circumference of the arc at point H. Join lines AH and DG so that each of the triangles AZH and GZD becomes set square [number] 5. The original triangle ADG is "set square [number] 6," as has been shown. The original triangle AHG from different set squares becomes [number] 15.[a]

a The text is composed of ten horizontal lines below the figure.

در عمل گونیا ۵ بفتح نصف قطر از گونیا ۶ و آنچنان باشد که بر خط آ ج آ ب نصف
دائره آ د ج بگرداند بر مرکز ب آنگه نقطهٔ آ مرکز سازد و قوس ه ب بگرداند و باز نقطهٔ
ج مرکز سازد و بر محیط قوس نقطه د معلوم کند و خط آ د بپیوندد تا متقاطع قوس ه
ب شود بنقطه ز و خط ز ج بکشد تا منتهی شود بمحیط قوس بنقطهٔ ح و خط آ ح د
ج بپیوندد که هر یک از مثلث آ ز ج و مثلث ج ز د گونیا ۵ باشد و اصل مثلث آ د ج
گونیا ۶ بود چنانک نموده شد و اصل مثلث آ ح ج از گویناها مخالف ۱۵ باشد.

CONSTRUCTION 14 (fol. 184v [14])[1]

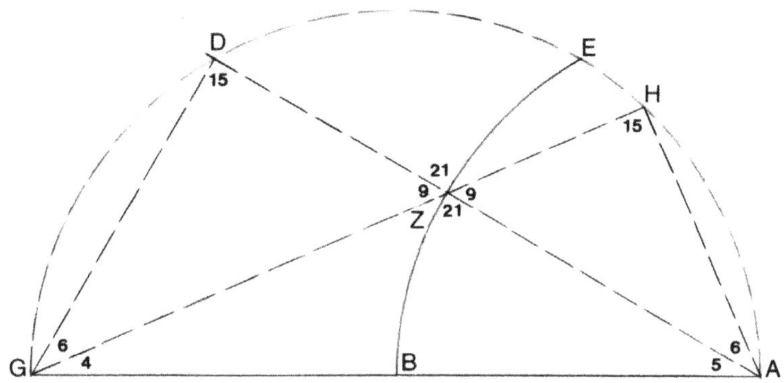

Pl. 14.1. Construction of set squares inscribed in a semicircle.

Like the previous two constructions, this one is mostly drawn in red ink and concerns triangles that are defined by angle modules. The objective is to establish set squares (*gūnyā*), that is, right-angled triangles, inscribed in the same semicircle. The text describes the construction of "set square [number] 5" (*gūnyā-5*) with a compass opening obtained from half the diagonal of "set square [number] 6" (*gūnyā-6*). The two triangles AZH and GZD are "set square [number 5]," while triangle ADG is "set square [number] 6." The sum of angles in triangle AHG (excluding the right angle) has 15 modules, which is also true for angle ADG. In this system of angle modules, each module corresponds to 6°, so that 90° becomes 15 modules, 36° becomes 6, and 30° becomes 5. While *gūnyā-5* (with 36°-54°-90°) was used for constructing pentagons and decagons, *gūnyā-6* (with 30°-60°-90°) was used for equilateral triangles, hexagons, and dodecagons. The construction proposed for *gūnyā-5* is a close approximation, even though correct construction methods for the pentagon were known.

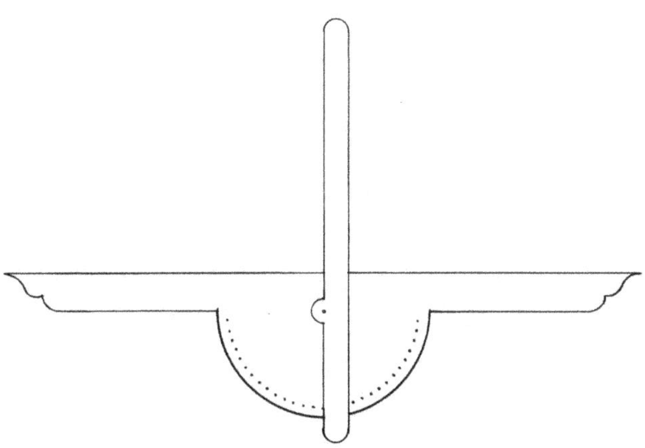

Pl. 14.2. The T-square, with a movable alidade and semicircular sextant pierced with holes, described and drawn in Badiʿ al-Zaman ibn al-Razzaz al-Jazari's treatise *al-Jāmiʿ bayn al-ʿilm wa'l-ʿamal al-nāfiʿ fī ṣināʿat al-ḥiyal* (A Compendium of Science and Useful Practice in the Mechanical Arts), written in Diyarbakır circa 1206. With a T-square, the center-point of three points on a sphere or a circle could be determined, as well as angles, whether acute or obtuse.[2]

1 Özdural, "Analyses" (unpublished chapter), 73–77.
2 Al-Jazarī 1974, 196–98.

Translation and Transcription (fol. 184v [15])

Fol. 184v [15]

No accompanying text.

Construction 15 (fol. 184v [15])[1]

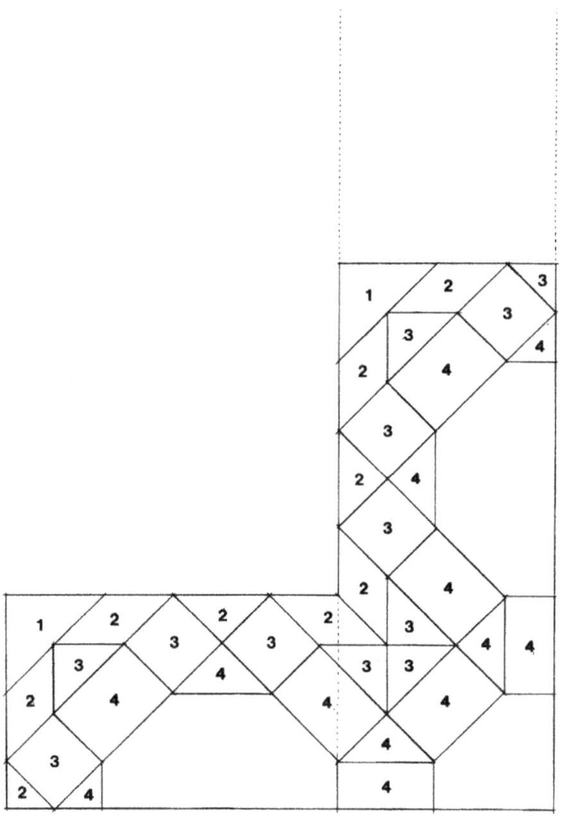

Pl. 15.1. Construction of the orthogonal projection of a quarter-*muqarnas* vault.

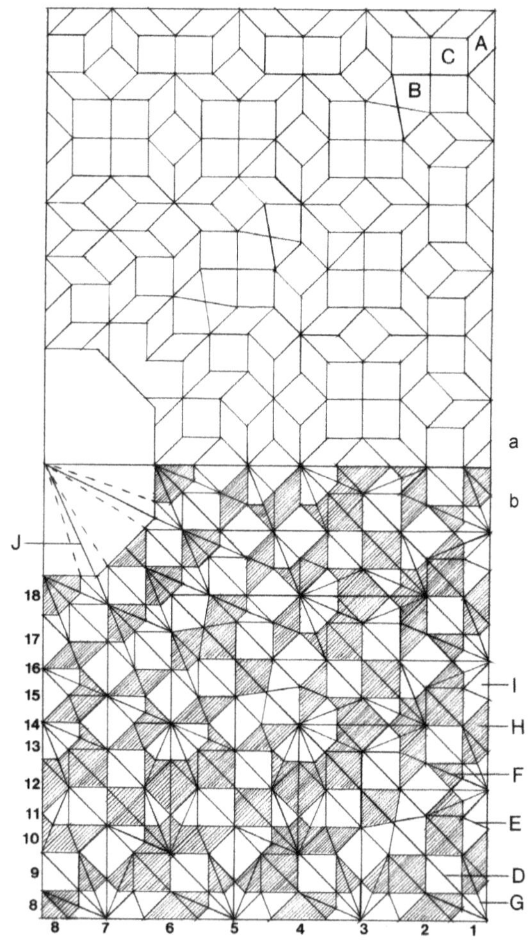

Without any accompanying text, this L-shaped figure is subdivided into squares, rhombi, right-angled isosceles triangles, and rectangles. These inaccurately drawn divisions form a network around two half-octagons and a quarter-octagon. Two fixed lengths, *a* and *b*, determine the lengths of all the subdivisions, numbered 1 to 4. Özdural proposed that these numbers correspond to the corbelling order of *muqarnas* tiers on successive horizontal planes.

Pl. 15.2a–b. The upper half of the drawing (a) depicts a 50-centimeter plaster slab with incised lines discovered at the Takht-i Sulayman Palace in Iran, built by Abaqa Khan in the 1270s; the slab features the orthogonal projection of a quarter-*muqarnas* vault. In the lower half of the drawing (b), the slab's successive *muqarnas* tiers are reconstructed. The simple type of *muqarnas* depicted in it resembles that of Construction 15 (pl. 15.1), and corresponds to a type described by the Timurid mathematician Jamshid al-Kashi in 1427 as having only three shapes: a rhombus (A), a rhomboid (B), and a rectangle (C) (i.e., quadrangle, a term including the square). This type of *muqarnas*, in which "the heights of the tiers are not uniformly equal, and sometimes two or three tiers are found with ceilings, but without [vertical] facets," was more flexible. This rare type was seen by al-Kashi only in the old buildings of Isfahan.[2]

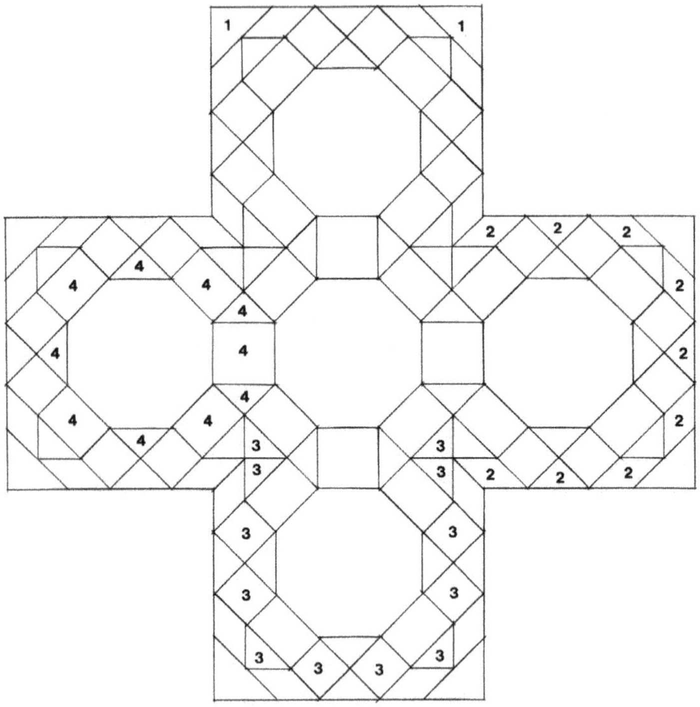

Pl. 15.3. Unlike the Takht-i Sulayman slab above, Construction 15 has numbers that differentiate *muqarnas* tiers. Özdural's drawing reconstructs the full version of the quarter vault and its coding system of successive corbelled tiers, numbered from 1 to 4, which increase from the corners toward the medial octagons.

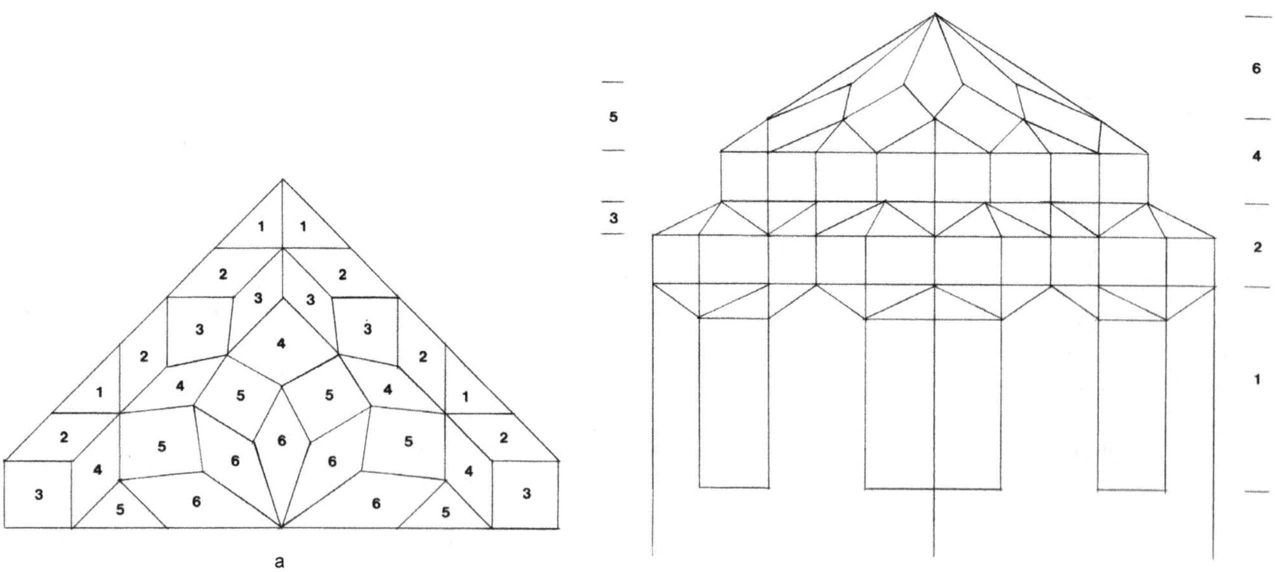

Pl. 15.4a–b. Plan and elevation of a *muqarnas* squinch of carved plaster at the Ilkhanid-period shrine of Imamzada Yahya in Varamin, Iran (1307); it is composed of rectangles, rhombuses, and rhomboids.[3] In Özdural's drawing, the consecutive tiers are numbered, starting from the lowest tier and increasing upwards.

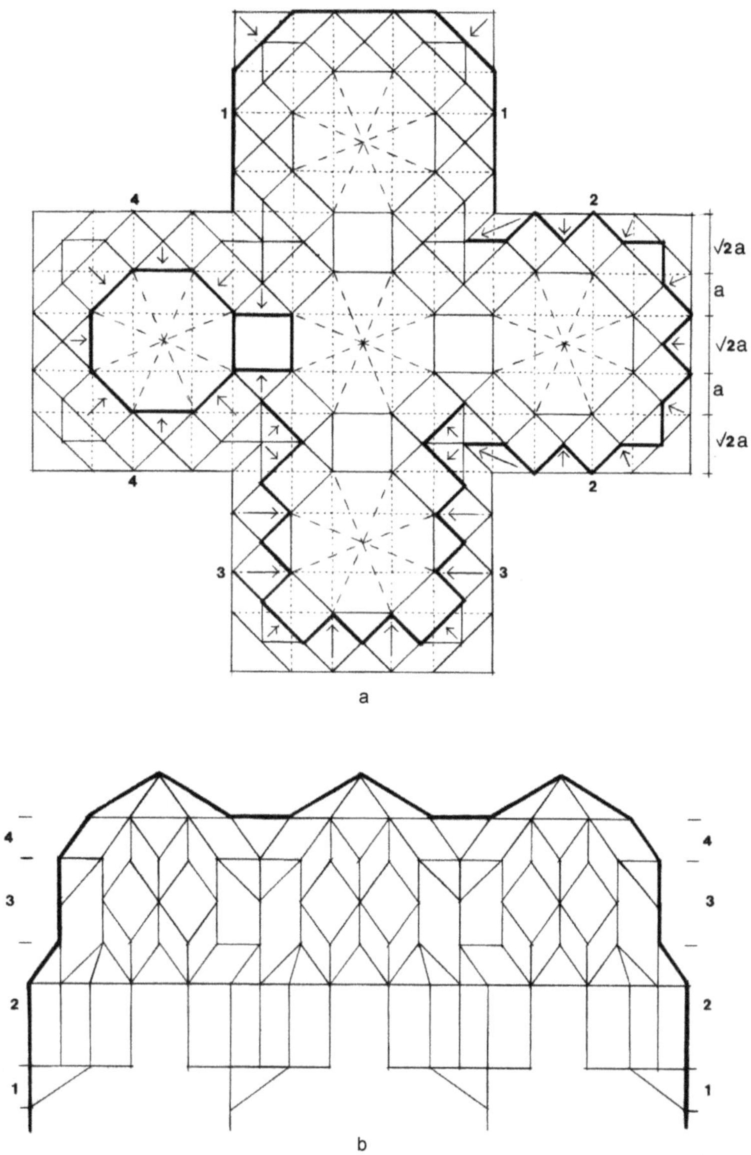

Pl. 15.5a–b. A reconstruction plan and elevation proposed by Özdural for the cross-shaped *muqarnas* vault, only a quarter plan of which is provided in Construction 15. He noted the similarity between it and its counterpart at Varamin, each with a "strong prismatic character."

---

1 Özdural, "Analyses" (unpublished chapter), 77–87.

2 Özdural, "Analyses" (unpublished chapter), 77–87; the drawing is adapted from the *muqarnas* plan on the plaster slab first published in Harb 1978, pl. 1. On al-Kashi's discussion of this type of early *muqarnas*, which Özdural translates as a "muddly *muqarnas*," due to its ambiguity, see Özdural, "Analyses" (unpublished chapter), 79–85, 178, 213–14nn75–77. The same term is translated as the "clay-plastered" *muqarnas* in Golombek and Wilber 1988, 1:166; Dold-Samplonius 1992, 226; "clay-covered" in Necipoğlu 1995, 353; "mudded" in Özdural 1990, 37.

3 Özdural, "Analyses" (unpublished chapter), 86; for pictures, see Pope and Ackerman 1938, 6: 531; Wilber 1969, pl. 6. Examples of the "simple" and "curved" type of *muqarnas* coexisted in Ilkhanid buildings. For those listed below, see Özdural, "Analyses" (unpublished chapter), 85, 215nn87–89. The "simple" type is found on the entrance portals of the Masjid-i Jami' of Ashtarjan (1315–16), the mausoleum of Baba Qasim in Isfahan (1340–41), and the Masjid-i Jami' of Kirman (1349). The "curved" type is seen on the tomb of Shaykh 'Abd al-Samad in Natanz (1307) and the shrine of Shaykh Bayazid at Bistam (1313).

Translation and Transcription (fol. 185r [16])

Fol. 185r [16]

Triangle AK[B] is a right-angled triangle in which the ratio of the difference between the shortest side and the hypotenuse to the difference between the [former] difference and the shortest side [is the same as the ratio of the intermediate side to the shortest side].[a]

Section (*faṣl*): The procedure is this: [We draw an arbitrary line on which the required points A and B will lie. On it we mark a given magnitude GD.] By means of GD[b] we erect perpendicular GE [being equal to GD]. We make point E the center and then [with compass opening] EG we inscribe arc ZH on the rectilinear extension of B. We bisect GD [at point T]. We put one arm of the compass fixed on point E, and we place the ruler [at changing positions] so that its edge is always touching point E. [At every position it cuts simultaneously arc ZH at a point K and the line at a point A. With the other arm of the compass we compare the changing magnitudes TK and TA.] We move the ruler so that magnitudes TK and TA become equal. [At this position] we mark points K, A, and B [by taking magnitude DB equal to magnitude GA]. When lines AK and KB are drawn, triangle AKB is the required right-angled one. [In this triangle,] AD is equal to BG and BK. When we subtract BL, which is equal to BD, from BK, KL is equal to GD. God knows best.

---

a The text is composed of nine horizontal lines below the figure. The text is defective in the critical opening lines of the enunciation of the problem.

b Text: ABD.

مثلث آ گ [ب] قایم الزاویه مثلثی است درو نسبت مابین ضلع اقصر و وتر با مابین مابین و ضلع اقصر.

فصل طریق آنست که بواسطهٔ آ ب د عمود ج ه اخراج کنیم و نقطهٔ ه مرکز سازیم و ببعد ه ج قوس ز ح بگردانیم از طرف ب و ج د تنصیف کنیم بنقطهٔ ه و پای پرگار بر نقطه ثابت داریم و حرف مسطره از نقطهٔ ه منفک بگردانیم و چندان تحریک دهیم که مقدار ط گ ط آ متساوی شوند نقطهٔ آ گ ب نشان کنیم و خط آ گ گ ب بکشد که مثلث آ گ ب قایم الزاویه مطلوب باشد و آ د مثل ب ج ب گ بود و چون ب ل بمثل ب د از ب گ نقصان کنند گ ل مثل ج د بود والله اعلم.

CONSTRUCTION 16 (fol. 185r [16])[1]

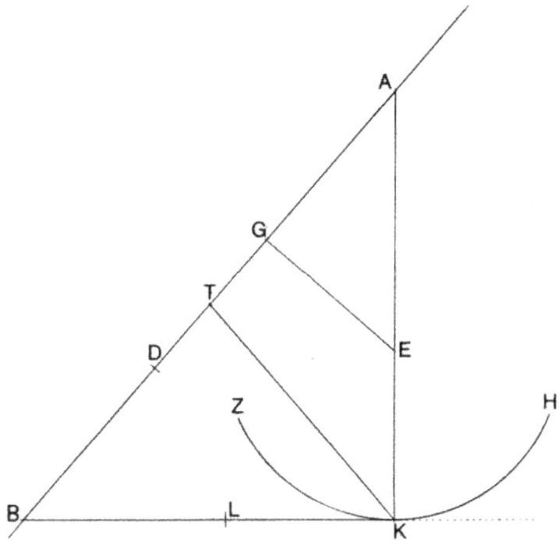

Pl. 16.1. Construction of a right-angled triangle containing a kite.

The description of the "ratio/proportion" (*nisbat*) of this triangle is followed by its construction procedure, which involves a "verging construction" equivalent to a cubic equation. One arm of the compass is placed on a fixed point [T], with the edge of a moving ruler always touching point E, around which it moves. In this triangle the ratio of AG to GD is the same as AK to KB.

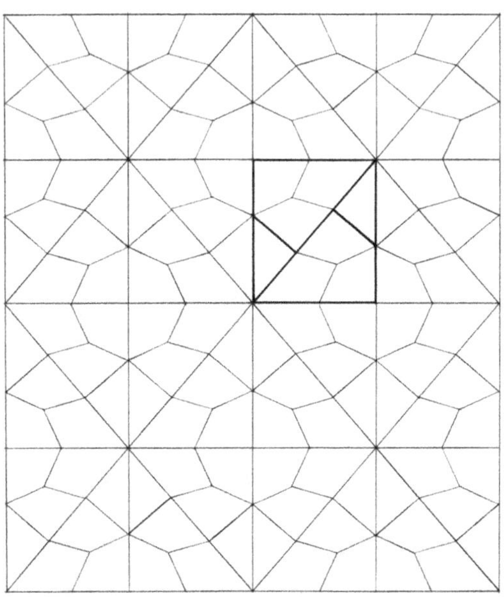

Pl. 16.3. This ornamental composition is generated by replicating the rectangular repeat-unit above (pl. 16.2b).

a

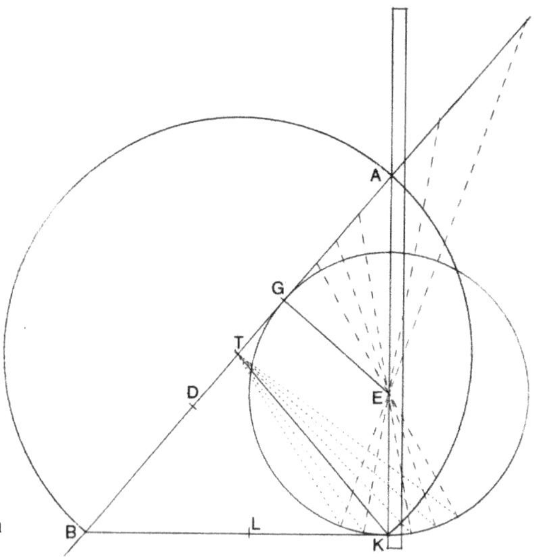

b

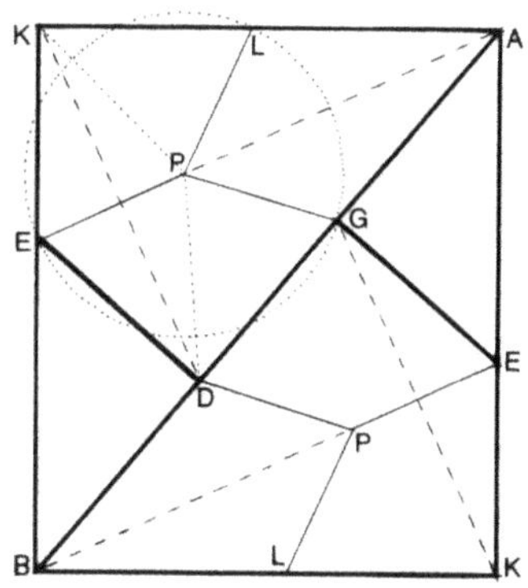

Pl. 16.2a. A drawing that shows the movement of the ruler around the fixed point E in the "verging construction" mentioned above.

Pl. 16.2b. The main objective of the verging construction above is to create this "repeat-unit" (*ʿaqd*) with a set of interrelated ratios between its interlocking parts. The rectangular repeat-unit is composed of two right-angled triangles (such as the one described in pl. 16.1), which are joined along diagonal AB. When the small congruent triangles BDE and AGE are cut off from the main ones, there remain congruent "isosceles quadrangles" or kites BKEG and AKED facing opposite directions and sharing segment GD. Each of the kites is further subdivided into three kites with particular proportional properties. Kites play a prominent role in Islamic ornamental patterns and recur throughout the *Anonymous Compendium*.

1 Özdural, "Analyses" (unpublished chapter), 87–94.

Translation and Transcription (fol. 185v [17])

Fol. 185v [17]

The proportion of this other repeat-unit is based on the figure that is established on the backside of this folio (i.e., fol. 185r [16]).[a]

Section (*faṣl*): In this repeat-unit we need a right-angled triangle, such that if we cut off from the hypotenuse of the triangle towards the shortest side [a magnitude equal to the shortest side] and erect a perpendicular on the point of cutting, it cuts off the intermediate side at a point so that [the distance] from it to the right angle will be equal to the perpendicular that cuts off. Obtaining a triangle of this sort is difficult. It falls outside Euclid's *Elements* and concerns the science of conics. If the magnitude of the perpendicular is assumed, the construction is achieved by means of a moving ruler, as in the (following) example (*mis̱āl*):[b]

---

a On the figure in the upper left corner of the page two words are written: "hypotenuse" and "side." The text consists of two parts. The first part, under this figure, is composed of two oblique lines, facing downwards, from lower right to upper left.

b The second part of the text is composed of seven oblique lines, from lower right to upper left, on the upper right side of the page.

نسبت این عقد اخیر موقوف برین شکل است که بر ظهر ورق مثبت است.

فصل [در] این عقد احتیاج بمثلثی قایم الزاویه داریم که چون از وتر مثلث در طرف ضلع اقصر فصل کنیم و از مقام فصل عمودی اخراج کنیم تا ضلع اوسطرا قطع کند و از موضع قطع تا زاویهٔ قائمه مثل عمود مقطوع باشد و حصول این چنین مثلثی مشکل است و از اصول اقلیدسی خارج می‌افتد و آن در علم مخروطاتست و بعمل تحریک مسطره حاصل می‌آید چون مقدار عمود مفروض باشد برین مثال.

Construction 17 (fol. 185v [17])[1]

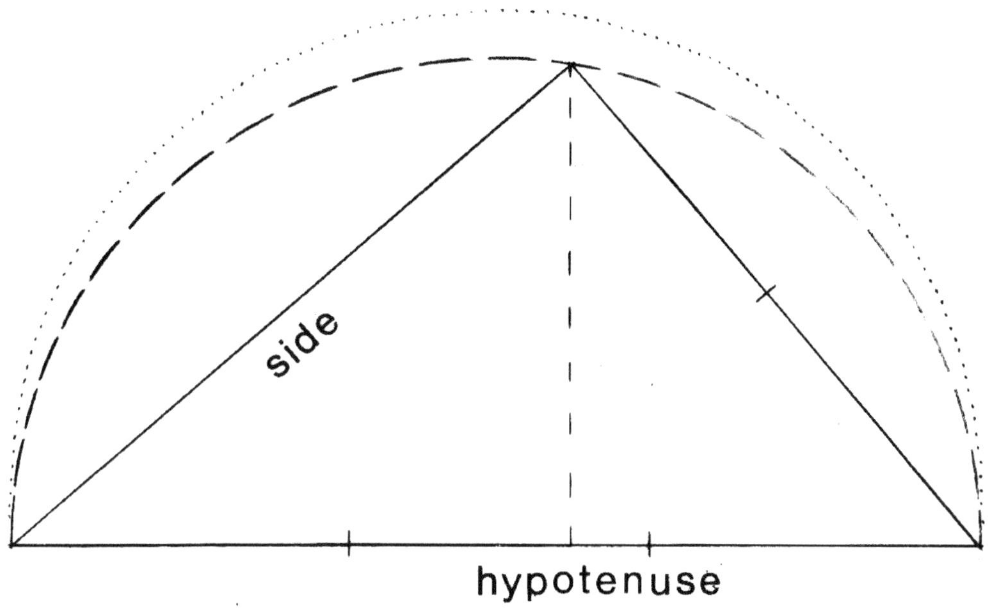

Pl. 17.1. A right-angled triangle inscribed in a semicircle.

The triangle is drawn in black ink, with the semicircle in red. Its components, written in black ink, read: "side" (*żilʿ*) and "hypotenuse" (*quṭr,* equal to the diagonal of the circle). The text explains that the "ratio/proportion" (*nisbat*) of "this other repeat-unit" (Construction 18 below) is based on the construction that is established on the backside of this page (i.e., Construction 16). The hypotenuse and shortest side of the triangle are divided by marks into three and two parts, respectively. These marks correspond to the main divisions of the triangular repeat-unit under consideration. In other words, Construction 17 succinctly illustrates the proportion that governs Constructions 16 and 18: a right-angled triangle with sides 2, $\sqrt{5}$, and 3.

1 Özdural, "Analyses" (unpublished chapter), 95–96 (labeled as "Construction 18" by Özdural).

Translation and Transcription (fol. 185v [18])

Fol. 185v [18]

On half the assumed side AB, construct square AGED. Draw line GET. With compass opening AB inscribe quarter circle BTZ in order to find on perpendicular GT point T. Join lines TA and TZ so that TA intersects line ED at point I. Then from point Z, with compass opening TI, cut off magnitude ZK [on line ZT]. [By means of] line KA, find point L on the circumference of quarter circle ATH. This is what was required.[a]

a The text is composed of eight horizontal lines on the left side of the figure.

بر نصف ضلع آ ب مفروض مربع آ ج ه د عمل کند و خط ج ه ط اخراج کند و بفتح آ
ب ربع ب ط ز بگرداند تا بر عمود ط ج نقطه ط دریابند و خط ط آ ط ز بپیوندد تا ط
آ ه د را قطع کند بنقطهٔ ی. آنگه از نقطهٔ ز بفتح ط ی مقدار ز ک فصل کند که خط ک
آ بر محیط ربع آ ط ح [د] نقطهٔ ل دریابد و این مطلوبست.

Construction 18 (fol. 185v [18])[1]

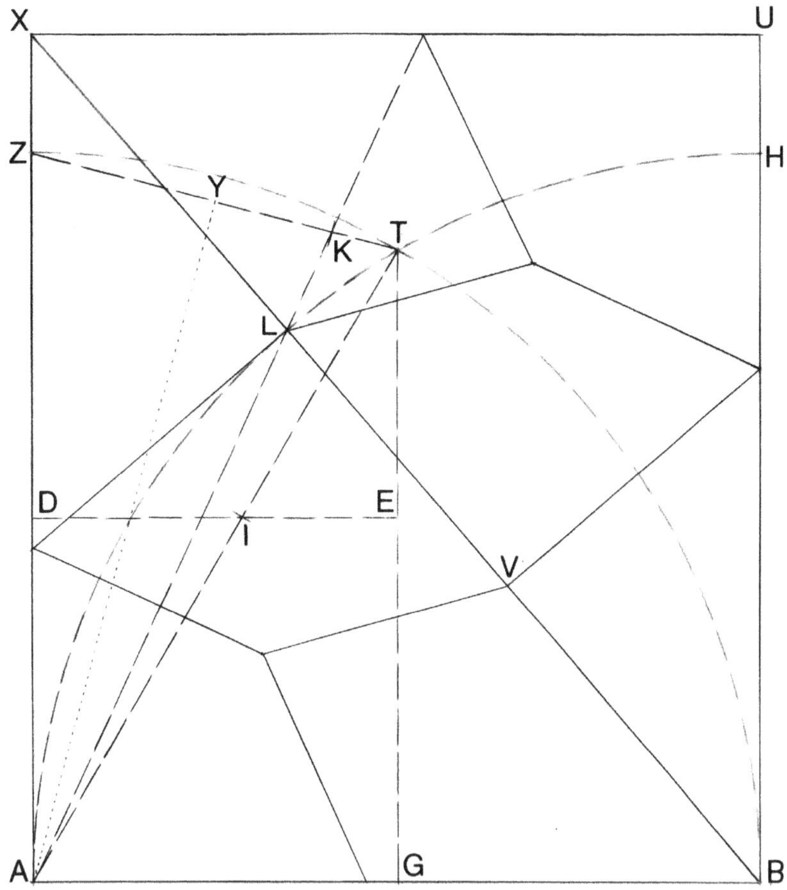

Pl. 18.1. A rectangle composed of two right-angled triangles containing two congruent kites, with further subdivisions.

This figure is an approximation of Construction 16. Auxiliary construction lines are drawn in red ink to distinguish the black ink repeat-unit. Özdural suggests that this close approximation was provided because "verging constructions" may have been considered inadmissible. Its author attempted to solve a cubic equation by way of ruler and compass, even if it is not theoretically correct.

1 Özdural, "Analyses" (unpublished chapter), 94–95 (labeled as "Construction 17" by Özdural).

Translation and Transcription (fol. 186r [19])

Fol. 186r [19]

Construction of an equilateral and equiangular pentagon with a [fixed compass] opening [equal] to the altitude.[a]

In this figure line AB is assumed to be equal to the altitude. The intersections G and D[b] are obtained from points A and B. Points Z and H are obtained from point E, point E from point G, and points K and L on lines AZ and BH from point T [i.e., the midpoint of AB]. Both points M and N are also obtained from point T. Points S and O are obtained from K and L. Points Q and J are obtained from point C, and point C is obtained from point F on line SO. All these points are obtained by compass opening AB. From lines MQ, MJ, NS, and NO, the equilateral and equiangular pentagon MX is obtained by compass opening AB, which is equal to altitude TM,[c] as is illustrated.[d]

---

a The heading of the construction is written in three oblique lines, from lower right to upper left, on the upper left corner of the page.

b Point D is not in the original Persian figure.

c Text: LM.

d The text is composed of six horizontal lines underneath the main figure.

عمل مخمس متساوی الأضلاع والزوایا بفتح العمود

درین شکل خط آ ب مثل عمود مفروضست و تقاطع ج د از نقطه آ ب حاصل میشود و نقطهٔ ز ح از نقطهٔ ه و نقطهٔ ه از نقطهٔ ج و نقطه ک ل بر خط آ ز ب ح از نقطهٔ ط حاصل میشود و هم چنین نقطهٔ م ن هم از ط حاصل میشود و نقطه س ع از ک ل حاصل میشود و نقطه ق ش از نقطهٔ ص حاصل میشود و نقطهٔ ص از ف و نقطهٔ ف از خط س ع. چون این نقطها حاصل شده باشد جمله بفتح آ ب از خط م ق م ش ن س ن ع مخمس م خ متساوی الأضلاع و الزوایا بفتح آ ب که مثل عمود ط م[1] است حاصل شود چنانک مصور است.

---

[1] ط م] ل م

Construction 19 (fol. 186r [19])[1]

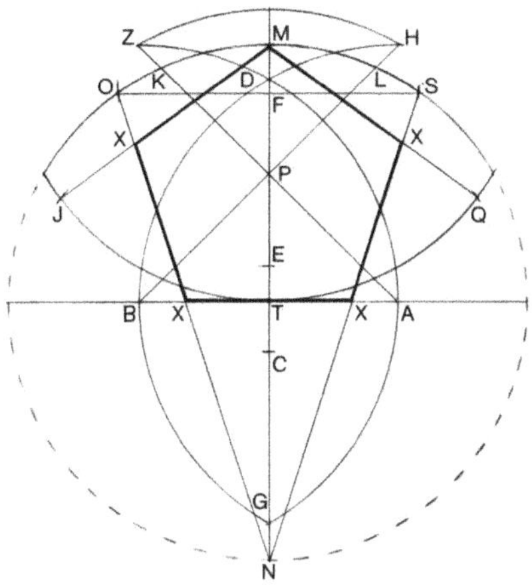

Pl. 19.1. Construction of a regular pentagon with a fixed compass opening equal to the altitude.

This construction, like most of the following five constructions of the pentagon, is based on an approximation method, even though correct methods were provided by Euclid and Abu'l-Wafa'. In it, all lines are in black ink, but the pentagon itself is highlighted by an additional dotted line. The text gives a step-by-step explanation of the construction procedure.

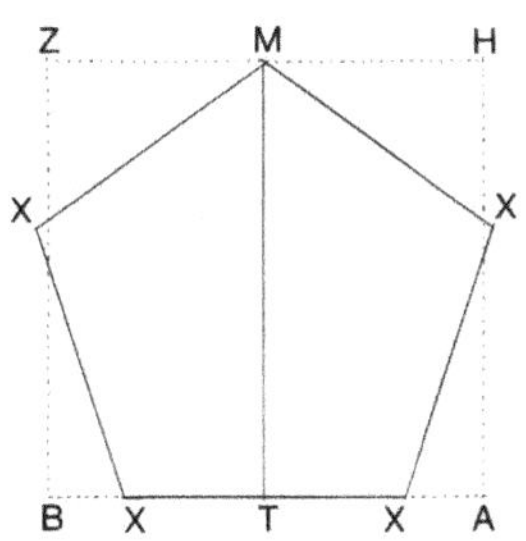

a

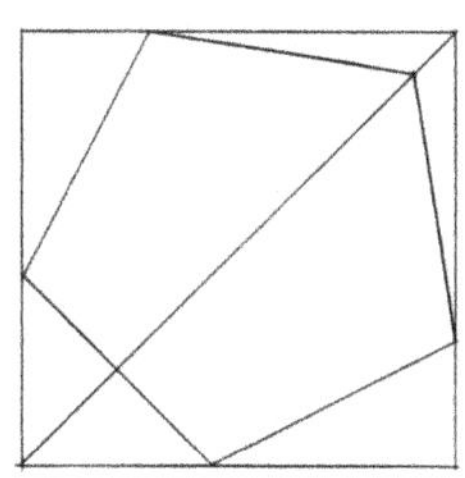

b

Pl. 19.2a. In the construction above (pl. 19.1), the pentagon has the appearance of fitting in a square. Since only three points are touching the square, it approximates constructing a pentagon in a square.

Pl. 19.2b–c. Two ways in which a regular pentagon can be inscribed in a square; in both cases only four points can lie on the square. Case (b) was solved in Abu'l-Wafa''s *Geometrical Constructions*, in which the pentagon's fifth point lies on the diagonal of the square.[2] In the second case (c), the diagonal of the pentagon corresponds to the side of the square.[3]

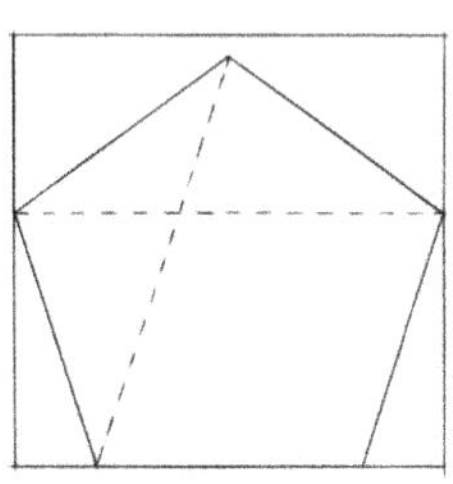

c

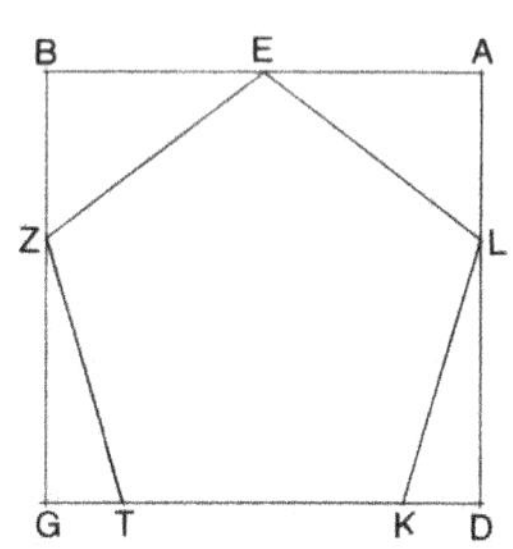

d

Pl. 19.2d. Since all points of a regular pentagon cannot possibly lie on a square, the condition of equiangularity was sometimes dropped. One such example is seen in the work of the mathematician Abu Sahl al-Kuhi (ca. 940–ca. 1000), who flourished in Buyid Baghdad and treated the problem in his "Treatise on the Construction of an Equilateral Pentagon in a Known Square," which was based on conic sections.[4] Later on, Ibn al-Haytham (960–ca. 1040) also wrote a now-lost treatise on this problem.[5] Construction 19 (see pl. 19.2a) recalls al-Kuhi's figure, on which Ibn al-Haytham's construction may have been based. In this case, the required pentagon is both equilateral and equiangular, since it was achieved by approximation, a solution perfectly suitable for ornamental geometry.

1 Özdural, "Analyses" (unpublished chapter), 96–99.
2 Abū'l-Wafā' al-Būzjānī, Ms. Ayasofya 2753, 34 (7.20).
3 In this case, the diagonal of the pentagon corresponds to the hypotenuse of the right triangle whose intermediate side is the altitude. Thus the altitude is shorter than the hypotenuse, which is equal to the side of the square.
4 See Hogendijk 1985, 104–5n10.
5 Hogendijk 1985, 105.

Translation and Transcription (fol. 186r [20])

Fol. 186r [20]

Construction of this equilateral and equiangular pentagon [is] made with a [fixed compass] opening [equal] to the side.[a]

---

a The text along the right margin of the page is composed of two vertical lines, from top to bottom. On the small figure are written three words: "side," "chord," and "altitude."

ضلع وتر عمود

عمل اين مخمس متساوى الأضلاع والزوايا بفتح الضلع

Construction 20 (fol. 186r [20])[1]

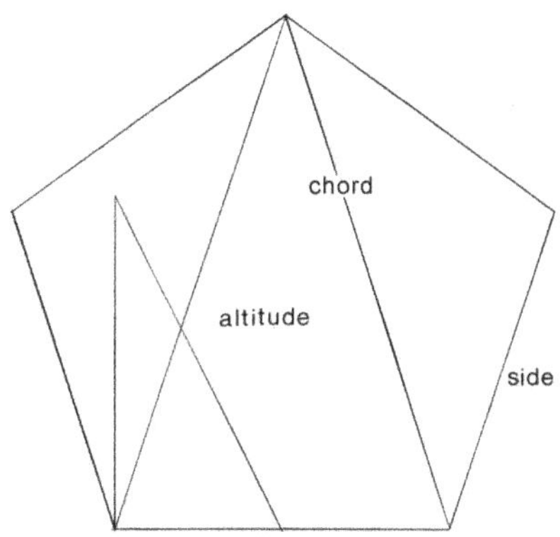

Pl. 20.1. Construction of a regular pentagon with a fixed compass opening equal to the side.

The small figure, without any explanatory text, consists of a pentagon that circumscribes an isosceles triangle, with a small right-angled triangle on its base. The dimensions of the pentagon are labeled "altitude" (*ʿamūd*), "chord" (*watar*), and "side" (*żilʿ*).

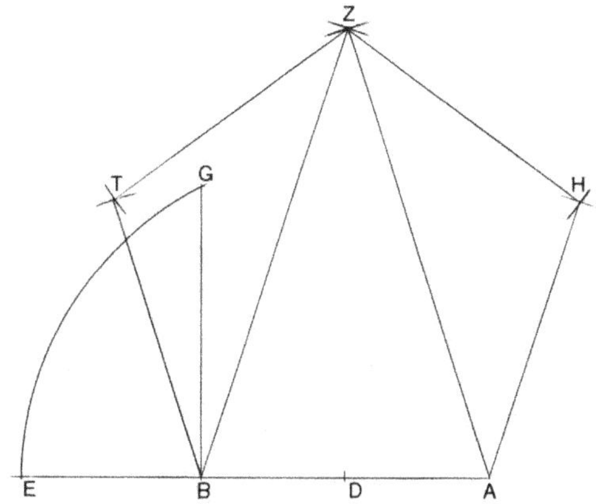

Pl. 20.2. A correct construction method (but not with a fixed compass opening) was explained in Abu'l-Wafa''s *Geometrical Constructions*.[2] It involves the construction of the isosceles "triangle of the pentagon," in which angles A and B are each the double of angle Z. Once these three points of the triangle are established, the remaining two (T and H) are easily found as points equidistant from the known ones. Abu'l-Wafa''s construction was based on setting the compass opening to two different lengths (DG = AB and AE).

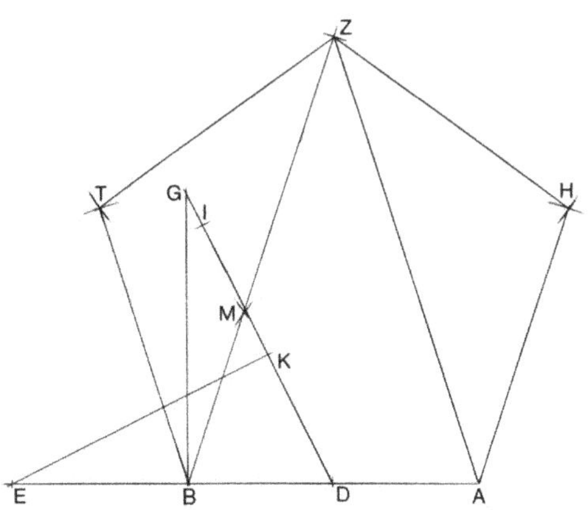

Pl. 20.3. This is another method offered by Abu'l-Wafa' for constructing a regular pentagon on an assumed side with a fixed compass opening.[3] The construction procedure is essentially the same as the previous one, except for an increased number of steps due to the fixed compass opening, which allows greater control and is more reliable.

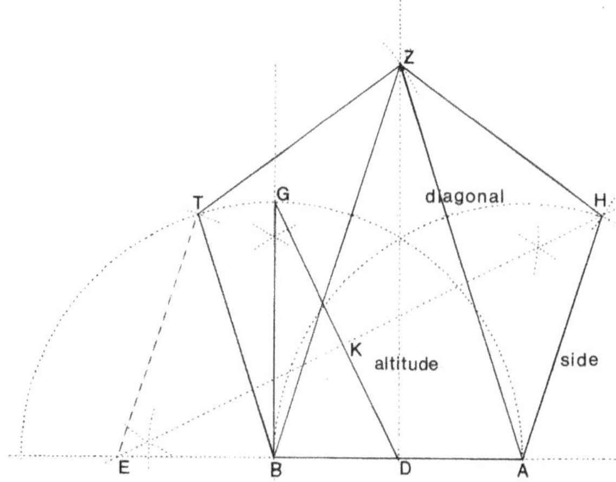

Pl. 20.4. Construction 20 (pl. 20.1) is an improved version of Abu'l-Wafa''s previous construction (pl. 20.3) with a fixed compass opening on an assumed side, without going through the intermediary point M.

1 Özdural, "Analyses" (unpublished chapter), 99–102.
2 Abū'l-Wafāʾ al-Būzjānī, Ms.Ayasofya 2753, 16–17 (3.3).
3 Abū'l-Wafāʾ al-Būzjānī, Ms. Ayasofya 2753, 17 (3.4).

Translation and Transcription (fol. 186v [21])

Fol. 186v [21][a]

Section (*faṣl*): On the construction of an equilateral and equiangular pentagon with a [fixed compass] opening equal to the altitude.[b]

We want to construct an equilateral and equiangular pentagon such that its altitude is equal to the assumed line AB. The procedure is as follows: Let the compass opening be AB and [with that] on line AB from points A and B inscribe arcs HAG and TBG. Obtain the intersections D and G. Draw line GDI to find intersection E. From intersection E find the center, I. Inscribe the greatest circle [with center I].

Then from point G find point Z, and from the latter inscribe arc TH to find points T and H. From these two points find intersection C. On the rectilinear extensions of lines TC and HC and on the circumference of the circle find points Q and J. From these two points, again on the circumference, find points t[c] and h opposite to one another, that is to say, from point Q [find] point h and from point J [find] point t.[d] Keep these two points fixed. Then, from intersection E, mark points K and L on the circumference of the arc. From line KL find intersection M [i.e., midpoint of KL]. On the rectilinear extension of lines BMT[e] and AMH find points N[f] and S on the circumference of the circle. From these two points, also on the circumference, find points O and F. Again, on the rectilinear extensions of lines tM and hM find points X and V on the circumference of the circle. Then draw lines IO, IF, hVz, and tXd. Thus emerges pentagon Iλzdg. This is the desired pentagon, equilateral and equiangular, as was required, to construct and to explain. God is all-knowing and all-wise.

---

a This is the explanatory text of Construction 21 (found on the next page [fol. 186bis(r)]). It is composed of nineteen horizontal lines that fill the entire page.

b This heading is listed in Arabic.

c Text: Q.

d Text: MtMT. According to Hogendijk, the Persian text is probably a corruption of: *nuqṭa* T, i.e., point T.

e Text: NMT.

f Text: X.

فصل فی عمل المخمس المتساوی الأضلاع والزوایا بفتح العمود

میخواهیم که مخمسی متساوی الأضلاع والزوایا عمل کنیم چنانک عمود او مثل خط آ بٓ مفروض باشد. طریق آن بود که پرگار
بر فتح آ بٓ و بر خط آ بٓ از نقطهٔ آ بٓ قوس حٓ آ جٓ طٓ بٓ جٓ بگرداند و تقاطع دٓ جٓ حاصل کند و خط جٓ دٓ یٓ بکشد و
تقاطع هٓ معلوم کند و از تقاطع هٓ مرکز یٓ معلوم کند و دائره اعظم بگرداند آنگه از نقطهٔ جٓ نقطهٔ زٓ معلوم کند و ازین نقطه
قوس طٓ حٓ بگرداند و نقطهٔ طٓ حٓ معلوم کند و ازین هر دو نقطهٔ تقاطع صٓ معلوم کند و بر استقامت خط طٓ صٓ حٓ صٓ
بر محیط دائره نقطه قٓ شٓ معلوم کند و ازین هر دو نقطه هم بر محیط نقطه تٓ ثٓ بر خلاف یکدیگر یعنی از نقطهٔ قٓ نقطهٔ
ثٓ و از نقطهٔ شٓ مٓ تٓ مٓ طٓ معلوم کند و این هر دو نقطه‌را نگاه دارد آنگه تقاطع هٓ نقطهٔ کٓ لٓ بر محیط قوس نشان کند
و از خط کٓ لٓ تقاطع مٓ معلوم کند و بر استقامت خط نٓ مٓ طٓ آ مٓ حٓ بر محیط دایره نقطه خٓ سٓ معلوم کند و ازین هر دو
نقطه هم بر محیط نقطهٔ عٓ فٓ معلوم کند. باز بر استقامت خط تٓ مٓ ثٓ مٓ بر محیط دائره نقطهٔ خٓ ذٓ معلوم کند آنگه خطوط
یٓ عٓ، یٓ فٓ، ثٓ ذٓ، ظٓ تٓ، خٓ ضٓ بکشد مخمس یٓ لآ ظٓ ضٓ غٓ ظاهر گردد و این مخمس مطلوب باشد متساوی الأضلاع
والزوایا و چنانک خواهد عمل کند و در بیان آورد والله اعلم احکم.

Translation and Transcription (fol. 186bis(r) [21])

Fol. 186bis(r) [21][a]

As in this example:[b]
And in this construction, line KL does not need to be drawn since intersection M is known from lines AH and BT. This construction is deduced, not transmitted.[c]

---

a The explanation for Construction 21 is on the previous page (fol. 186v).
b This heading is written above the main figure, which occupies most of the page.
c The text is composed of two oblique lines surrounding the lower part of the figure. The first line, from upper right to lower left, is on the right side of the figure. The second line is written on the left side of the same figure, from lower right to upper left.

برین مثال

و درین عمل اگر خط یٓ لٓ نکشد شاید چه از خط آ حٓ تقاطع مٓ معلوم میشود.

و این عمل استنباطی است، نه منقول.

Construction 21 (fol. 186bis(r) [21])[1]

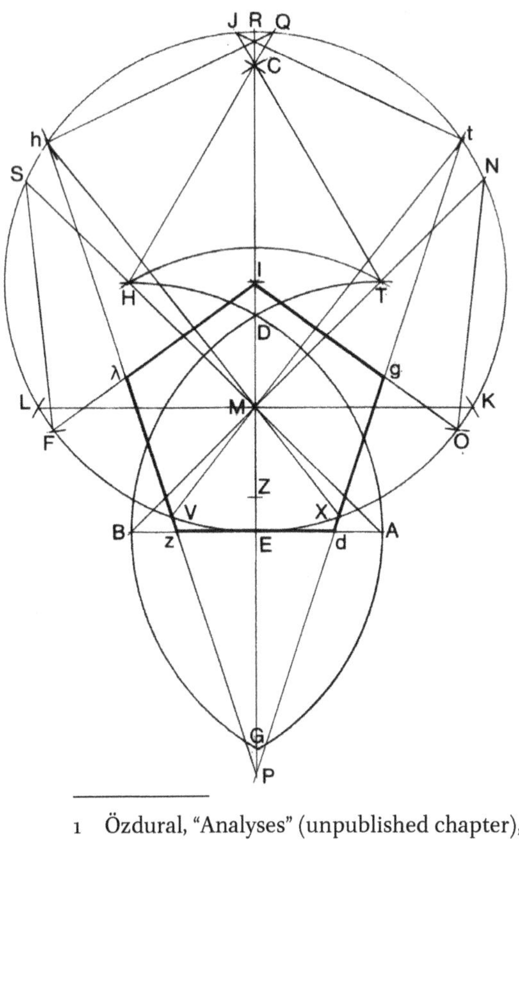

Pl. 21.1. Construction of a regular pentagon with a fixed compass opening equal to the altitude.

Drawn within a large square, the pentagon is surrounded by a network of arcs and lines. This is a more elaborate alternative to Construction 19 (pl. 19.1), both being approximations based on the pentagon's altitude.

---

1 Özdural, "Analyses" (unpublished chapter), 102–4.

Translation and Transcription (fol. 186bis(r) [22])

Fol. 186bis(r) [22]

No accompanying text.

Construction 22 (fol. 186bis(r) [22])[1]

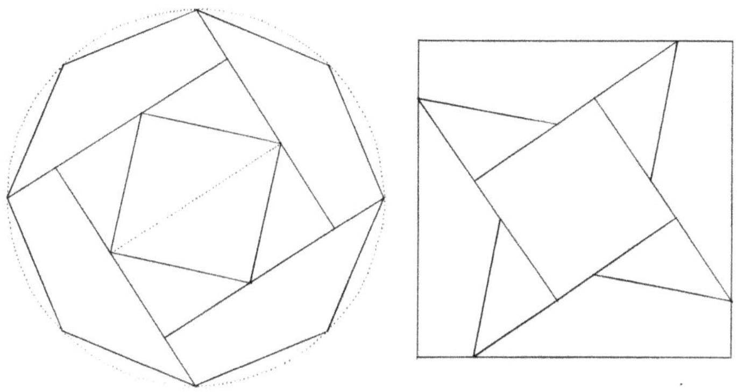

Pl. 22.1. Construction of a square and octagon of equal areas.

In this textless "cut-and-paste" construction, an octagon is transformed into a square. The octagon on the left is dissected into nine pieces, which are then reassembled into the square on the right. The uninked incised construction lines provide clues about the construction procedure.

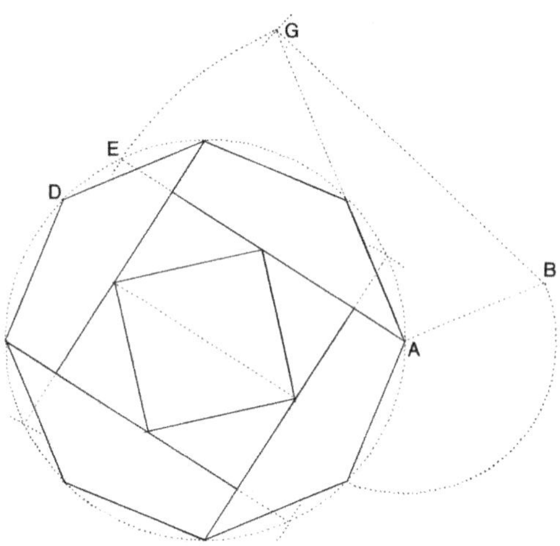

a

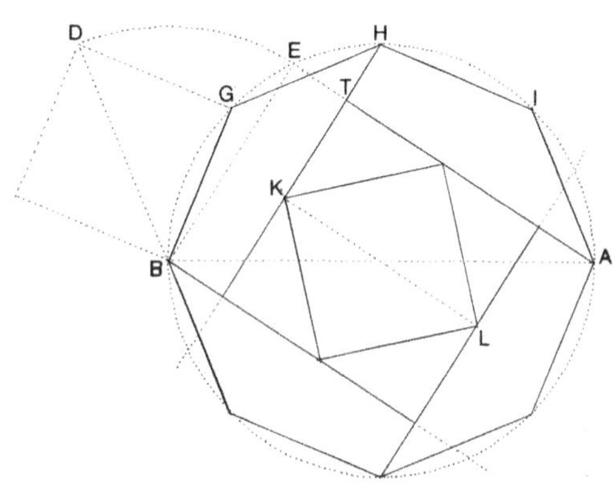

b

Pl. 22.2a. Perhaps tne solution in Construction 4 (see pl. 4.1) was used here as well.

Pl. 22.2b. Since it does not make sense to produce a nine-piece dissection by using the same method that produced an elegant five-piece dissection in Construction 4, a different method may be assumed. According to Özdural, Construction 22 appears to have been offered as an alternative to Construction 4. In the previous case, the side of the required square was obtained from the diameter of the inscribed octagon. Here, the required dimension was obtained from the diameter of the circle circumscribing the octagon. Though not as economical as the former case, according to Özdural, this construction is "equally clever."

1 Özdural, "Analyses" (unpublished chapter), 104–5.

Translation and Transcription (fol. 186bis(v) [23])

Fol. 186bis(v) [23]

On the construction of an equilateral and equiangular pentagon with a [fixed compass] opening equal to the side.[a]

If one wants to construct an equilateral and equiangular pentagon with a [fixed compass] opening equal to side AB, the procedure is as follows: On side AB, with center A, inscribe arc BG, and with center B, arc AD, until they intersect at point E. Then, with center E inscribe arc DHZG. From points G and D on arc DG,[b] find points Z and H. Draw lines ZA and HB. Then, from points Z and H on lines AZ and BH find points T and K. Draw lines AKM and BTL so that each of them ends at the circumferences of arcs AD and BG at points L and M. Then, with points L and M as centers obtain intersection N. Draw lines AL, BM, LN, and MN so that the equilateral and equiangular pentagon ABMNL is produced. When arc ML is inscribed from intersection N, it should pass through intersections M and L. If that does not happen, the construction is erroneous, as we wanted to explain. God knows best.[c]

---

a This heading is written in two horizontal lines under the figure.
b Text: DS.
c The text is composed of nine horizontal lines on the lower half of the page.

فی عمل المخمس المتساوی الأضلاع والزوایا بفتح الضلع

چون خواهد که عمل مخمسی متساوی الأضلاع والزوایا کنند بفتح ضلع آ ب طریق آن بود که بر ضلع آ ب از مرکز آ قوس ب ج و از مرکز ب قوس آ د بگرداند تا متقاطع شود بنقطهٔ ه آنگه از مرکز ه قوس د ح ز ج بگرداند و از نقطه ج د بر قوس د ج نقطه ز ح معلوم کند و خط ز آ ح ب بکشد آنگه از نقطهٔ ز ح بر خط آ ز ب ح نقطهٔ ط ک معلوم کند و خط آ ک م ب ط ل بکشد تا هر یک منتهی شوند بمحیط قوس آ د ب ج بنقطهٔ ل م آنگه از مرکز ل م تقاطع ن حاصل کند و خط آ ل ب م ل ن م ن بکشد تا مخمس آ ب م ن ل متساوی الأضلاع والزوایا حاصل شود. باید که چون از تقاطع ن قوس م ل بگرداند از قوس تقاطع م ل بگذرد و اگر غیر ازین آید عمل خطا باشد و چنانکه خواهد بیان کند والله اعلم.

Construction 23 (fol. 186bis(v) [23])[1]

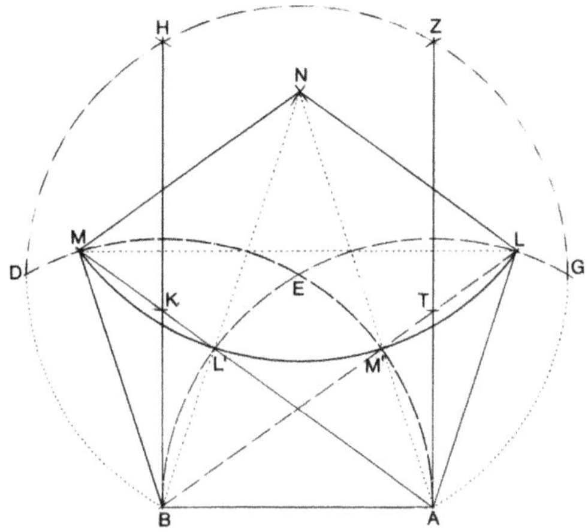

Pl. 23.1. Construction of a regular pentagon with a fixed compass opening equal to the side.

The text explains the approximate construction procedure, which is less complicated than the ones based on the altitude and quite close to the theoretical value. The description ends with a method for checking the accuracy of the execution.

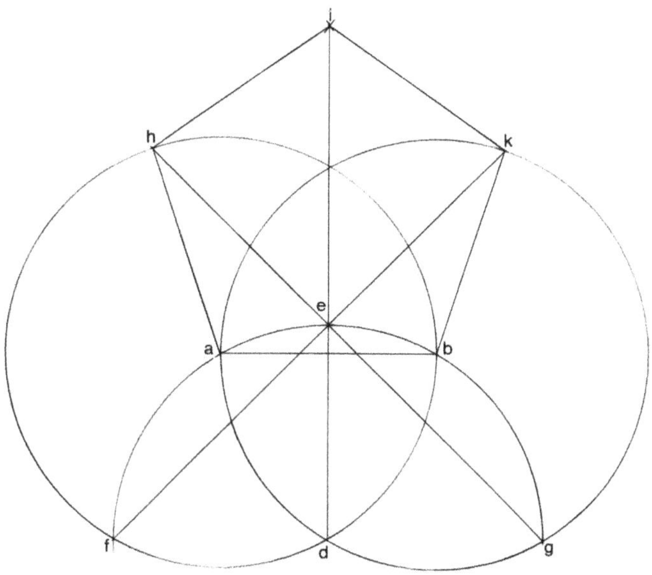

Pl. 23.2. A comparable approximate construction of the pentagon with a fixed compass opening corresponding to one side, from the fifteenth-century master mason Mathes Roriczer's *Geometria deutsch*, which is a cruder approximation.[2] Albrecht Dürer repeated the same method in the sixteenth century.[3] According to Özdural, the authors of the pentagon constructions in the *Anonymous Compendium* seem to have been informed about the correct ways of constructing the regular pentagon. Although they sought alternative approximation methods to achieve the same goal, they always paid attention to the accuracy of their constructions.

1 Özdural, "Analyses" (unpublished chapter), 106–8.

2 Shelby 1977, 116–17 (figs. 26, 27), 180–81n4.

3 Albrecht Dürer, *Vnderweysung der Messung, mit dem Zirckel vñ Richtscheyt, in Linien ebnen vnnd gantzen Corporen* (Nüremberg: Hieronymus Andreae, 1525), fol. E iiii, fig. 16.

Translation and Transcription (fol. 186bis(v) [24])

Fol. 186bis(v) [24]

On the construction of the equilateral and equiangular pentagon with compass opening equal to the side.[a]

a The heading, consisting of two horizontal lines, is written under the figure, with no accompanying explanatory text.

فى عمل المخمس المتساوى الأضلاع والزوايا بفتح الضلع

CONSTRUCTION 24 (fol. 186bis(v) [24])[1]

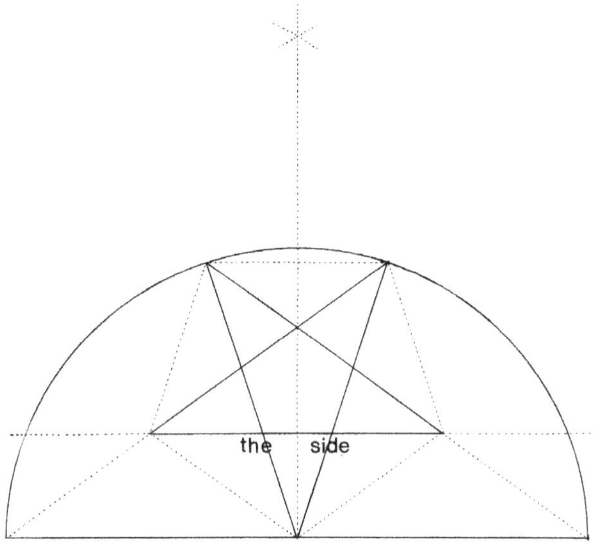

Pl. 24.1. Construction of a regular pentagonal star inscribed in a semicircle, with a compass opening equal to the side.

The pentagonal star is inscribed in a semicircle in such a way that one of its vertices is at the center on the diameter, while the two opposite vertices are touching the circumference. This drawing, unlike the previous one drawn on its left side, uses a correct method. Uninked incised lines connect the pentagon's angular points, so as to define a pentagon around it. The sides of the pentagonal star correspond to the diagonals of the invisible pentagon circumscribing it, which are extended as uninked lines. Underneath the horizontal diagonal is written "the side" (*al- żilʿ*), referring to the pentagram's side. This textless construction, in which the compass opening was not fixed, is based on the division of a line segment in extreme and mean ratio.

a

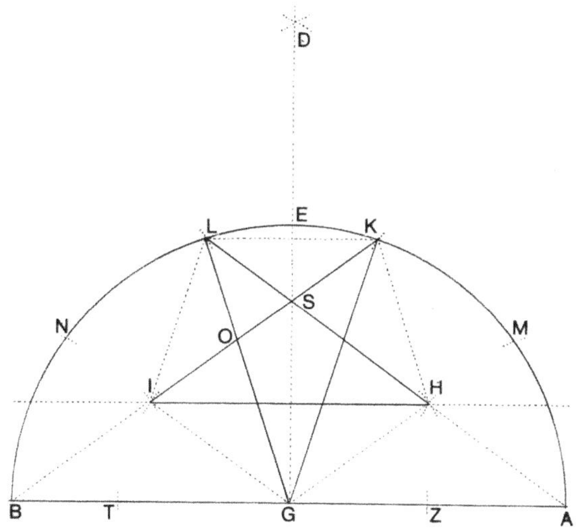

b

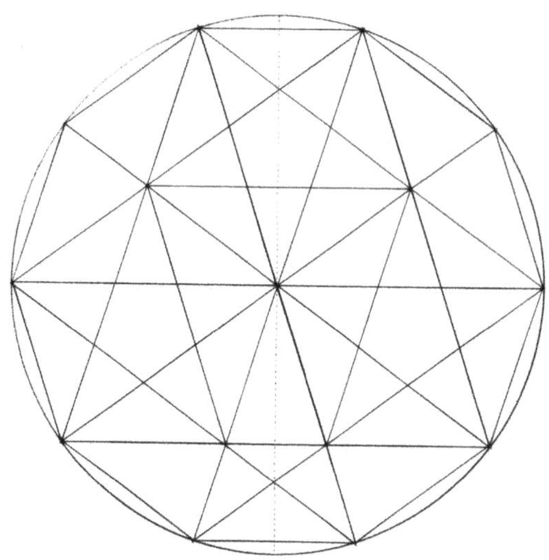

Pl. 24.2a. Özdural's reconstruction of the construction procedure of Construction 24, based on uninked construction lines.

Pl. 24.2b. When three pentagonal stars are inscribed in a circle, they form a rich network of interrelated segments divided in extreme and mean ratio, which, according to Özdural, may have been the "reason for the conception of Construction 24."

1 Özdural, "Analyses" (unpublished chapter), 108–10.

Translation and Transcription (fol. 187r [25])

Fol. 187r [25]

Construction of a pentagon with a [fixed compass] opening equal to the diagonal, by Abu Bakr al-Khalil al-Tajir.[a]

If line AB, which is the diagonal of a pentagon, is known, and one wants to construct a pentagon with that compass opening: Construct on it the elliptical figure AGBD [from points A and B]. Draw diagonal DGZ. Put the compass on point D and mark point [E], then put it on point E and mark point Z. Put it on point Z and take points H and T on the circumference of the elliptical figure. From these two points find intersection K. Put it [the compass] on point K and locate point L. From point L locate points M and N. Then pentagon ALBNM becomes equilateral and equiangular. God knows best.[b]

---

a The heading is composed of two oblique lines, from lower right to upper left and from upper right to lower left, surrounding the upper part of the figure.

b The text is composed of fifteen horizontal lines on the left side of the page.

عمل مخمس بفتح الوتر لأبي بكر الخليل التاجر

چون خط آ بٓ که وتر مخمس است معلوم باشد و خواهد که بدان فتح مخمسی عمل کند
بر وی بیضی آ جٓ بٓ دٓ عمل کند و قطر دٓ جٓ زٓ اخراج کند و پرگار بر نقطه دٓ نهد و نقطه
نشان کند و بر نقطهٔ هٓ نهد و نقطهٔ زٓ نشان کند و بر نقطهٔ زٓ نهد و نقطهٔ حٓ طٓ بر محیط
بیضه بگیرد و ازین هر دو نقطه تقاطع کٓ معلوم کند و بر نقطهٔ کٓ نهد و نقطهٔ لٓ بگیرد و از
نقطهٔ لٓ نقطه مٓ نٓ بگیرد که مخمس آ لٓ بٓ نٓ مٓ متساوی الأضلاع والزوایا باشد والله اعلم.

CONSTRUCTION 25 (fol. 187r [25])[1]

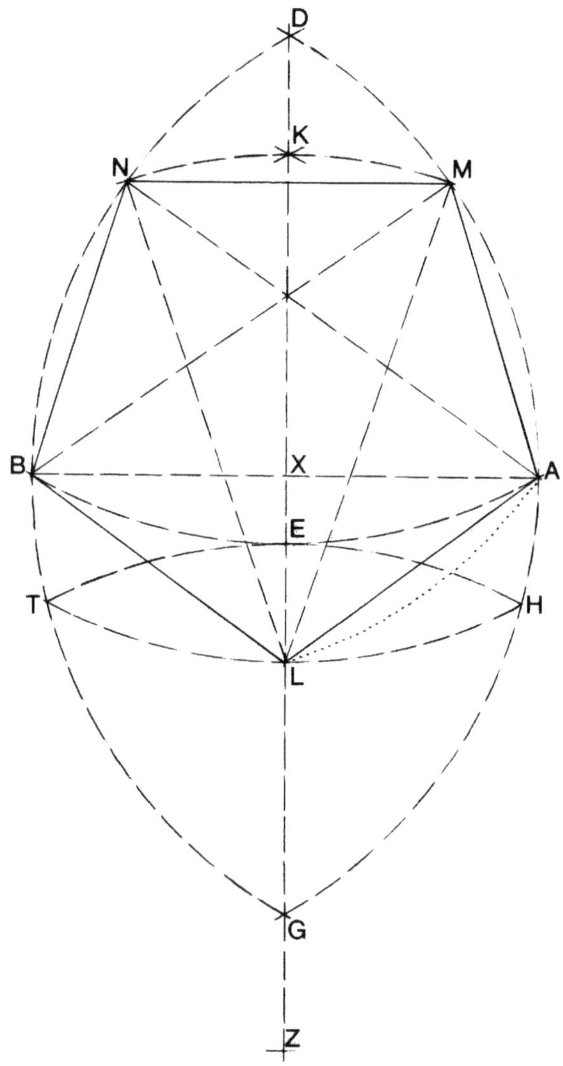

Pl. 25.1. Construction of a regular pentagon with a fixed compass opening equal to the diagonal, by Abu Bakr al-Khalil al-Tajir.

An elliptical figure circumscribing a regular pentagon, whose construction procedure is explained in the text. Özdural observes that the named author "appears [to have been] a mathematician who was competent in approximation techniques. The method he proposed here is simple and easy to follow; it also produces a fairly accurate result." This is the last construction in the pentagon series, only two of which are correctly drawn by means of a segment cut in extreme and mean ratio.

1 Özdural, "Analyses" (unpublished chapter), 110–14.

Translation and Transcription (fol. 187v [26])

Fol. 187v [26]

Simple proportion.[a]

If angle A is a right angle and [magnitude] AB is assumed, the construction should be made as follows: Bisect angle A[b] by line AD and inscribe quarter-circle BDG so that point D is known. Then make point B the center and with compass opening BA inscribe arc BH and with [compass opening] BD inscribe arc DE parallel to arc BH. Then make point E the center and with [compass opening] EB inscribe arc BHZ to intersect arc BH at point H. Draw line[c] ZHT so that magnitude GT becomes equal to ZH, as this is the proportion of the required triangle of the repeat-unit, which has been drawn. God knows best.[d]

Some artisans draw this repeat-unit such that its length is in seven parts and the width is six parts. This is extremely close [to the required ratio]. God knows best.[e]

---

a This note is written obliquely inside the figure, from upper right to lower left.

b Text: AB.

c The phrase "and line" is repeated twice.

d The text is composed of six oblique lines, from upper right to lower left, along the diagonal of the figure.

e Annotation belonging to Construction 26: four horizontal lines facing downwards on the right side of the page, inserted between the accompanying texts of Construction 26 and Construction 27. It could also relate to Construction 27.

نسبت سهل

چون زاویه آ قایمه باشد و جزو مفروض آ ب عمل چنان باید کرد که زاویهٔ آ ب دو قسم کند بخط آ د و ربع دائره ب د ج بگرداند تا نقطهٔ د معلوم شود.آنگه نقطهٔ ب مرکز سازد و بفتح ب آ قوس آ ح بگرداند و ببعد ب د قوس د ه بگرداند موازی قوس آ ح.آنگه نقطهٔ ه مرکز سازد و ببعد ه ب قوس ب ح ز بگرداند تا متقاطع قوس آ ح گردد بنقطهٔ ح و خط[١] ز ح ط بکشد چنانک مقدار ج ط مثل ز ح باشد که این نسبت مثلث مطلوب است درین عقد که کشیده شده است والله اعلم.

بعضی از صناع این عقدرا چنان می‌کشند که طول او هفت قسمت می‌گیرند و عرض شش قسمت و عظیم نزدیکست والله اعلم.

---

[١] و خط] و خط و خط

Construction 26 (fol. 187v [26])[1]

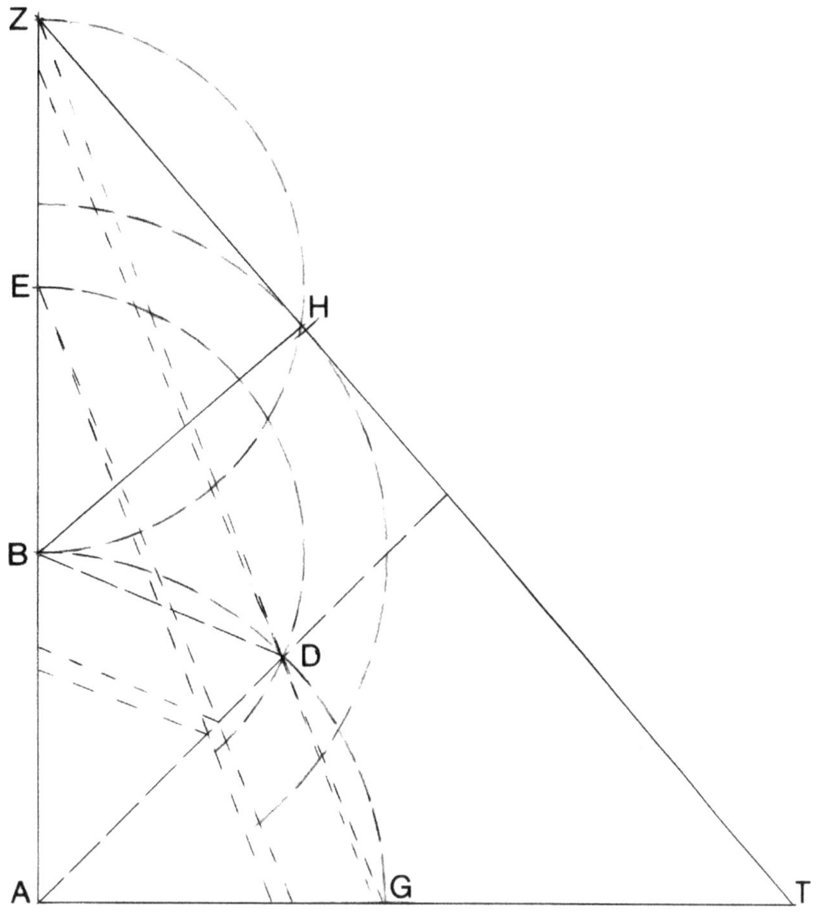

Pl. 26.1. Construction of a kite in a right-angled triangle.

For this triangular repeat-unit an accurate verging construction and an approximation have already been offered in Constructions 16 and 17. The main text explaining the construction procedure makes a cross-reference to the proportion of the required triangle in the repeat-unit that was drawn earlier. Özdural interprets this explicit cross-reference as an indication that Construction 26 is an alternate solution to Constructions 16 and 17. According to the property stated in Construction 16, AB equals BH, which is also equal to AG. Unlike these correct properties, drawn with an advanced verging procedure, Construction 17 was executed by means of a close approximation. Construction 26, labeled "simple proportion," provides an easier and more practical approximation. Both constructions have the same objective, namely, to produce equality between lengths GT and ZH.

1 Özdural, "Analyses" (unpublished chapter), 114–17.

Translation and Transcription (fol. 187v [27])

Fol. 187v [27]

The ratio of the length and width of this repeat-unit is determined by geometrical method from the octagon, such that its length [AS] is half the diagonal of the octagon plus twice the side, and its width [AB] is half the diagonal of the octagon plus the square root of the difference between the square root of half the diagonal and the square of twice the side. The diagonal [BS] of this repeat-unit is half the diagonal of the octagon plus twice the above-mentioned square root of the difference. God knows best.[a]

---

a The text is composed of four vertical lines, from top to bottom, on the right side of the figure.

نسبت طول و عرض این عقد بطریق هندسه از مثمن یافته می‌شود که طول او نصف
قطر مثمن باشد و دو ضلع و عرض او نصف قطر مثمن باشد با [1]جدر مابین مربع نصف
قطر و مربع دو ضلع و قطر این عقد نصف قطر مثمن باشد با ضعف جدر مابین مذکور
والله اعلم.

---

1 A dot is missing in the word جدر, meaning "square root"; it is obviously meant to be read as جذر. This applies to all subsequent uses of the term below.

Construction 27 (fol. 187v [27])[1]

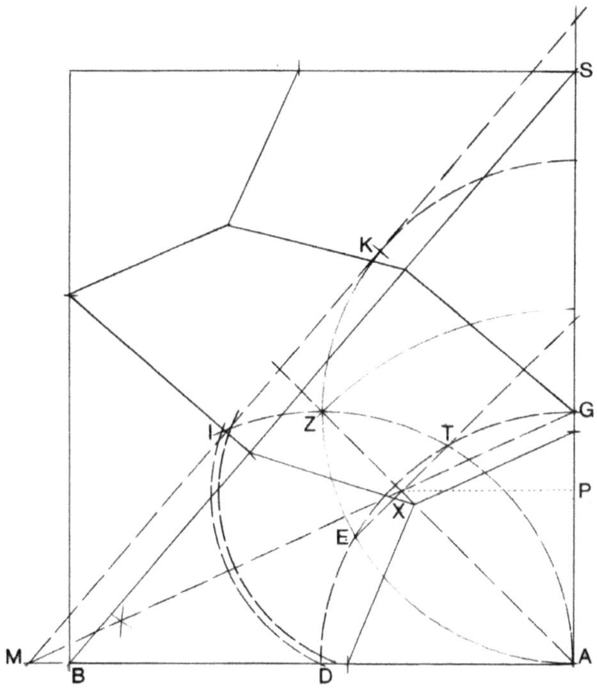

Pl. 27.1. Construction of two equivalent kites in a rectangle.

This construction provides a more accurate approximate version of Construction 26. It starts with a quarter-octagon that is not equilateral. Although no uninked incised lines are visible, auxiliary lines are drawn in red ink so that the black ink pattern itself is clearly distinguished. Özdural states that the method proposed here as an alternative for Construction 17 "answers the needs of the artisans satisfactorily."

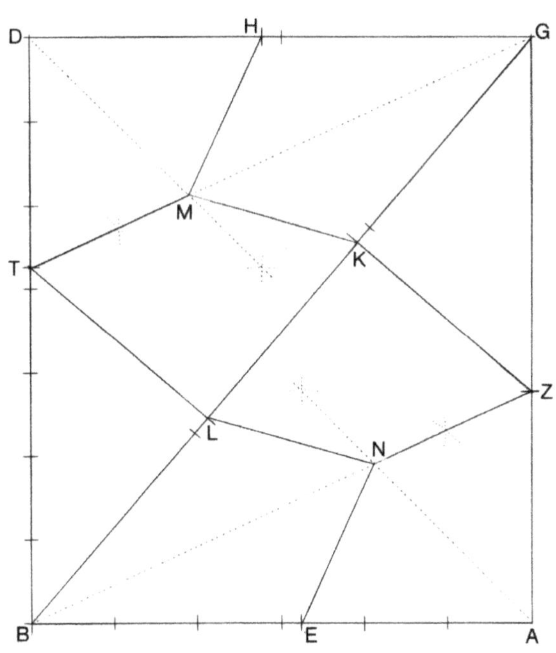

Pl. 27.2.[2] A short note between Constructions 26 and 27 explains that some artisans draw this repeat-unit such that its length is in seven parts and the width is six parts. That missing figure is reconstructed here by Özdural. It is much simpler than the previous constructions of the same pattern, but "happens to be the most accurate of all." The text indicates that artisans were familiar with "numerical ratios" for expressing dimensions obtained from geometrical constructions.

1 Özdural, "Analyses" (unpublished chapter), 117–18.

2 This missing construction is reconstructed and numbered "28" by Özdural, "Analyses" (unpublished chapter), 118–19. Necipoğlu renumbered it as 27.2, and accordingly renumbered the following constructions as well, to make them correspond with the numbering of the individual constructions in the *Anonymous Compendium*.

Translation and Transcription (fol. 188r [28])

Fol. 188r [28]

In this figure there are two kinds of [procedures to establish the desired] proportion; one is circular and the other rectilinear. The first, circular one is this: Inscribe quarter-circle AMK and semicircle LMK. From their intersection M, [drop] perpendicular MN, which is equal to side LS. This is what is required. The other [procedure] is this: Construct a triangle that is half the square [on an arbitrary line] BG. [Take] magnitude EB equal to hypotenuse GD, and EZ equal to EG. And [take] magnitude DH equal to ZB, and HT equal to DH. Then draw line ZTI and line KO parallel to IZ. That magnitude ZO is the required side.[a]

Another kind of [procedure to establish] the properties of the proportion is this: Find point F at the middle of side AL. From magnitude LC that is equal to half the diagonal, find point C. Then between these two points, the required point Q is found [as the midpoint].[b]

---

a The text is composed of five vertical lines, from bottom to top, along the left side of the figure.
b The text is composed of seven oblique lines, from lower right to upper left, on the left side of Construction 29, but with no relation to it.

درین شکل از دو نوع نسبت است یکی دوری و دیگر خطی. اول دوری آنست که ربع دائره آ مٓ کٓ بگرداند و نصف دائره لٓ مٓ کٓ که از تقاطع مٓ عمود مٓ نٓ مثل ضلع لٓ سٓ بود و این مطلوبست. و دیگر آنک مثلث نصف مربع بٓ جٓ عمل کند و مقدار هٓ بٓ مثل وتر جٓ دٓ و هٓ زٓ مثل هٓ جٓ و مقدار دٓ حٓ مثل زٓ بٓ و حٓ طٓ مثل دٓ حٓ و خط زٓ طٓ یٓ بکشد و کٓ عٓ موازی یٓ زٓ که مقدار [زٓ] عٓ ضلع مطلوب بود.

نوعی دیگر از خواص نسبت آنست که نقطه فٓ وسط ضلع آ لٓ معلوم کند و از مقدار لٓ صٓ که مثل نصف قطر است نقطهٔ صٓ معلوم کند. مابین این هر دو نقطه نقطه قٓ که مطلوبست معلوم شود.

Construction 28 (fol. 188r [28])[1]

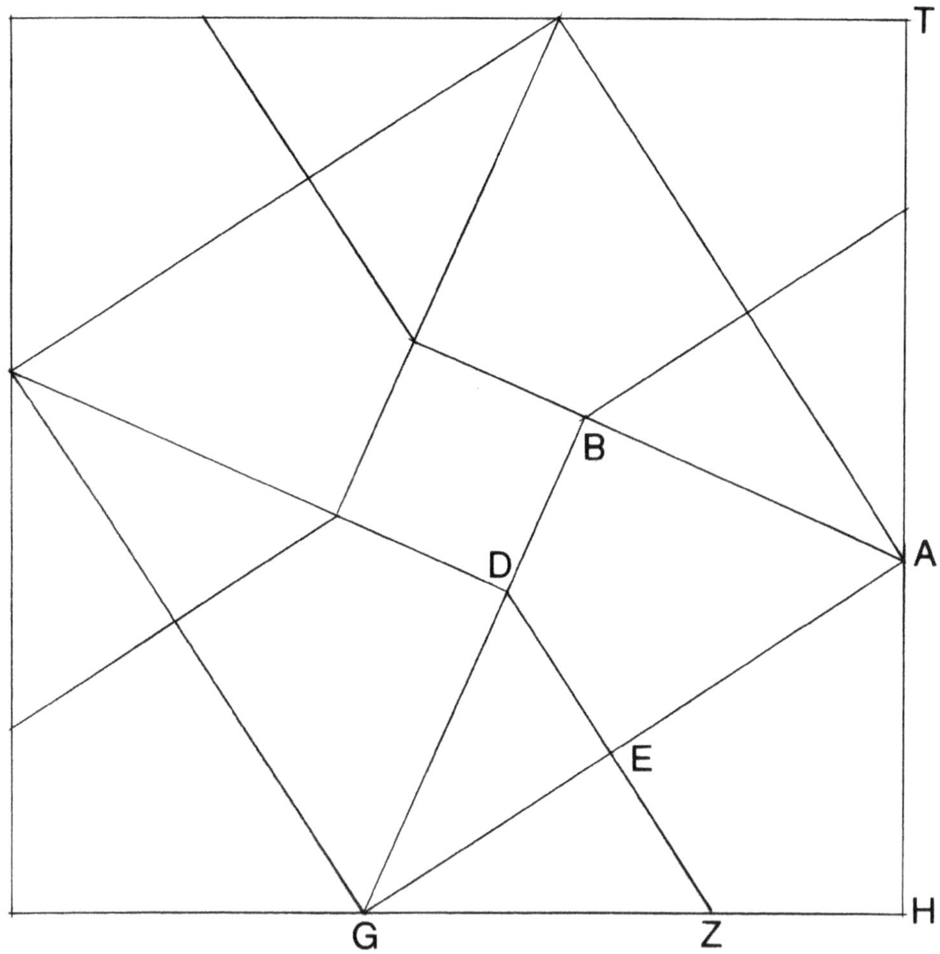

Pl. 28.1. Construction of a square containing two rotated squares, four kites, and triangles.

This construction and the following ones explore a special ornamental pattern with four rotating right-angled triangles around a rotated central square. It is found in Abu'l-Wafa''s *Geometrical Constructions* (see pls. 2.4a–d and 2.5a–c). On the basis of that scheme a potential ornamental pattern was envisaged here: four isosceles quadrangles (kites) and triangles are rotated around a rotated central square, generating a pattern of interlocking shapes within a large square, whose side is equal to AB plus BG. If the outermost square is given, the problem can be reduced to determining the position of point A on side HT in order to establish the proportion of the pattern.

1 Özdural, "Analyses" (unpublished chapter), 120–22 (labeled as "Constructions 29, 30, and 31" by Özdural).

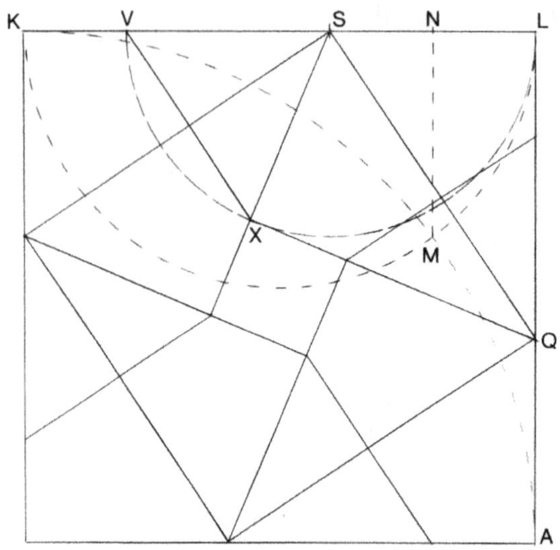

Pl. 28.2. An alternative solution to the same problem. The text explains that the figure provides two kinds of procedures for establishing the desired proportion, one "circular" and the other "rectilinear." For the circular method, inscribe quarter-circle AMK and semicircle LMK; from the intersection M, drop perpendicular MN, which is equal to side LS. The idea here is to divide the side of a given square into two parts by a simple ratio, so that the angle of rotation can be established. The approximation process amounts to determining the point that divides the side of the square into two-thirds (MN:KL = 2/5, SL:SK = 2/3).

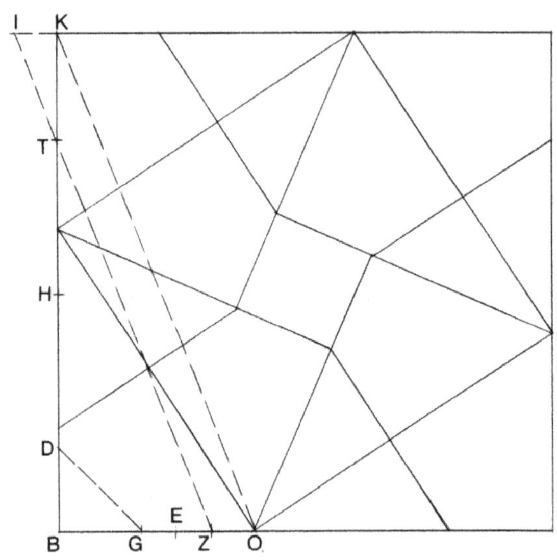

Pl. 28.3. The objective of the second method for the same construction is to divide the side of the given square into two parts so that their ratio corresponds to the required angle of rotation. The author first determines right angle ZBH, which approximates the required triangle; he then adds the shortest side to the intermediate side to find point O, by way of parallel line KO. Here, "length BO is the desired side." Özdural points out that this rectilinear method provides a closer approximation of the required angle than the previous circular method.

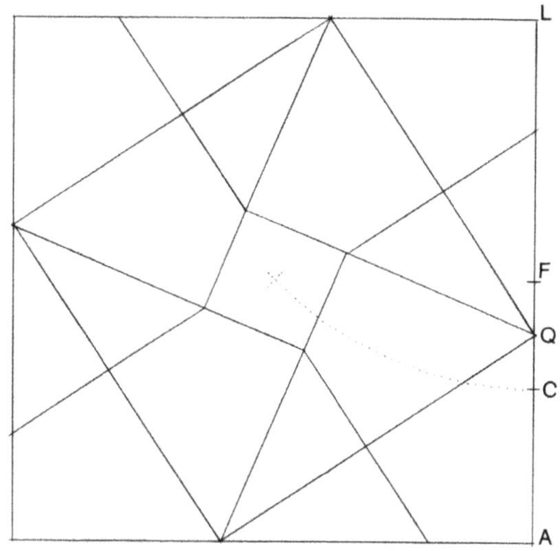

Pl. 28.4. The third approximation for constructing the same repeat-unit is derived from the "properties of the proportion." The required point Q is estimated between half the side of a given square and half of its diagonal. All three constructions (pls. 28.2, 28.3, and 28.4) use approximation methods to find the points where the vertices of the intermediate rotated square touch the outer edges of the square repeat-unit. Construction 28.3 yields a more satisfactory result.

Translation and Transcription (fol. 188r [29])

Fol. 188r [29]

No accompanying text.

Construction 29 (fol. 188r [29])[1]

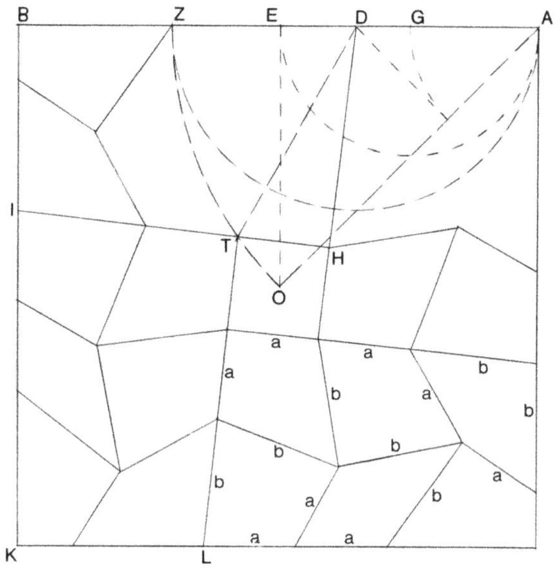

Pl. 29.1. Construction of a square with four quadrangles surrounding a rotated central square, each subdivided into a series of quadrangles.

There is no textual explanation, but the procedure is shown in the upper right corner of the figure by means of auxiliary construction lines. In this construction, line DT must be determined so that points H, T, and I lie on a straight line.

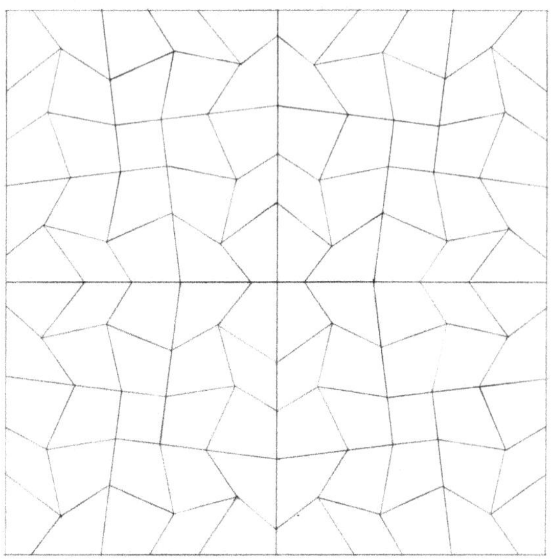

Pl. 29.2. A replication of the repeat-unit in four directions, which does not generate an attractive composition.

1 Özdural, "Analyses" (unpublished chapter), 123–24 (labeled as "Construction 32" by Özdural).

Translation and Transcription (fol. 188v [30])

Fol. 188v [30]

Some practitioners of geometry held that [if] we want the distance between two parallel squares to be equal to the side of the square in the middle, then it is necessary that magnitude SN[a] be equal to [R]J. The proportion of it [the repeat-unit] is such that, since the square of OF is assumed, they draw this repeat-unit in another way. All parts F′C, CQ, QR, RJ, Jt t[h and h]X are equal. Then line Qt is the square root of 5 and the side of the inner square is two units plus the square root of 5. The side of the outer square is four units plus the square root of 5. If one of these two squares is known, the other is found by means of the proportion [of the repeat-unit]. God knows best.[b]

---

a Text: AN.
b The text is composed of six horizontal lines below the figure.

بعضی از اصحاب هندسه بران بودند که مابین المربعین المتوازیین مثل ضلع مربع وسط می‌خواهیم. پس چنین باشد باید که مقدار آن مثل شٓن باشد. نسبت وی آنست که چون مربع عٓف مفروض بود این عقدرا بطریقی دیگر میکشند اقسام فٓص صٓق قٓر رٓش شٓت تٓث ثٓخ[1] جمله متساوی شود. پس خط قٓت جدر ۵ باشد و ضلع مربع داخل ۲ باشد صحیح و جدر ۵ و ضلع مربع خارج چهار باشد صحیح و جدر ۵. چون ازین هر دو مربع یکی معلوم باشد دیگری بنسبت وی معلوم شود والله اعلم.

---

[1] تث ثخ] تخ

CONSTRUCTIONS 30 and 31 (fol. 188v [30–31])[1]

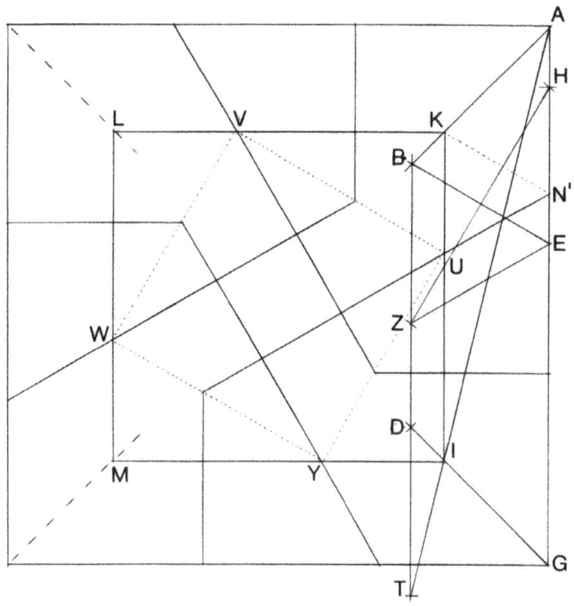

Pl. 30.1. Construction of two parallel squares containing a rotated central square surrounded by rotating kites and chevrons.

Constructions 30 and 31 share a single figure. It recalls Abu'l-Wafa's square, containing a small rotated central square surrounded by four rotating triangles (see pls. 2.4d and 2.5a–c), which may have inspired Constructions 28 and 29. According to the text of the present construction, some practitioners of geometry established its proportions as being based on the distance between the two parallel outer squares, which is equal to the side of the innermost rotated third square. If that is desired, then SN = RJ. In case the outer square OF is assumed as given (rather than the intermediate parallel square IKLM), they draw this construction in another way (as is explained in Construction 31).

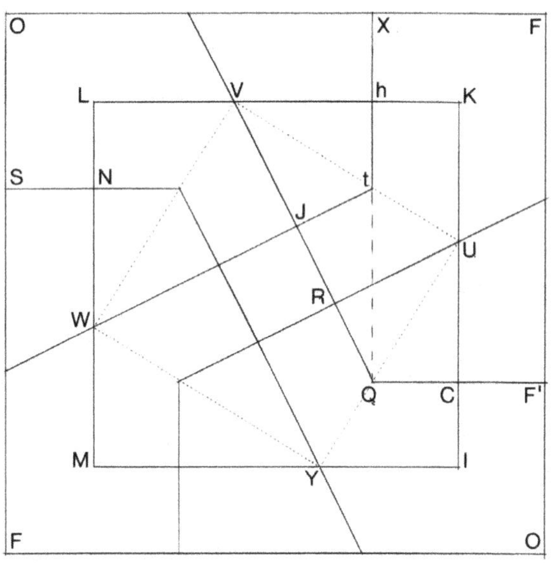

Pl. 31.1. This repeat-unit, described in the text of Construction 31, is illustrated in the previous figure accompanying Construction 30. Here, the outer square is assumed as given, rather than the intermediary, second square (IKLM). The text of Construction 31 explains that its objective is to determine the distance between the two parallel outer squares. If that distance corresponds to the altitude of an equilateral triangle (BEZ), then the side of the innermost square becomes two altitudes plus the side of that equilateral triangle. Özdural suggests that the author, whom the text identifies as Abu Bakr al-Khalil, placed both constructions on the same figure in order to compare the differing proportions of two variants of the same pattern. The kite, identified in the text as "*ṭurunj-6*" is composed of two right-angled triangles. These paired triangles, with 30°- and 60°-angles, correspond to the set square described in Construction 14 (see pl. 14.1), where it is referred to as "set square [number] 6" (*gūnyā-6*). Özdural infers that kites were named after the right-angled triangles from which they were created.

1 Özdural, "Analyses" (unpublished chapter), 124–32 (labeled as "Constructions 33–34" by Özdural).

Translation and Transcription (fol. 189r [31])

Fol. 189r [31][a]

The procedure for drawing and [determining] the proportion (*nisbat*) of this repeat-unit is as follows: From an assumed square with angle A and angle G, draw forth line[s] [number] four equal to line AB [and] GD. Mark AB equal to GD. Draw line BD parallel to AG. Then, between the two parallel lines construct equilateral triangle BEZ from point B and base BZ, such that the apex [E] of the triangle ends on the parallel line [AG]. Then put one arm of the compass on point E, the apex of the triangle, and with opening EZ, which is the side of the triangle, mark point H on side AG. Draw line HZ. Again, put the compass on point Z, and with opening ZH mark point T on [the rectilinear extension of] line BD. Draw line AT to intersect line GD at point I. This is the required point. Draw line IK parallel to AG and complete square IKLM. Once this square becomes known, put in it four [number] six kites (*turunj*),[b] and complete the construction.[c] Masters examine the proportion of this repeat-unit. Abu Bakr al-Khalil examined it in a few ways. The constructions were made. One of these [constructions] is the one that has been explained. God knows best.

---

a No separate figure is provided for Construction 31. This construction is superimposed on Construction 30 on the previous page (fol. 188v). The text is composed of twelve horizontal lines in the upper part of the page.

b A "kite [number] 6" corresponds to two right-angled triangles of 30°-, 60°-, and 90°-angles.

c That is, the construction of the desired inner square formed by the hypotenuses of the triangles in the kites.

طریق کشیدن و نسبت این عقد آنست که در مربع مفروض از زاویه آ و زاویهٔ جَ خط چهار مثل خط آ بَ جَ دَ اخراج کند و آ بَ مثل جَ دَ نشان کند و خط بَ دَ بکشد که موازی آ جَ باشد. آنگه مابین الخطین المتوازیین از نقطهٔ بَ بر قاعده بَ زَ مثلث بَ هَ زَ متساوی الأضلاع عمل کند چنانک راس مثلث بخط موازی منتهی شود. آنگه رجل پرگار بر نقطهٔ هَ راس مثلث نهد و بفتح هَ زَ که ضلع مثلث است بر ضلع آ جَ علامت حَ نشان کند و خط حَ زَ بکشد و باز پرگار بر نقطهٔ زَ نهد و بفتح زَ حَ بر خط بَ دَ نقطه طَ نشان کند و خط آ طَ بکشد تا متقاطع خط جَ دَ شود بنقطه یَ و این نقطه مطلوبست. خط یَ کَ موازی آ جَ بکشد و مربع یَ کَ لَ مَ تمام کند و چون این مربع معلوم شد در وی چهار ترنج ۶ بنهد و عمل تمام کند و استادان نسبت این عقد امتحان کنند و ابی بکر الخلیل امتحان کرده بچند وجه و عمل کرده شد و یکی از آن جمله اینست که شرح داده شد والله اعلم.

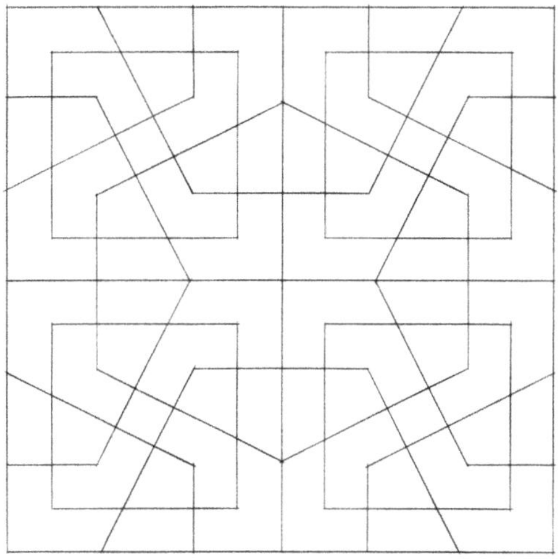

Pl. 30.2. Özdural observes that this is a version of the ornamental motif introduced in Construction 2 (pl. 2.1 and pl. 2.6). When repeated in four directions, Construction 30 generates this geometric scheme.

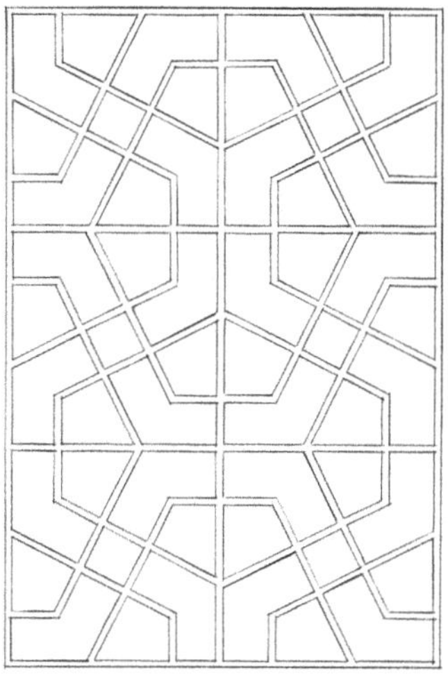

a

Pl. 30.3a. When the middle square IKLM is repeated without the outer borders, it forms a composition more common in the history of Islamic ornamental patterns. This particular example is found on a carved wooden door of the shrine of Imamzada ʿAbdallah and Imamzada Fazlallah, dated 1473, in Rawshanabad (Iran).[2]

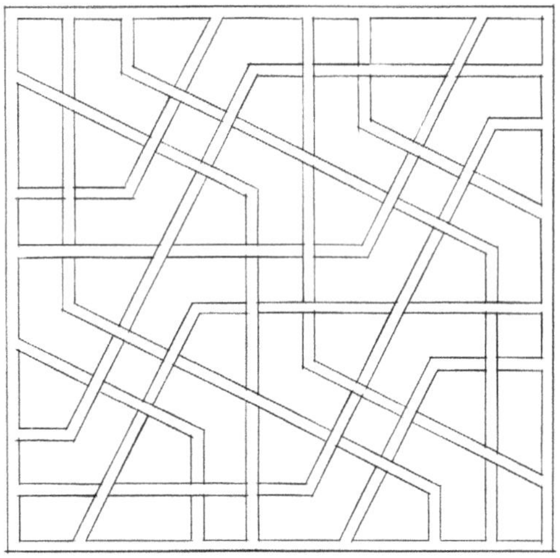

b

Pl. 30.3b. When the border, whose width is the same as the side of the innermost rotated third square, is included, the pattern gains more versatility in generating ornamental schemes. This drawing is based on a mosaic faience panel over the bazaar entrance of the Gawhar Shad Mosque in Mashhad (1418), where a slightly different interpretation of that property is explored.

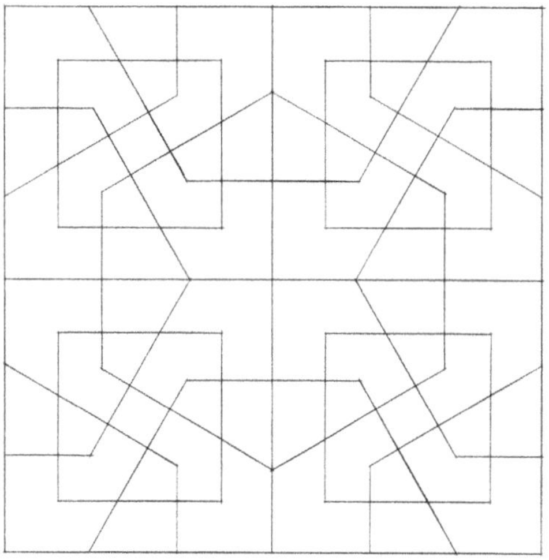

Pl. 31.2 When the repeat-unit in Construction 31 is replicated, it generates this ornamental scheme, which is a variant of pl. 30.2. The difference between them is the result of applying two different proportional systems to the same repeat-unit, the latter based on $\sqrt{5}$ and the former on $\sqrt{3}$.

2 The figure is drawn from the picture published in Golombek and Wilber 1988, 2: fig. 477.

Translation and Transcription (fol. 189r [32])

Fol. 189r [32]

The side AB is two parts. The length BG and the line ED are each the square root of 5.[a]

a The text is composed of three oblique lines, from lower right to upper left, on the left side of the figure.

ضلع آ بٓ دو قسم است و طول بٓ جٓ و خط هٓ دٓ هر یک جدر ۵.

CONSTRUCTION 32 (fol. 189r [32])[1]

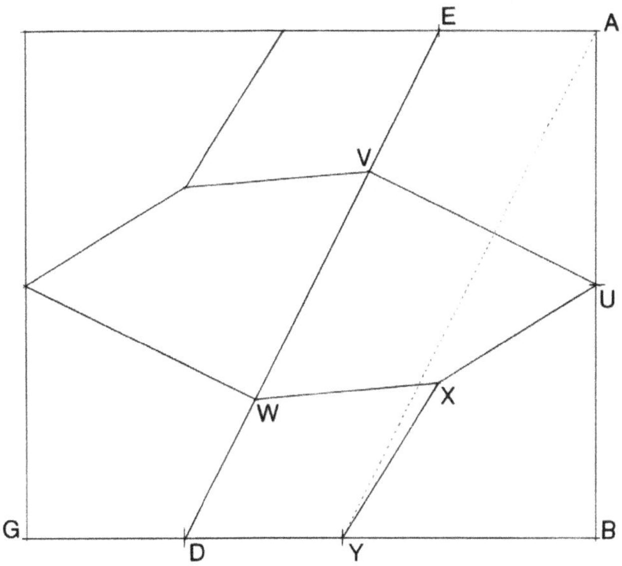

Pl. 32.1. Construction of kites within a rectangle.

In this construction the sides of the rectangular frame have the proportion 2: √5. The text explains that side AB of this rectangle is bisected into two parts, while the other side, BG, and the length of diagonal ED are both √5. Özdural notes that this pattern recalls Construction 16 (see pl. 16.2b), in which a diagonal divides the rectangular frame into two identical right-angled triangles, each containing a kite and a triangle, with each kite further subdivided into three quadrangles. In Construction 32, diagonal DE bisects the rectangle into two identical quadrangles, each containing a large and small kite. The two large kites are further subdivided into three quadrangles. In Constructions 16 and 32, the diagonal axis of symmetry is divided into three parts by two points. The points on the diagonal of Construction 32 are related to one another in the extreme and mean ratio.

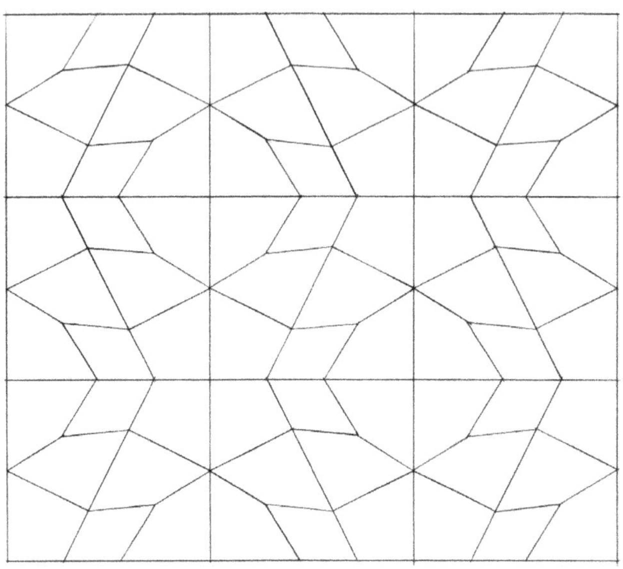

Pl. 32.2. The interlocking large and small kites of Construction 32 also show similarities to Constructions 30 and 31, where they rotate around a central square. But unlike the latter two, this one is not in and of itself successful as a decorative pattern according to Özdural.

1 Özdural, "Analyses" (unpublished chapter), 132–33 (labeled as "Construction 35" by Özdural).

Translation and Transcription (fol. 189v [33])

Fol. 189v [33]

Line AD is the diagonal of a square. The magnitudes of AB and BG are equal, and AD is equal to AB. When line GD is extended rectilinearly, point E will be found. Then EZ and ZH are each equal to AG. Join [G and H as] line GH, and from angle K line KL parallel to GH is drawn to find point L. What is required is obtained. God knows best.[a]

a The text is composed of four horizontal lines below the figure.

خط ا̅ د̅ قطر مربع است و مقدار ا̅ ب̅ ب̅ ج̅ متساویست و ا̅ د̅ مثل ا̅ ب̅. چون خط ج̅
د̅ بر استقامت اخراج [کند] نقطهٔ ه̅ دریابد باز ه̅ ز̅ ز̅ ح̅ هر یک مثل ا̅ ج̅ بود. خط ج̅
ح̅ بپیوندد و از زاویهٔ ک̅ خط ک̅ ل̅ موازی ج̅ ح̅ اخراج کند و نقطهٔ ل̅ دریابد مقصود
حاصل شود والله اعلم.

Construction 33 (fol. 189v [33])[1]

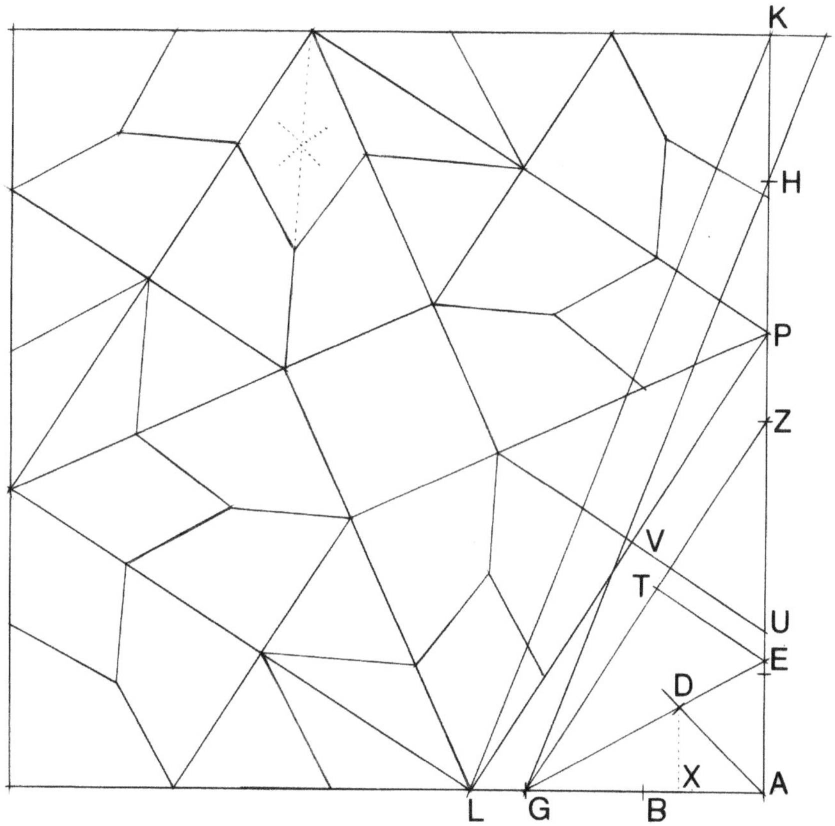

Pl. 33.1. Construction of a square with two rotated squares and four kites subdivided into smaller units.

This construction is related to Constructions 30, 31, and 32. As in the previous examples, its proportionally related components are determined by way of approximation. The four kites are here divided by their diagonals into two congruent triangles, each triangle containing a triangle and a kite, while each kite is further subdivided into three quadrangles. Özdural observes that in this case the angle of rotation is closest to the correct solution. In order to understand the gist of this construction, it is crucial to construct kite AGTE in the lower right corner, on the base of a right-angled triangle. Line GZ is parallel to line LP of the rotated intermediary square, within which is a smaller rotated central square, whose four extended sides touch the four vertices of the rotated intermediary square.

1 Özdural, "Analyses" (unpublished chapter), 134–35 (labeled as "Construction 36" by Özdural).

Translation (fol. 190r [34])*

Fol. 190r [34]

We construct the proportion of this repeat-unit with a T-square (*gūnyā miṣtar*). We say that in this repeat-unit (*ʿaqd*), which is defined by the compartment (*khāna*, lit., cell) AB[G]D, it is required that kite (*ṭurunj*) ABZT be the same as kite DGEH, in such a way that BZ equals GE, and EZ is shared by both. Thus, BE equals GZ. The other [condition] is that, since in kite EGDH the side EG equals GD and EH equals HD, by necessity, angle E equals angle D and will be right [-angled]. And EZ is equal to EH. As this preliminary (*muqaddima*) is known now, let us assume that the side GD of the repeat-unit's compartment (*khāna-i ʿaqd*) is known, but the indefinitely extended side DB is unknown [that is, DB is drawn but point B has not been defined]. Then we take the ruler and [on the edge of it we mark point D′ as the intersection of the ruler's two perpendicular edges]. From point D′ on the ruler, with any compass opening we want, we mark magnitudes D′K′ and D′T′ equal to each other. Then, in the same manner that we have marked the magnitudes on the ruler, we mark, on the repeat-unit's compartment, from angle D′ points M′ and N′ on the sides DG and DB. We again take the ruler and position the letter T′ on the perpendicular arm at the letter N′, so that both points correspond to each other. Then we give motion to the ruler pivoted on this point from left to right until [the position is reached at which] magnitudes S′D′ and K′O′ on either side of the ruler become equal. [While doing this,] point T′ should never be separated from point N′. [At this position] we draw line S′O′. From point G′, which is known, and parallel to S′O′, we draw line GB [to define the repeat-unit's rectangle]. We divide line BG by the proportion (*nisbat*) of S′D′, D′K′, and K′O′. On the ruler, which is parallel to GB, magnitude S′D′ is equal to K′O′, and K′D′ is equal to D′T and to D′T′. Therefore, on line GB magnitude BE will be equal to ZG, and ZE equal to EH and to HD. This is what is required. God knows best.[a]

However, we have found a construction by approximation. When we divide the right angle into 9 equal parts, four parts of it will constitute angle AGB, and 5 parts will constitute angle BGD. This is a very close approximation.[b]

a The text is composed of fifteen horizontal lines on the lower part of the page, under the figure.

b Annotation: Four oblique lines, from lower right to upper left, on the right side of the figure.

* For the Persian transcription, turn to page 268.

Construction 34 (fol. 190r [34])[1]

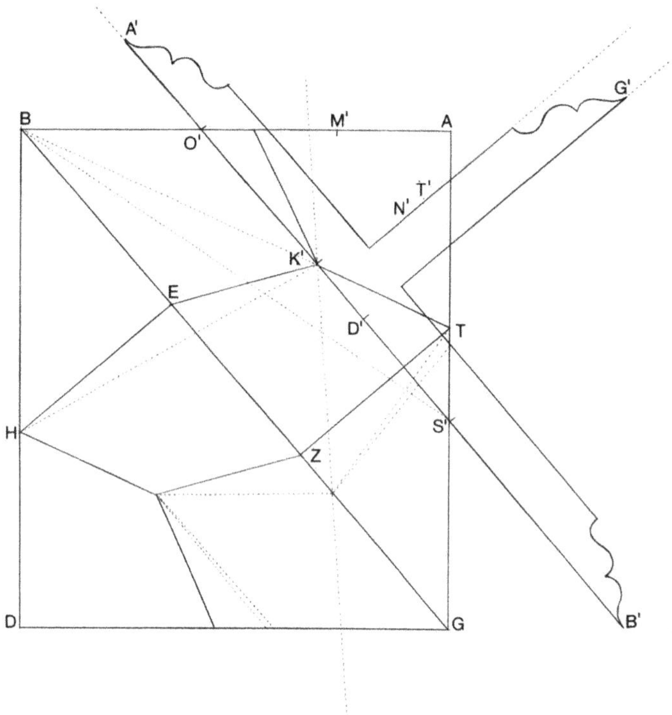

Pl. 34.1. Construction with a T-square of congruent kites along the diagonal of a rectangle.

This construction resolves the problem previously dealt with in Constructions 28, 29, 30–31, and 33, which were based on approximations. This one is solved with a tool devised to aid constructions involving conic sections. Construction 34 concerns the repeat-unit dealt with intermittently in Constructions 16–18, 26, 27, and 32. Like Construction 16, it is based on a verging construction, but the moving ruler is replaced here with a "T-square" (*gūnyā misṭār*) shown laid over the pattern. Özdural observes that the letters mentioned in the detailed textual commentary do not always match those designating points in the figure (those of the figure are drawn in black ink, with points of the verging procedure rendered in red ink). The undecipherable uninked incised lines led him to infer that the fifteenth-century copyist was perplexed while trying to redraw this construction.

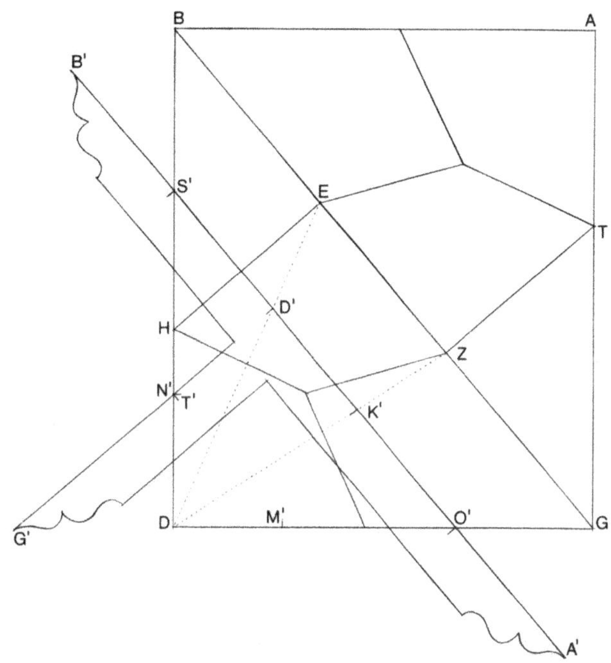

Pl. 34.2. Özdural believed that the T-square was misplaced in the original drawing. He placed it differently, distinguishing between the two sets of letters by adding a prime symbol (as in "D′") over the red letters. The text indicates that "kite" (*turunj*) ABZT should be congruent to kite DGEH, in addition to embodying other proportional correspondences. The shorter side of the rectangle (GD) is assumed as given, and the problem is reduced to finding its longer side. As in the verging construction of the same repeat-unit (pl. 16.2a) the exercise amounts, in modern terms, to solving the cubic equation $x^3 + 2x^2 = 2x + 2$.[2]

1 Özdural, "Analyses" (unpublished chapter), 135–43. (labeled as "Constructions 37 and 38" by Özdural).
2 Özdural, "Analyses" (unpublished chapter), 139.

Transcription (fol. 190r [34])*

عمل نسبت این عقد بگونیا مسطر گوئیم که درین عقد که در خانۀ آ بٓ دٓ است می‌باید که طرنج آ بٓ زٓ طٓ مثل طرنج دٓ جٓ هٓ حٓ باشد. پس چون چنین باشد بٓ زٓ مثل جٓ هٓ بود و هٓ زٓ مشترک میان هر دو. پس بٓ هٓ مثل جٓ زٓ بود. و دیگر چون در طرنج هٓ جٓ دٓ حٓ ضلع هٓ جٓ مثل جٓ دٓ و هٓ حٓ مثل حٓ دٓ باشد بضرورت زاویۀ هٓ مثل زاویۀ دٓ بود و قائمه باشد و هٓ زٓ مثل هٓ حٓ است. پس چون این مقدمه معلوم شد فرض کنیم که از خانۀ عقد ضلع جٓ دٓ معلوم است و ضلع دٓ بٓ مالایتناهی مجهول. آنگه مسطره برداریم و از نقطۀ دٓ بر مسطره مقدار دٓ کٓ و دٓ طٓ بهر فتح که خواهیم متساوی نشان کنیم و حرف بزنیم. آنگه بهمین مقدار که بر مسطره نشان کرده ایم در خانۀ عقد از زاویه دٓ نقطه مٓ نٓ نشان کنیم بر ضلع دٓ جٓ دٓ بٓ. آنگه مسطره برداریم و حرف طٓ که بر طرف قایمه است بر حرف نٓ نهیم منطبق نقطه بر نقطه. آنگه برین نقطه مسطره‌را از چپ و راست حرکت میدهیم چندانک از طرف مسطره مقدار سٓ دٓ مثل کٓ عٓ گردد و نقطۀ طٓ از نقطۀ نٓ مفارقت نکند. آنگه خط سٓع بکشیم و در سطح از نقطۀ جٓ که معلوم است خط جٓ بٓ موازی سٓ عٓ بدر بریم و خط بٓ جٓ نسبت سٓ دٓ دٓ کٓ کٓ عٓ قسمت کنیم که چون بر مسطره که موازی جٓ بٓ مقدار سٓ دٓ مثل کٓ عٓ است و کٓ دٓ مثل دٓ طٓ طٓ دٓ است. از خط جٓ بٓ مقدار بٓ هٓ مثل زٓ جٓ بود و هٓز مثل هٓ حٓ و حٓ دٓ و مقصود اینست والله اعلم.

اما عملی بتقریب یافته‌ایم که هرگاه که زاویۀ قائمه‌را به ۹ قسم متساوی کنیم چهار قسم ازان زاویۀ آ جٓ بٓ باشد و ۵ قسم زاویه بٓ جٓ دٓ بود و این غایت تقریب است.

* For the English translation, see page 266.

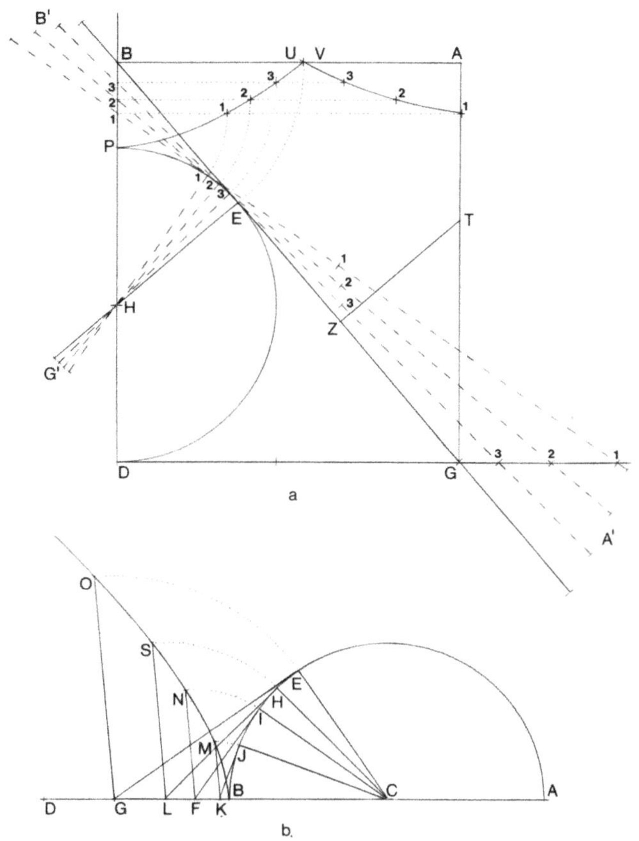

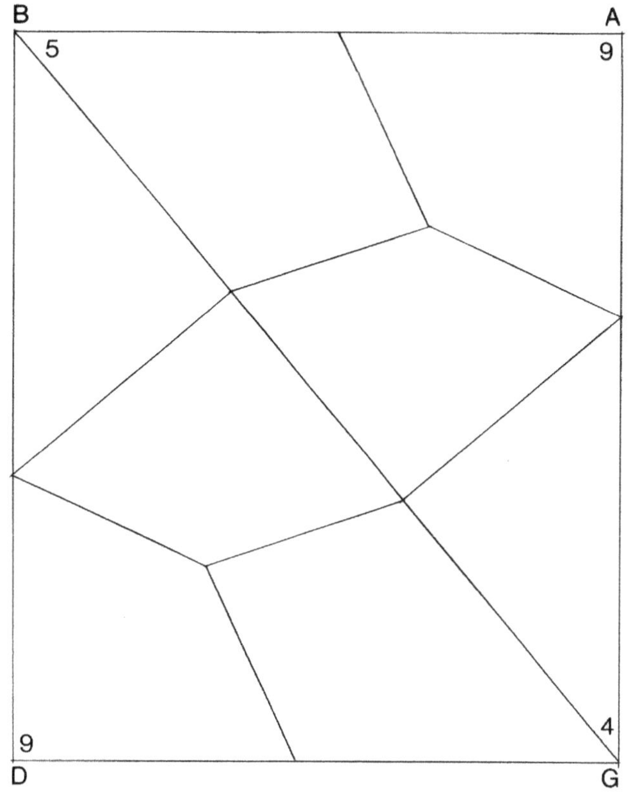

Pl. 34.3a. Construction 34 is a more practical and correct version of Construction 16. The rotational movement of the T-square reveals, according to Özdural, how a construction by means of the T-square can be the direct equivalent of a solution by means of conic sections, in this case two intersecting hyperbolas. It involves the use of tangents to a semicircle (PED), by means of which the equilateral hyperbola is drawn.

Pl. 34.3b. This way of drawing was not invented by the author of Construction 34, who may have borrowed it from Ibrahim ibn Sinan (d. 946), the author of a treatise on drawing the three conic sections.[73] Özdural speculates that the author of the construction must have been a mathematician familiar with the theory of conic sections, which he applied to the practical problem of a construction by means of the T-square.

Pl. 34.4. Özdural adapted this construction from pl. 34.1, based on a short text written in diagonal lines next to the figure corresponding to Construction 34. It states that a construction by approximation was found, which involves the division of the right angle of this pattern's repeat-unit into nine equal parts. Four parts constitute angle AGB and five parts form angle BGD, with each angle module representing 10°. Özdural hypothesizes that this "anticlimactic" approximation was discovered incidentally, perhaps by the fifteenth-century copyist, who had been perplexed by the preceding construction (see pl. 34.1).

3 Edited and translated into French in Ibn Sinān 2000, 292–95. See Özdural, "Analyses" (unpublished chapter), 140.

Translation and Transcription (fol. 190v [35])

Fol. 190v [35]

Section: The procedure of producing this repeat-unit is as follows: On line AB erect two perpendiculars, AL and BM, parallel to each other. Draw from angle A a line [number] 4, and from angle B a line [number] 6, so that they intersect each other at point G. Then, from point A with compass opening AG, mark point D on side AB. From point B with compass opening BD, mark point E on line BE. Join lines GD and GE. On line GE construct heptagon GETZH, such that line GE becomes its base. From point H draw line HK parallel to AL, so that it is equal to EI, that is to say, to half of ED. From point K draw line LKM parallel to AB. Thus, the construction is completed. God knows best.[a]

If one wants a tetragonal star (*khātim-i chahār*) to also be included in the repeat-unit, then from point T draw a line [number] four at the two points M on perpendicular BM and draw line ML parallel to AB. God knows best.[b]

---

a The text is composed of two parts. This first part is written in six horizontal lines on the lower part of the page.

b The second part of the text is composed of two vertical lines, from bottom to top, written along the edge of the figure.

فصل طریق اخراج این عقد آنست که بر خط آ بَ قائمتین آ لَ بَ مَ موازی یکدیگر اخراج
کند و از زاویه آ خط چهار و از زاویهٔ بَ خط ۶ اخراج کند تا متقاطع شوند بنقطهٔ جَ.
آنگه از نقطهٔ آ بفتح آ جَ بر ضلع آ بَ نقطهٔ دَ نشان کند و از نقطهٔ بَ بر خط بَ هَ نقطهٔ
هَ نشان کند بفتح بَ دَ و خط جَ دَ جَ هَ بپیوندد و بر خط جَ هَ مسبع جَ هَ طَ زَ حَ عمل
کند چنانک خط جَ هَ قاعدهٔ او باشد و از نقطهٔ حَ خط حَ كَ موازی آ لَ اخراج کند
چنانک مثل هَ یَ باشد یعنی نصف هَ دَ و از نقطهٔ كَ خط لَ كَ مَ موازی آ بَ بکشد
که عمل تمام شود والله اعلم.

و اگر خواهد که خاتم چهار نیز درین عقد باشد از نقطهٔ طَ خط ۴ بر دو نقطهٔ مَ بگیرد
بر قائمه بَ مَ و خط مَ لَ موازی آ بَ بکشد والله اعلم.

Construction 35 (fol. 190v [35])[1]

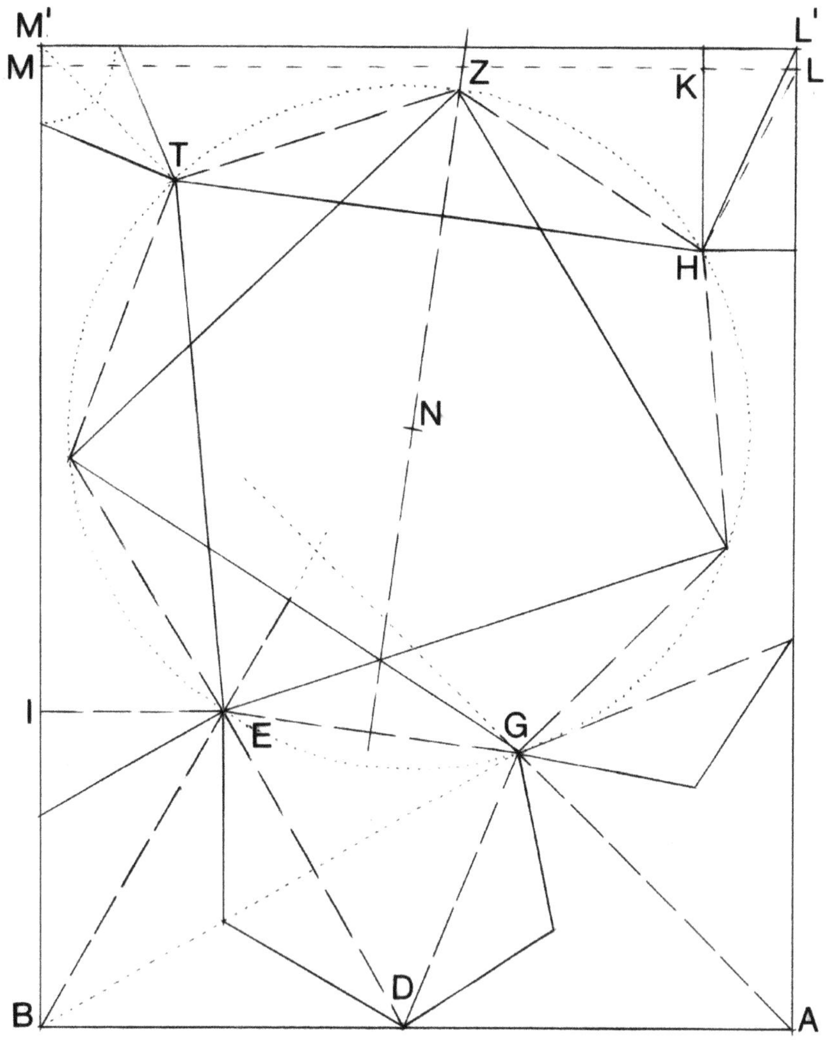

Pl. 35.1. Construction of a rectangle combining hexagonal, heptagonal, and octagonal stars.

The text of this repeat-unit explains that the aim is to obtain a heptagonal star two of whose vertices coincide with the vertices of a quarter-octagonal star (with center A) and of a quarter-hexagonal star (with center B). The line AG, referred to as a "line [number] 4" (which delineates 45°) intersects with line BG, referred to as "line [number] 6" (which defines 30°). These lines determine the angular point of the quarter-octagon in the lower right corner. The side of the octagon (GD) not only is equal to the side of the heptagon (EG), but is also correlated with the vertex of the hexagonal star (defined by BE). Thus, all three polygons and their stars are, in an ingenious way, proportionally correlated with one another, even though they geometrically belong to different categories.

The author of the textual commentary suggests another option: If one wants, a "tetragonal star" (*khātim-i chahār*) can be added to this repeat-unit from point T. This requires drawing a "line [number] four" (corresponding to 45°) at two points (M) on perpendicular BM and drawing line ML parallel to AB.

1 Özdural, "Analyses" (unpublished chapter), 143–47 (labeled as "Construction 39" by Özdural).

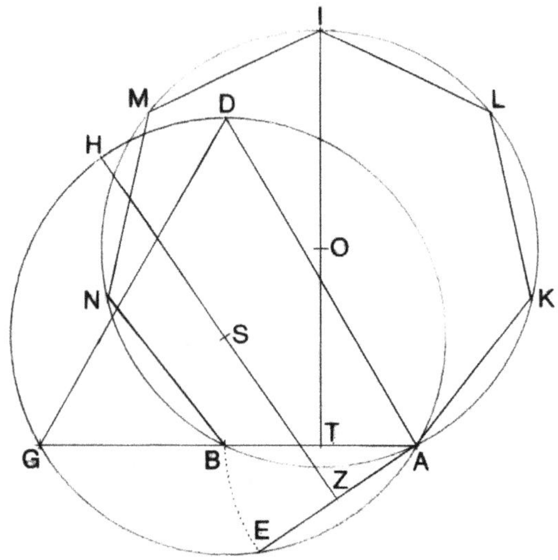

a

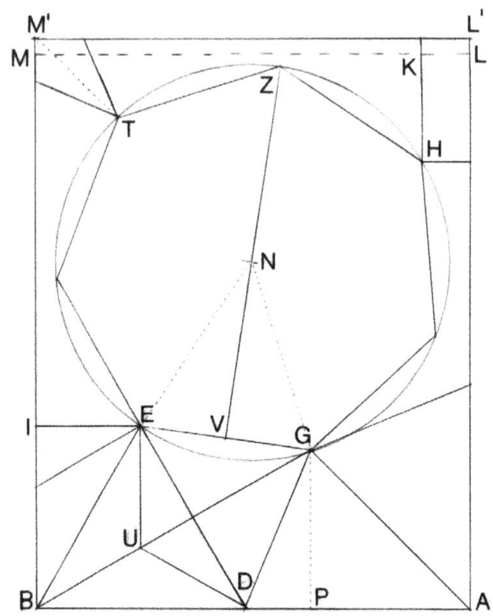

b

Pl. 35.2a. Abu'l-Wafa' proposed an approximation method for constructing the regular heptagon within a circle (see pl. 12.2d), which in reverse order "answers exactly" the case of Construction 35. However, other than the resemblance of the perpendicular (TI) in his construction to the perpendicular in pl. 35.1 (on the midpoint of GE) and the circle circumscribing the heptagon, there is nothing else to suggest that his method was applied here.

Pl. 35.2b. Construction lines indicate that the circumference of the circumscribing circle was divided into seven arcs equal to GE. The center of the circle (N) was found from points G and E with a compass opening equal to GU. Point G is the common vertex of the heptagon and octagon, while point U defines the concave angular point of the hexagonal star.

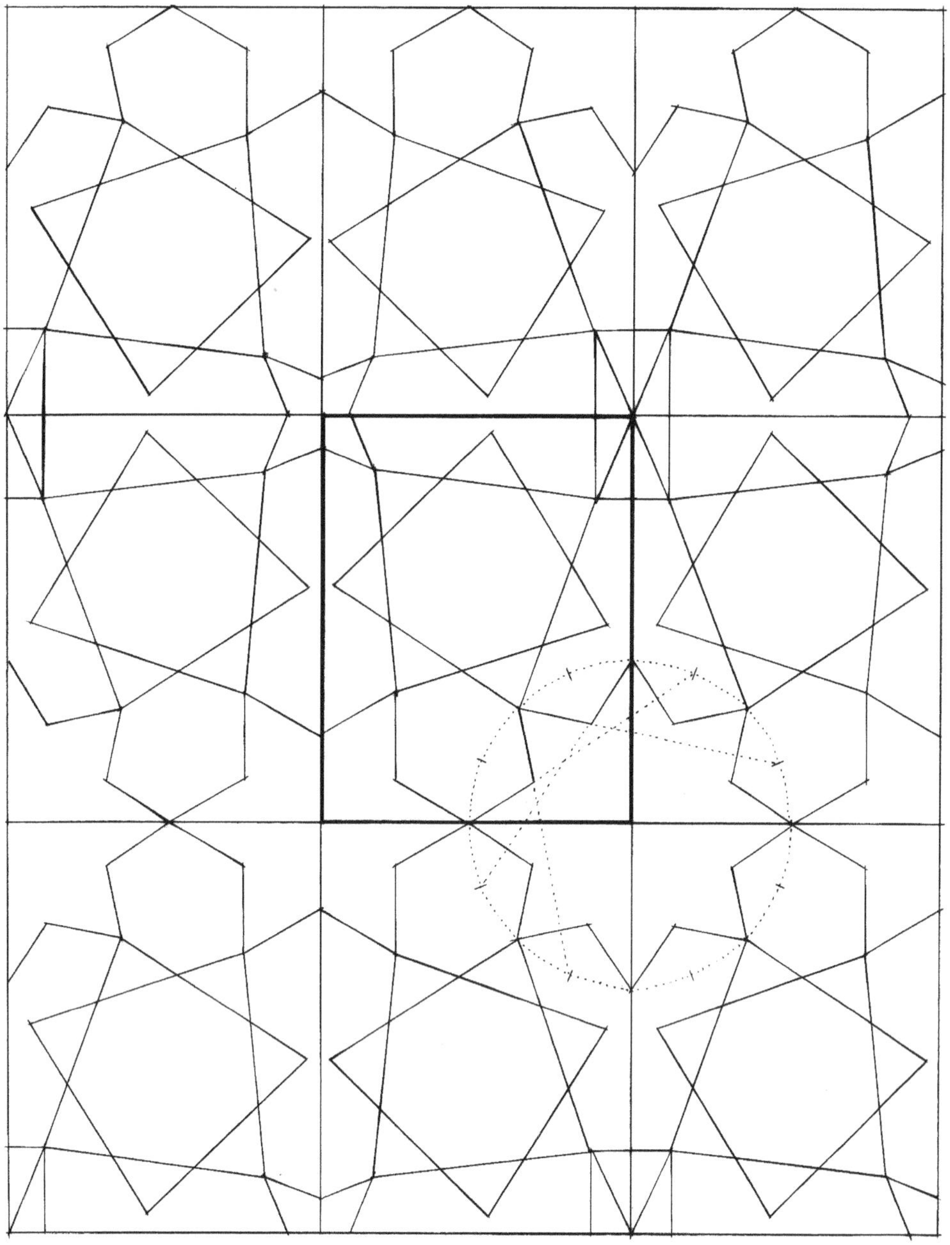

Pl. 35.3. In Construction 35.1 the central heptagon is married with the two polygons in the lower right and left angles of the rectangular repeat-unit. Realizing that the upper corners have been treated insufficiently, the author attempts to compensate for this imbalance by mentioning the option of adding a "tetragonal star" (*khātim-i chahār*, lit. tetragonal seal) to this repeat-unit. Özdural's drawing shows an unattractive pattern obtained by the repetition of the repeat-unit, featuring tetragonal stars fitted in the interstices of regular octagons. He assumed that the repeat-unit was meant for an ornamental design, rather than being an exercise that explores the methods for connecting the vertices of stars with differing numbers of points. Thus, the pattern generated by repeating Construction 35, which integrates four different types of star polygons as a geometrical exercise, is not, according to Özdural, "a great accomplishment in terms of the ornamental arts."

Translation and Transcription (fol. 191r [36])

Fol. 191r [36]

The proportion of this repeat-unit is also [derived] from "conics" (*makhrūṭāt*). In it the objective is to construct a right-angled triangle of which the altitude plus the shorter side is equal to the hypotenuse. Ibn-i Haytham [i.e., Ibn al-Haytham] has composed a treatise on the construction of this triangle, and his construction is, by means of conic sections, a hyperbola and a parabola. However, the objective can be achieved here with the aid of this T-square. As we said in the preceding "preliminary" (*muqaddima*), the objective of our repeat-unit is four right-angled "pinecone figures" (*shakl-i ṣanawbarī*) surrounding a right-angled equilateral and equiangular quadrilateral [i.e., square], like the "pinecone-like quadrilaterals" (*murabbaʿ-i ṣanawbarī*) AIHK, GHMN, DMLS, and BLKO that surround the square KHML. Now, as angle H of the square and both [angles] of the [pinecone] figure[s] are right angles, by necessity lines K[G] and HD are straight. Thus, triangle AKG is right-angled and the same as triangle GHD. Since this triangle is right-angled, it is inscribed in a semicircle. Then point H has to be sought on arc GE.[a] Subsequently, when we have angle T of the ruler at a right angle, and its edge AB straight, [then] side AB of the square and of the ruler will coincide with each other. God knows best.[b]

a Text: AE.
b The text is composed of eleven horizontal lines below the figure.

نسبت این عقد نیز هم از مخروطاتست که در اینجا مقصود عمل مثلثیست قایم الزاویه
که عمود آن مثلث با ضلع اقصر منضم مثل وتر زاویه بود و ابن هیثم در عمل این مثلث
رسالهٔ گفته است و عمل او در نسبت قطاع مخروطات زاید و مکافی گفته اما اینجا ازین
مسطره گونیا مقصود حاصل می‌توان کرد چنانک گوئیم در مقدمهٔ اول مقصود این عقد مارا
چهار شکل است صنوبری قایم الزاویه[۱] محیط بر یک مربع متساوی الأضلاع قایم الزاویه
مثل مربع صنوبری آ ی̄ ح̄ ک̄، ج̄ ح̄ م̄ ن̄، د̄ م̄ ل̄ س̄، ب̄ ل̄ ک̄ ع̄ محیط بر مربع ک̄ ح̄
م̄ ل̄. اکنون چون زاویهٔ ح̄ در مربع و شکل هر دو قایمه است بضرورت خط ک̄ [ج̄] ح̄ د̄
مستقیم باشد و مثلث آ ک̄ ج̄ قائمة الزاویه بود و مثل مثلث ج̄ ح̄ د̄ باشد، و این مثلث
چون قائمه است در نصف قوس واقع شود پس نقطهٔ ح̄ بر قوس ج̄[۲] ه̄ طلب باید کرد. پس
هرگاه که مارا در مسطرهٔ زاویهٔ ط̄ قایمه باشد و طرف آ ب̄ مستقیم و منطبق بر ضلع آ
ب̄ از مربع و مسطره‌را والله اعلم.

۱ قایم الزاویه] قایم الزاویتین
۲ ج̄] آ

CONSTRUCTION 36 (fol. 191r [36])[1]

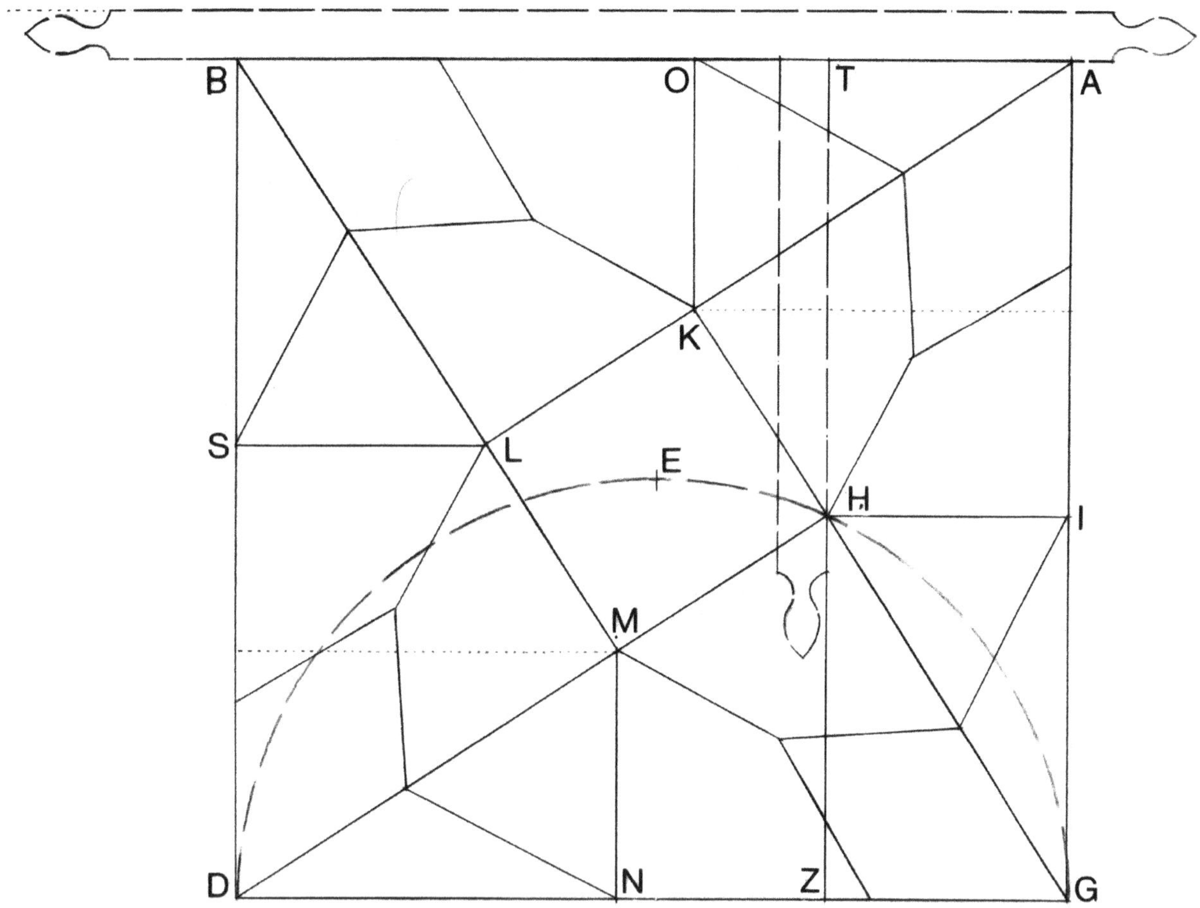

Pl. 36.1. Construction with a T-square of a square containing a rotated central square surrounded by four right-angled triangles, which are divided into four kites, triangles, and smaller components.

As in Construction 34, a T-square is used in the construction of this pattern. It is placed along the upper edge of a square frame. The T-square and auxiliary construction lines are drawn in red ink, so as to distinguish them from the main pattern in black ink. The text explains that the proportion of the repeat-unit is derived from conics, and involves the construction of a special right-angled triangle on which Ibn al-Haytham had composed a treatise. The object is to obtain, by means of the T-square, four kites (pinecone figures, *shakl-i ṣanawbarī*) surrounding the rotated central square KHML. According to Özdural, this repeat-unit, previously drawn with approximation methods in Constructions 28, 29, 30–31, and 33, reaches its "climax" here.

1 Özdural, "Analyses" (unpublished chapter), 147–65 (labeled as "Constructions 40, 41, and 42" by Özdural).

Translation and Transcription (fol. 191v [36])

Fol. 191v [36]

The true nature of the proportion of this [preceding] repeat-unit belongs to conics, [but] we can draw it with the aid of a tool called a T-square (*gūnyā misṭar*), which is a tool by means of which many difficult repeat-units formed by conics can be drawn. This is presently the deduction of the scribe (*kātib*); whether or not it is so is not known. That is to say, one makes the ruler (*misṭara*) in the same way as the alidade of an astrolabe and at the middle of it erects a perpendicular ruler (*misṭara*) similar to the vertex of the alidade of the boat astrolabe (*usṭurlāb-i zawraqī*). They call this the [right-angled] ruler of the set square (*siṭāra-i gūnyā*),[a] as in the ruler ABGD, which consists of ruler AB and the perpendicular [leg] GD. It is necessary to give the [horizontal] ruler the shape of a trapezium, with AB as the longer side [and ZE its shorter side], corresponding to the trapezoidal shape of the alidade of an astrolabe. Similarly, the perpendicular (arm) GD should be in the shape of a trapezium with the longer side GD, as in the way of the clearly defined [shape of] the alidade. Point D at the end of the alidade should be distinctly apparent and angle GDA should be a very correct right angle, so that by means of this ruler many extraordinary, difficult proportions can be obtained. First, the procedure for constructing the proportion of this repeat-unit:[b]

---

a The term *siṭāra* is given in the *Burhān-i qāṭiʿ* dictionary as an "implement for drawing straight lines, long and narrow about two fingers wide, made of steel, wood, or bone, etc., and in Arabic it is called *misṭar*." Hence, it is the Persian word for a *misṭar*.

b The accompanying text is composed of ten horizontal lines in the upper part of the page.

حقیقت نسبت این عقد از مخروطاتست و به آلتی که آنرا گونیا مسطر خوانیم توان کشیدن
و آن آلتی باشد که بسیار عقود مشکل مخروطاتی بدین آلت توان کشید و این حالیا
استنباط کاتب است اگر بوده است معلوم نیست و آنچنان باشد که مسطرهٔ بسازد مانند
عضادهٔ اسطرلاب و بر وسط آن مسطر قایمه راست کند مانند سهم عضاده اسطرلاب
زورقی، و این‌را سطارهٔ گونیا خوانند مثل مسطرهٔ آ ب ج د که از وی مسطرهٔ آ ب بود
و قائمه ج د و باید که از مسطره طرف آ ب منحرف بود. چون انحراف عضادهٔ مجیَّب و
هم چنین از قائمهٔ طرف ج د منحرف باشد و از قائمه خط ج د بر روی عضاده مبین باشد
و نقطه د بر طرف عضاده معین و زاویهٔ ج د آ قائمه عظیم درست یابد که بدین مسطره
بسیار نسبتهاء غریب مشکل توان کرد. اول در طریق عمل نسبت این عقد.

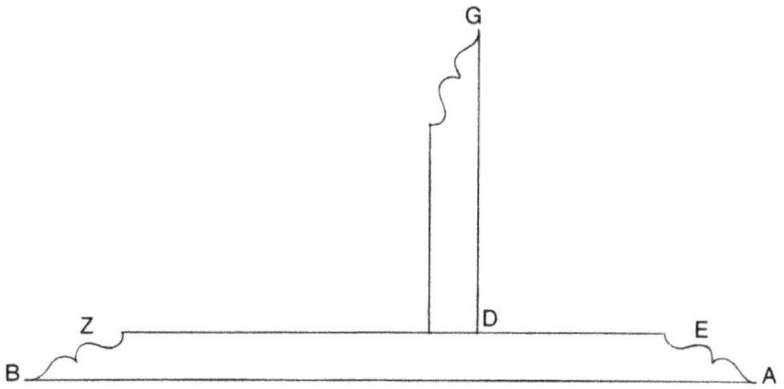

Pl. 36.1′. Construction of the T-square.
Özdural's redrawing of the T-square described verbally and graphically on fol. 191v [36]. Used in Constructions 34 and 36, this drafting tool must have attracted the attention of mathematicians and artisans alike.

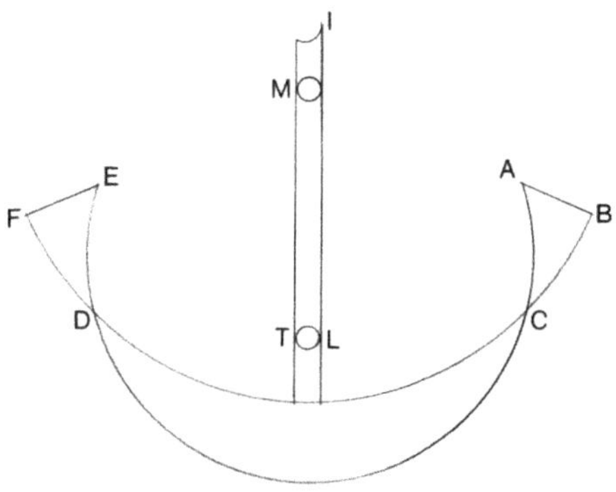

Pl. 36.3. The textual description compares the perpendicular upper ruler (alidade) of the T-square to that of the "boat astrolabe" (*usṭurlāb-i zawraqī*). Özdural points out that the astrolabe developed by al-Sijzi (ca. 980) belonged to this category. The drawing here is based on an extant undated Indian "boat astrolabe."[2]

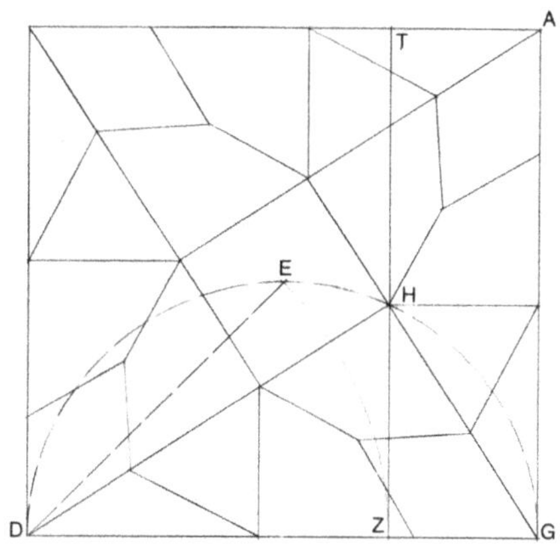

Pl. 36.2. Özdural's redrawing of Construction 36 (see pl. 36.1), in which he suggests that the construction lines in red ink work out an alternative approximation method for the same repeat-unit, shown graphically on the already drawn figure. He suggests that an artisan possibly added this "easy but sufficiently accurate approximation," as in the contribution of an artisan to Construction 27 (see pl. 27.2). The approximation here is based on the observation that diagonal DE corresponds closely to segment DZ, which is cut off from side DG by the extension of perpendicular TH in the preceding construction (see pl. 36.1).

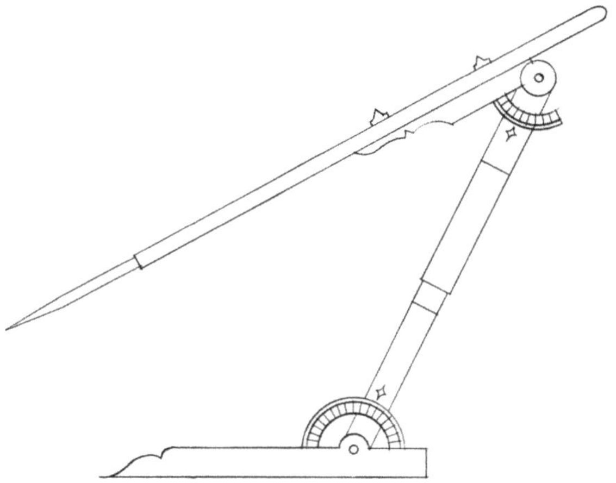

Pl. 36.4. The "complete compass" or "perfect compass," which could draw conic sections and other lines, was described by such celebrated tenth-century mathematicians as al-Kuhi and al-Sijzi.[3]

---

2 Discussed in Özdural, "Analyses" (unpublished chapter), 156. The figure here is adapted from fig. 112 in Gunther 1932, 218. An undated Indian example of an extant "boat astrolabe" is illustrated in Gunther 1932, pls. XLVIII and XLIX. Özdural cites al-Sijzi's treatise on the construction of the astrolabe in Istanbul, Topkapı Sarayı Library, Ms. A. 3342/9.

3 These treatises on the "perfect compass" are published in Woepcke 1874. See also Rozenfel'd and Iushkevich 1996, 458–59.

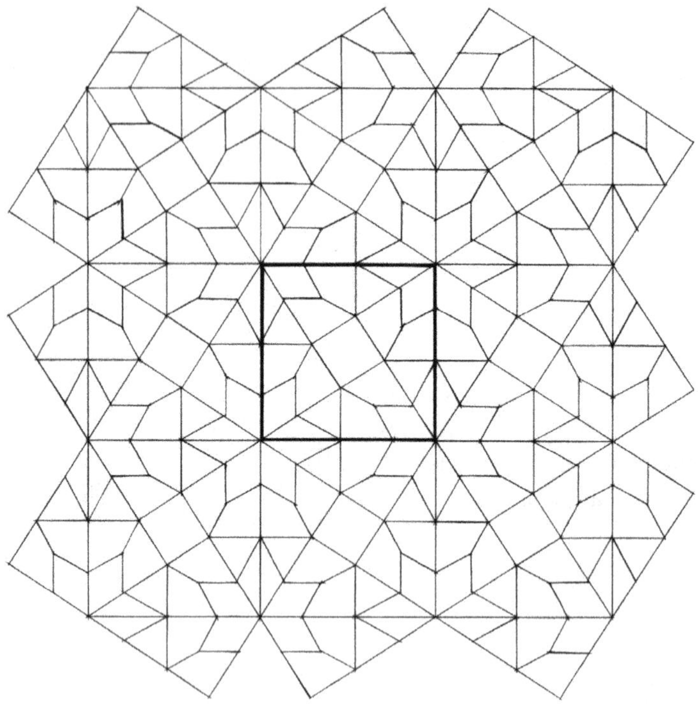

Pl. 36.5. Replicating the repeat-unit generates this geometric scheme, which is, according to Özdural, a simpler version of Abu'l-Wafa''s solution for composing a square from several squares (see pls. 2.4a–d, 2.5a–c).

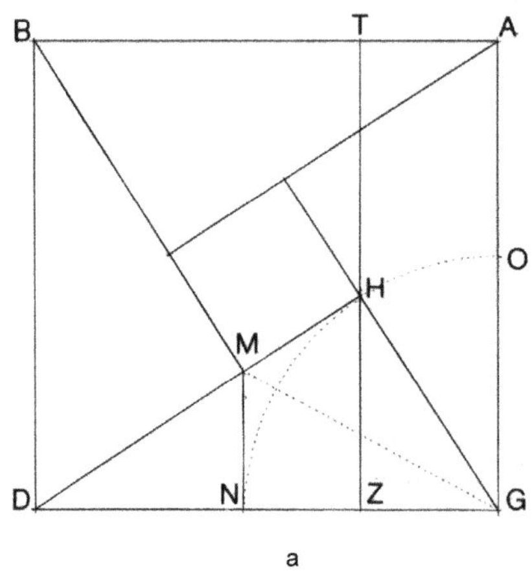

a

Pl. 36.6a. The previous four solutions of this special pattern, composed of four rotating triangles around a rotated central square, were based on approximations. In Construction 36, the problem was approached from a new perspective: how to construct a right-angled triangle (GDH) so that the altitude (HZ) plus the shortest side (HG) is equal to the hypotenuse (GD).[4]

---

4 See Özdural, "Analyses" (unpublished chapter), 151, where he attributes the triangle to ʿUmar Khayyam; he suggests it was misattributed to Ibn al-Haytham by the scribe: Omar Khayyām 1963a, 326; French translation in Omar Khayyām 1981, 76. On Özdural's questionable attribution, see the chapters by Hogendijk and Necipoğlu in this volume.

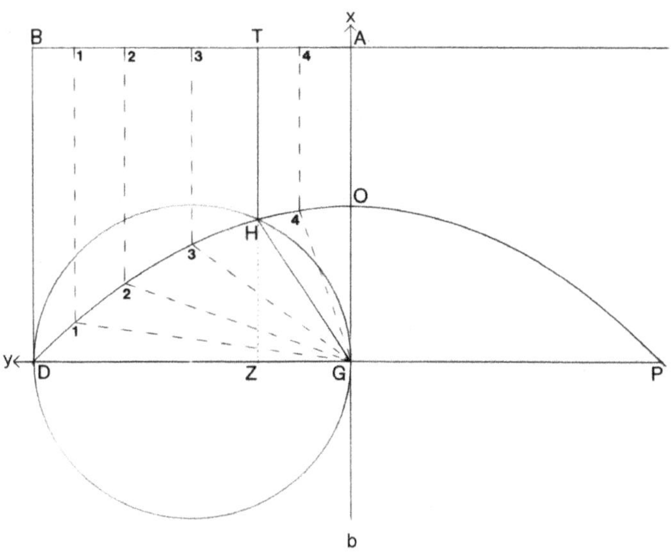

Pl. 36.6b. This drawing analyzing Construction 36 geometrically shows that the pattern is "a shortcut translation of a solution by means of conic sections," involving a parabola and semicircle that intersect at point H. The solution of the problem corresponds to, in modern terms, a cubic equation.

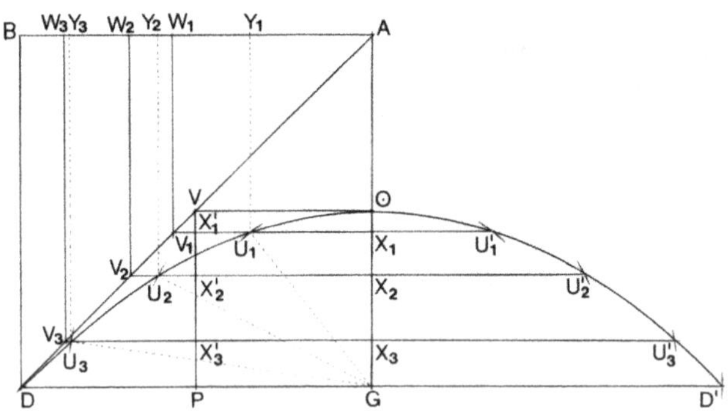

a

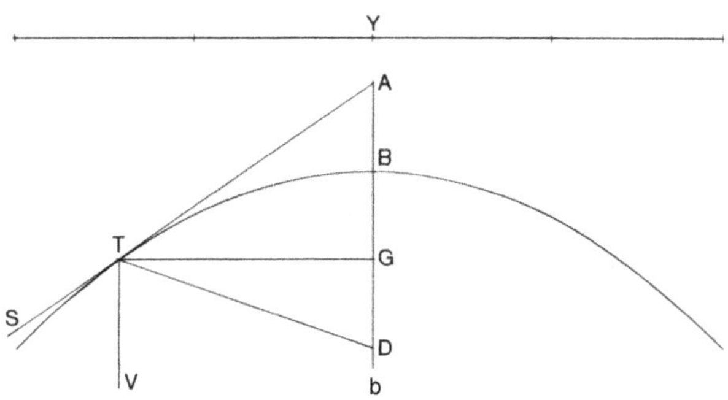

b

Pl. 36.7a. Construction of the parabola attributed to Diocles (ca. 190–180 B.C.), whose treatise was probably translated into Arabic in the tenth century for al-Kuhi. It is similar to that of Construction 36.[5]

Pl. 36.7b. Ibn al-Haytham provides correct proofs, in response to the "unsatisfactory" proofs given by the ancients, for the focal property of the parabola in his treatise on parabolic mirrors. This indicates his familiarity with Diocles's work.[6] In Construction 36, the sliding movement of the T-square translates properties of the parabola into an instrumental construction. Özdural attributes Construction 36 to the author of Construction 34, who used the same tool for constructing the hyperbola.

5 See Toomer, in Diocles 1976, 1–17, 62–66.
6 See Diocles 1976, 18–23.

Translation and Transcription (fol. 192r [37])

Fol. 192r [37][a]

Make angle BAG three-sevenths of the right angle. Bisect line AG at point D. Cut off BE equal to AD. Draw line EZ parallel to AG. Draw line TI parallel to BE. Bisect TE at point H and mark [point I so that] TI is equal to TH. Extend line EI until it intersects line AB at point K. Draw KL parallel to BE. From the center Z, inscribe arc KMN, such that [it intersects line EZ at point M and] the arc KM becomes equal to the arc MN. Extend LN[b] until point S is obtained on line AF. This becomes the center of [the circle that circumscribes] the heptagon, and [the rest of] the construction should be easy, God, may He be exalted, willing. Alternatively, angle ELN[c] is constructed equal to angle ELK, and the center is taken on line LNS. Alternatively, EO is cut off equal to EL, so that point O becomes the center of the heptagon. Then line OS becomes parallel to GA and equal to GA,[d] and again point S becomes the center of the other heptagon, and also GO equals AS. God knows best.

---

a The textual commentary for Construction 37 is continuous with that of the T-square illustrated on the preceding page (fol. 191v [36]). The text for the figure on the upper part of this page is composed of nine horizontal lines written below the figure; part of the first sentence is on the previous page (i.e., fol. 191v [36], where the last line is incomplete and reads: "First, the procedure for constructing the proportion of this repeat-unit:".).

b Text: KMN.

c Text: EAN.

d Text: AD.

زاويهٔ ب آ ج ثلثة اسباع قائمه عمل كند و خط آ ج تنصيف كند بنقطهٔ د و ب ه مثل آ د جدا كند و خط ه ز موازى آ ج برون برد و خط ط ى موازى ب ه بكشد و ط ه تنصيف كند بنقطه ح و ط ى مثل ط ح نشان كند و خط ه ى اخراج كند تا متقاطع خط آ ب شود بنقطهٔ ک و ک ل موازى ب ه بيرون آورد و از مركز ز قوس ک م ن بگرداند چنانک قسم ک م مثل م ن باشد. ک م ن اخراج كند تا بر خط آ ف نقطهٔ س بگيرد و اين مركز مسبع باشد و عمل آسان گردد ان شاء الله تعالى. و إلّا زاويهٔ ه ل[١] ن مثل زاويهٔ ه ل ک عمل كند و بخط ل ن س مركز بگيرد و إلّا ه ع مثل ه ل فصل كند كه نقطهٔ ع مركز مسبع باشد و خط ع س موازى ج آ و مثل ج آ و مثل ج آ[٢] و باز نقطهٔ س مركز مسبع آخر و ديگر ج ع مثل آ س باشد والله اعلم.

---

١ ل] آ

٢ ج آ] آ د

Construction 37 (fol. 192r [37])[1]

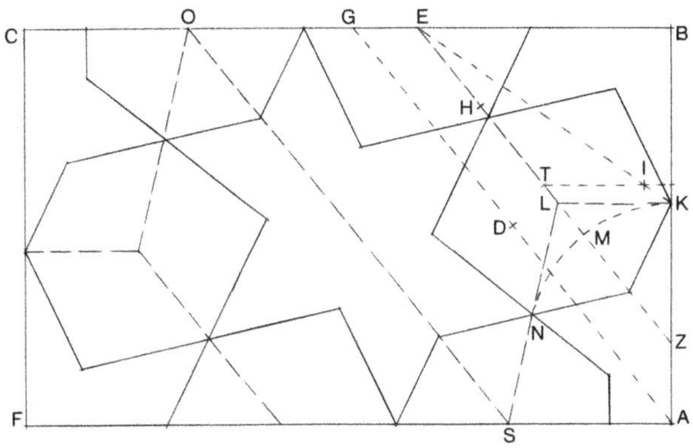

Pl. 37.1. Construction of a rectangle with regular heptagons, irregular hexagons, and two types of six-pointed stellate figures.

This construction is based on the division of the right angle into seven parts. Angle BAG is three-sevenths of the right angle that defines the lower right corner of the rectangular repeat-unit. Özdural suggests that a set square having 38.6°/51.4°-angles was probably used, given the absence of uninked incised lines.[2] Points S and O become the centers of half-heptagons in this figure, subdivided into two congruent quadrilaterals by the diagonal axis OS. The central portion of the pattern has an equilateral, but not equiangular, hexagon, drawn in broken lines, which circumscribes the central six-pointed stellate figure bisected by OS. Angles B and F, as well as C and A are occupied by two different types of quarter-six-pointed stellate figures, which interlock with irregular hexagons.

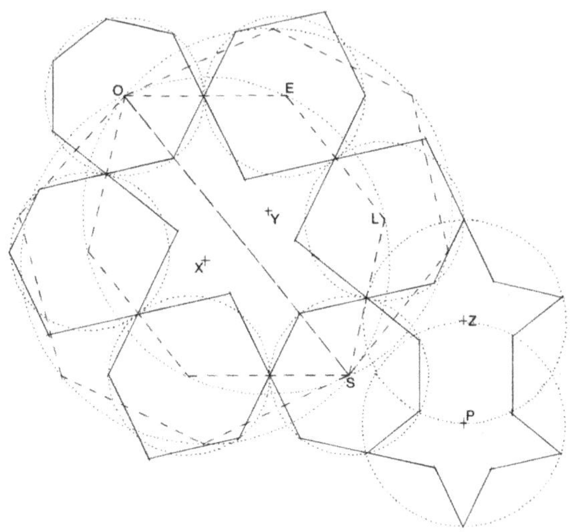

Pl. 37.2. It is only when the rectangular repeat-unit is expanded that the central six-pointed stellate figure, bisected by OS, becomes surrounded by two regular heptagons and four irregular hexagons. These polygons interlock with the two different types of six-pointed stellate figures.

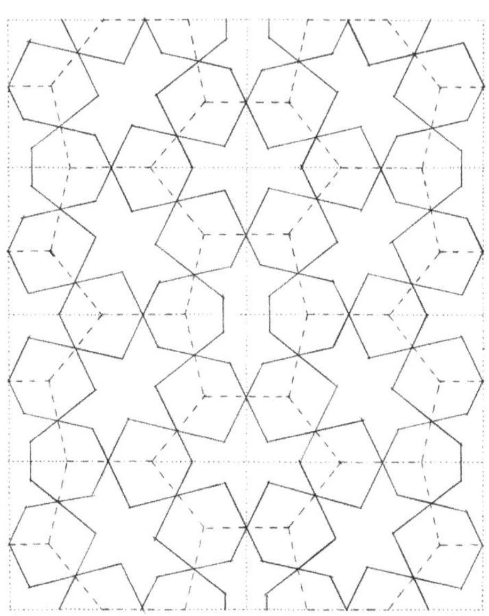

Pl. 37.3. Replication of the rectangular repeat-unit generates this ornamental composition. Two types of equilateral hexagons in contact, which are drawn in red ink (here in broken lines), form a double-layered scheme. These hexagons are superimposed over a network of two types of six-pointed stellate figures, irregular hexagons, and regular heptagons, which constitute the actual pattern in black ink. The lines in red ink (i.e., the aforementioned broken lines) highlight the underlying generative grid of the main pattern in black ink. The main pattern, from which these generative construction lines are to be omitted, is created by joining the midpoints of the hexagons.

1 Özdural, "Analyses" (unpublished chapter), 165–67 (labeled as "Construction 43" by Özdural).

2 See Necipoğlu's chapter in this volume for a different interpretation. She suggests that the T-square, which is depicted on the previous page (fol. 191v [36]), was probably used for this construction, since the first line of the textual commentary for this repeat-unit starts on the page that depicts the T-square.

Translation and Transcription (fol. 192r [38])

Fol. 192r [38]

No accompanying text.

Construction 38 (fol. 192r [38])[1]

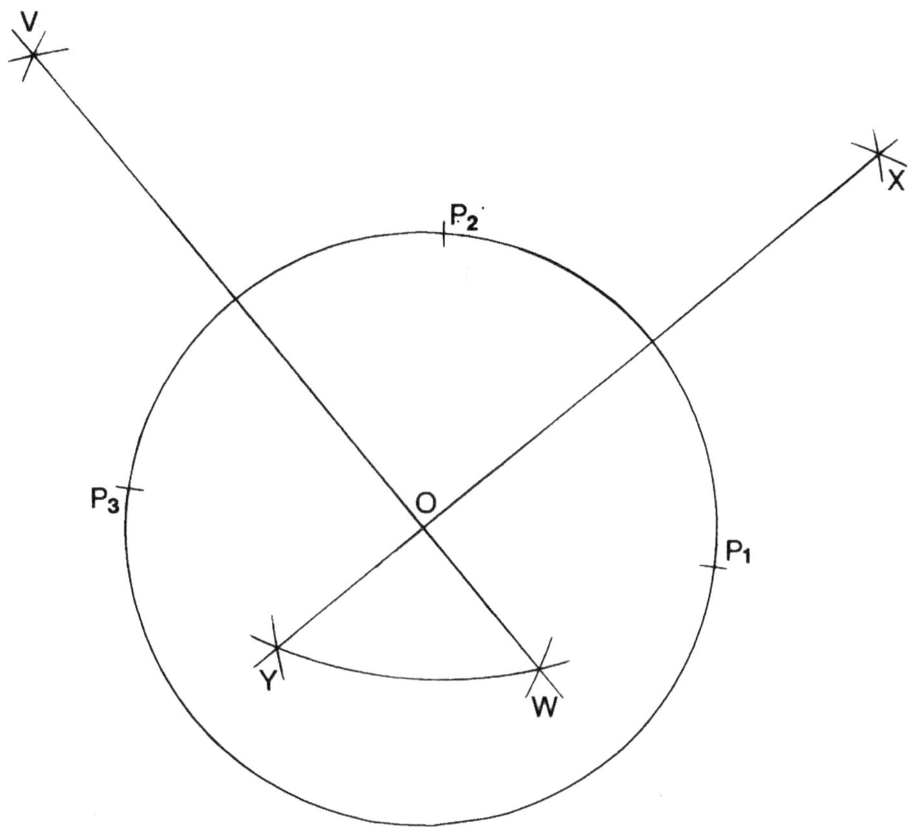

Pl. 38.1. Construction of a circle passing through three points not on a straight line.

This figure, which lacks an accompanying text, is drawn at the bottom of the page containing Construction 37, which precedes it. A construction widely known since ancient times, it describes a circle passing through three points (P1, P2, and P3) that are not on a straight line. Perpendiculars XY and VW are drawn so that they bisect the imaginary chords between P1, P2 and P2, P3. The intersection point O of these perpendiculars is the center of the circle, defined by the given points. The proof of this construction is given in Euclid, *Elements*, 4:5.[2]

1 Özdural, "Analyses" (unpublished chapter), 167–68 (labeled as "Construction 44" by Özdural).
2 Euclid 1956, 2:88–90: "About a given triangle to circumscribe a circle."

Translation and Transcription (fol. 192v [39])

Fol. 192v [39][a]

No accompanying text.

---

a This folio contains four drawings, Constructions 39, 40, 41 and 42, accompanied by laconic texts. The text written in two parts on the right sides of Constructions 39 and 40 has no relation to these figures; instead, it relates to Construction 41. The first part of the text accompanying Construction 40 is composed of two vertical lines, from top to bottom, and the figure breaks a sentence in the middle. The continuation of the second part of that sentence is written next to Construction 39, in three vertical lines from top to bottom.

Construction 39 (fol. 192v [39])[1]

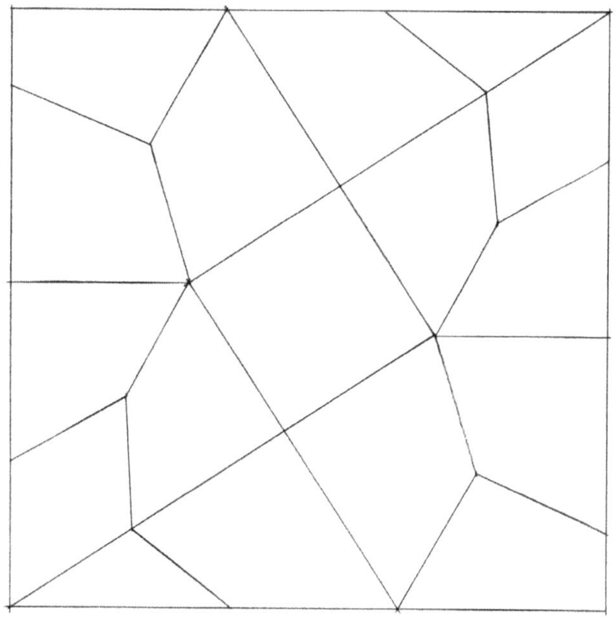

Pl. 39.1. Construction of a square with a rotated central square surrounded by kites and triangles that are further subdivided.

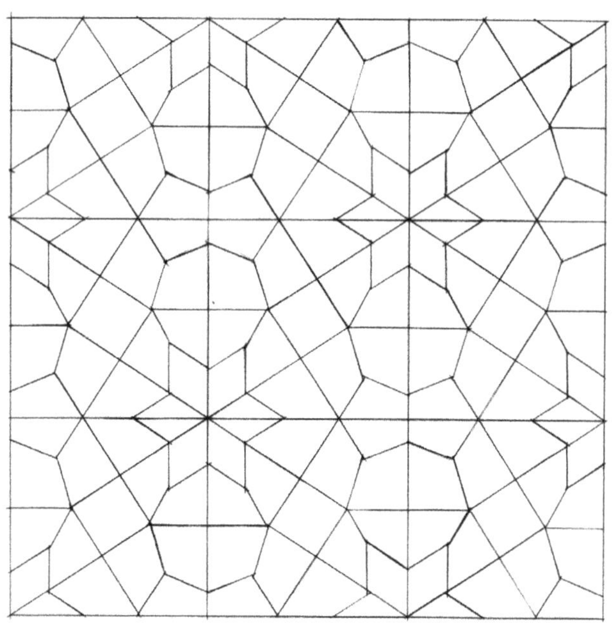

Pl. 39.2. Repeating this construction generates an ornamental pattern composed of hexagonal stars inscribed in a hexagon, which interlock with diverse polygons (recalling a type of composition seen in pl. 36.2).

1 Özdural, "Analyses" (unpublished chapter), 168–69 (labeled as "Construction 45" by Özdural).

Translation and Transcription (fol. 192v [40])

Fol. 192v [40]

No accompanying text.

CONSTRUCTION 40 (fol. 192v [40])[1]

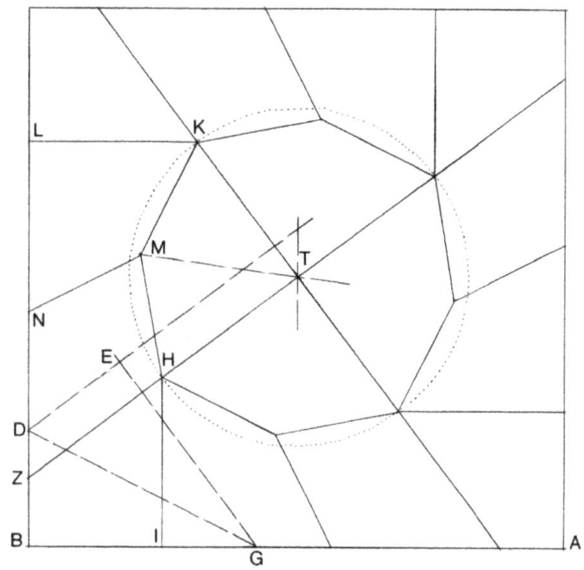

Pl. 40.1. Construction of a square with four kites rotating around a central octagon.

In this construction a square is composed of four kites in such a way that they rotate around the center point (T) of the square. Each corner is occupied by an irregular quadrilateral. Every kite is subdivided into three quadrilaterals. Those around the center form an octagon, which is equilateral but not equiangular. The method of construction is indicated in red ink on the lower left corner, showing how to obtain kites GBDE and TKLZ.

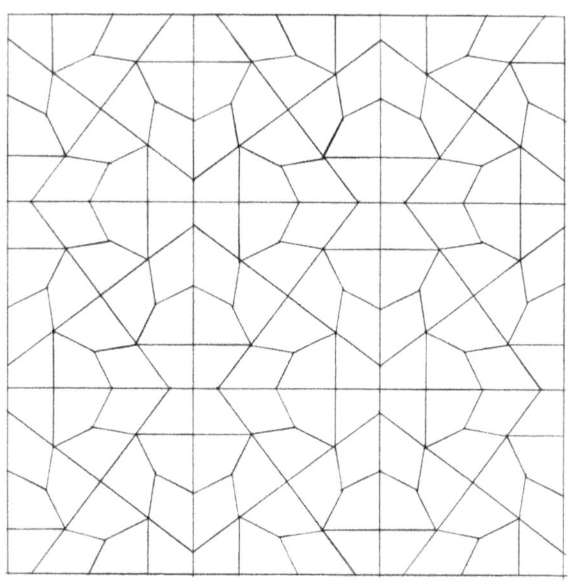

Pl. 40.2. When the pattern is repeated it generates this composition.

1 Özdural, "Analyses" (unpublished chapter), 169 (labeled as "Construction 46" by Özdural).

TRANSLATION AND TRANSCRIPTION (fol. 192v [41])

Fol. 192v [41]

The side of this square is three plus the square root of 7, because in the circle ABGD circumscribing the quadrilateral[a] ABGD side AB is 2, side AG is 2, side GD is one, and side DB is the square root of 7. It is from this situation that the small and large proportions are determined. God knows best.[b]

---

a The two-part text corresponding to the figure of Construction 41 is detached from it. This is the first part of the text, written on the right side of Construction 40.

b Written on the right side of Construction 39, this second part of the text completes the interrupted sentence above, which concerns Construction 41.

ضلع این مربع سه است و جذر ۷ چنانک در دایرۀ آ بٓ جٓ دٓ ذو اربعه اضلاع.

آ بٓ جٓ دٓ ضلع آ بٓ ۲ باشد و ضلع آ جٓ ۲ و ضلع جٓ دٓ یکی و ضلع دٓ بٓ جذر ۷ و از
اینجا نسبت کوچک و بزرگ معلوم شود والله اعلم.

Construction 41 (fol. 192v [41])[1]

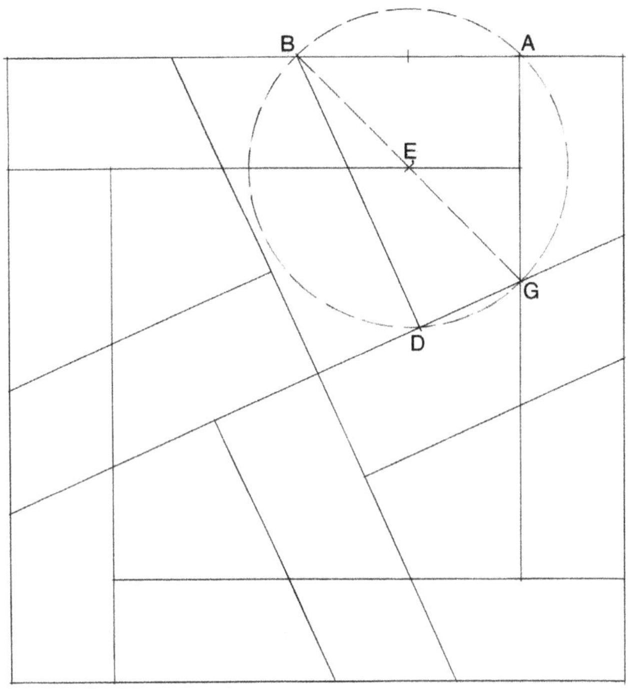

Pl. 41.1. Construction of a square containing a parallel square with extended sides and four rotating kites.

The accompanying text explains that the side of this square is three plus √7, since in circle ABGD circumscribing quadrilateral ABGD, side AB = 2, AG = 2, GD= 1, and DB = √7. The "small and large proportions" are derived from these properties. The rotational symmetry is based on a swastika motif at the center and two similar sets of kites: the four smaller ones within the inner square, and the four larger ones within the outer square. Özdural suggests that the "small and large proportions" refer to the proportions of the small and large kites.

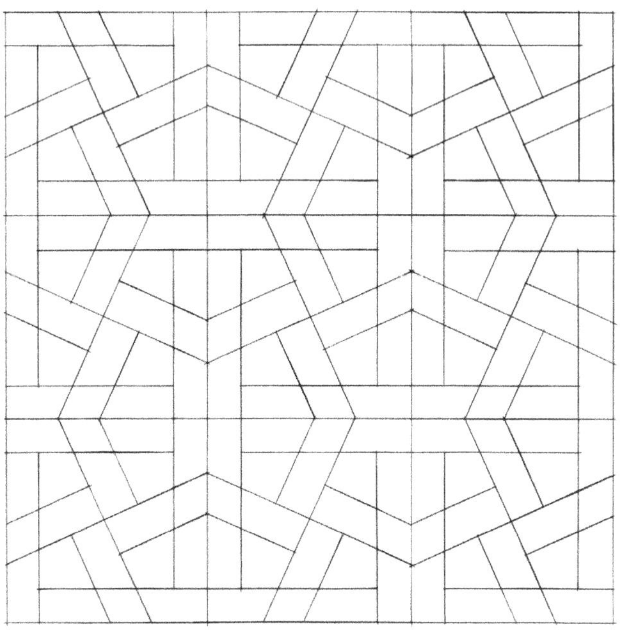

Pl. 41.2. This geometric scheme is generated by repeating the pattern. The composition with two parallel squares forming a border recalls Construction 30–31 (see pls. 30.2 and 31.2), but the proportion of the present one is based on the square root of 7 rather than the square root of 5.

1 Özdural, "Analyses" (unpublished chapter), 170–71 (labeled as "Construction 48" by Özdural).

Translation and Transcription (fol. 192v [42])

Fol. 192v [42]

Points A, [B,] and G are assumed so that AB and BG are equal to each other. From centers A and G, with compass opening AG, intersection D is found. On the rectilinear extension of side BG from point G mark point E with [compass opening] twice AG. Then, from point E with compass opening EG and from point D with compass opening DG, find the intersection Z. Place the ruler exactly on points B and Z, and obtain point H. Draw HT parallel to ZE to obtain what is required. God knows best.[a]

---

a The text is composed of eight vertical lines, from top to bottom, on the lower right side of Construction 42.

نقطه آ ج̅ فرض کند چنانک آ ب̅ ب̅ ج̅ متساوی باشد و از مرکز آ ج̅ بفتح آ ج̅ تقاطع د̅
معلوم کند و بر استقامت ضلع ب̅ ج̅ از نقطهٔ ج̅ بضعف آ ج̅ نقطهٔ ه̅ نشان کند. آنگه از
نقطهٔ ه̅ بفتح ه̅ ج̅ و از نقطهٔ د̅ بفتح د̅ ج̅ تقاطع ز̅ معلوم کند و مسطره بنقطهٔ ب̅ ز̅ راست
بنهد و نقطهٔ ح̅ بگیرد و ح̅ ط̅ موازی ز̅ ه̅ بکشد که مطلوبست والله اعلم.

CONSTRUCTION 42 (fol. 192v [42])[1]

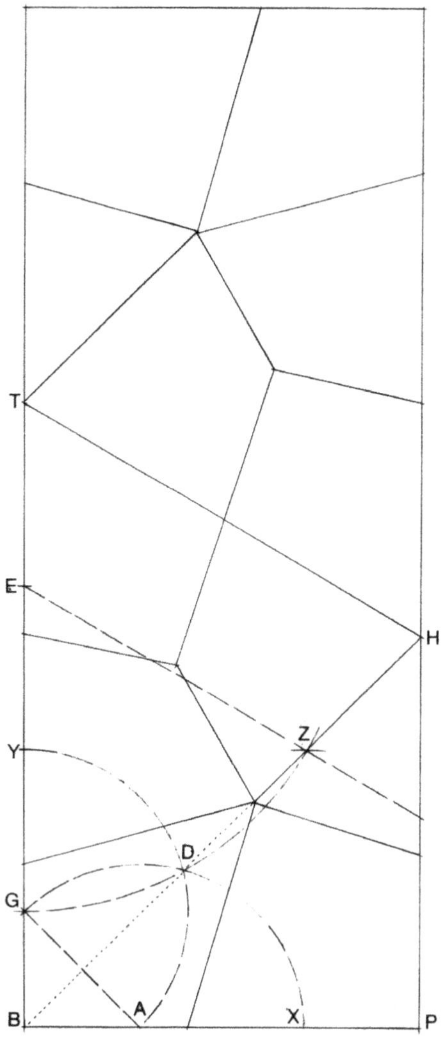

Pl. 42.1. Construction of a long rectangle containing two large kites with subdivisions.

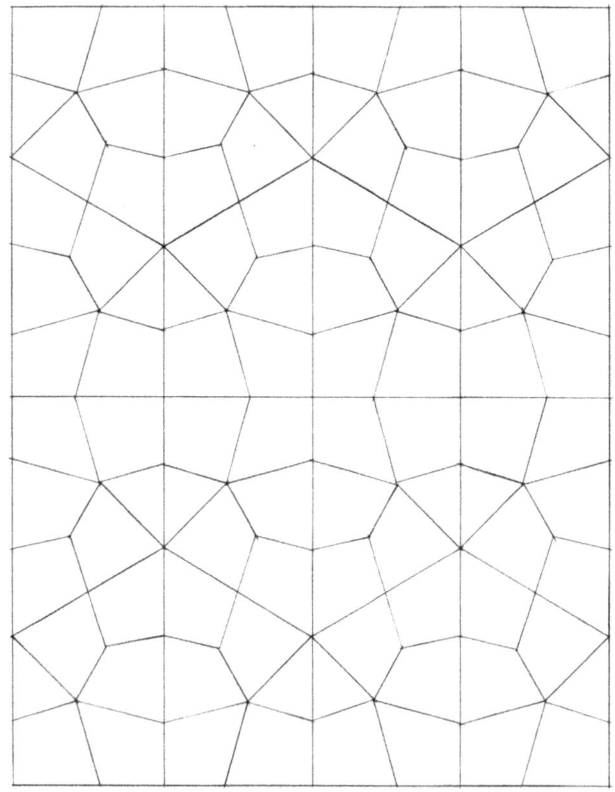

Pl. 42.2. When the two large kites are subdivided into smaller components, their centers are aligned along a straight line. According to Özdural, this appears to have been the main objective of the configuration, which produces this ornamental scheme when repeated.

In this construction two large kites facing in opposite directions are joined along line HT, which constitutes their shared side as well as the pattern's main axis of symmetry. Each of these large kites is divided into three smaller quadrangles. At both ends of the rectangular frame two quadrangles and a triangle are fitted. The proportion is first established by the small drawing in red ink at the lower left corner, and then transferred to the actual size by a sliding technique. The text explains the essential points of the construction procedure.

1 Özdural, "Analyses" (unpublished chapter), 169–70 (labeled as "Construction 47" by Özdural).

Translation and Transcription (fol. 193r [43])

Fol. 193r [43]

In set square ABG [i.e., set square [number] 6] mark AD and AE with any compass opening. Join DE and cut off DZ equal to DA. Join line AZ and draw ZH parallel to AG, so that ZH is twice AZ. Place the ruler on points H and A and obtain point T on line BG. Draw TK parallel to GA and bisect it at point L. Draw line LM [parallel to] DZ so that LM becomes equal to LT. Construct angle LMN equal to angle TLM and extend line MNS.[a] Complete the construction, so that angles O are right [angles]. God knows best.[b]

---

a Text: TNS

b The text is composed of eight horizontal lines on the lower part of the page.

در گونیا آ ب ج بهر فتح که خواهد آ د آ ه نشان کند و د ه بپیوندد و د ز مثل د آ جدا کند و خط آ ز بپیوندد و ز ح موازی آ ج اخراج کند چنانک ز ح ضعف آ ز باشد و مسطره بر نقطهٔ ح و آ بنهد و بر خط ب ج نقطهٔ ط بگیرد و ط گ موازی ج آ بیرون آرد و تنصیف کند بنقطهٔ ل و خط ل م د ز اخراج کند چنانک ل م مثل ل ط باشد و زاویهٔ ل م ن مثل زاویه ط ل م عمل کند و خط م[١] ن س اخراج کند و عمل تمام کند که زوایا ع قوایم باشند والله اعلم.

---

١ م] ط

Construction 43 (fol. 193r [43])[1]

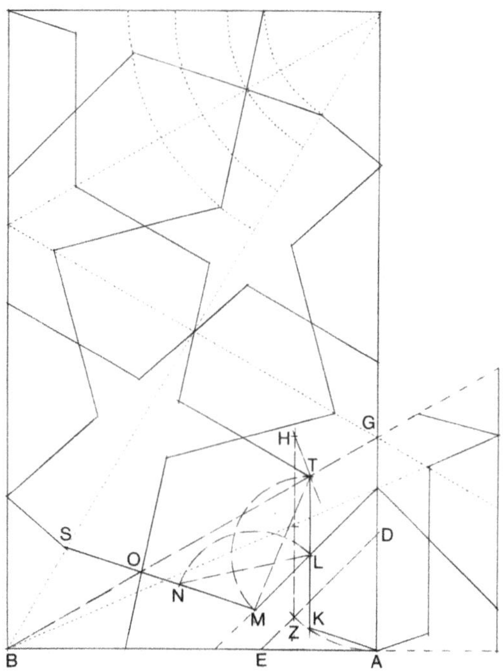

Pl. 43.1. Construction of a rectangle containing irregular hexagons and six-pointed stellate figures.

According to Özdural, the author seems to have been disturbed by the irregularities of the polygons interlocking with the stellate six-pointed forms (for instance, in hexagon TLMO the sides and angles are unequal). He attempts to correct this irregularity in the amended design by assuming the standard pattern. The amendments made in the lower right quadrant make the polygon TLMO look more regular. Since this necessitates altering the six-pointed stellate figure around point G, a small portion of it is drawn outside the frame of the pattern. The two shorter sides of the irregular half-hexagon are made equal by connecting the angular point to B with an incised uninked line.

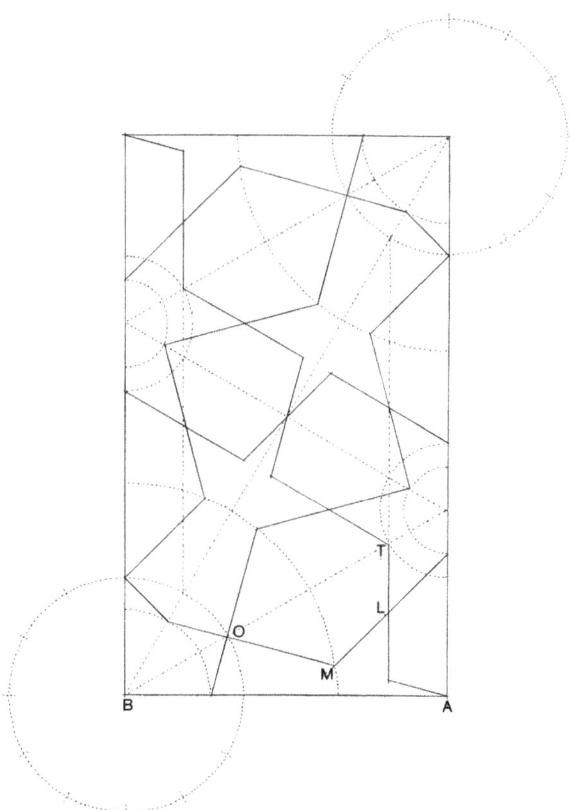

Pl. 43.2. This construction grows from two quarter-hexagonal stars at the lower left and upper right corners. Özdural interprets the design as an amendment made to a common pattern already existing since Seljuk times. The construction procedure explained in the text is related to a small detail of this well-known pattern, but uninked incised lines allow the reconstruction of the whole pattern as follows: the standard formula starts with drawing a "line [number] 6" as the main axis of symmetry connecting angle B to the opposite angle of the rectangle. The secondary diagonal axis is established by drawing another "line [number] 6," which bisects the former line perpendicularly at the center. The tertiary axes are determined by drawing two other "lines [number] 6" from the same opposite angles to the ends of the secondary axis (i.e., line BT). From the center, a pair of lines are drawn to define the radii of the two concentric circles from the opposite angles. The outer circles are divided into 12 equal parts and alternating points are joined to the third one in the sequence, on either side. In the hexagonal star thus formed, the vertices are right angles. The sides of the stars are extended until the opposite ones are joined. The pattern is completed with the aid of the incised uninked diagonal axes of symmetry and the pair of concentric circles (at the upper right and lower left corners defining the quarter-hexagonal stars, and along the secondary axis defining the irregular half-six-pointed stellate figures).

1 Özdural, "Analyses" (unpublished chapter), 171–73 (labeled as "Construction 49" by Özdural).

TRANSLATION (fol. 193v [44])*

Fol. 193v [44]

It should be known that the procedure for drawing this repeat-unit starts from the center. It is as follows: As the median line AB is assumed, bisect it at midpoint G. With this point as the center inscribe circle DZEH. At two-fifths of arc ED mark points Z and H on the circumference. Draw line ZH so that it becomes a line [number] five on line AB. Then divide angle ZGE, that is, arc ZE, into four [equal] parts, and draw the [dividing] line GTI. Take the magnitude of BI and make GL and GK equal to that. Draw line KL. From line KL and line LG point M is found [with the compass centered on points G and L, respectively]. Draw line M[K] and on KM erect the perpendicular SO. Take O as the center and complete the construction. If one wants to put the arm of the compass at point K′ and with the compass opening K′M′ obtain point N, then in this repeat-unit angle LGK is equal to angle GK′F, and each of them is equal to one right angle plus one-tenth of the right angle. God knows best.[a]

If magnitude M′C′ is made equal to half of GK′, then a reliable (*ma'mūn*) repeat-unit comes out. If it is not a *ma'mūn* repeat-unit, magnitude M′C′ is less than half of GK′.

Remarks on the properties [of this repeat unit's proportion] by means of approximation: First, perpendicular SQ traversing the pentagon is equal to half of BG. Second, [constructing] the width of this repeat-unit as two parts, and its length as three parts, is a very close approximation.[b]

---

a The text is in two parts. This is the first part, composed of eight horizontal lines below the figure.

b The second part of the text is composed of four vertical lines, from top to bottom, on the right side of the figure.

---

* For the Persian transcription, turn to page 296.

Construction 44 (fol. 193v [44])[1]

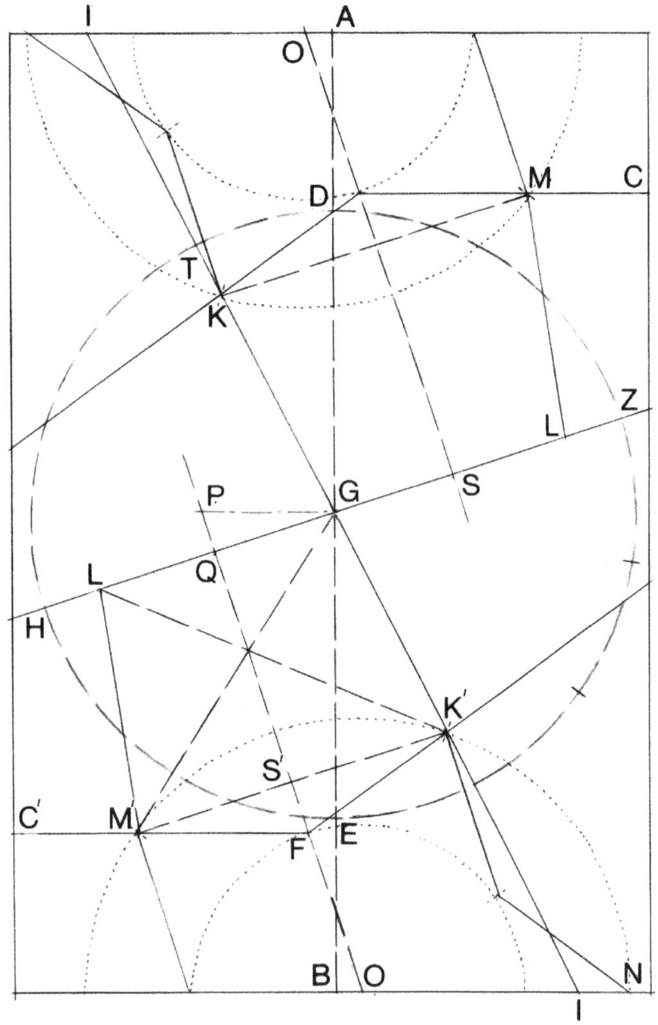

Pl. 44.1. Construction of a rectangle with pentagonal stars and irregular pentagons.

The half-pentagonal stars along the upper and lower sides of the rectangle are separated by irregular pentagons on either side of a diagonal axis of symmetry. Auxiliary construction lines drawn in red ink are distinguished from the main pattern in black ink. Two pairs of uninked concentric semicircles incised by a divider facilitate the drawing of the two pentagonal stars. The main text explains that the construction procedure of this repeat-unit starts from the center. The assumed middle line (AB) is bisected at midpoint G, forming the center of circle DZEH. Points Z and H are marked on the circumference so that arcs ZD, EH are two-fifths of arc ED. Line ZH becomes a "line [number] 5" with regard to line AB. The diagonal lines intersect at point G, and constitute dynamic axes of symmetry. Angles HGK and GKF are equal, each being one right angle plus one-tenth of the right angle.

1 Özdural, "Analyses" (unpublished chapter), 173–76 (labeled as "Construction 50" by Özdural).

Transcription (fol. 193v [44])*

باید[1] دانست که طریق کشیدن این عقد از وسط است و آن چنان باشد که چون خط آ بٓ وسط معلوم بود بدو قسم کند بنقطهٔ جٓ در وسط و برین نقطه دائره دٓ زٓ هٓ حٓ بگرداند و بر دو خمس قوس هٓ دٓ نقطه زٓ حٓ بر محیط نشان کند و خط زٓ حٓ بکشد که این خط پنج باشد بر خط آ بٓ. آنکه زاویه زٓ جٓ هٓ یعنی قوس زٓ هٓ چهار قسم کند و خط جٓ طٓ یٓ بکشد و مقدار بٓ یٓ بردارد و جٓ لٓ جٓ کٓ مثل وی سازد و خط کٓ لٓ بکشد. آنکه از خط کٓلٓ و خط لٓ جٓ نقطهٔ مٓ معلوم و خط مٓ بکشد و از کٓ مٓ عمود سٓ عٓ اخراج کند و مرکز عٓ بگیرد و عمل تمام کند. و اگر خواهد پای پرگار بر نقطه کٓ نهد و بفتح کٓ مٓ نقطه نٓ بگیرد و درین عقد زاویه لٓ جٓ کٓ مثل زاویه جٓ کٓ فٓ بود و هر یک قائمه و عشر قائمه باشند والله اعلم.

و اگر مقدار مٓ صٓ مثل نصف جٓ کٓ سازد عقد مأمونی آید و اگر عقد مأمونی نبود مقدار مٓ صٓ کمتر از نصف جٓ کٓ بود.
ذکر الخواص بالتقریب اول عمود سٓ قٓ از مخمس مخالف مثل نصف بٓ جٓ بود و دیگر عرض این عقد دو قسم و طول سه قسم عظیم تقریب است.

[1] باید] بباید

* For the English translation, see page 294.

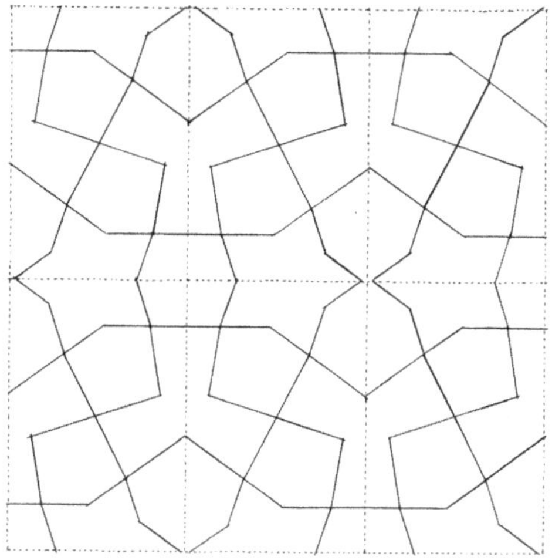

a

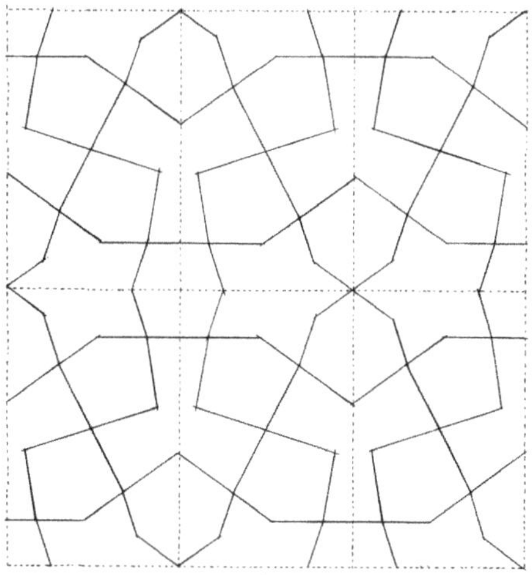

b

Pl. 44.2a. Another textual commentary explains that if line MC is made equal to half of GK, then "a reliable" (*ma'mūn*) repeat-unit comes out, but if MC is less than half of GK then it is unreliable. The text also remarks on two properties of approximating this repeat-unit: (1) SQ = half of BG; (2) the width of the repeat-unit is two parts and its length is three parts, which is a "very close approximation."[2] The distinction sheds light on aesthetic criteria. This drawing illustrates the "reliable" geometric scheme generated when MC equals half of GK and the rectangle thus formed is repeated, creating a "stable" impression according to Özdural. In this version, the three-armed polygon, surrounded by irregular pentagons and hexagons, has equal sides all around, but a slight gap appears between the horizontal vertices of the two pentagonal stars.

Pl. 44.2b. If MC is taken as less than half of GK, the gap between the horizontal vertices of the pentagonal stars disappears, but now the three-armed polygon is no longer equilateral. According to Özdural, this composition does not "give a well-fitted impression because of the irregularity of the three-armed polygons."

2 Özdural mistranslated "a very close approximation" as "too coarse an approximation": see Özdural, "Analyses" (unpublished chapter), 175.

Translation and Transcription (fol. 194r [45])

Fol. 194r [45]

The procedure for drawing this repeat-unit is as follows: From angle A [with an arbitrary compass opening], place point B on the diagonal of the square and point G on the side, so that AB is equal to AG. Draw line BG. Then, from the [same] angle, with compass opening BG, construct [equilateral] triangle ADE. Put the compass on the apex of the triangle at point D and with compass opening AB find point Z on the diagonal. Join line ZE. From the center draw line IH parallel to EZ and obtain point H. [From angle A] inscribe arc HT. Points H and T are what was required. Complete the construction. God knows best.[a]

---

a The text is composed of seven horizontal lines below the figure.

طریق کشیدن این عقد آنست که از زاویهٔ آ بر قطر مربع نقطهٔ ب نهد و بر ضلع نقطهٔ ج
چنانک آ ب مثل آ ج بود و خط ب ج بکشد. آنگه از زاویه مثلث آ د ه بفتح ب ج عمل
کند و پرگار بر راس مثلث بر نقطهٔ د نهد و بفتح آ ب بر قطر نقطهٔ ز بگیرد و خط ز ه
بپیوندد و از مرکز خط ی ح موازی ه ز بکشد و نقطهٔ ح بگیرد و قوس ح ط بگرداند
که نقطه ح ط مطلوب بود. عمل تمام کند والله اعلم.

Construction 45 (fol. 194r [45])[1]

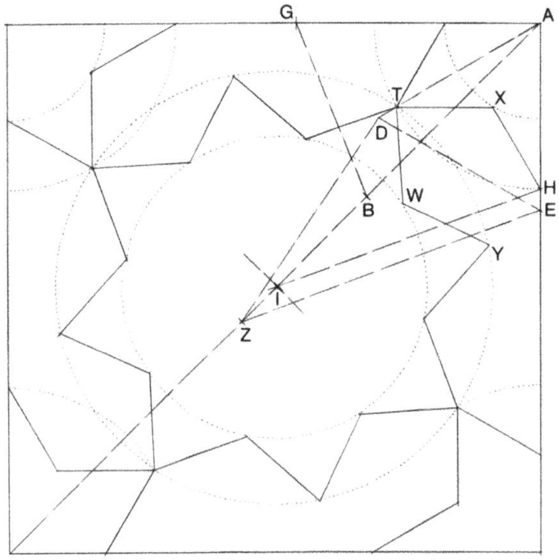

Pl. 45.1. Construction of a square combining hexagonal and octagonal stars with composite polygons.

The text explains the "procedure for drawing this repeat-unit" in such a way that the central regular octagonal star is joined harmoniously with regular quarter-hexagonal stars at each of the four corners of the square frame and with composite polygons (paired pentagons joined with a heptagon). The construction starts from angle A, from which point B is marked on the diagonal of the square and point G on the side, so that AB = AG. Then, from angle A an equilateral triangle ADE is drawn. Intermediary steps show how to obtain points H and T; these "are what is required," since they define the vertices of the quarter-hexagonal star at the upper right corner. The rest of the construction is not explained in the text. Pairs of incised uninked concentric circles and arcs define the stars, while other lines are drawn in red ink, from which the construction procedure can be deduced by those familiar with drawing star polygons.

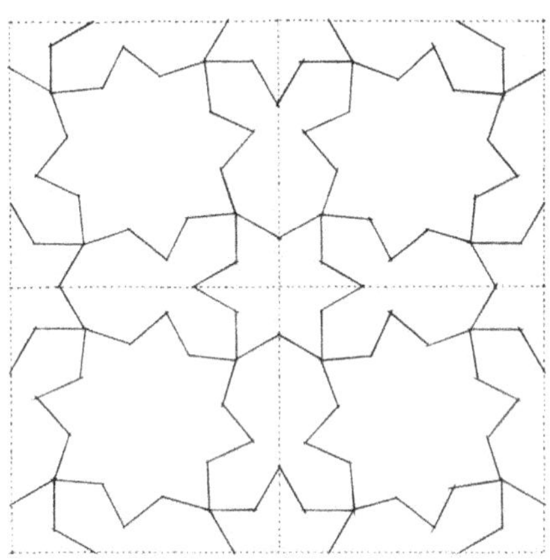

a

Pl. 45.2a. According to Özdural's reconstruction, first the side of the octagon is obtained and made equal to the side of the hexagon at the vertex where they touch. Hence, the sides of the octagonal star and hexagonal star are equal. Once the required proportion is determined, it is transferred to the actual size of the pattern by a sliding technique. While the hexagonal stars are laid on perpendicular axes, the octagonal star is slightly tilted. Repetition of the square repeat-unit generates this geometric scheme in which the sides of all the different interlocking shapes are equal.

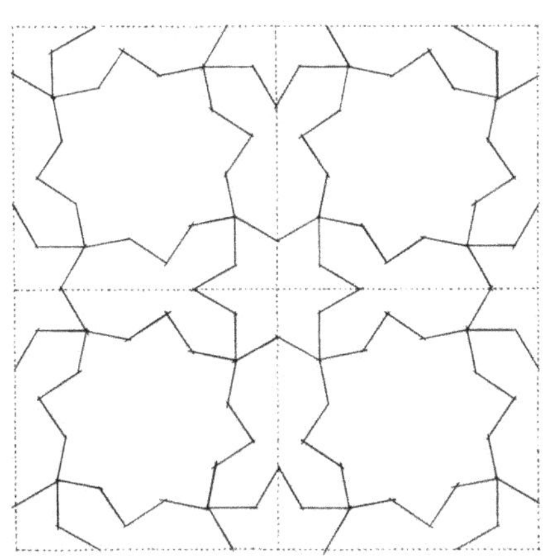

b

Pl. 45.2b. This scheme, which Özdural generated by using a standard octagonal star, is, according to him, "aesthetically inferior" because the composite polygons are no longer equilateral, unlike those in pl. 45.2a. (The composite polygons combine two pentagons and a heptagon.) The superior previous version, depicted in the *Anonymous Compendium*, shows the designer's ability to devise subtly refined ways of integrating different geometric systems, harmoniously united so as to become undistinguishable.

1 Özdural, "Analyses" (unpublished chapter), 176–77 (labeled as "Construction 51" by Özdural).

Translation and Transcription (fol. 194r [46])

Fol. 194r [46]

No accompanying text.

Construction 46 (fol. 194r [46])[1]

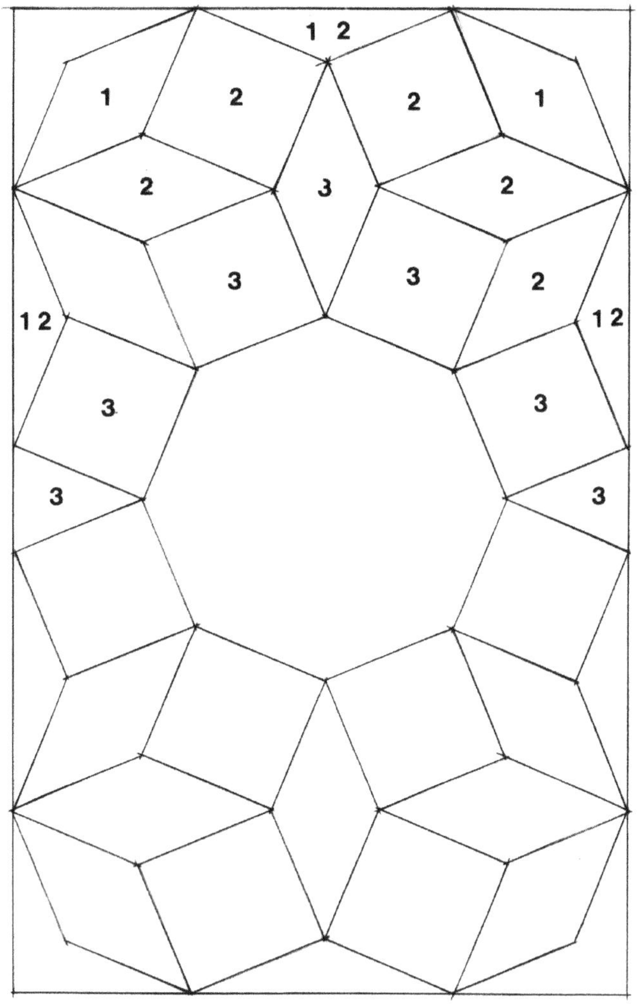

Pl. 46.1. Construction of the orthogonal projection of a rectangular *muqarnas* vault.

The textless construction is composed of squares, rhombuses, and half-rhombuses. Some of these are marked by numbers 1 to 3, which increase from the corners towards the central octagon. As in Construction 15 (see pl. 15.1–5), the enumeration seems to represent the tiers of a simple *muqarnas*, reaching the highest level around the perimeter of the octagon. Özdural did not propose a reconstructed section showing the three-dimensional layout because he did not find any extant rectangular vault like it. The rectangular frame of the construction indicates the shape of the space to be covered. This is unusual for *muqarnas* vaults, which generally have square plans repeated twice or four times.

---

1 Özdural, "Analyses" (unpublished chapter), 178 (labeled as "Construction 52" by Özdural).

Translation and Transcription (fol. 194r [47])

Fol. 194r [47]

No accompanying text.

Construction 47 (fol. 194r [47])[1]

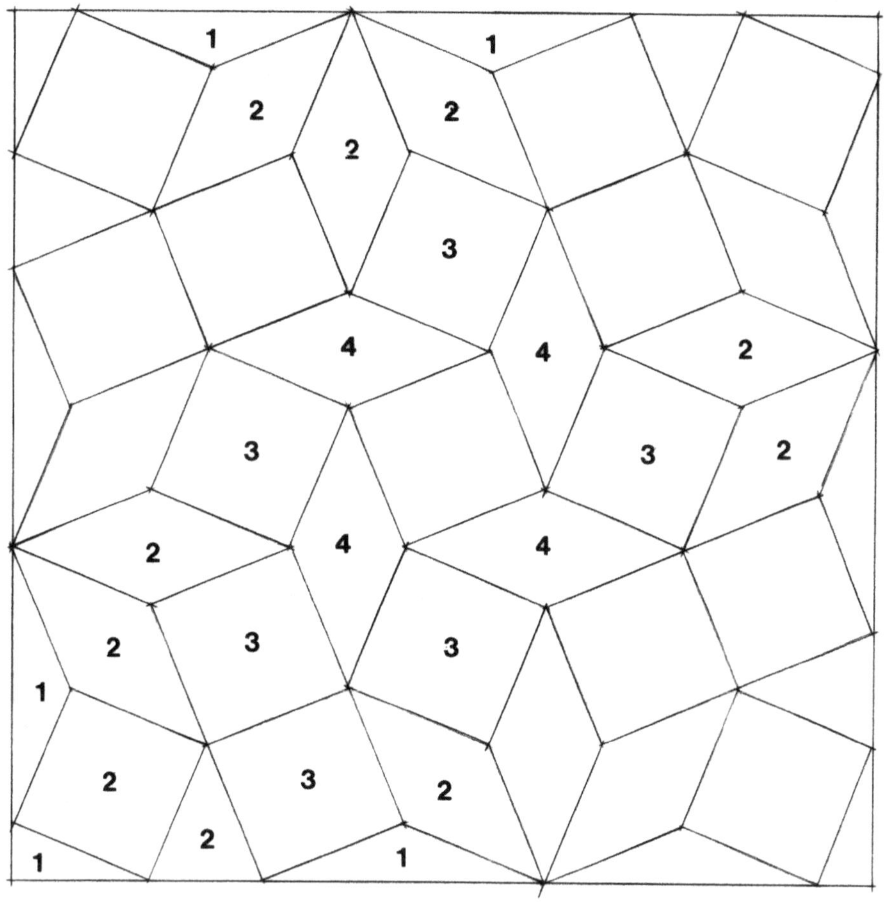

Pl. 47.1. Construction with the orthogonal projection of a square *muqarnas* vault.

This textless construction is subdivided into squares, rhombuses, and half-rhombuses, marked by numbers 1 to 4. Increasing from the corners towards the central rotated square, these numbers presumably represent the number of tiers in the square *muqarnas* vault. The units revolving around the rotated square at the center create a dynamic effect. According to Özdural, had it ever been realized, this type of rotational symmetry would have been "revolutionary" for *muqarnas* vaulting. As in the previous case (see pl. 46.1), he found no extant example to reconstruct its three-dimensional configuration.

1 Özdural, "Analyses" (unpublished chapter), 179 (labeled as "Construction 53" by Özdural).

Translation (fol. 194v [48])*

Fol. 194v [48]

The procedure for drawing this repeat-unit and its proportion is as follows. Assign point A anywhere on one side of an assumed square. Draw lines [number] seven AB and AG on AE. From point B drop perpendicular BD[a] onto line AG. On perpendicular BD[b] construct rectangle DE, so that its length is twice its width. This is perpendicular BD.[c] Then check whether point E′ of the rectangle falls on the midpoint of the assumed square, or not.[d]

If it is on the midpoint of the square, then this is the required proportion. But if it [E] is not on the midpoint of the square, from point E draw line AE to the assigned point A. From the center of the square draw another line parallel to line AE, in order to find the center of the heptagon, which is required, on the side of the square. God knows best.

Another kind [of construction procedure] pertaining to the proportion of this repeat-unit is the following: Mark point A anywhere on the side of the assumed square [repeat-unit], and draw lines [number] seven AB and AZ on [line] AE. [From point Z] draw a line [which will become ZT] perpendicular to AZ. From angle Z, which is a right angle, draw line [number] four until it intersects line AB at point B. Then put the compass on letter B, and with compass opening BZ find point H on perpendicular ZT. Again, open the compass and set it as opening AZ, and put it on point H. Mark point T on the perpendicular ZT, so that HT is equal to AZ. Then check whether point T falls on the side of the assumed square or not. If it does, this is the required proportion, and if it does not, it is not.[e]

---

a Text: AD.
b Text: AD.
c Text: AD.
d The text, constituting one continuous sentence, is divided into two parts. This is the first part, composed of four vertical lines, from top to bottom, on the right side of the figure. The last few words are written at the beginning of the second part of the text, under the figure.
e The second part of the text is composed of nine horizontal lines below the figure.

---

* For the Persian transcription, turn to page 306.

Construction 48 (fol. 194v [48])[1]

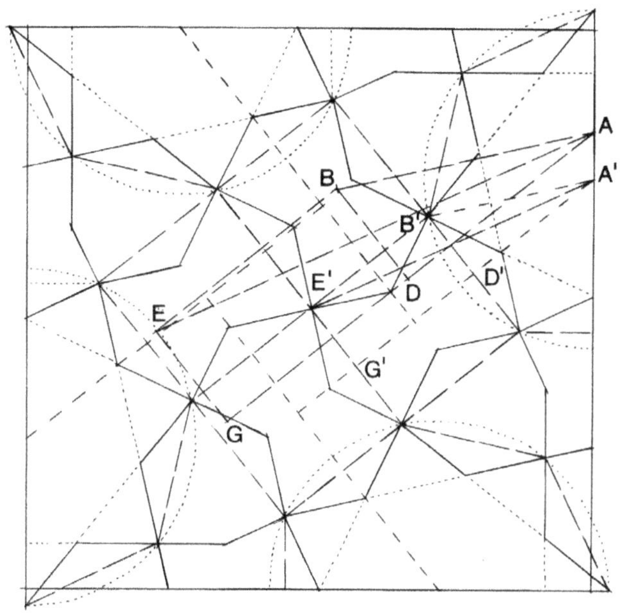

Pl. 48.1. Construction of a square with heptagonal stars, double-pentagons, and irregular hexagons.

The text explains the procedure and required proportion of this repeat-unit. It comprises four central, rotating composite polygons (double-pentagons) that form a swastika motif within a small, rotated central square whose extended sides coincide with the centers of four heptagonal half-stars lined up along each side of the repeat-unit. The four rotation axes are drawn as dotted lines in the original figure, while auxiliary construction lines are shown in red ink, and the incised construction lines of the heptagonal stars are left uninked. Like Construction 37, this pattern is based on heptagonal symmetry and was likely executed by means of set squares, a method referred to in the text as "lines [number] seven" (i.e., lines AB and AG on line AE). Once the desired proportion is established at an arbitrarily chosen scale, the sliding technique is used to translate it to the actual scale of a given square repeat-unit. This is one of two different methods of construction described and shown in the same figure.

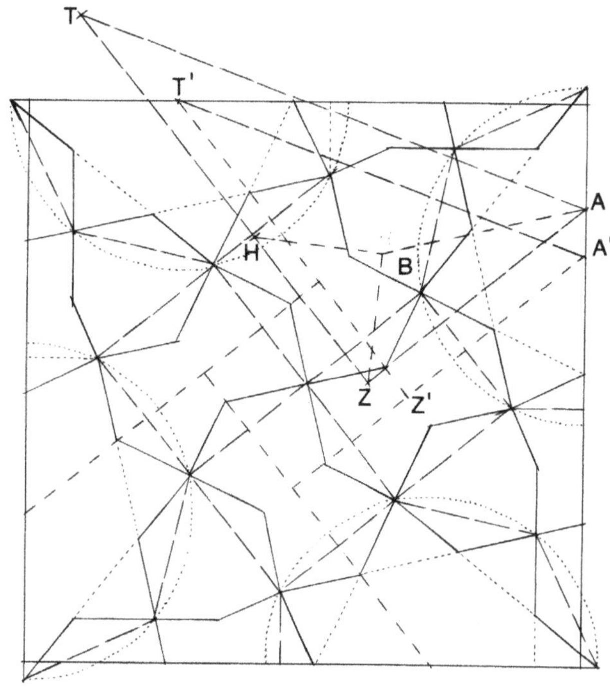

Pl. 48.2. The alternative method for the preceding construction consists of drawing the four axes (here broken lines) that meet the sides of the square repeat-unit at four-sevenths of a right angle. The lines that issue from the centers of the half-heptagonal stars and intersect each other perpendicularly form a small rotated square at the center, whose angles are equidistant from the centers of the heptagonal stars. In other words, the centers of the four half-heptagonal stars along the sides of the assumed square and the "lines [number] 7" issuing from those centers create four right-angled figures rotating around a small rotated central square. If the point T does not fall on the required position, a sliding technique must be applied in order to fit the established proportion into the assumed square. Özdural reconstructs this laborious sliding technique for finding the correct position of T′A′ (drawn parallel to TA), which was not explained in the text, probably because of its impracticality.

1 Özdural, "Analyses" (unpublished chapter), 179–83 (labeled as "Constructions 54 and 55" by Özdural). The figure on fol. 194v of the manuscript contains only points A′, B′, E′, G′, D′ in fig. 48.1, and points Z′, T′ in fig. 48.2, as well as a point H′ on Z′T′ corresponding to point H on ZT in fig. 48.2.

TRANSCRIPTION (fol. 194v [48])*

طریق کشیدن این عقد آنست و نسبت او که در مربع مفروض بر احد الأضلاع او نقطه آ کیف ما اتّفق وضع کند و خط آ ب̅ آ ج̅ بر آ ه̅ هفت بیرون کند و از نقطهٔ ب̅ بر خط آ ج̅ عمود آ د̅ اخراج کند و بر عمود آ د̅ سطح د̅ ه̅ عمل کند چنانک طول او ضعف عرض بود که آن عمود ب̅ د̅[1] است. آنگه بنگرد که نقطهٔ ه̅ از سطح مستطیل بر وسط مربع مفروض افتاده است یا نه. اگر بر وسط مربع بود و این نسبت مطلوبست و اگر بر وسط مربع نباشد از نقطه ه̅ بنقطه آ که موضوع است خط آ ه̅ بکشد و از مرکز مربع خطی دیگر موازی خط آ ه̅ بکشد تا بر ضلع مربع مرکز مسبع که مطلوبست معلوم شود والله اعلم.

نوعی دیگر هم در نسبت این عقد[2] آنچنان است که بر ضلع مربع مفروض نقطه آ کیف ما اتفق وضع کند و خط آ ب̅ آ ز̅ بر آ ه̅ هفت بیرون کند و بر خط آ ز̅ عمود ز̅ ط̅ هر جا که خواهد اخراج کند و از زاویه ز̅ که قائمه است خط چهار بیرون کند تا متقاطع خط آ ب̅ شود بنقطه ب̅. آنگه پرگار بر حرف ب̅ بنهد و بفتح ب̅ ز̅ بر عمود ز̅ ط̅ نقطهٔ ح̅ بگیرد و باز پرگار بگشایند و بفتح آ ز̅ آورد و بر نقطهٔ ح̅ نهد و بر عمود ز̅ ط̅ نقطه ط̅ نشان کند چنانک ح̅ ط̅ مثل آ ز̅ باشد. آنگه بنگرد که نقطهٔ ط̅ بر ضلع مربع مفروض افتاده است یا نه. اگر افتاده باشد این نسبت مطلوبست واگر چنانچ نیفتاده باشد نه.

---

۱ ب̅ د̅] آ د̅

۲ عقد] عقد و

---

* For the English translation, see page 304.

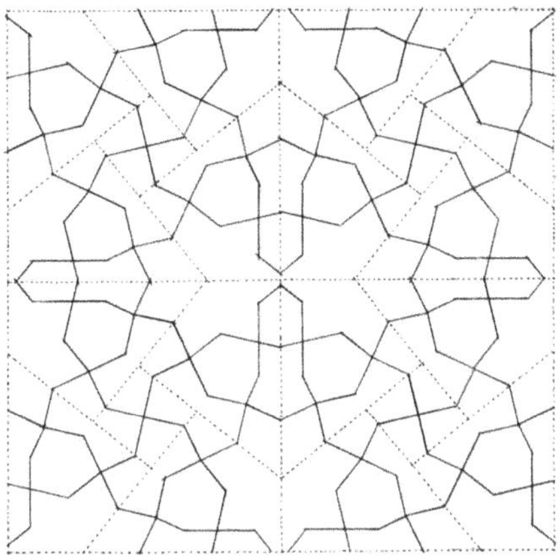

Pl. 48.3. This ornamental scheme is generated by replicating the "repeat-unit." The motif of four interlocking double-pentagons generated from polygonal stars was a favorite since Seljuk times, early examples of which usually display pentagonal, hexagonal, or octagonal symmetries. The heptagonal versions in pl. 48.1 and pl. 48.2 are the earliest instances according to Özdural, in keeping with the concentration on the heptagon in this document.

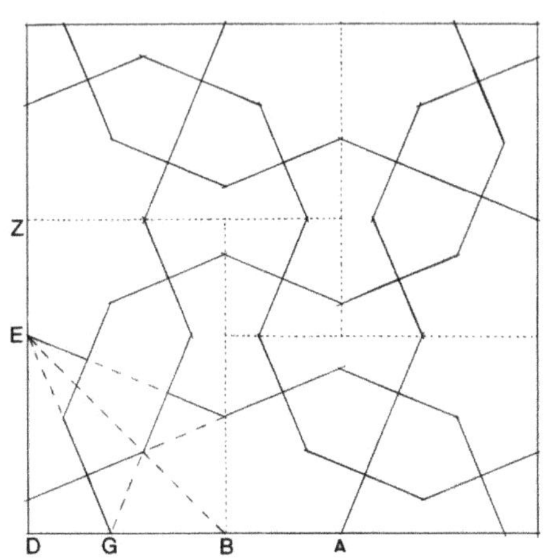

a

Pl. 48.4a. A simpler version of the same motif found in the Timurid-Turkmen *Topkapı Scroll* (late fifteenth or early sixteenth century), based on octagonal symmetry.

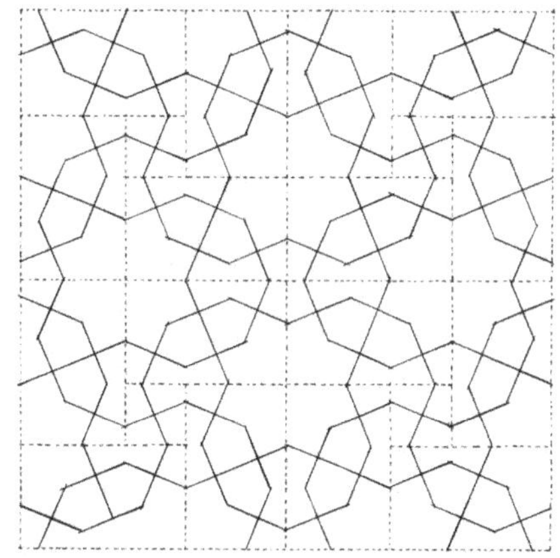

b

Pl. 48.4b. The ornamental pattern generated by the repetition of (a).[2]

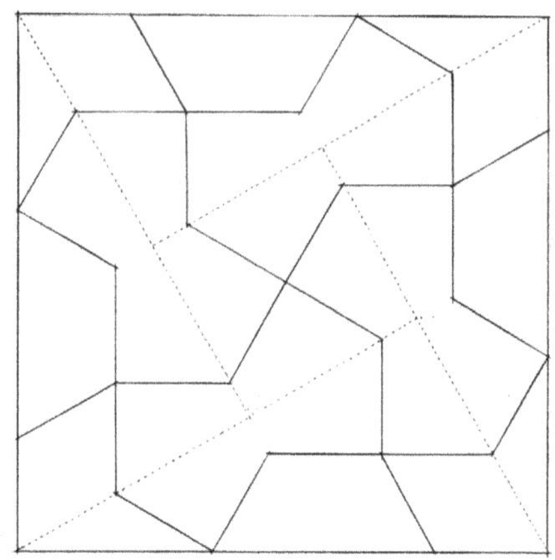

a

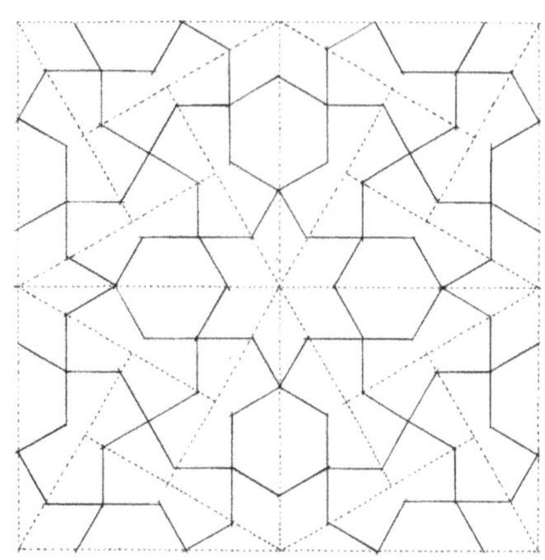

b

Pl. 48.5a–b. An even simpler version of the same motif found in the *Topkapı Scroll*, based on hexagonal symmetry (a) and the ornamental pattern generated by its repetition (b).[3]

2 Necipoğlu 1995, cat. no. 61.
3 Necipoğlu 1995, cat. no. 59.

Translation and Transcription (fol. 195r [49])

Fol. 195r [49]

In this figure, both GE and EB are equal to AD. If line KT perpendicular [to HV] is made equal to side ZH, then it will fit [the repeat-unit].[a]

a The text is composed of five short oblique lines, from upper right to lower left, below the figure.

درین شکل هر یک از جٓ هٓ هٓ بٓ مثل آ دٓ باشند و اگر عمود کٓط مثل ضلع زٓحٓ سازند شاید.

Construction 49 (fol. 195r [49])[1]

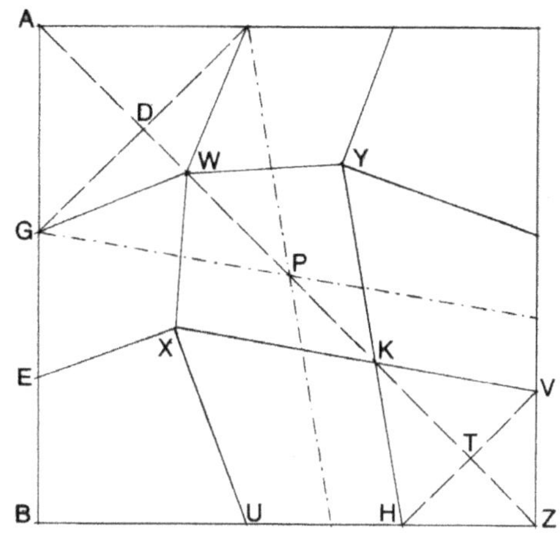

a

Pl. 49.1a. Construction of a square combining kites, octagons, and tetragonal stars.[2]

One of five square figures on the same page (Constructions 49–53) based on the derivatives of the square. Each of them is divided into different types of isosceles quadrangles (kites and rhombuses) in such a way that one type is always a quarter-octagon. Özdural suggests that these variations on a single theme seem to have been treated "as an academic exercise," rather than serving ornamental purposes. The construction procedures are not described in detail; the essential points are given as short notes, in addition to incised construction lines. Construction 49 has letters designating points, auxiliary construction lines in red ink, and a short note giving all the necessary information on its properties and construction procedure. The square frame is subdivided into proportionally interrelated quadrangles. Three aligned kites are bisected by the diagonal AZ of the square frame, the one at the upper left being a quarter-octagon, and the one at the lower right a quarter-tetragonal star. Moreover, two other axes of symmetry intersecting at point P subdivide the square frame into pairs of congruent quadrangles. According to the text, GE = EB = AD, and the components of the pattern fit together if KT = ZH.

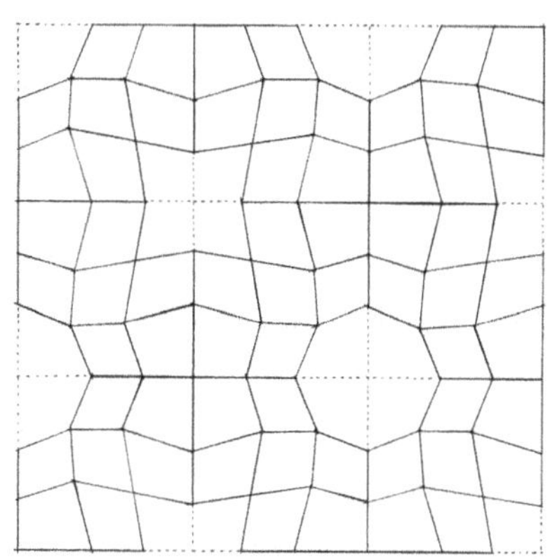

b

Pl. 49.1b. Repeating the square construction generates this pattern, which, according to Özdural, "is not a great success."

1 Özdural, "Analyses" (unpublished chapter), 183–85 (labeled as "Construction 56" by Özdural).
2 Özdural's drawing is flipped in comparison with the original drawing on the facsimile page.

Translation and Transcription (fol. 195r [50])

Fol. 195r [50]

In this figure EB, DE, and AD are equal to each other.[a]

a The text is composed of three horizontal lines below the left corner of the figure.

درين شكل هَبَ دَ هَ آ دَ متساوى باشند.

CONSTRUCTION 50 (fol. 195r [50])[1]

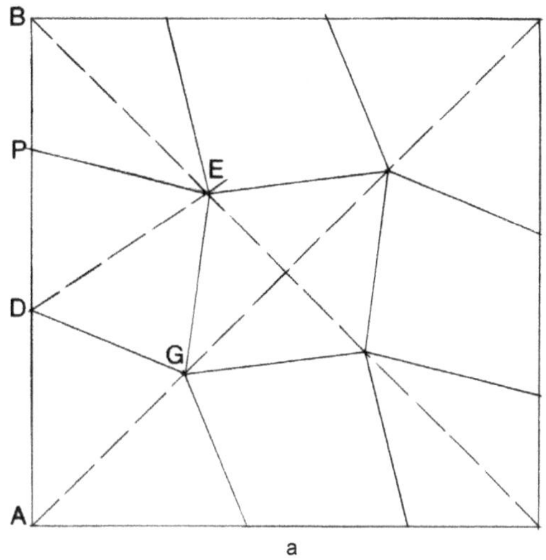

Pl. 50.1a. Construction of a square with a central rhombus, kites, octagons, and tetragonal stars.

This construction consists of a square subdivided into kites surrounding a central rhombus. The two intersecting diagonals of the square frame form axes of symmetry that bisect the isosceles quadrangles lined up along them. Quarter-octagons fill the upper right and lower left corners, while the other two corners contain quarter-tetragonal stars. Auxiliary construction lines in red ink and a short note provide the necessary information, summed up as DE = AD.

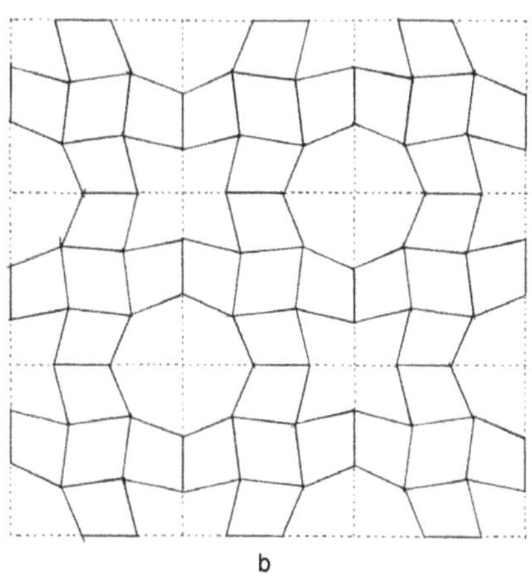

Pl. 50.1b. Repeating the square construction generates this geometric composition.

1 Özdural, "Analyses" (unpublished chapter), 185 (labeled as "Construction 57" by Özdural).

Translation and Transcription (fol. 195r [51])

Fol. 195r [51]

In this figure, DE should be equal to AD, or BZ [should be] equal to one-third of AB.[a]

---

a The text is a vertical line, from bottom to top, along the left side of the figure.

درین شکل دٓ هٓ مثل آ دٓ باید یا بٓ زٓ ثلث آ بٓ.

Construction 51 (fol. 195r [51])[1]

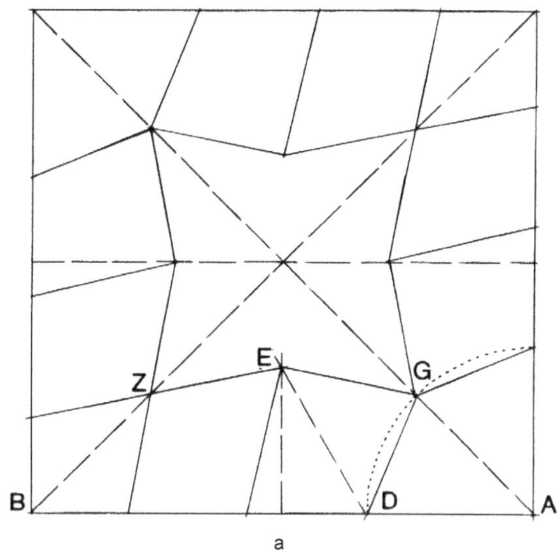

Pl. 51.1a. Construction of a square combining kites, octagons, and tetragonal stars.

This construction consists of a square frame with intersecting diagonals, which constitute axes of symmetry that bisect the central tetragonal star, as well as the pairs of quarter-octagons and quarter-tetragonal stars in opposite corners. Auxiliary construction lines in red ink are complemented by an uninked, incised quarter-circle that circumscribes the quarter-octagon in the lower right corner. A short note explains the required property of the square construction, whose side (AB) is three times the unit length AD = BZ = DE.

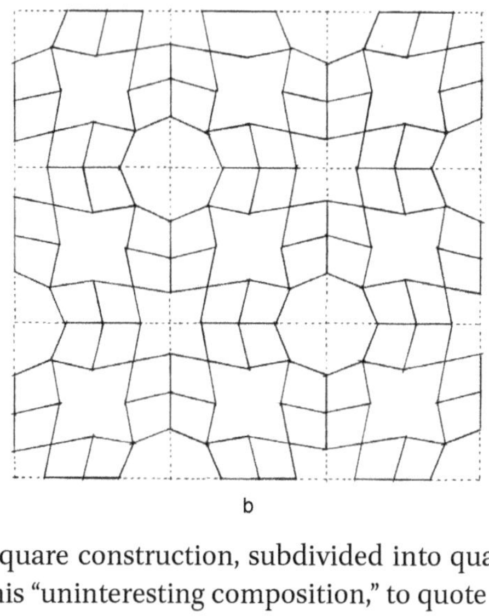

Pl. 51.1b. Repeating the square construction, subdivided into quadrangles not all of which are isosceles, generates this "uninteresting composition," to quote Özdural.

1 Özdural, "Analyses" (unpublished chapter), 185–86 (labeled as "Construction 58" by Özdural).

Translation and Transcription (fol. 195r [52])

Fol. 195r [52]

No accompanying text.

Construction 52 (fol. 195r [52])[1]

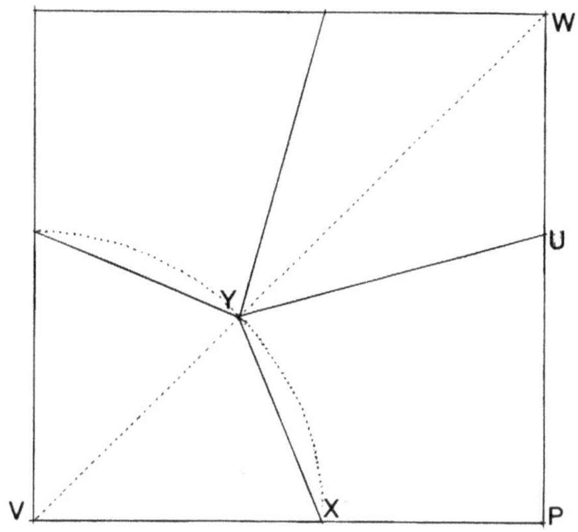

a

Pl. 52.1a. Construction of a square combining kites, octagons, and tetragonal stars.

In this textless square construction, the diagonal (VW) constitutes the axis of symmetry; on it point Y forms the shared vertex of two kites. A quarter-circle in the lower left corner (V) defines a quarter-octagon, while the opposite angle (W) has a quarter-tetragonal star. There are no auxiliary lines, but the construction procedure can be deduced from the uninked incised guiding lines.

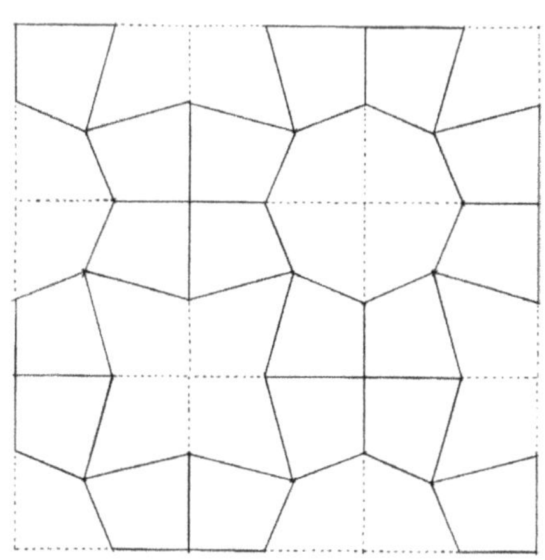

b

Pl. 52.1b. Repeating this economical square construction generates this "equally neat and crisp" geometric scheme, in Özdural's words.

1 Özdural, "Analyses" (unpublished chapter), 186 (labeled as "Construction 59" by Özdural).

Translation and Transcription (fol. 195r [53])

Fol. 195r [53]

In this figure, line GZ bisects angle DGE, on line GD.[a]

a The text is composed of two vertical lines, from bottom to top, along the left side of the figure.

درين شكل خط جٓ زٓ زاويۀ دٓ جٓ هٓ را بدو نيم كند بخط جٓ دٓ.

Construction 53 (fol. 195r [53])[1]

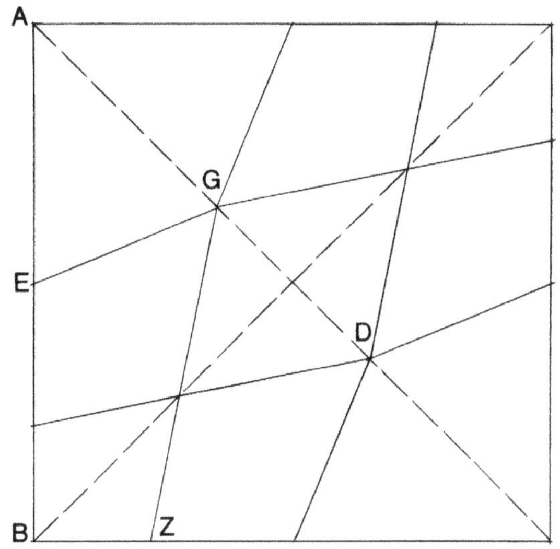

a

Pl. 53.1a. Construction of a square with a central rhombus, combining kites, octagons and tetragonal stars.

This square construction is subdivided into kites surrounding a central rhombus, which is connected to two quarter-octagons at both corners, along one diagonal. The other two corners contain quarter-tetragonal stars. The intersecting diagonals of the square, drawn in red ink as auxiliary construction lines, are accompanied by a brief text explaining that line GZ bisects angle DGE on line GD. It does not mention the assumed unit length, which is AE (half of side AB of the square), defining the radius of the quarter-octagon at the upper left corner.

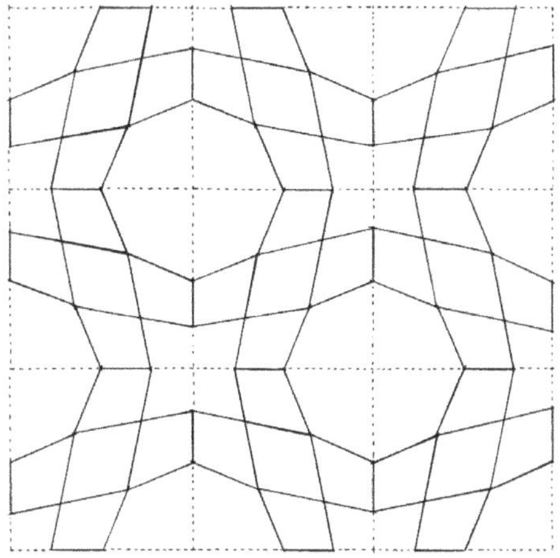

b

Pl. 53.1b. Repeating the square construction generates what Özdural calls this "simple yet interesting" composition.

1 Özdural, "Analyses" (unpublished chapter), 186–87 (labeled as "Construction 60" by Özdural).

Translation and Transcription (fol. 195v [54])

Fol. 195v [54]

With any compass opening mark [on] lines [DE] AD, AK.[a] From point D erect perpendicular DE on BG, and make EZ equal to DE. Draw DZ and inscribe arc HET, so that ET is one-third of EH. Join line DT and from it cut off DL equal to DK. On the rectilinear extension of AL obtain point M. Draw MN parallel to DT. Then from point N line NG becomes a line [number] 6. Thus obtain the compartment (*khāna*) [of the repeat-unit] and complete it. God knows best.[b]

---

a The lines AD, AK are lines [number] 5.
b The text is composed of four horizontal lines below the figure.

بهر فتح که خواهد بر خطها دٓ هٓ آ دٓ آ کٓ نشان کند و از نقطهٔ دٓ عمود دٓ هٓ بر بٓ جٓ آرد
و هٓ زٓ مثل دٓ هٓ سازد و دٓ زٓ بکشد و قوس حٓ هٓ طٓ بگرداند چنانک هٓ طٓ ثلث هٓ حٓ
باشد و خط دٓ طٓ بپیوندد و از وی دٓ لٓ مثل دٓ کٓ فصل کند و از استقامت آ لٓ نقطهٔ
مٓ بگیرد، و مٓ نٓ موازی دٓ طٓ بکشد و از نقطهٔ نٓ خط نٓ جٓ خط ۶ باشد. خانه بگیرد و
تمام کند والله اعلم.

CONSTRUCTION 54 (fol. 195v [54])[1]

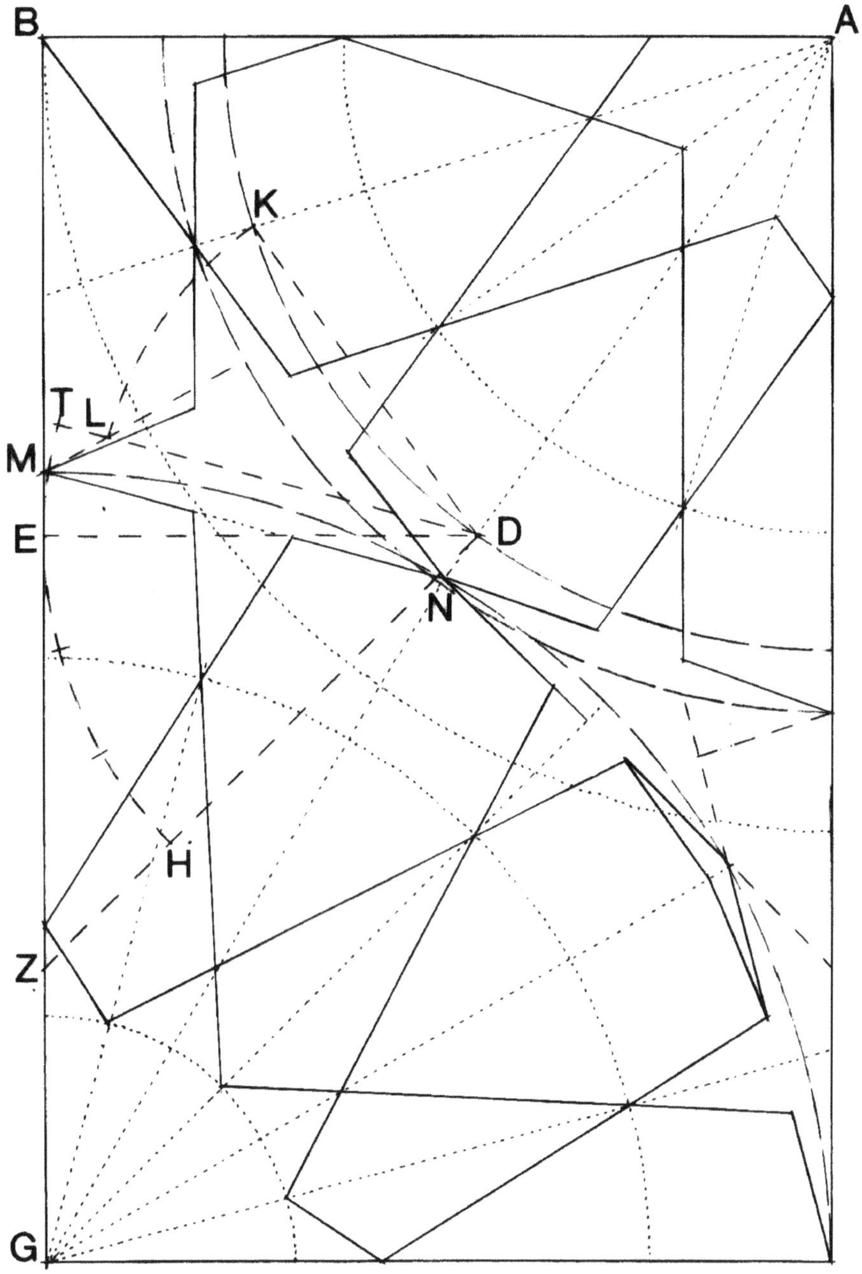

Pl. 54.1. Construction of a rectangle with pentagonal, decagonal, and dodecagonal stars.

1 Özdural, "Analyses" (unpublished chapter), 187–89 (labeled as "Construction 61" by Özdural).

For the commentary on Construction 54, turn to the next page.

Pl. 54.1. See previous page for redrawing.

This construction with radial axes involves the division of opposite angles of a rectangle into five and six parts. It combines pentagonal, decagonal, and dodecagonal stars with irregular hexagons and kites. Özdural observed that, as in the case of Construction 43, this construction concerns a repeat-unit that was well known at the turn of the thirteenth and fourteenth centuries. The proposed amendment here is an attempt to integrate two different systems, namely pentagonal and hexagonal symmetries. The popular traditional pattern that inspired this construction is analyzed by Özdural below, in pl. 54.2a–b. The author of the present construction devised a method for regularizing the interstitial pentagonal stars formed between the quarter-decagonal and dodecagonal stars in the upper right and lower left corners, respectively. The uninked incised concentric quarter-circles and radial lines illustrate the standard procedure for constructing the decagonal and dodecagonal stars separately. The proposed alteration is shown by auxiliary construction lines in red ink and explained in the text. This amounts to modifying the pentagonal star into a slightly irregular interstitial pentagonal star, so that it can be adapted to hexagonal symmetry. In the new position of point N, diagonal MN of the pentagonal star corresponds to the side of the dodecagon and also becomes equal to the side of the decagon. The new pentagonal star, which is slightly irregular, provides an agreeable link between the two parent stars. Özdural observed that the dodecagon was a new type devised for this particular pattern, with the angles at its vertices close to those of the regular pentagonal star. However, the drawing was not finished (a full and a half-pentagonal star on the right side are not defined in solid lines), suggesting that the problem remained unsolved in this construction.

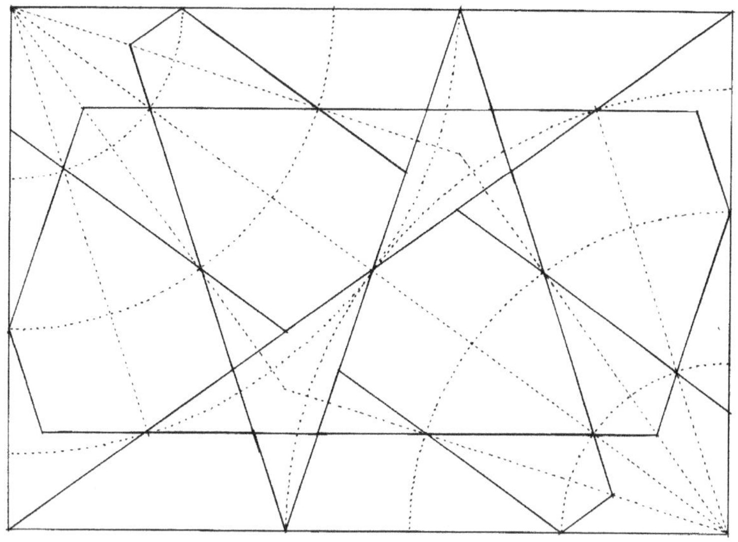

a

Pl. 54.2a. The traditional pattern that inspired Construction 54 is composed of a pair of quarter-decagonal stars at diagonally opposite angles of a rectangular frame, so that two regular pentagonal stars are formed between their vertices. The arrow-shaped six-sided polygons on the radial axes, generated from the extended arms of the decagonal stars, form pentagonal stars when joined together. In this configuration, the arms of the pentagonal and decagonal stars have the same angle (36°).

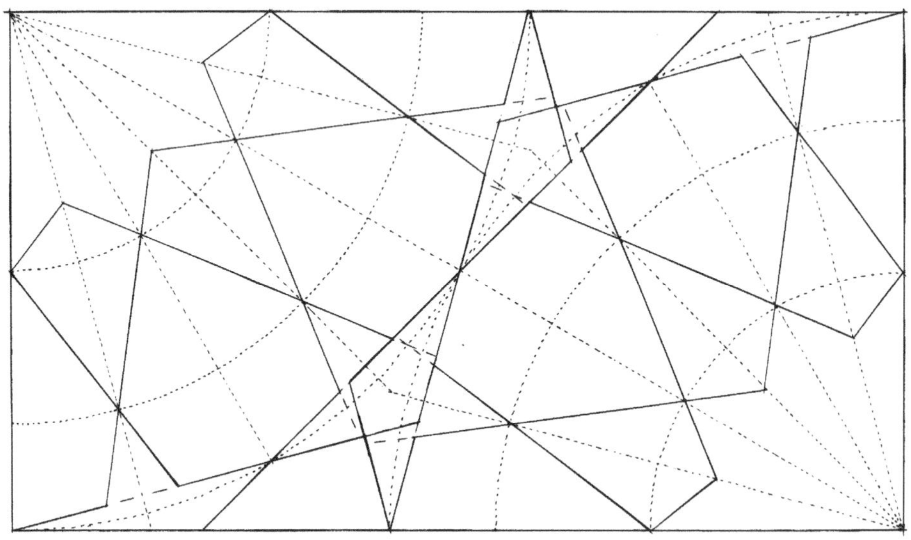

b

Pl. 54.2b. Similar schemes were known since Seljuk times. For instance, when the arms of a pair of quarter-dodecagons are extended similarly, the two opposite arrow-like polygons join together to form pentagonal stars. However, these interstitial stars are irregular and do not provide a harmonious link between the parent stars.

Translation and Transcription (fol. 196r [55])

Fol. 196r [55]

The most correct proportions [of this repeat-unit] are as follows: Let AT be one-fourth of AB. BG is a line [number] four and EB is a line [number] 8. EZ and EH are made equal to AW, which is half of AG. What is required is thus achieved.[a]

a The text is composed of three vertical lines, from bottom to top, on the left side of the figure.

اصح نسبتها آنست که آ طَ ربع آ بَ باشد. بَ جَ خط چهارست و هَ بَ خط ۸، آ وَ نصف
آ جَ هَ زَ. هَ حَ مثل آ وَ سازد مطلوب حاصل باشد.

Construction 55 (fol. 196r [55])[1]

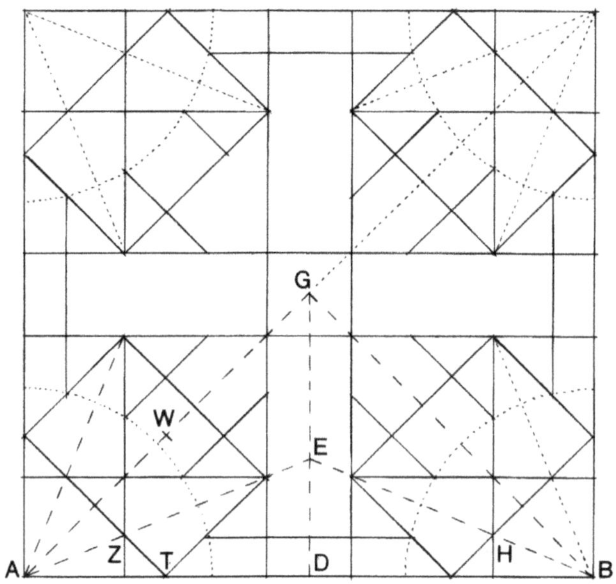

Pl. 55.1. Construction of a square with a central latticework of eight arms.

In this construction, the central lattice originates from quarter-octagonal stars in each corner. This pattern, with four-fold symmetry, is based on the derivatives of a square. Dotted auxiliary construction lines illustrate the construction procedure explained in the text. Uninked incised guiding lines, which consist of quarter-circles and radial lines issuing from all angles of the square, indicate how to complete the construction. The text explains the "most correct proportions" of this pattern, in which AT is one-fourth of AB. In it, BG is "line [number] 4" and EB "line [number] 8." Also, AW equals half of AG, EZ, and EH. When the required proportion is achieved, all the horizontal, vertical, and diagonal arms of the latticework have the same width.

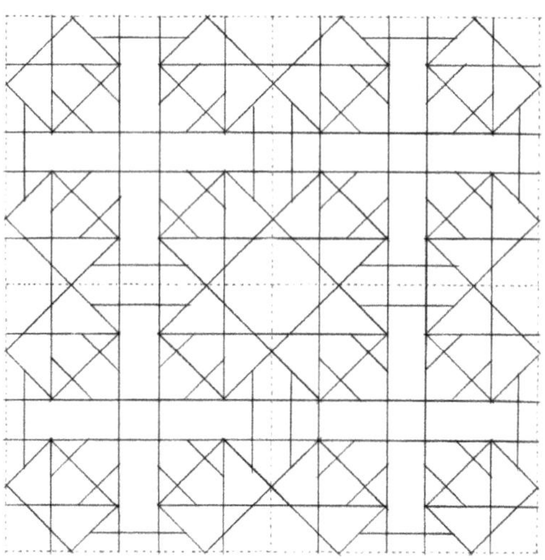

Pl. 55.2. Repeating the square pattern generates this geometric scheme, where octagonal stars are surrounded by a latticework all of whose widths are equal.[2]

1 Özdural, "Analyses" (unpublished chapter), 189–90 (labeled as "Construction 62" by Özdural).
2 Özdural's drawing only indicates the main outlines of the lattice, without the final pattern formed by the octagonal stars. For early examples of this pattern appearing on mid-eleventh- to early twelfth-century brick monuments in Khurasan, and subsequently on the stone mihrabs and portals of late twelfth- to thirteenth-century monuments in Anatolia, see references in Necipoğlu's chapter, n. 96.

Translation and Transcription (fol. 196r [56])

Fol. 196r [56]

No accompanying text.

Construction 56 (fol. 196r [56])[1]

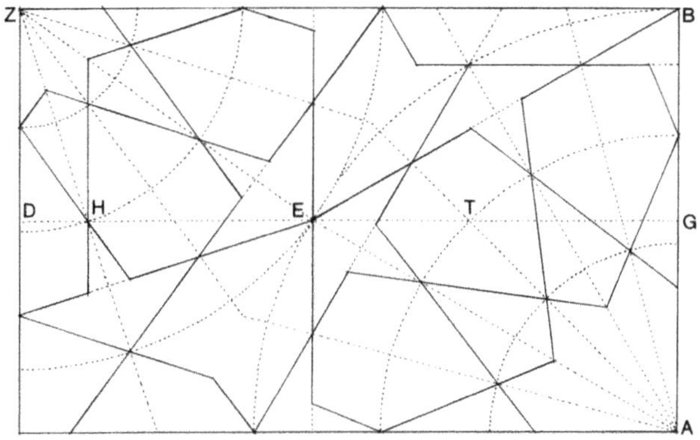

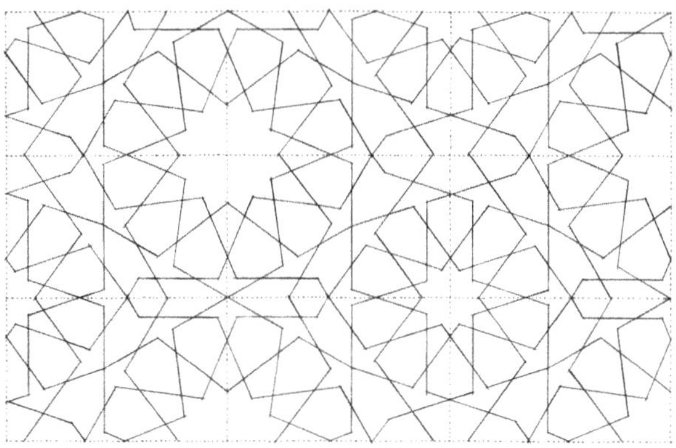

a

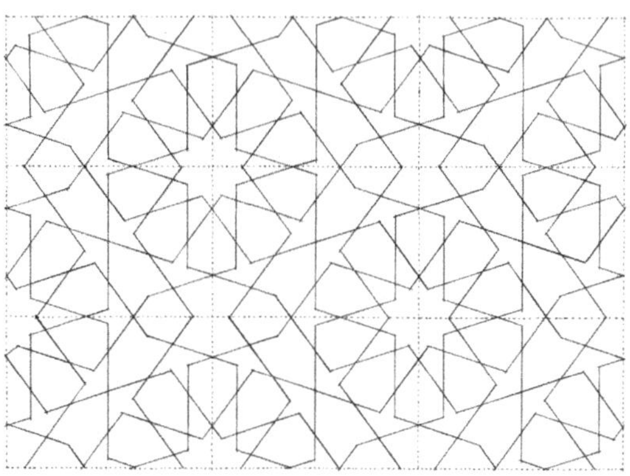

b

Pl. 56.1. Construction of a rectangle combining decagonal and dodecagonal stars.

This textless construction with a system of radial axes involves the division of opposite angles of the rectangle into six and five parts. It is a successful variant of Construction 54, which failed in integrating decagonal and dodecagonal stars in opposite corners with intermediate pentagonal stars. Based on uninked incised lines, Özdural deduced the construction procedure. Instead of modifying the pentagonal stars of Construction 54, the author linked both geometric systems by two similar-looking arrow-shaped polygons facing each other. This is accomplished by changing the position of the decagonal and dodecagonal stars so that from point A a "line [number] 6" is drawn (making a 60°-angle with the horizontal), until it intersects GD at point E. Then a "line [number] 5" issuing from point E is drawn; it intersects BZ at point Z, so as to determine the length of the outer rectangular frame. These two lines are joined at point E, where the two arrows meet. The main objective here is to obtain similar arrows that belong to different geometric systems. This can be achieved when the size and shape of the two stars, drawn with the aid of concentric quarter-circles and radial lines, have the correct relationship.

Pl. 56.2a–b. By repeating the rectangular construction, an elaborate geometric scheme (a) is generated that became one of the favorite ornamental compositions in Islamic art. When this scheme is compared with another one based on the standard pattern, composed of two similar decagonal stars (b), one sees how it gained sophistication by the integration of pentagonal and hexagonal symmetries.[2]

1 Özdural, "Analyses" (unpublished chapter), 190–91 (labeled as "Construction 63" by Özdural).

2 Özdural states that integrating patterns with different geometric systems, seen already in the Seljuk period (e.g., the mosque at Barsian in Iran, 1134), was a trend that gained wider recognition in Ilkhanid times: see Özdural, "Analyses" (unpublished chapter), 191, 254n222. He points out that the pattern composed of two similar decagonal stars is also found in the *Topkapı Scroll*: Necipoğlu 1995, cat. no. 73. For Timurid-Turkmen examples of patterns integrating different geometric systems in the same scroll, he cites Necipoğlu 1995, cat. nos. 30, 39, 42, and 44.

Translation and Transcription (fol. 196v [57])

Fol. 196v [57]

Example by length: The magnitude of TZ is one perpendicular. ZK is one and a half sides.[a] KF is two perpendiculars. Thus, the total is 3 perpendiculars plus one and a half sides.

Example by width: Each [magnitude] is one and a half sides. ZS stays as one and a half sides. We look at SO and we say: MN is one and a half sides. ML is one perpendicular. Then LN remains one and a half sides minus one perpendicular, and this [LN] equals SO. Thus, the total EO amounts to four and a half sides minus one perpendicular. God knows best.[b]

Length: 3 perpendiculars plus one and a half sides.[c]

Width: 4 and a half sides minus one perpendicular.[d]

---

a That is, a side of the rotated square ZL.
b The text is composed of five horizontal lines below the figure. In three additional lines at the bottom of the page, one finds Arabic letters and their numerical *abjad* values, which are used in designating points in the constructions of the *Anonymous Compendium* (see page 182).
c Annotation: The first of two sets of two slanted lines, from upper right to lower left, on the right side of the figure.
d Annotation: The second of two sets of two slanted lines, from upper right to lower left, on the right side of the figure.

مثال طول مقدار طٓ زٓ یک عمودست و زٓ کٓ یک ضلع و نیم و کٓ فٓ دو عمود. پس مجموع
۳ عمود باشد و یک ضلع و نیم.
مثال عرض هر یک ضلع و نیم است و زٓ سٓ یک ضلع و نیم بماند. سٓ عٓ نگاه داریم و گویم
مٓ نٓ یک ضلع و نیم است و مٓ لٓ عمودی. پس بماند لٓ نٓ یک ضلع و نیم إلّا عمودی و این
مثل سٓ عٓ باشد و مجموع هٓ عٓ چهار ضلع و نیم بود إلّا عمودی والله اعلم.

طول ۳ عمود و یک ضلع و نصف ضلعی

عرض ۴ ضلع و نیم إلّا عمودی

Construction 57 (fol. 196v [57])[1]

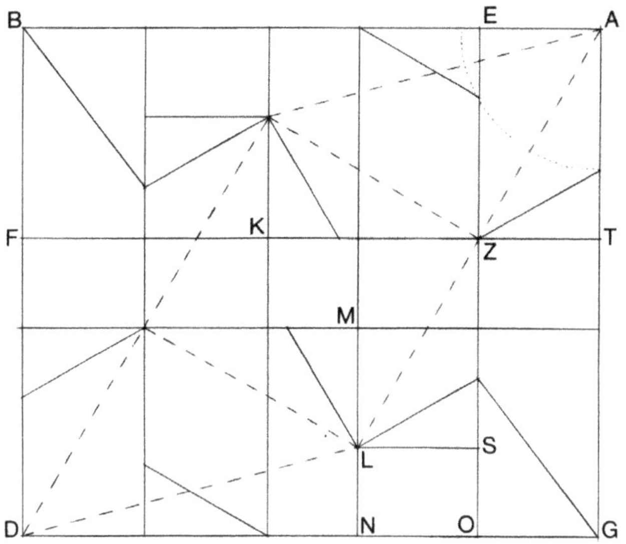

Pl. 57.1. Construction of a rectangle with hexagonal stars, hexagons, and composite pentagonal polygons.

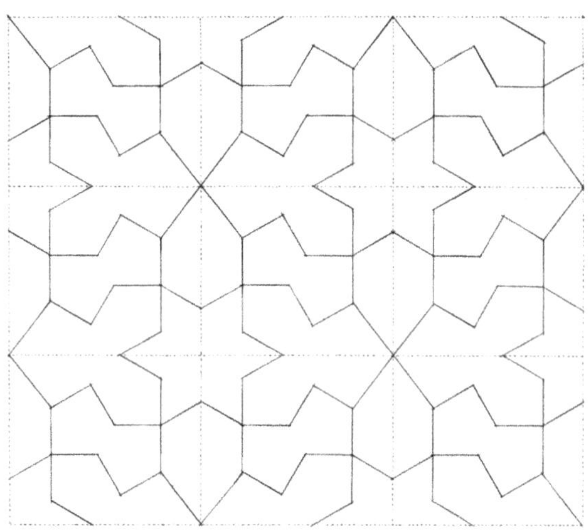

Pl. 57.2. When the rectilinear grid is removed, the pattern itself generates this ornamental scheme.

In this construction of a nearly square rectangular figure, two quarter-hexagonal stars at opposite angles (upper right and lower left corner) generate interlocking hexagons and composite polygons (double-pentagons and triple-pentagons). The simple pattern is based on hexagonal symmetry. The double-equilateral pentagon is similar to the ones in Construction 48 (see pl. 48.1–5). Also occupying the center in the present configuration, this shape is contained in a rotated small rectangle drawn in dotted lines. The vertices of four of the double-pentagons coincide with the radial dotted lines, issuing from the center of the quarter-hexagonal stars. Özdural notes that the construction is superimposed on a rectilinear grid whose perpendicular lines may have been drawn with a set square. The construction procedure is thus based on a modular system derived from the grid. The main text describes the dimensions in terms of two types of modules. The same information is summarized in marginal annotations, which explain that the length is 3 perpendiculars plus 1½ sides, while the width is 4½ sides without a perpendicular. The side corresponds to the side of the hexagonal star, that is, to the side of the equilateral triangle whose "perpendicular" is ZT. The side of the hexagonal star is the main module that determines all the dimensions, the secondary module being the "perpendicular" that corresponds to $1/2\sqrt{3}$. Özdural points out that this is the only construction in this document where the proposed method aims at explaining geometric properties in terms of numerical dimensions based on two modules.[2]

1 Özdural, "Analyses" (unpublished chapter), 191–93 (labeled as "Construction 64" by Özdural).

2 Judging by the letters of the Arabic alphabet and their corresponding numerical values written at the bottom of the fol. 196v, Özdural speculates that Construction 57 marks the end of the *Anonymous Compendium*. He suggests that the two extra folios with constructions of a "different character" beyond those studied so far were added later by the mid-fifteenth-century copyist, presumably Abu Ishaq Kubanani Yazdi. See Özdural, "Analyses" (unpublished chapter), 193. On p. 255n224, he cites Golombek (private correspondence), who believed that the last drawing was added later, and Necipoğlu 1995, 180n113, where it is also proposed that the last drawing might have been added during the Timurid period. See Özdural, "Analyses" (unpublished chapter), 197, where he notes that the next treatise in Ms. Persan 169, on "finger-reckoning" (*angusht shumārī*), was by the Timurid scholar Sharaf al-Din ʿAli Yazdi, an expert on puzzles (*al-muʿammāʾī*), who, together with Abu Ishaq Kubanani Yazdi, may have participated in a gathering where the latter took notes on, or was perhaps the author of, Constructions 58–60.

Translation and Transcription (fol. 197r [58–60])

Fol. 197r [58–60]

No accompanying text.

Construction 58 (fol. 197r [58])[1]

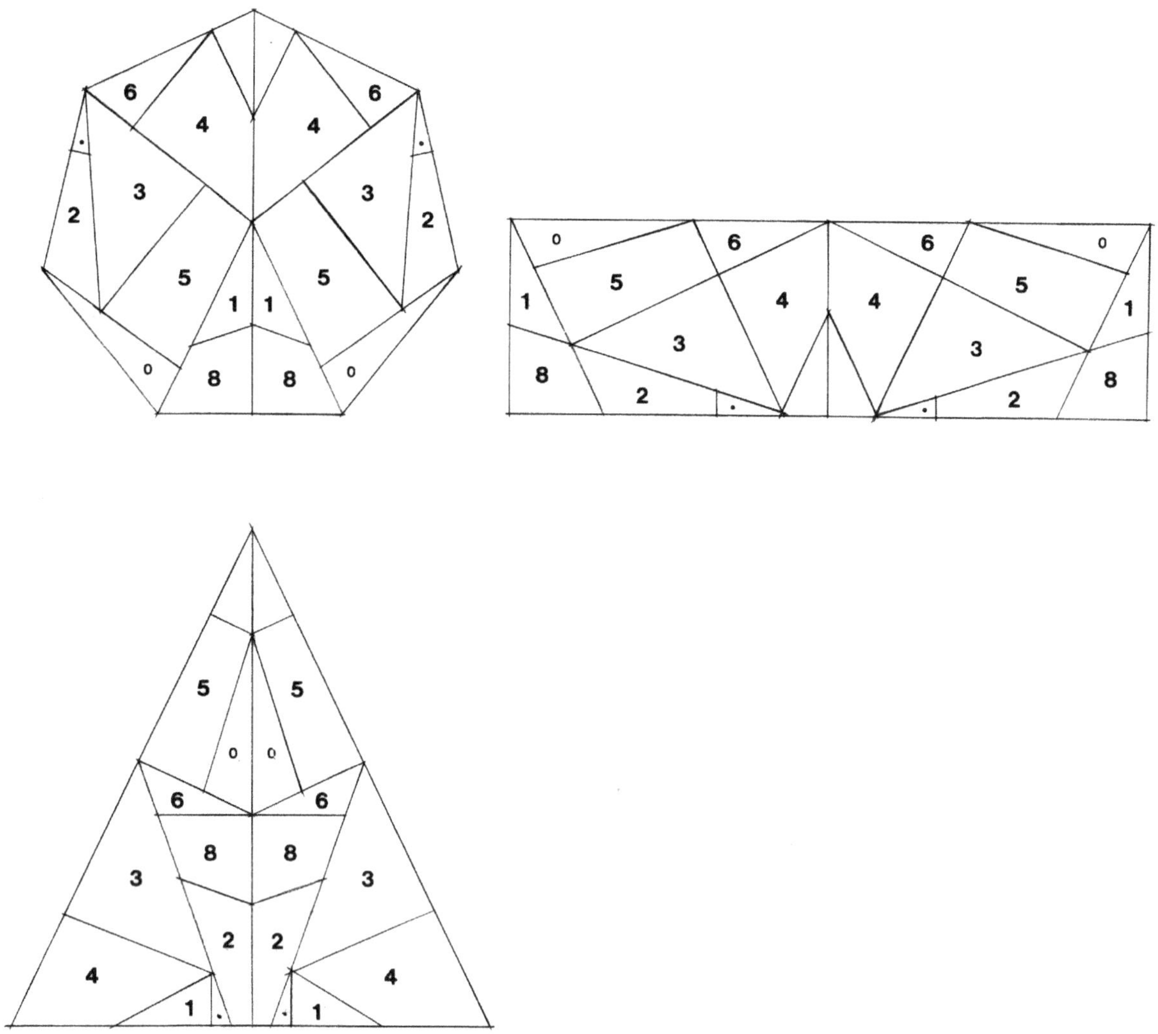

Pl. 58.1. Construction of a heptagon, rectangle, and triangle with equivalent areas.

This page contains three constructions without texts, each consisting of two or three figures (Constructions 58, 59, and 60). Özdural observes that these "cut-and paste" constructions look "much more complicated" than their earlier counterparts in this document. Resembling jigsaw puzzles, they serve no ornamental purposes and "give the impression of being solutions to some recreational problems." Construction 58 consists of three figures: a heptagon, a rectangle, and a triangle. The objective of these figures is to transform the heptagon into a rectangle and a triangle of equal areas. First the heptagon is dissected into triangular and quadrangular pieces, designated by numbers 1 to 8, and by some signs. These are then reassembled into a rectangle and then into a triangle.

1 Özdural, "Analyses" (unpublished chapter), 193–95 (labeled as "Construction 65" by Özdural).

Construction 59 (fol. 197r [59])[1]

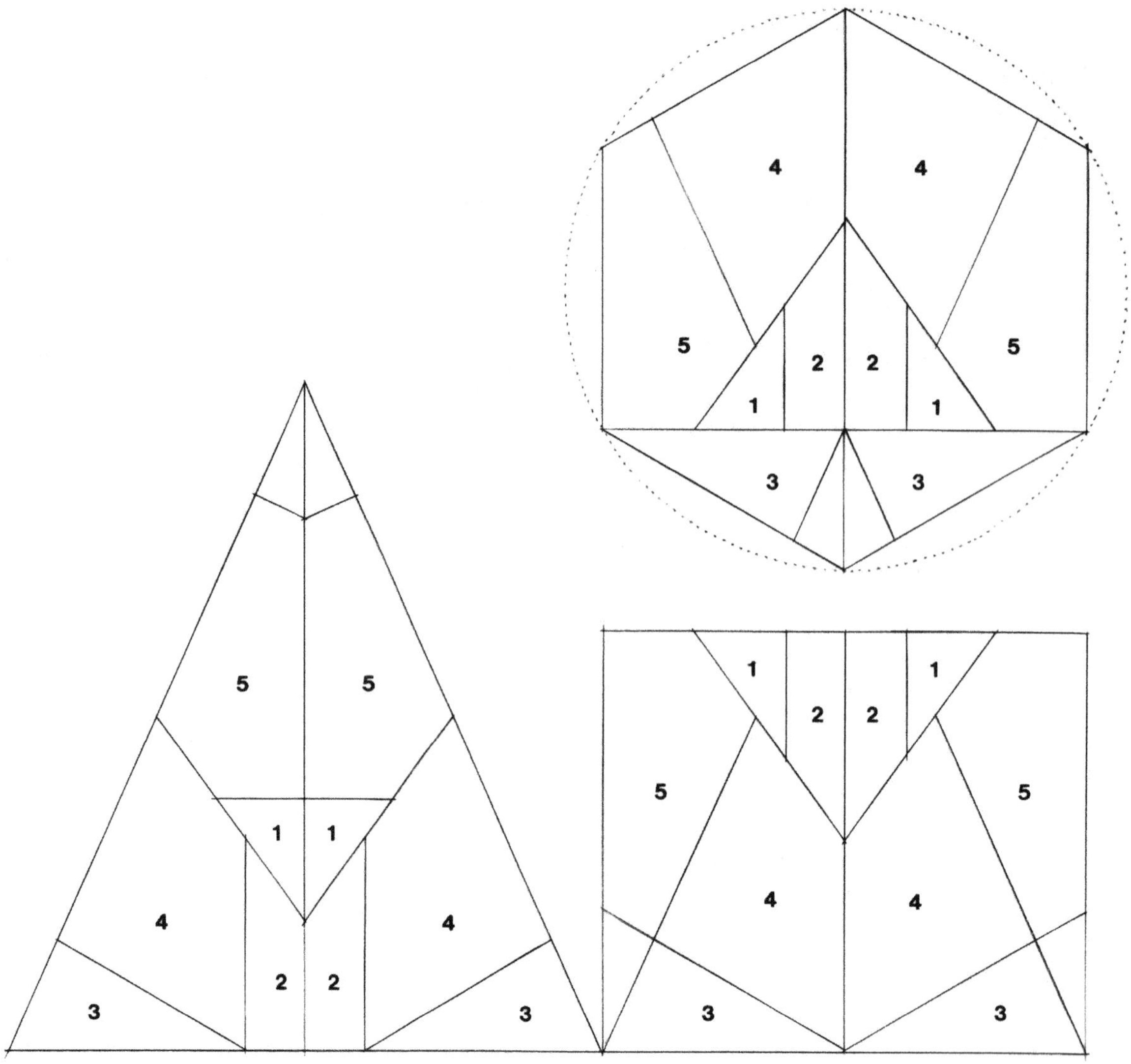

Pl. 59.1. Construction of a rectangle, triangle, and hexagon with equivalent areas.

This construction consists of three figures: a regular hexagon that is transformed into a rectangle and a triangle. The hexagon, inscribed in an uninked incised circle, is dissected into pieces arranged in pairs, numbered from 1 to 5, along with one pair without any sign. The transformation of the hexagon into the rectangle recalls Construction 8, in which a hexagonal star is transformed into a square by way of a rectangle. The present case is more complicated because further subdivisions are needed to transform the hexagon into a triangle.

1 Özdural, "Analyses" (unpublished chapter), 195–96 (labeled as "Construction 66" by Özdural).

Construction 60 (fol. 197r [60])[1]

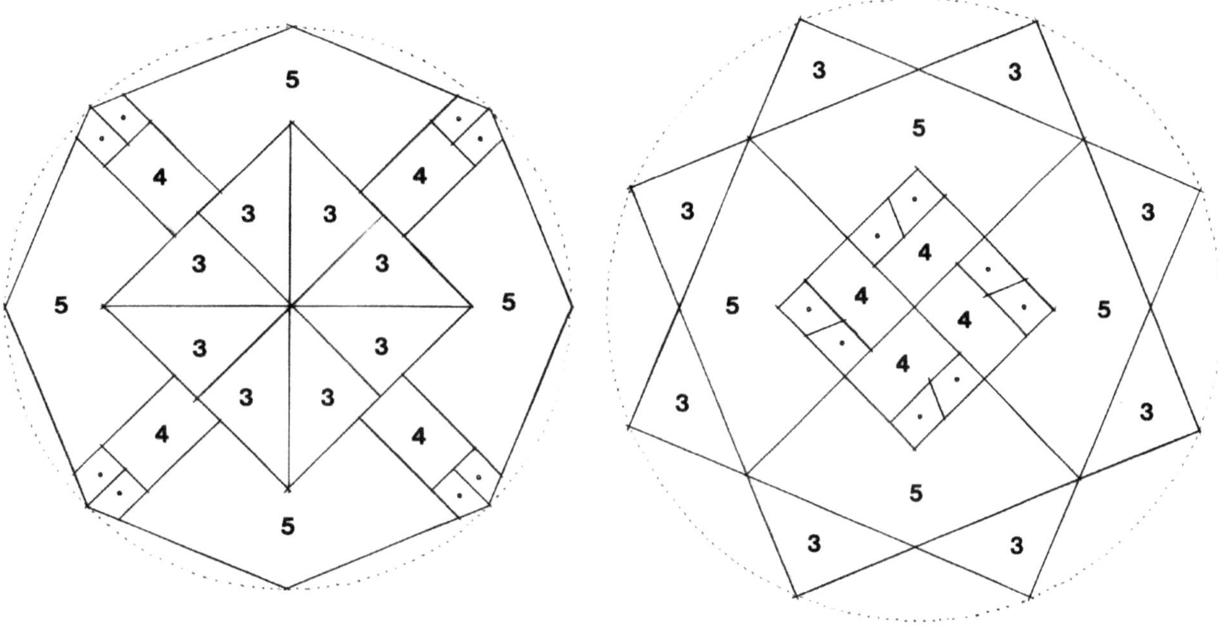

Pl. 60.1. Construction of an octagon and octagonal star with equivalent areas.

In this construction, which consists of two figures, an octagon is transformed into an octagonal star, or vice versa. Both figures are dissected into four sets of pieces, inscribed with numbers from 3 to 5, and a dot, so that the total number of pieces is 24. In comparison with the more elegant and economical transformations of the octagon or octagonal star into a square in Constructions 4, 10, and 22 (see pls. 4.1, 10.1, and 22.1), this construction reveals a trend toward unnecessary complication with the increased number of pieces. After fol. 197v, on which an apparently unrelated couplet is written, there is a sheet with no drawing or writing on either side (fols. 198r and 198v). On the following page, Construction 61 constitutes the last figure of this document.

1 Özdural, "Analyses" (unpublished chapter), 196–97 (labeled as "Construction 67" by Özdural).

Translation and Transcription (fol. 199r [61])

Fol. 199r [61]

No accompanying text.

Pl. 61.1. Construction of a rectangle combining pentagonal and decagonal stars with rotating swastikas.

This construction is based on radial axes combining pentagons, decagons, and squares. The pattern is generated by a pair of quarter-decagonal stars at diagonally opposite angles of the rectangular construction. The space in between the stars is filled with swastika motifs in squares revolving around small pentagonal stars. The uninked incised lines and quarter-stars reveal the construction procedure, drawn so that the diagonal AB becomes a "line [number] 5" and is divided into four equal parts. The diagonal AB and the line passing from points D and I, which intersect at the center of the composition (D), constitute the main axes of symmetry. The drafting technique and general character of this construction differ from previous ones, in that it occupies a whole page while the other side of the folio (fol. 199v) is empty. The figure is drawn in red ink, with some details highlighted in a brownish-green wash, unlike any other pattern in the *Anonymous Compendium*, although colors were commonly used in scrolls such as the *Topkapı Scroll*. Özdural noted that this textless figure resembles "drawings on separate sheets of paper that are compiled in later scrolls," and suggested that it was likely "produced in the Timurid world around the mid-fifteenth century."

Construction 61 (fol. 199r [61])[1]

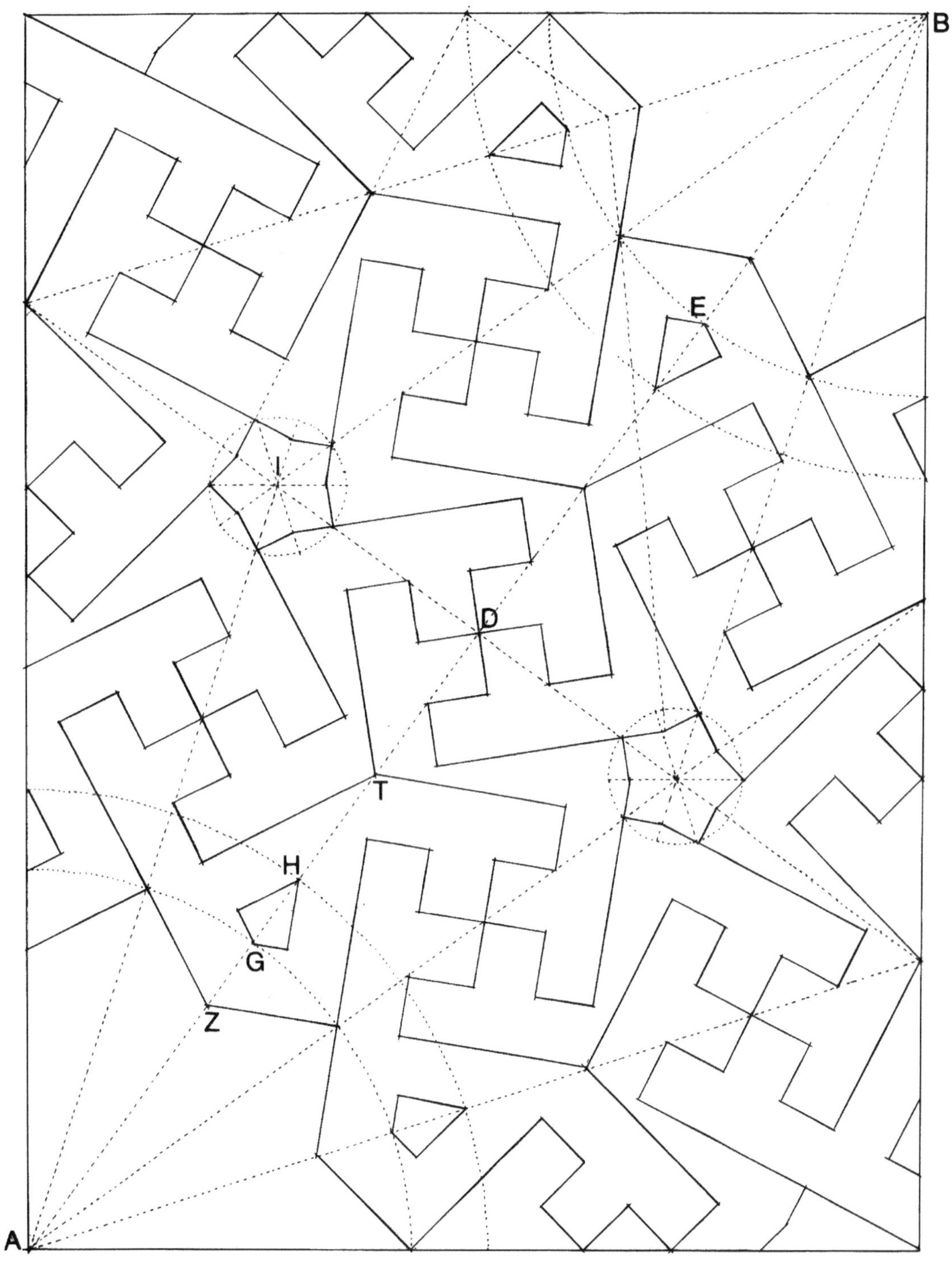

1 Özdural, "Analyses" (unpublished chapter), 197–99 (labeled as "Construction 68" by Özdural).

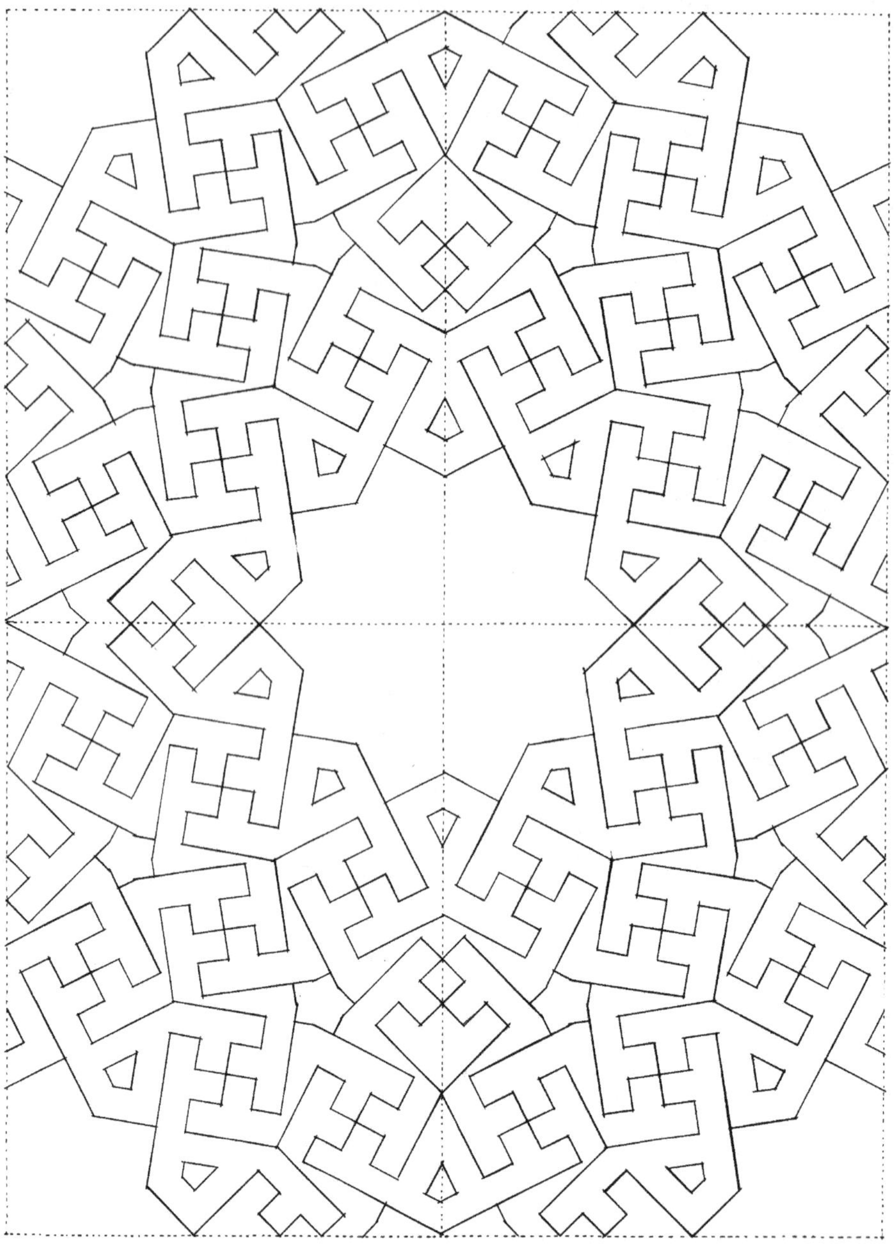

Pl. 61.2. According to Özdural, while comparable swastika motifs interlocking with polygons and stars appeared in Anatolian Seljuk monuments, this pattern gained popularity in Iran during the fifteenth century, as is seen in five patterns in the *Topkapı Scroll.* He hypothesizes that Construction 61, probably intended for a scroll, "seems to have fallen somehow in the hands of Abu Ishaq Yazdi," who attached it to the end of the *Anonymous Compendium* "at the last stage" of production, during the "binding process." Repetition of the pattern generates "this interesting composition of swastikas whirling around the decagonal and pentagonal stars."

End of facsimile of

Fī tadākhul al-ashkāl al-mutashābiha aw al-mutawāfiqa,

Bibliothèque nationale de France, Ms. Persan 169, folios 180r–199r.

The reduced-scale facsimile begins on page 376.

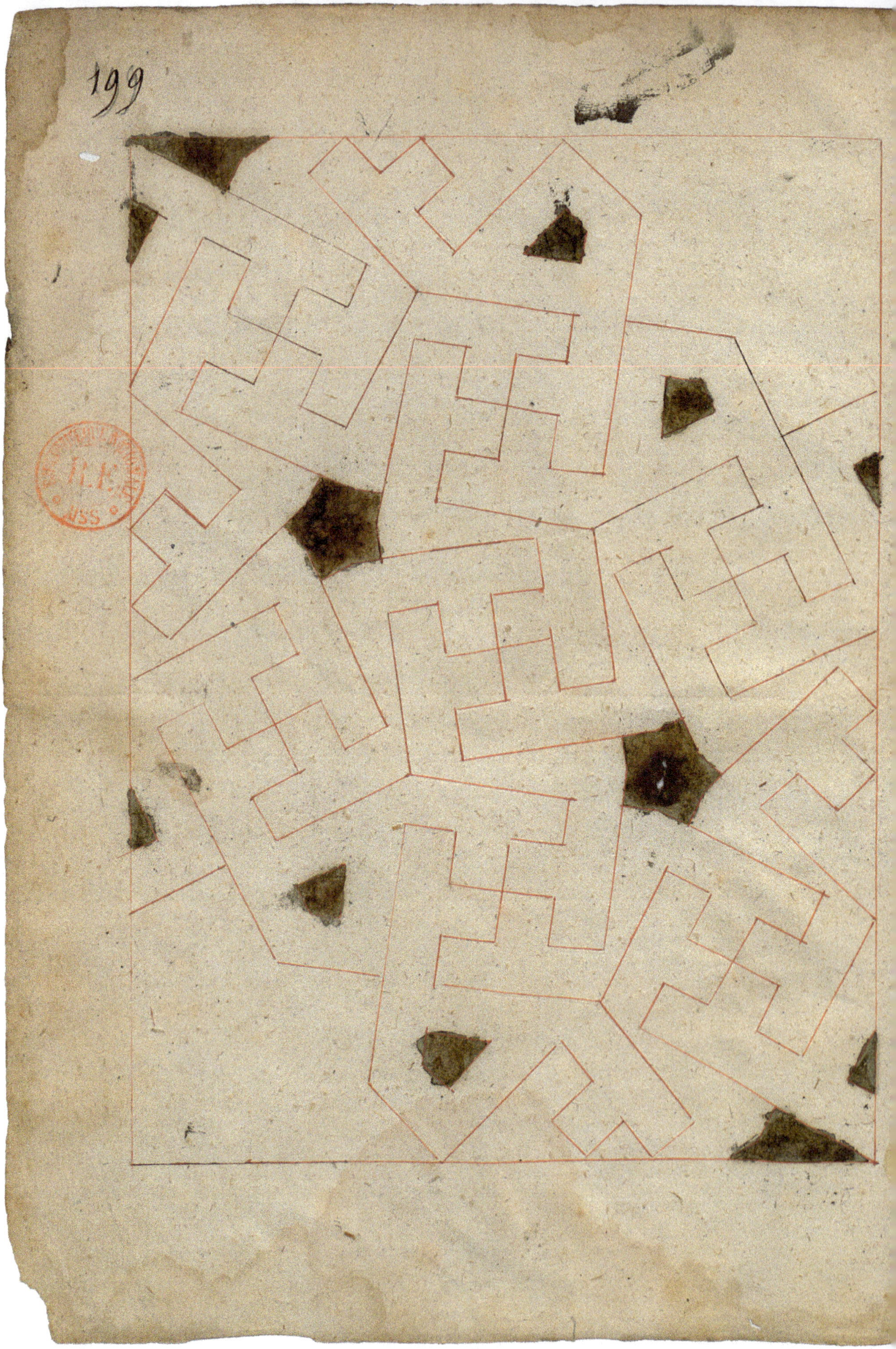

Fol. 199r [61]

Folio 197v contains a Persian couplet; folios 198r and 198v are blank.
For more description of the manuscript, see Richard 1989, 187.

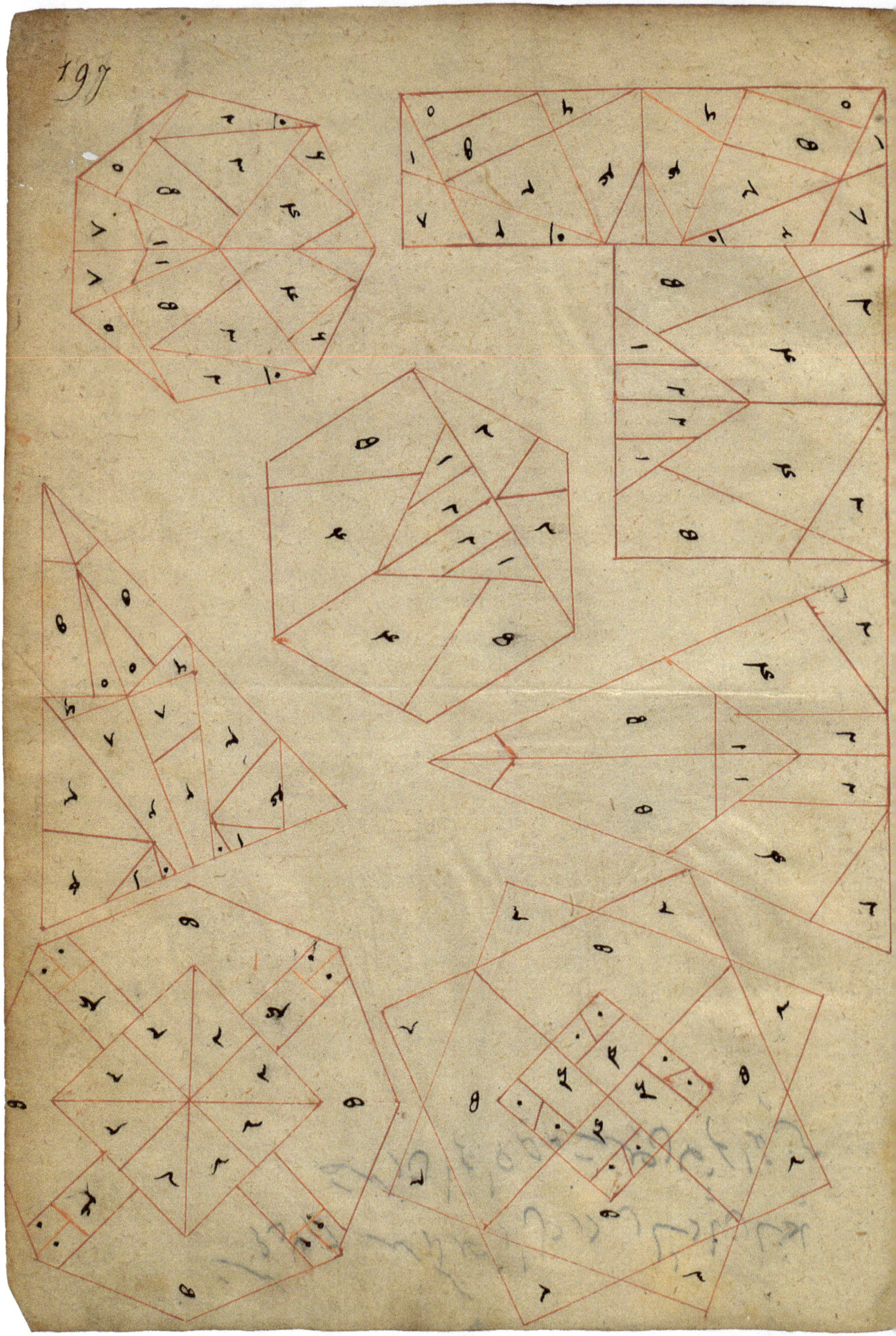

Fol. 197r [58 (center left, top left, top right), 59 (center, center right), and 60 (bottom left and right)]

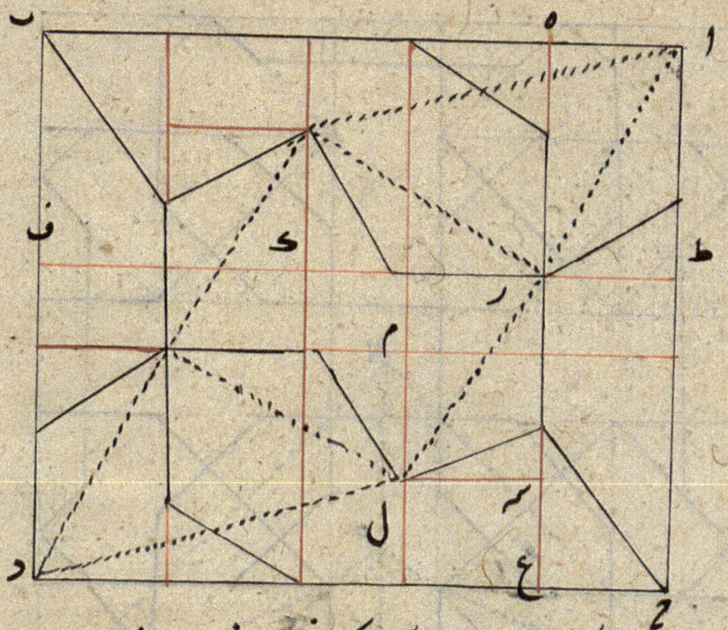

طول سه عمود و یک ضلع و نصف ضلع

عرض ع ن ضلع و نیم الا عمودی

مثال طول مقدار طز یک عمود ست و زی یک ضلع و نیم و ک ف دو عمود پس
مجموع طه سه عمود باشد و یک ضلع و نیم مثال عرض مربع یک ضلع و نیم است و ر مه
یک ضلع و نیم باشد سه ع نگاه داریم و گوییم م ن یک ضلع و نیم است و م ل عمودی پس
باشد ل ن یک ضلع و نیم الا عمودی و این مثل سه ع باشد و مجموع ن ع چهار
ضلع و نیم بود الا عمودی والله اعلم

ابجد هوز حطی کلمن سعفص قرشت ثخذ ضظغ

ا ب ج د ه و ز ح ط ی ک ل م ن س ع ف ص ق ر ش ت ث خ ذ ض ظ غ
۱ ۲ ۳ ۴ ۵ ۶ ۷ ۸ ۹ ۱۰ ۲۰ ۳۰ ۴۰ ۵۰ ۶۰ ۷۰ ۸۰ ۹۰ ۱۰۰ ۲۰۰ ۳۰۰ ۴۰۰ ۵۰۰ ۶۰۰ ۷۰۰ ۸۰۰ ۹۰۰ ۱۰۰۰

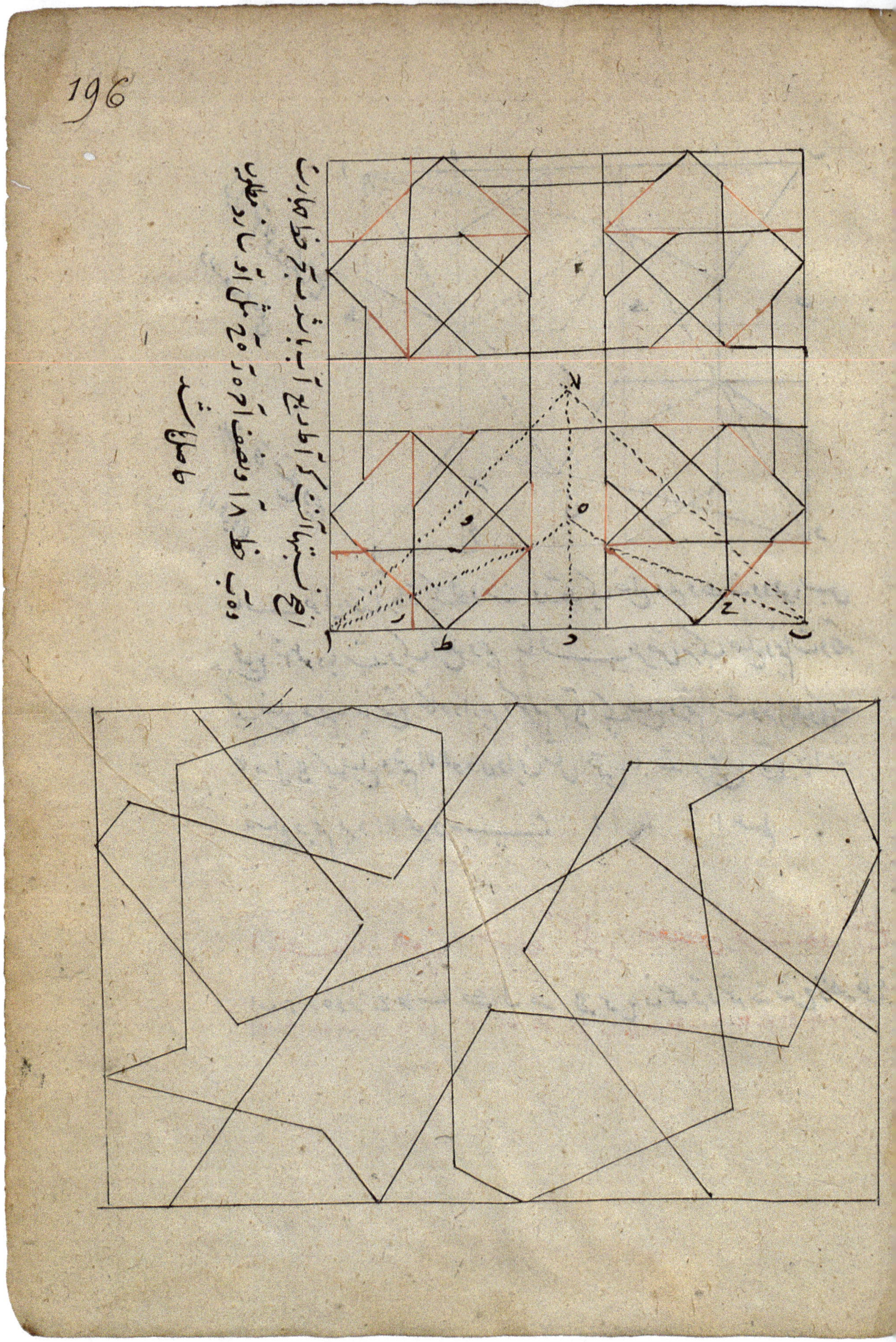

Fol. 196r [55 (top) and 56 (bottom)]

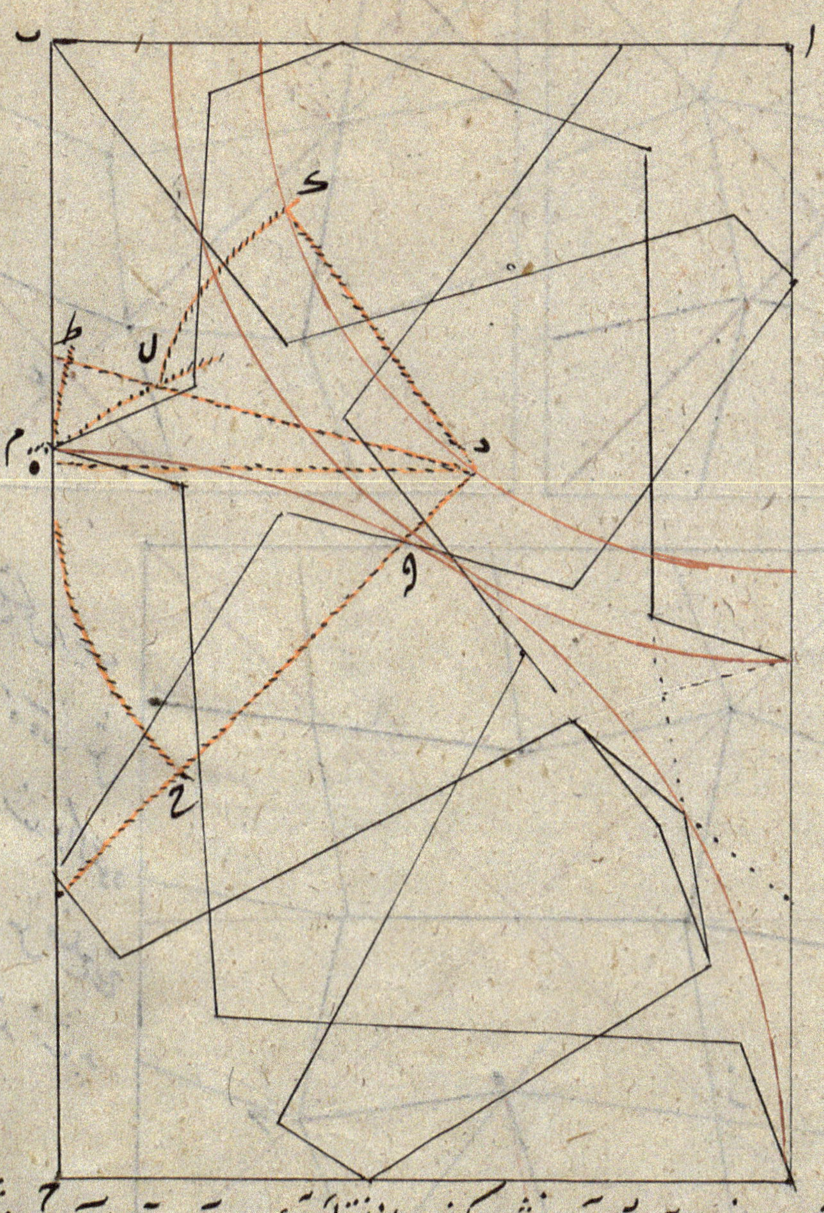

بهر فسیح که خواهد بر خطها ده آد آی نشان کند و از نقطه د عمود ده بر بح ارد و ده ر مثل
ده سازد و دز بکشد و قوس ح ه ط بگرداند چنانک ه ط مثل ه ح باشد و خط د ط ببوند د
و از وی د ل مثل د ک وصل کند و از استقامت آل نقطه م بگیرد و م ه موازی ح ط
بکشد و از نقطه ه خط ه ح خط ح ه باشد خانه بگیرد و تمام کند والله اعلم

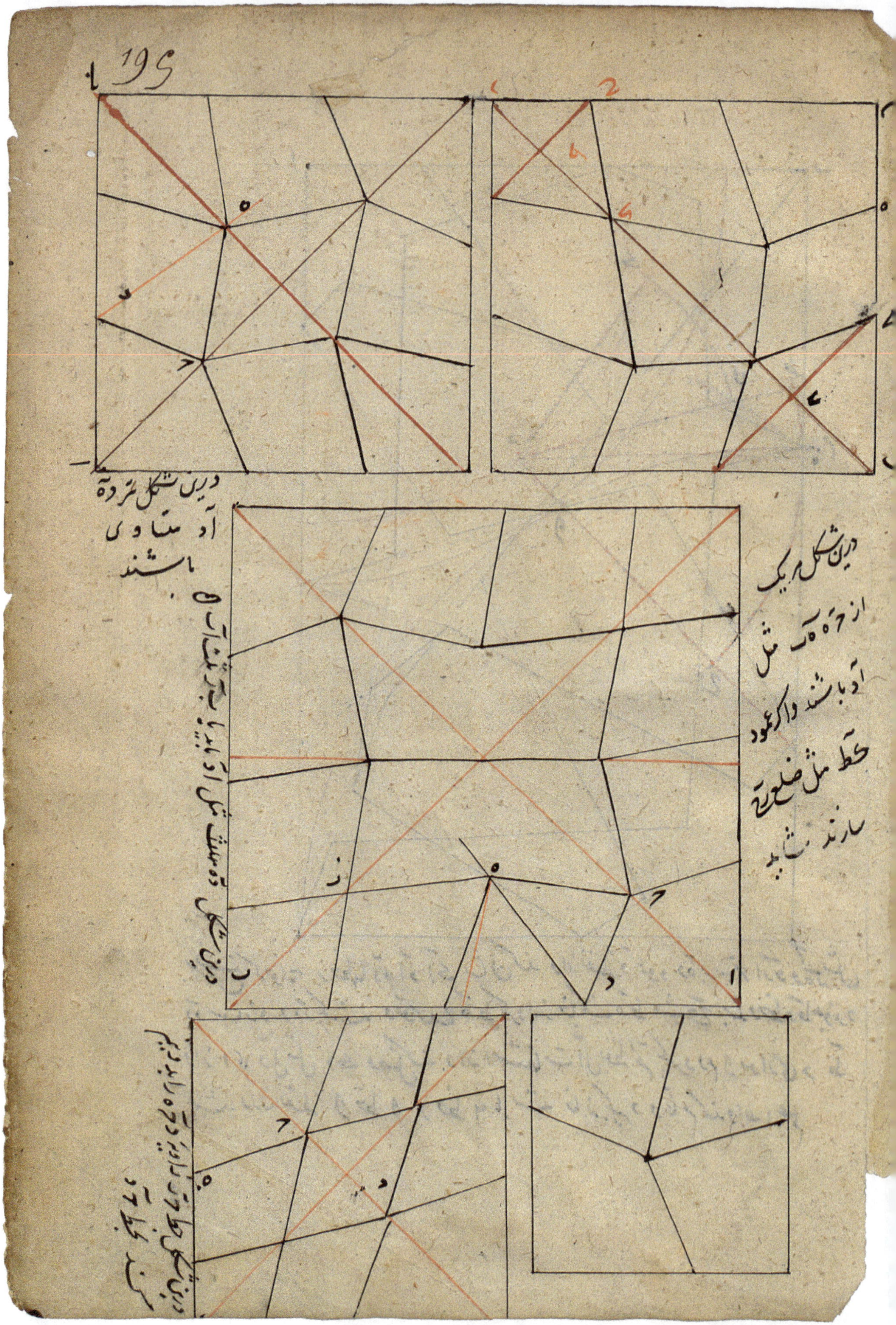

Fol. 195r [49 (top right), 50 (top left), 51 (center), 52 (bottom right), and 53 (bottom left)]

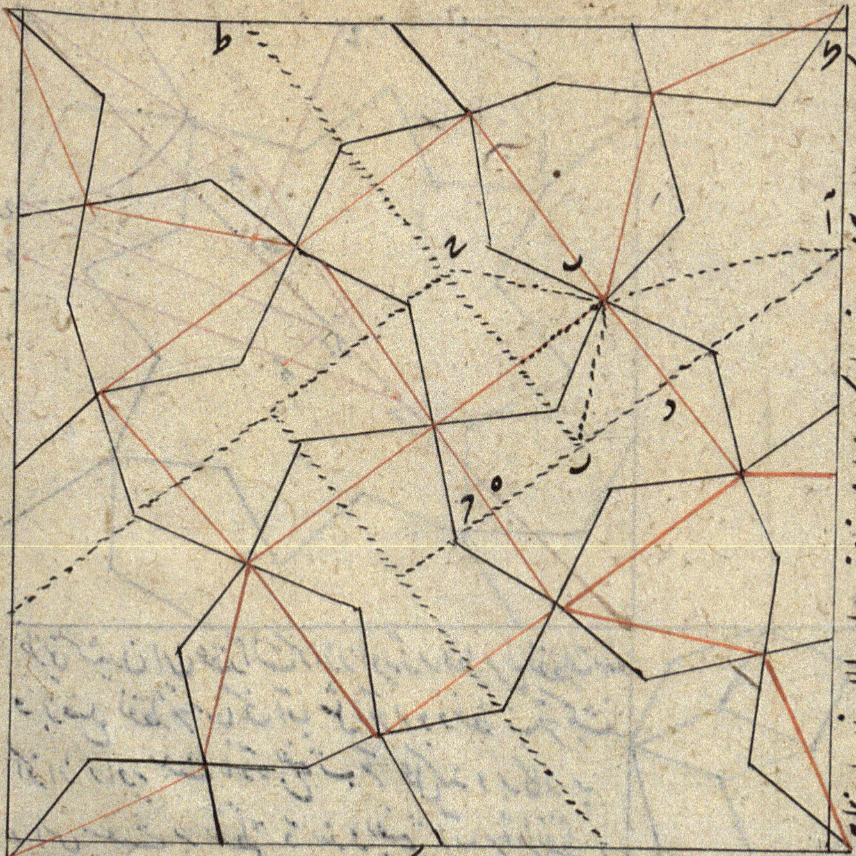

طریق کشیدن این عقد آنست و نسبت او که در مربع مفروض بر احد الاضلاع او نقطهٔ آ
کیف ما اتفق وضع کند و خط آب آد بر آه صفت بیرون کند و از نقطهٔ ب بر خط
آد عمود اد اخراج کند و بر عمود آد سطح ده کند چنانک طول او ضعف
عرض بود که آن عمود آد است انگه بنگرد که نقطهٔ ه از سطح مستطیل بر وسط مربع مفروض

افتاده است یا نه اگر بر وسط مربع بود این نسبت مطلوبست و اگر بر وسط مربع نباشد از نقطهٔ ه بنقطهٔ آ
که موضوع است خطی آه بکشد و از مرکز مربع خطی دیگر موازی خط آه بکشد تا بر ضلع مربع مرکز مربع که
مطلوبست معلوم شود والله اعلم نوعی دیگر هم در نسبت این عقد و آنچنانست که بر ضلع مربع مفروض
نقطهٔ آ کیف ما اتفق وضع کند و خط آب آد بر آه صفت بیرون کند و بر خط آد عمود ده بجا
که خواهد اخراج کند و از زاویهٔ ر که قائمه است خط چهار بیرون کند تا متقاطع خط آب شود بنقطهٔ
ب انگه پرکار بر حرف ب نهد و بفتح ب ر بر عمود ره بنقطهٔ ح بگیرد و باز پرکار بکشید
و بفتح آر آورد و بر نقطهٔ ح نهد و بر عمود ره نقطهٔ ط نشان کند چنانک ح ط مثل آر
باشد انگه بنگرد که نقطهٔ ط بر ضلع مربع مفروض افتاده است یا نه اگر افتاده باشد این نسبت
مطلوبست و اگر خارج نیفتاده باشد نه

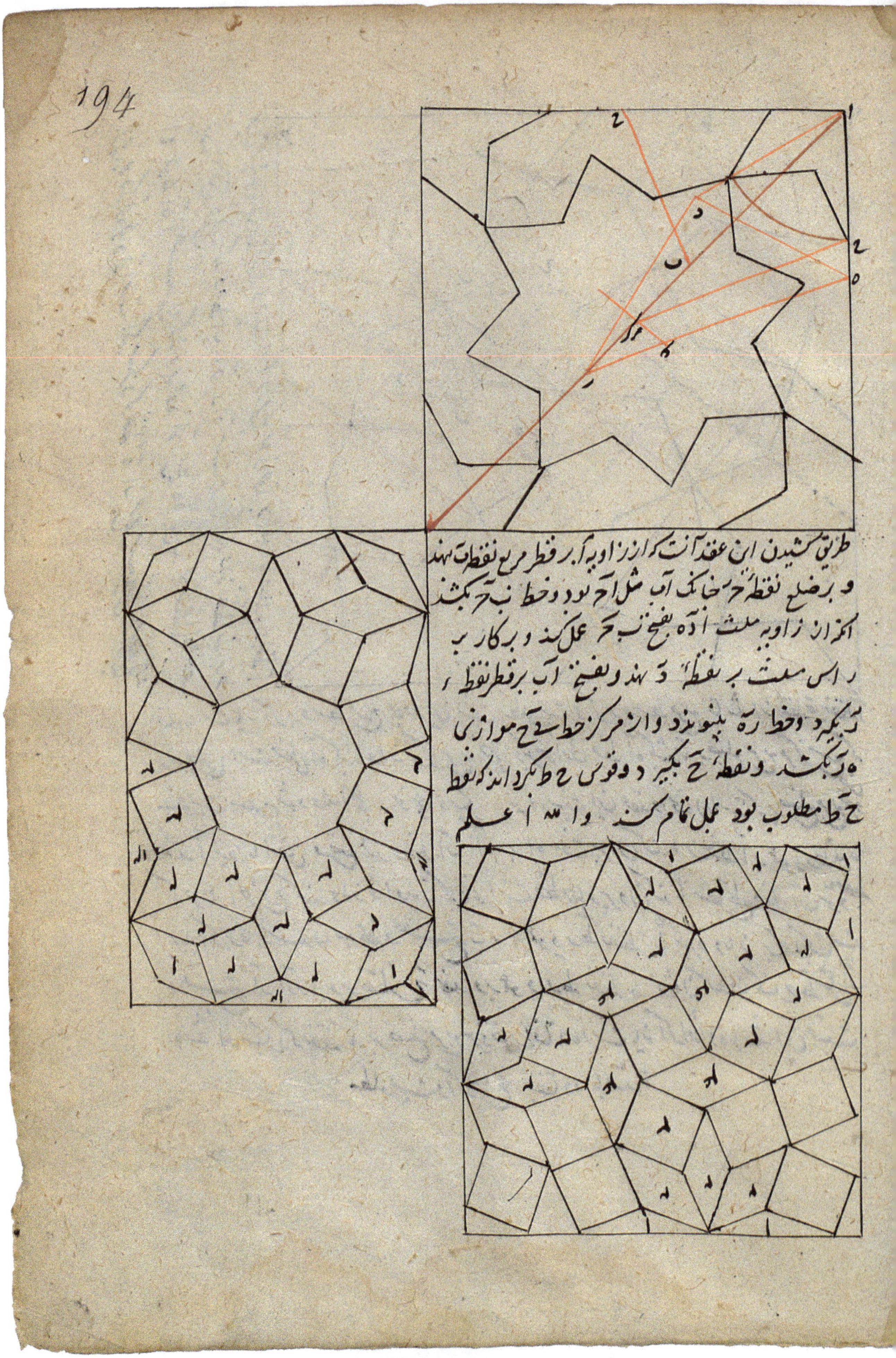

194

طریق کشیدن این عقد آنست که از زاویه آ. بر قطر مربع نقطه ب نهد
و بر ضلع نقطه ح چنانکه آب مثل آح بود و خط ب ح بکشد
انگه از زاویه مثلث آ ده بفتح ب ح عمل کند و برکار ب
راس مثلث بر نقطه د نهد و بفتح آب بر قطر نقطه ر
بگیرد و خط ر ه بپیوندد و از مرکز خط س ح موازی
ه ح بکشد و نقطه ح بگیرد و قوس ح ط بگرداند که نقطه
ح ط مطلوب بود عمل تمام کند والله اعلم

Fol. 194r [45 (top), 46 (center left), and 47 (bottom)]

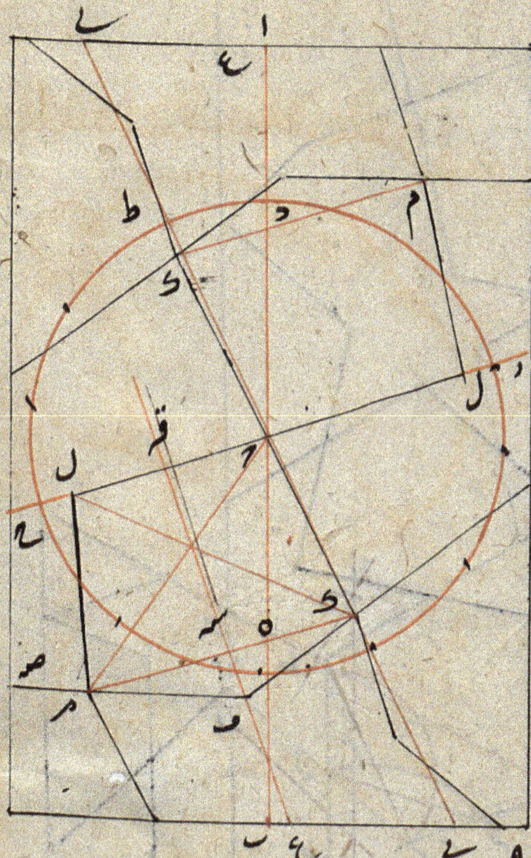

واگر مقدار م صه مثل نصف ح ک سازد عقد ما موینی آید و اگر عقد
ما موینی نبود مقدار م صه کمتر از نصف ح ک بود

ذکر الخواص بالتقریب اول عمود سه ته از مخمس مخالف مثل نصف ب ح
بود و دیگر عرض این عقد دو قسم و طول سه قسم عظیم تقریب است

بباید دانست که طریق کشیدن این عقد از وسط است وآن چنان باشد که چون خط آب وسط
معلوم بود بدو قسم کند بنقطهٔ ح در وسط و برین نقطه دایره در ه ح بگرداند و بر دو
خمس قوس ه د نقطه ز ح بر محیط نشان کند و خط ز ح بکشد که این خط نح باشد بر خط
آب اکنه زاویه زحه یعنی قوس زه چهار قسم کند و خط ح ط ... بکشد و مقدار ... بر
دارد و ح ل ح ک مثل وی سازد و خط ک ل بکشد اکنه از خط ک ل و خط ل ح نقطهٔ م معلوم
و خط م بکشد و از ک م عمود س ع اخراج کند و مرکز ع بگیرد و عمل تمام کند و اگر خواهد باید
پرکار بر نقطهٔ ک نهد و بفسحت ک م نقطهٔ و بگیرد و درین عقد زاویه ل ح ک مثل زاویه
ح ک ف بود و هر یکی قائمه و عشر قائمه باشند   والله اعلم

193

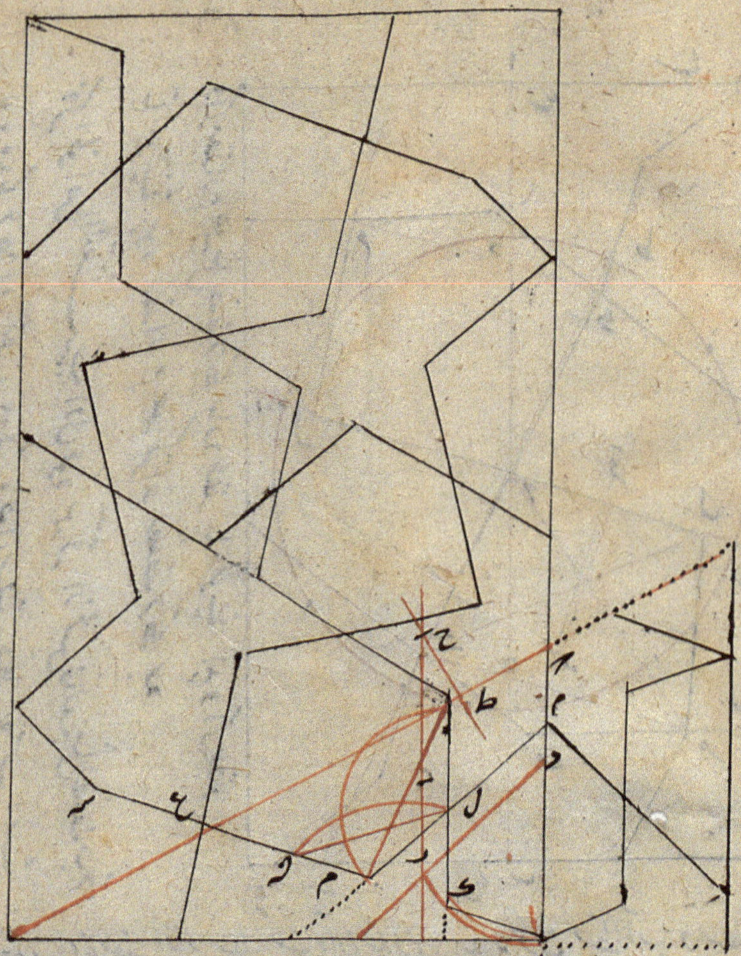

در گونیا اب‌ح بهر فسح که خواهد آد اه نشان کند وده بپیوندد ودر مثل دا حدا
کند وخط آد بپیوندد ورح موازی اح اخراج کند خابکند رح ضعف آر باشد
ومسطر بر نقطهٔ ح و آ نهد وبر خط بح نقطهٔ ط بگیرد وطک موازی حا
برون آرد و تنصیف کند بنقطهٔ ل وخط لم در اخراج کند خابکند لم مثل لط
باشد وزاویهٔ لم مثل زاویهٔ طلم عمل کند وخط طو سه اخراج کند وعمل
تمام کند که زوایا ع قوائم
باشند والله
اعلم

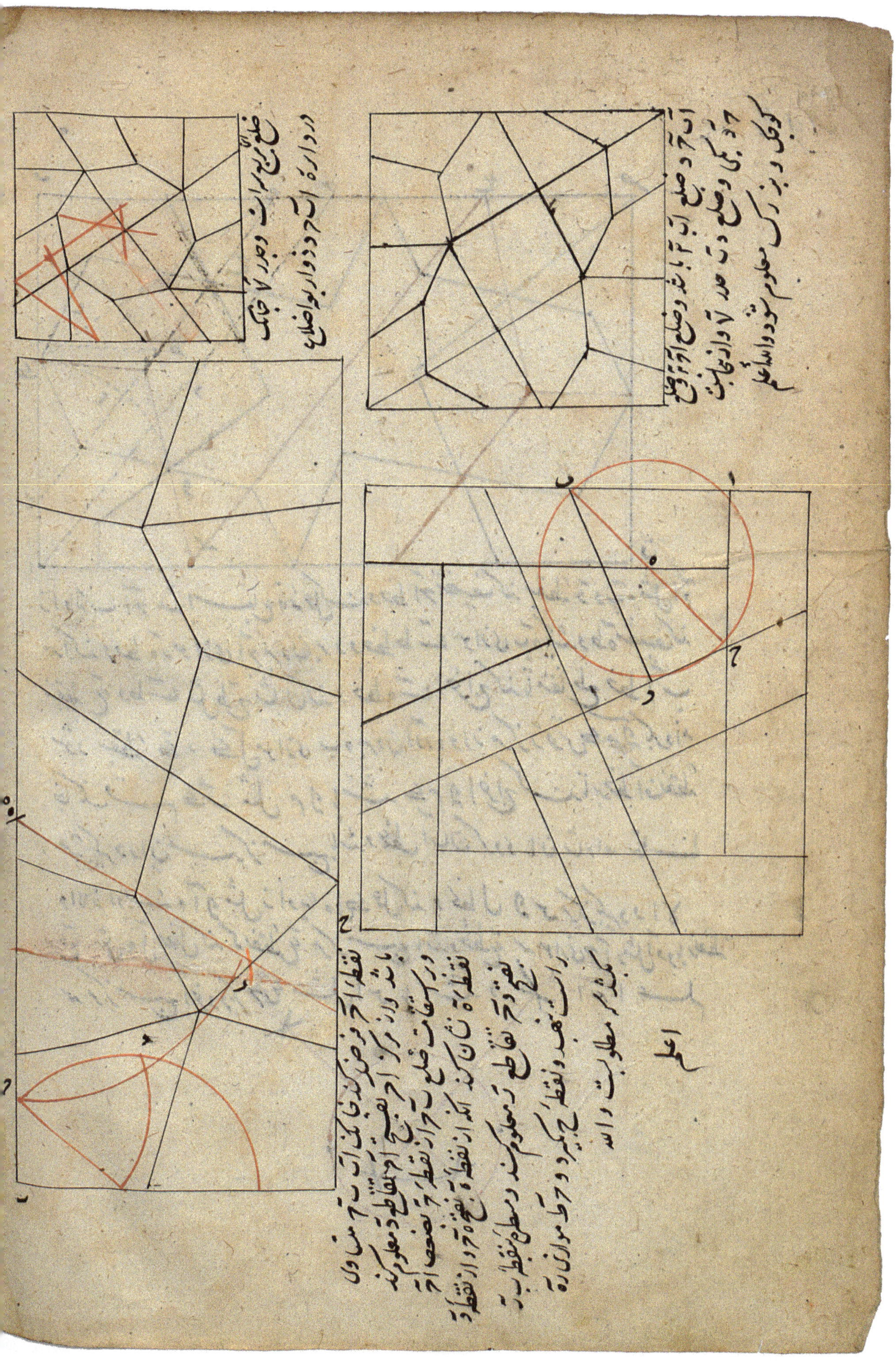

Fol. 192v [39 (top right), 40 (top left), 41 (center right), and 42 (bottom left)]

192

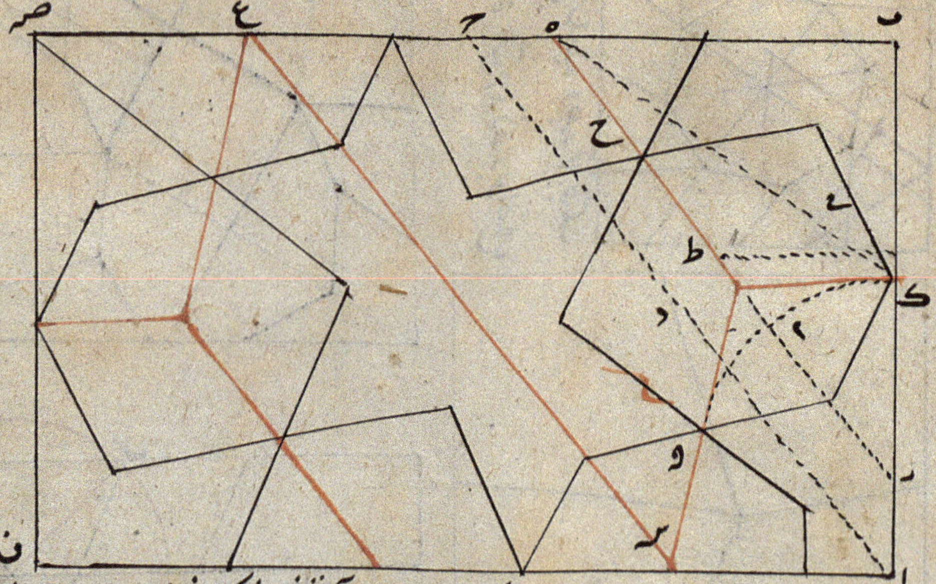

زاویهٔ ب اح مثلث اسباع قائمه عمل کند وخط اح تنصیف کند بنقطهٔ د و ب ه مثل اد
جدا کند وخط ه ز موازی اح برون برد وخط طی موازی ب ه بکشد وطه تنصیف کند
بنقطهٔ ح وطی مثل طح نشان کند وخط ه ی اخراج کند تا متقاطع خط اب
شود بنقطهٔ ک و ک ل موازی ب ه برون آورد واز مرکز ی قوس ک م ی بگرداند
جائیکه قسم ک م مثل م ا باشد ک م ا اخراج کند تا بر خط اف نقطهٔ
س بگیرد واین س مرکز مسبع باشد وغلط آسان گردد ان شاء الله تعالی

والا از زاویهٔ ه آ ک مثل زاویهٔ ه ل ک عمل کند وخط ل ا س مرکز بگیرد و ا لا
ع مثل ه ل فصل کند که نقطهٔ ع مرکز مسبع باشد وخط ع س موازی ح آ ومثل آ د و بار نقطهٔ
س مرکز مسبع آخر و دیگر ح ع مثل آ س باشد و الله اعلم

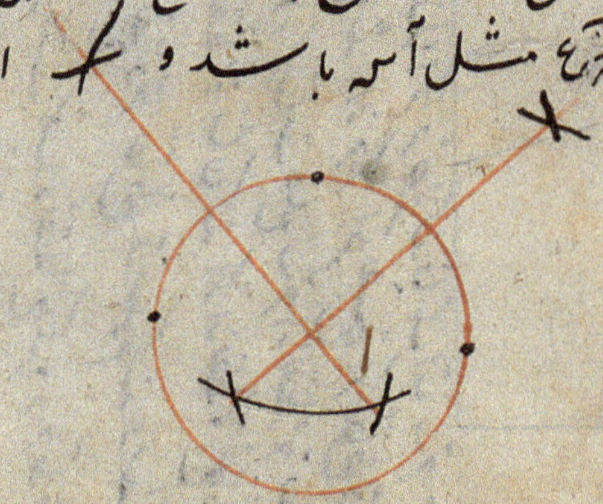

Fol. 192r [37 (top) and 38 (bottom)]

حقیقت نسبت این عقد از مخروطات است و به آلتی که آنرا کونیا مسطره خوانیم توان کشیدن
وآن آلتی باشد که بسیار عقود مشکل مخروطاتی بدین آلت توان کشید واین حالیا
استنباط کاتب است اگر بوده است معلوم نیست واینچنان باشد که مسطرهٔ بسازد
مانند عضادهٔ اسطرلاب و بر وسط آن مسطر قائمه راست کند مانند سهم عضادهٔ
اسطرلاب زورقی واین را اسطارهٔ کونیا خوانند مثل مسطرهٔ اب جد که
از وی مسطرهٔ اب بود و قائمهٔ جد و باید که از مسطره طرف اب منحرف
بود چون انحراف عضادهٔ مجیّب و سهم چنین از قائمهٔ طرف جد منحرف باشد
واز قائمه خط جد بر روی عضاده مبین باشد ونقطهٔ د بر طرف عضاده معین
وزاویهٔ جد آ قائمه عظیم درست باید که بدین مسطره بسیار نسبتهای
غریب مشکل توان کرد اول در طریق عمل نسبت این عقد

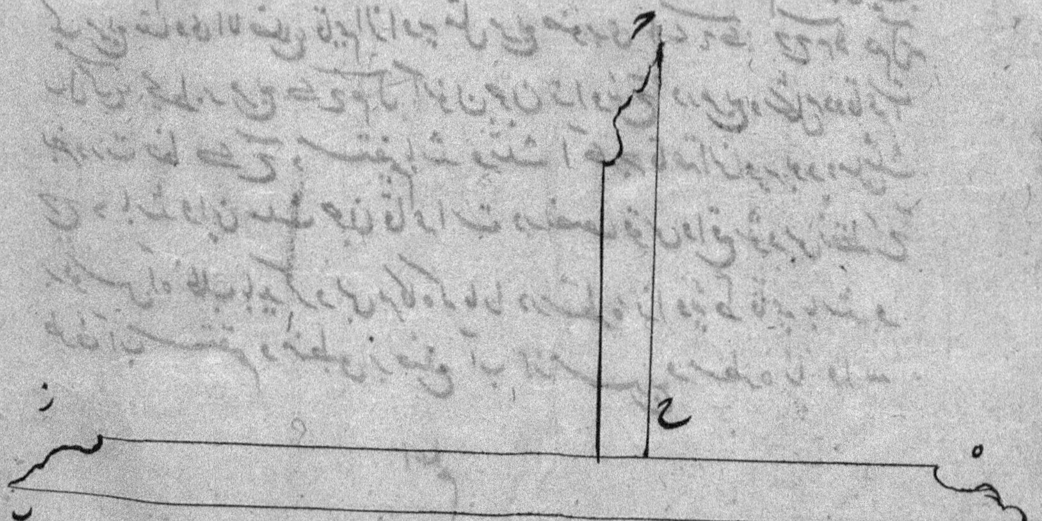

191

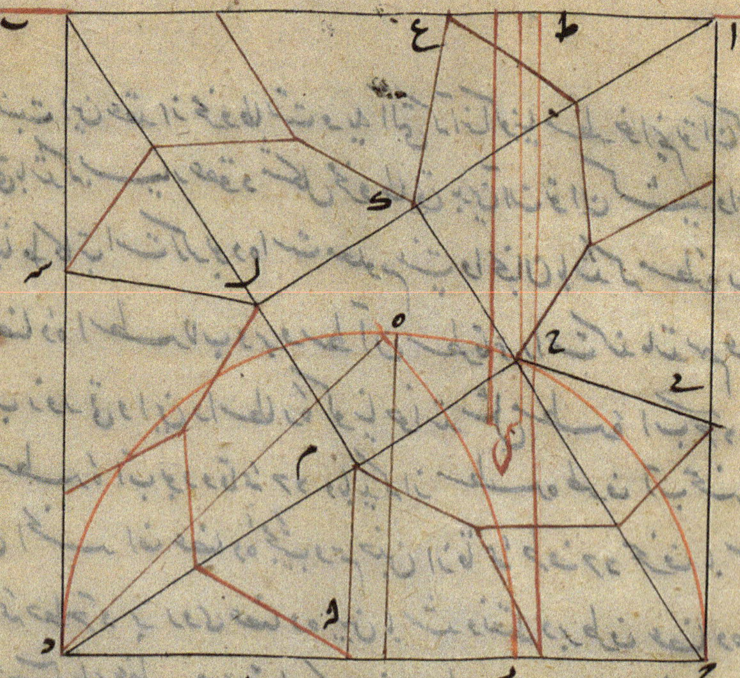

نسبت این عقد تیر هم از مخروطات است که در اینجا مقصود عمل مثلثیت قایم الزاویه که عمود آن مثلث
با ضلع اقصر منضم مثل وتر زاویه بود و ابن هیثم در عمل این مثلث رساله گفته است و عمل او در
منتسب قطاع مخروطات زاید و مکافی گفت اما اینجا ازین مسطره گونیا مقصود حاصل می توان کرد
چنانکه گوییم در مقدمهٔ اول مقصود این عقد مارا چهار شکل است صنوبری قایم الزاویتین محیط بر
یک مربع متساوی الاضلاع قایم الزاویه مثل مربع صنوبری ا ب ح ک ح ح م ک د م ل آ ه
ب ل ک آ محیط بر مربع ک ح م ل اکنون چون زاویهٔ ح در مربع و شکل هر دو قایمه است
بضرورت خط ک ح د مستقیم باشد و مثلث ا ک ج قایمة الزاویه بود و مثل مثلث
ح ج د باشد و این مثلث چون قایمه است در نصف قوس واقع شود پس نقطهٔ ح
بر قوس آه طلب باید کرد پس هرگاه که مارا در مسطره زاویهٔ ط قایمه باشد و
طرف آب مستقیم و منطبق بر ضلع آب از مربع و مسطره را والله

اعلم

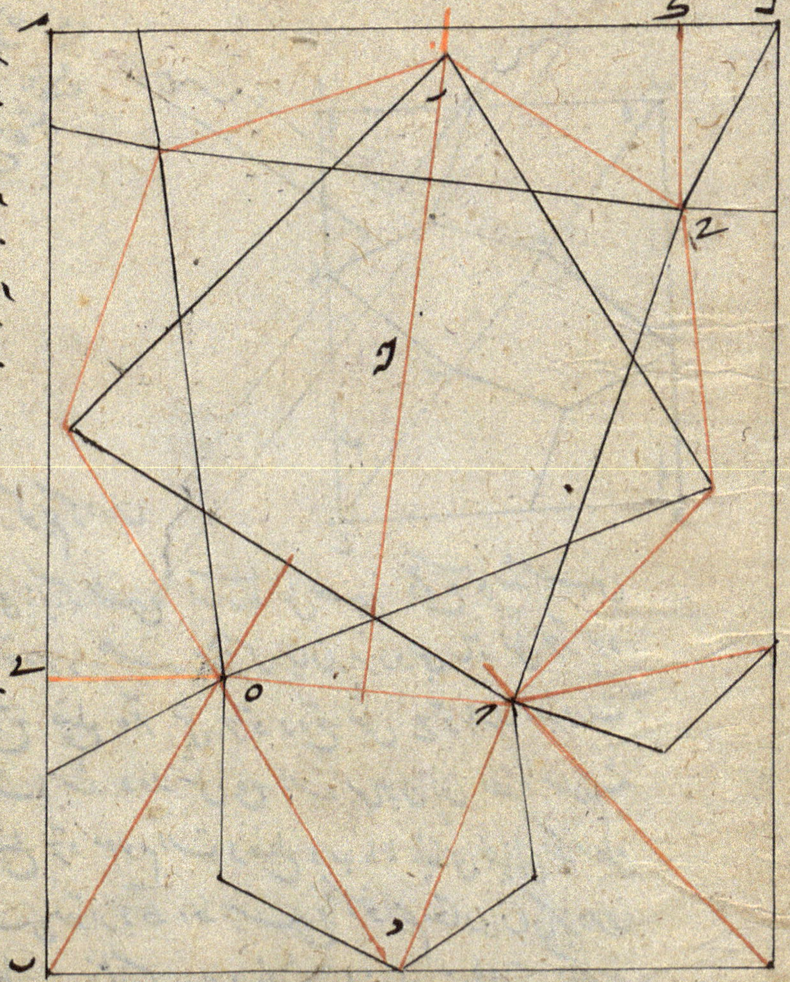

واگر خواهد که خاتم چهار بند درین عقد باشد از نقطهٔ ط خط ... بر دو نقطهٔ م
بگیرد بر قائمه ب ... و خط م ل موازی آب بکشد والله اعلم

فصل طریق اخراج این عقد آنست که بر خط آب قائمتین آل بم موازی یکدیگر اخراج کند واز زاویه آ
خط جهار رواز زاویه ب خط بح اخراج کند تا متقاطع شوند بنقطه ح انگه از نقطهٔ آ بفتح اح بر ضلع آب
نقطهٔ د نشان کند واز نقطهٔ ب بر خط به نقطهٔ ه نشان کند بفتح بد و خط حد حه بپیوندد
وبر خط حه مسبع حه طرح عمل کند چنانک خط حه قاعدهٔ او باشد واز نقطهٔ ح خط حی
موازی آل اخراج کند چنانک مثل ه ی باشد یعنی نصف ه د واز نقطهٔ ی خط لکم
موازی آب بکشد که عمل تمام شود والله اعلم

190

هرگاه که زاویهٔ قائمه را اما علی التقریب بانه [?] ابتدا
بخواهیم قسمت کنیم از ان زاویه اجب قسمت متساوی د ه
یعنی زاویه بجد بود و این غایت تقریب

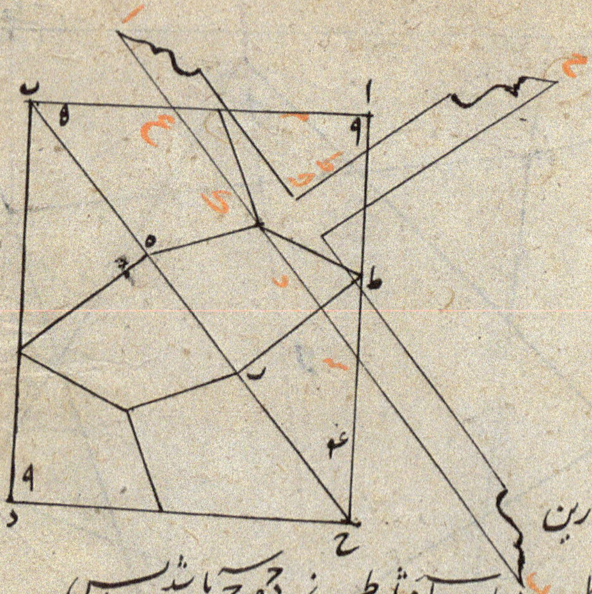

عمل بنت این عقد بکو بنا مسطره کوییم که درین
عقد که در خانهٔ اب‌جد است می باید که طرح اب رط مثل طرح ده‌ح باشد پس
چون چنین باشد سه ر مثل ح‌ه بود و هر مشترک میان هر دو پس سه مثل حز بود
و دیگر چون در طرح ه‌ح د‌ح ضلع ه‌ح مثل ح‌د و ه‌ح مثل ح‌د باشد بضرورت
زاویهٔ ه مثل زاویهٔ د بود دقیقه باشد و ه ر مثل ه‌ح است پس چون این مقدمه معلوم شد
فرض کنیم که از خانهٔ عقد ضلع ح‌د معلوم است و ضلع د ب ما لا تتناهی مجهول انگه مسطره
برداریم واز نقطهٔ د بر مسطره مقدار دک د ط به‌شرح که خواهیم متساوی نشان کنیم و حرف
بزنیم انگه بهمین مقدار که بر مسطره نشان کرده ایم در خانهٔ عقد از زاویهٔ د نقطهٔ م د نشان کنیم
بر ضلع دح دب انگه مسطره برداریم و حرف ط که بر طرف قائمه است بر حرف ه نهیم مطبق نقطه
بر نقطه انگه برین نقطه مسطره را از جب و راست حرکت میدهیم چندانکه از طرف مسطره مقدار
سه د مثل ک‌ع گردد و نقطهٔ ط از نقطهٔ ه مفارقت نکند انگه خط سع بکشیم و در سطح از نقطهٔ
ح که معلوم است خط حز موازی سع بدر بریم و خط ن‌ح مثبت سه د د‌ک ک‌ع
قسمت کنیم که چون بر مسطره که موازی حز مقدار سه د مثل ک‌ع است و ک د
مثل د ط ط د است از خط ح ب مقدار ب ه مثل ر ح بود و ه ر مثل ه‌ح و ح د و مقصود
اینست والله اعلم

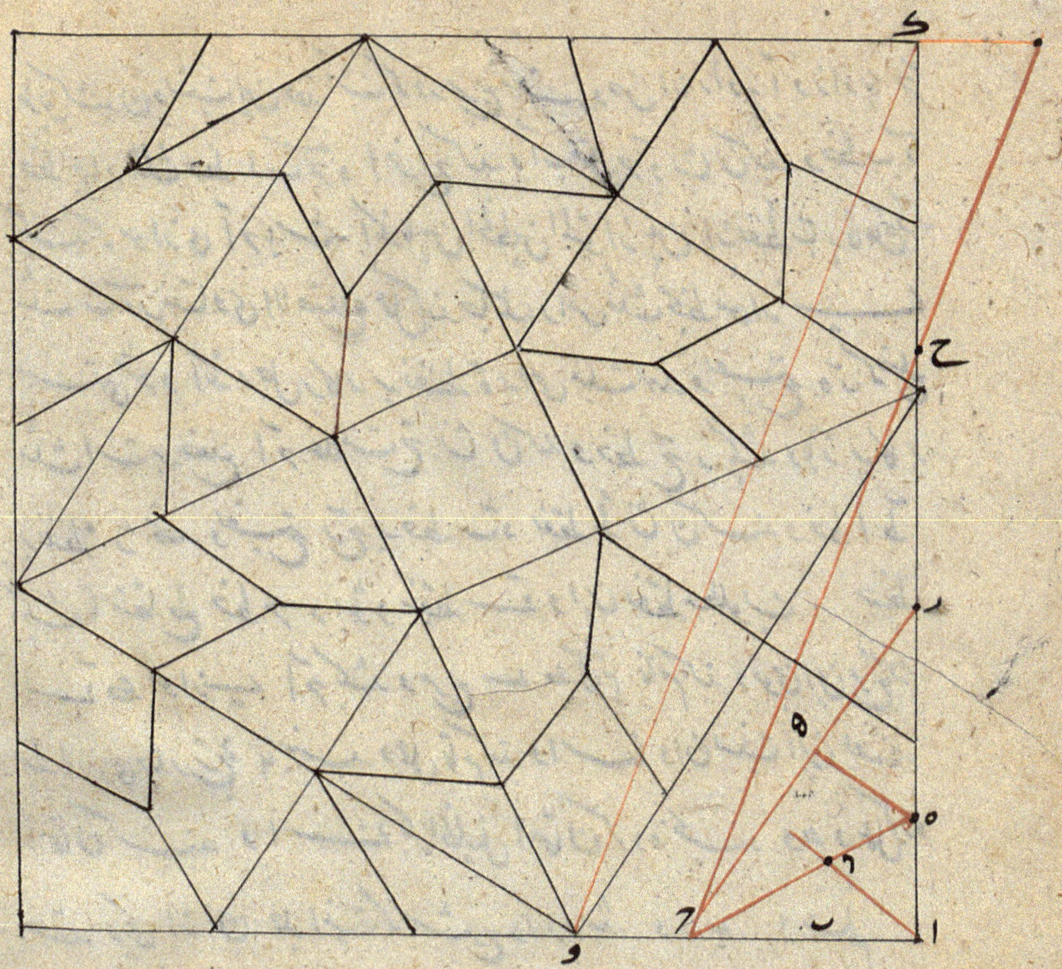

خط اد قطر مربع است ومقدار آب بج متساویست و آد مثل آب چون
خط آج در برابر استقامت اخراج نقطهٔ ه دریابد بازه ز زح مرکب مثل اح
بود خط حج بپیوندد و از زاویهٔ ک خط کل موازی حج اخراج کند و
نقطهٔ ل دریابد مقصود حاصل شود والله اعلم

189

طریق کشیدن ونسبت این عقد آنست که در مربع مفروض از زاویه آ و زاویه ج
خط جهار مثلث خط اب جد اخراج کند و آب مثل جد نشان کند وخط بد
بکشد که موازی آج باشد انکه مابین الخطین المتوازیین از نقطه ب بر قاعده بر
مثلث به ز متساوی الاضلاع عمل کند جانک راس مثلث بخط موازی
منتهی شود انکه رجل پرکار بر نقطه ه راس مثلث نهد وبفتح ه ز که ضلع
مثلث است بر ضلع آج علامت ح نشان کند وخط ح ز بکشد وباز پرکار
بر نقطه ز نهد وبفتح زح بر خط بد نقطه ط نشان کنند وخط اط
بکشد تا متقاطع خط ح د شود بنقطه ی واین نقطه مطلوبست خط
ی ک موازی آج بکشد ومربع ی ک ل م تمام کند وجون این مربع معلوم
شد دروی جهار ترنج آنها بنهند وعمل تمام کند واستادان نسبت این عقد
امتحان کنند واسنی بکر الخلیل امتحان کرده بجنند وجه وعمل کرده
نشد ویکی از ان جمله اینست که شرح داده شد والله اعلم

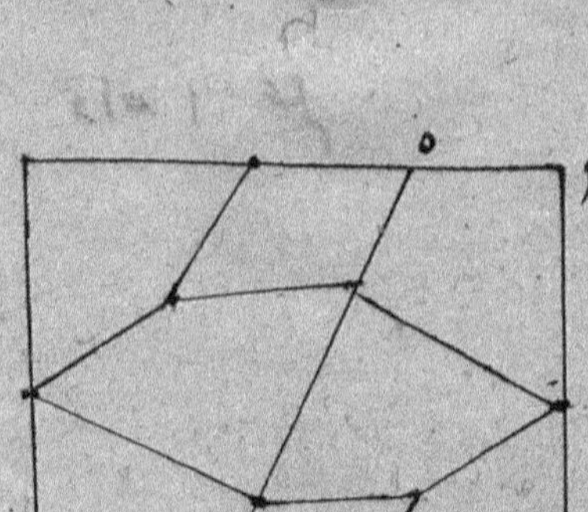

Fol. 189r [31 (text only) and 32 (bottom right)]

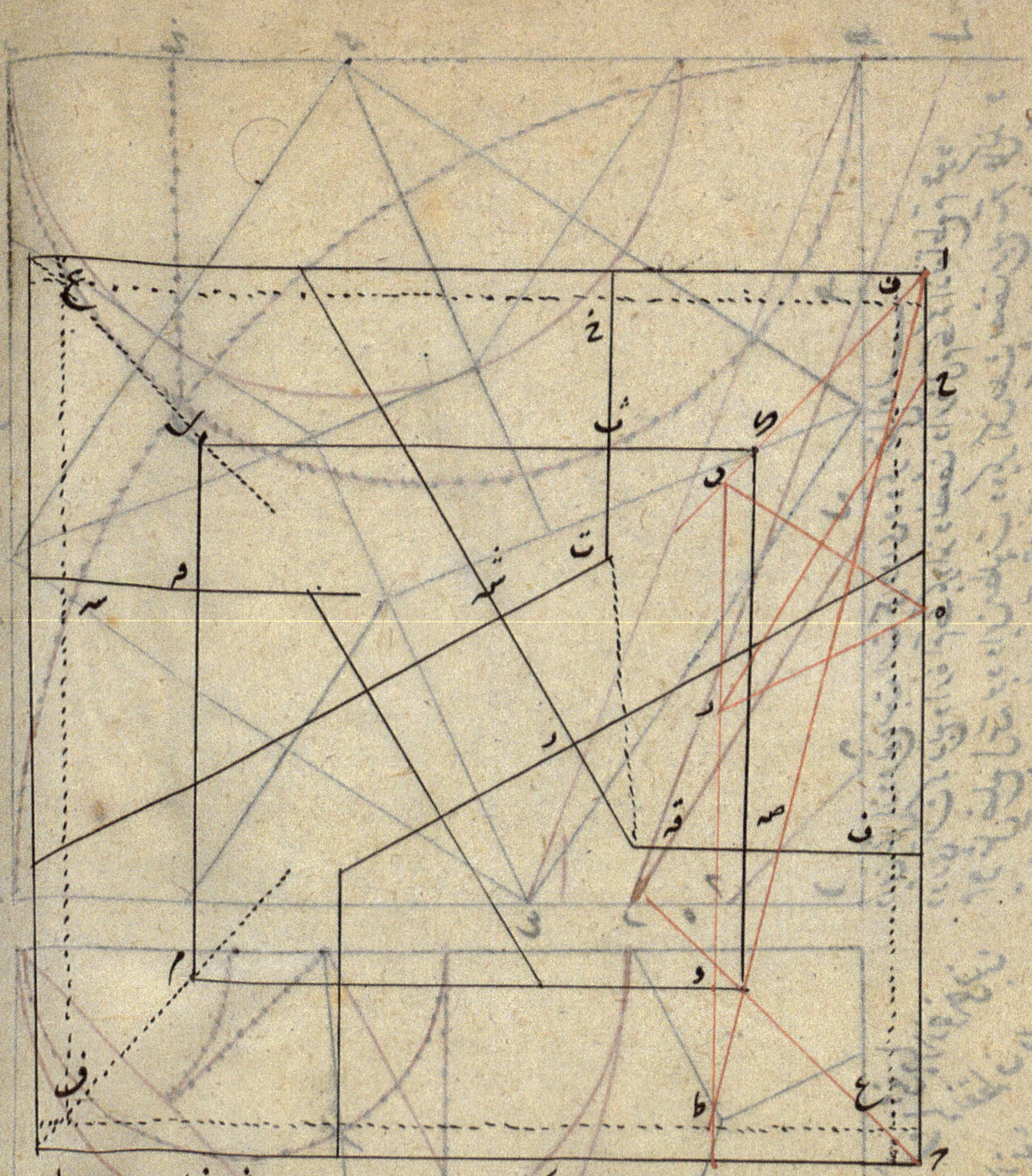

بعضی از اصحاب هندسه بران بودند که ما بین المربعین المتوازیین مثل ضلع مربع وسط
می خواهیم پس چنین باشد باید که مقدار آه مثل نشه باشد نسبت وی آنست که چون زبع
عف مفروض بود این عقد را بطریقی دیگر میکشند اقسام فصه صق قر رکبه
شت تخ جمله متساوی شود پس خط فق جذر آه باشد و ضلع مربع داخل تر باشد
صحیح وجذر آه و ضلع مربع خارج چهار باشد صحیح وجذر آه چون ازین
هر دو مربع یکی معلوم باشد دیگری نسبت وی معلوم شود والله اعلم

Fol. 188v [30 (with 31 superimposed)]

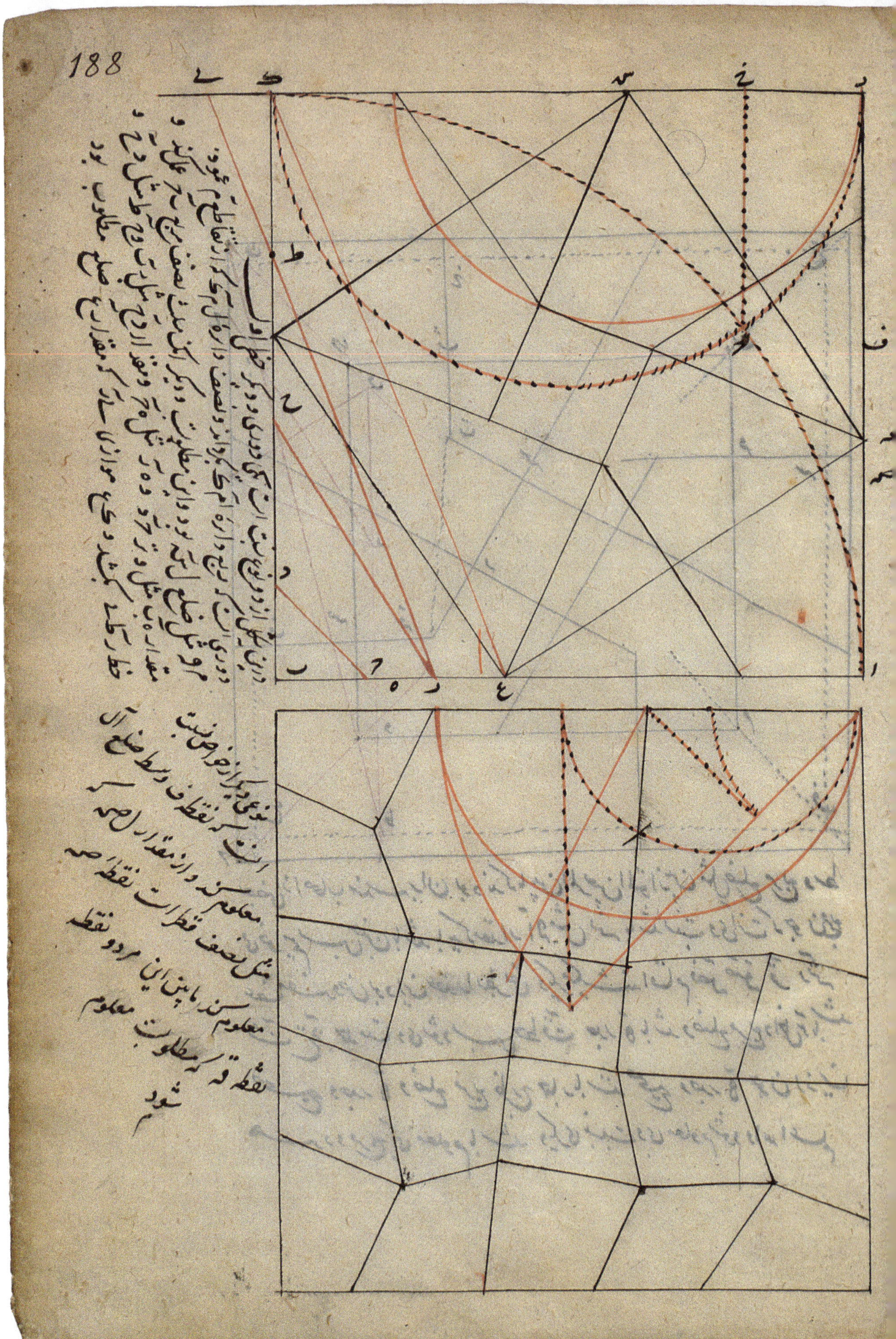

Fol. 188r [28 (top) and 29 (bottom)]

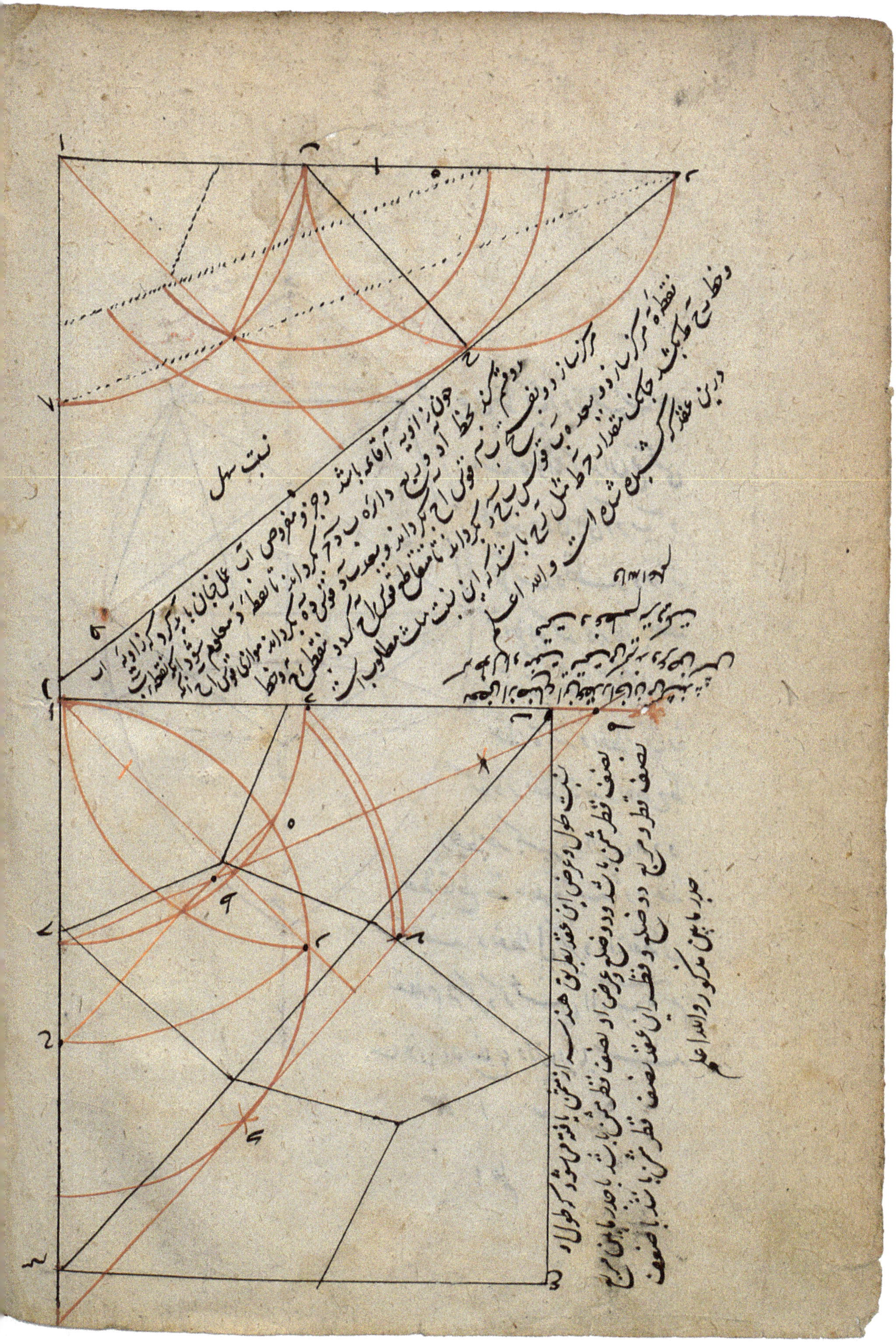

Fol. 187v [26 (top) and 27 (bottom)]

187

عمل مخمس منفتح الرؤس

لابی بکر الخلیل التاجر

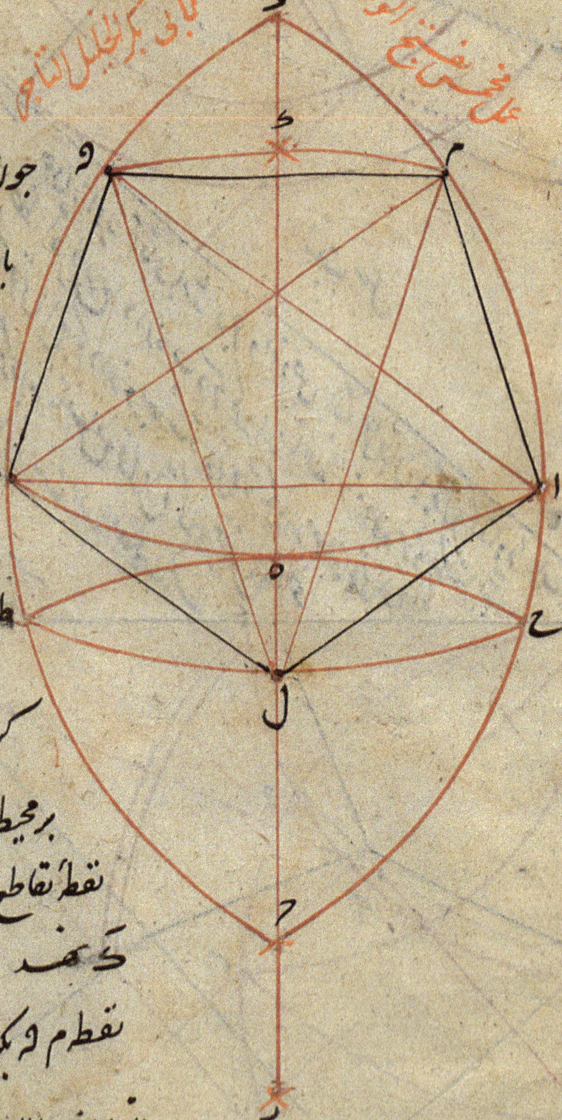

جون خط اب که وتر مخمس است معلوم
باشد وخواهد که بدان فتح مخمسی
عمل کند بروی بیضی اجب د
عمل کند وقطر دح ز
اخراج کند وبرکار بر نقطه
ز نهد ونقطهٔ ه نشان کند وبر
نقطهٔ ه نهد ونقطهٔ ز نشان
کند وبر نقطهٔ ز نهد ونقطهٔ ح ط
بر محیط بیضه بگیرد وازین هر دو
نقطهٔ تقاطع ک معلوم کند وبر نقطهٔ
ک نهد ونقطهٔ ل بگیرد واز نقطهٔ ل
نقطهٔ م ه بگیرد که مخمس ال ب ه م
متساوی الاضلاع والزوایا باشد

والله

اعلم

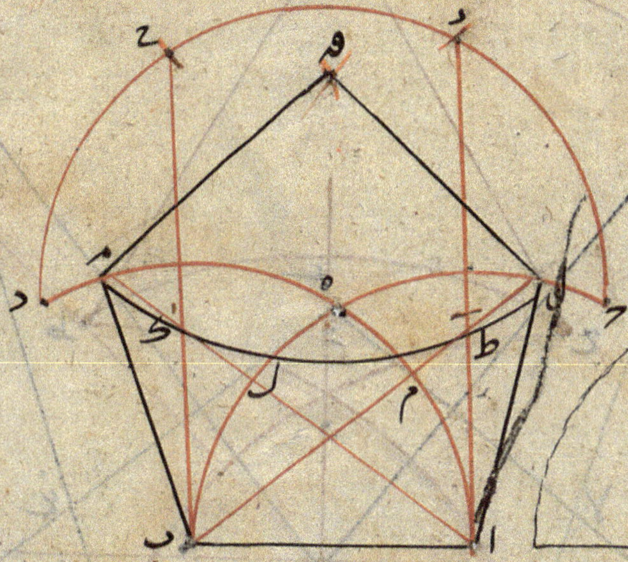

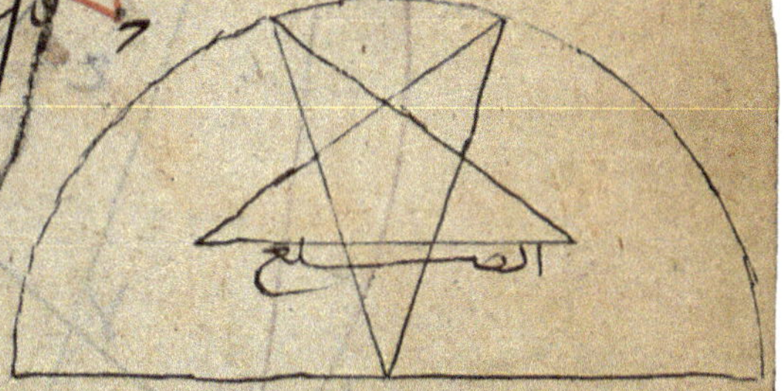

فی عمل المخمس المتساوی الاضلاع والزوایا
بفتح الضلع

فی عمل المخمس المتساوی الاضلاع والزوایا
بفتح الضلع

چون خواهد که عمل مخمس متساوی الاضلاع و الزوایا کند بفتح ضلع اب طریق آن بود که بر ضلع اب
از مرکز ا قوس ب ج و از مرکز ب قوس ا د بگرداند تا متقاطع شود بنقطه ه آنگه از
مرکز ه قوس د ح ر ج بگرداند و از نقطه ج د بر قوس د ج نقطه ر ح معلوم کند و خط
ر ا ح ب بکشد آنگه از نقطه ر ح بر خط ا د ب ج نقطه ط ک معلوم کند و خط ا ک م ب ط ل بکشد
تا هر یک منتهی شوند بمحیط قوس ا د ب ج بنقطه ل م آنگه از مرکز ل م
تقاطع ه حاصل کند و خط ل ب م ل ه م ا بکشد تا مخمس ا ب م ه ل متساوی
الاضلاع و الزوایا حاصل شود باید که چون از تقاطع که قوس م ل بگرداند از قوس تقاطع م ل
بدرد و اگر غیر ازین آید عمل خطا باشد و خاک خواهم پنهان کند و الله
اعلم

Fol. 186bis(v) [23 (left) and 24 (right)]

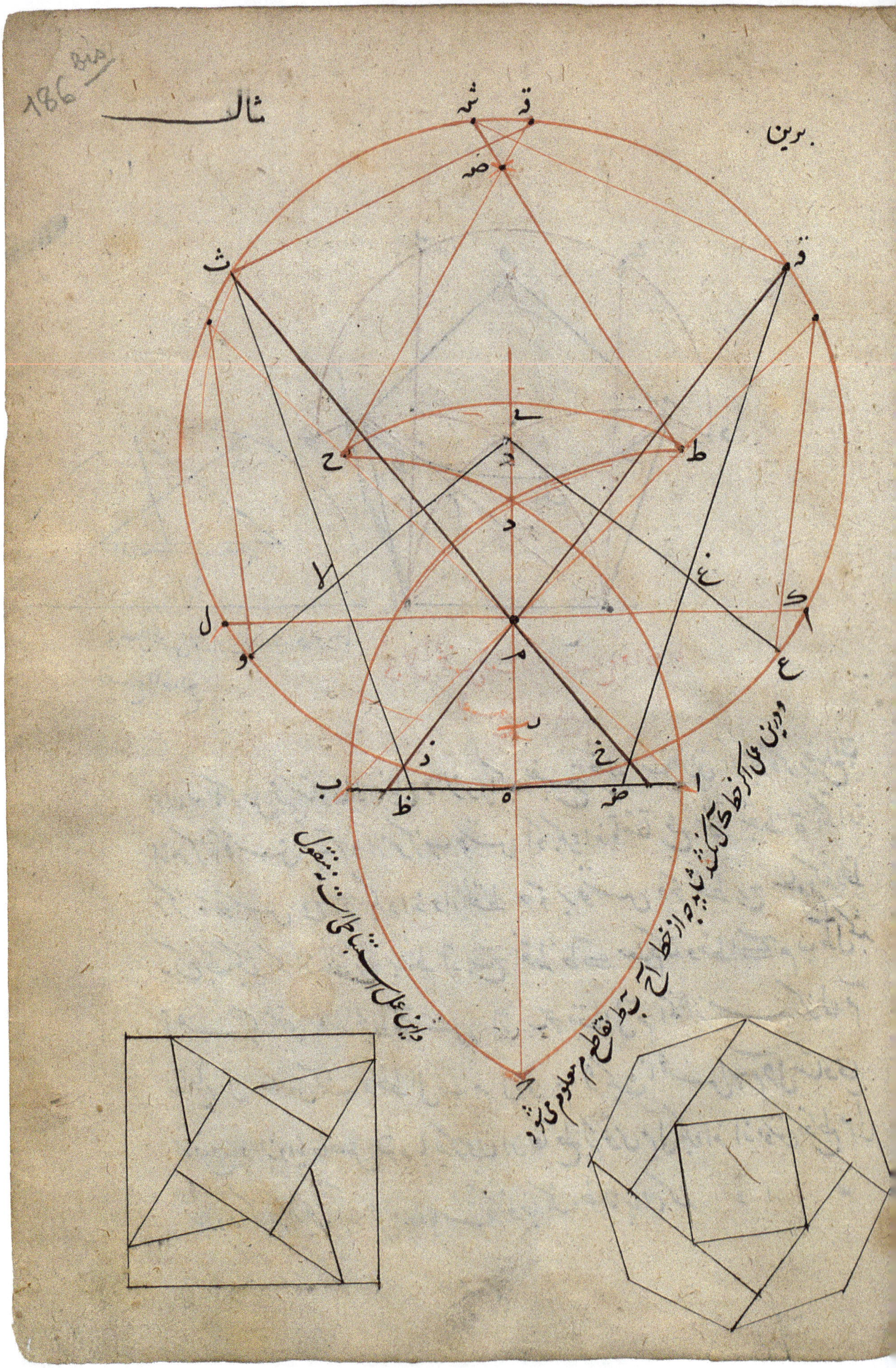

Fol. 186bis(r) [21 (top center) and 22 (bottom left and right)]

فصل فی عمل المخمس المتساوی الاضلاع والزوایا بفتح العمود

میخواهیم که مخمسی متساوی الاضلاع والزوایا عمل کنیم جانبش عمود او مثل خط اب مفروض باشد
طریق آن بود که برکار بر فتح اب و بر خط اب از نقطهٔ اب قوس ح اح ط ب ح بگردانند و
تقاطع د ح حاصل کند و خط ح د ی بکشد و تقاطع ه معلوم کند و از تقاطع ه مرکز ی معلوم
کند و دایرهٔ اعظم بگرداند آنکه از نقطهٔ ح نقطهٔ ز معلوم کند و ازین نقطه قوس ط ح
بگرداند و نقطهٔ ط ح معلوم کند و ازین هر دو نقطهٔ تقاطع ص معلوم کند
و بر استقامت خط ط ص ح ص بر محیط دایره نقطه ق ثه معلوم کند و ازین
هر دو نقطه هم بر محیط نقطه ت ث بر خلاف یکدیگر یعنی از نقطهٔ ق نقطه ث و
از نقطهٔ ثه م ت م ط معلوم کند و این هر دو نقطه را نگاه دارد آنکه تقاطع
ه نقطهٔ ک ل بر محیط قوس نشان کند و از خط ک ل تقاطع م معلوم کند و
بر استقامت خط ه م ط ا م ح بر محیط دایره نقطه خ ثه معلوم کند و ازین
هر دو نقطه هم بر محیط نقطهٔ ع ف معلوم کند باز بر استقامت خط ث م ث م
بر محیط دایره نقطهٔ خ ذ معلوم کند آنکه خطوط ی ع ی ف ث ذ ط ت خ ص
بکشد مخمس ی لا ظ ض غ ظاهر گردد و این

مخمس مطلوب باشد متساوی الاضلاع و
الزوایا و جانبش خواهد
عمل کند و در بیان
آورد والله اعلم
اعلم

Fol. 186v [(text refers to) 21]

186

عمل مخمس متساوی الاضلاع والزوایا نفتح العمود

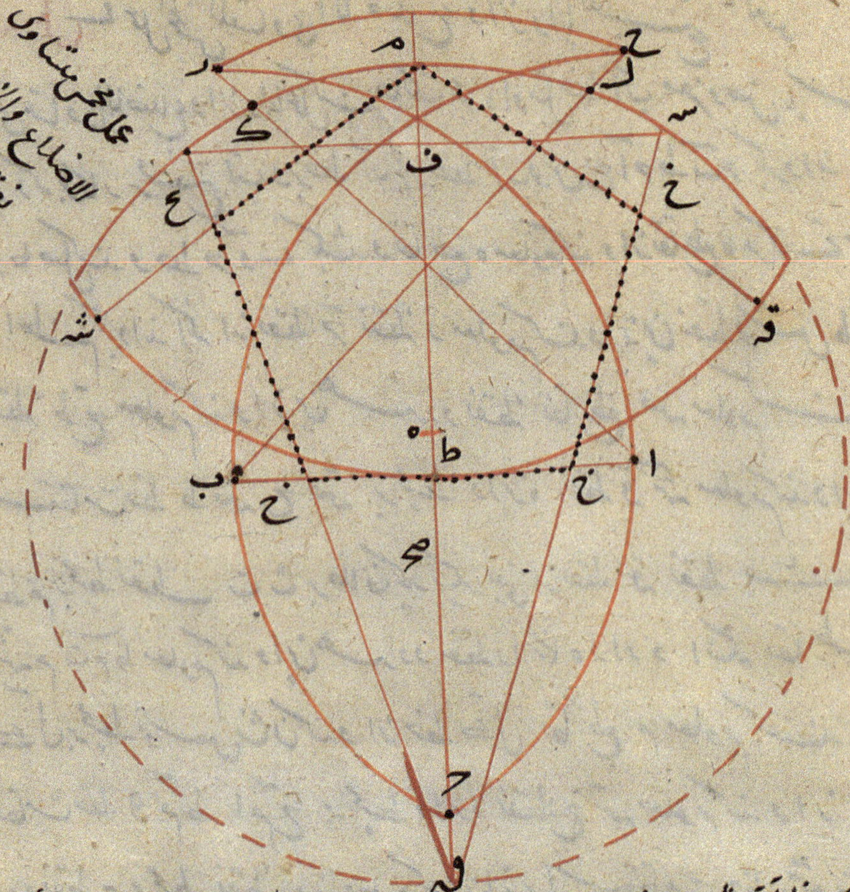

درین شکل خط آت مثل عمود مفروض است و تقاطع حرّد از نقطه آب حاصل میشود و نقطه رح
از نقطه ة و نقطه ة از نقطه ح و نقطه کل بر خط ار سح از نقطه ط حاصل میشود و همچنین
نقطه م د بم از ط حاصل میشود و نقطه سع از کل حاصل میشود و نقطه قش از نقطه
صه حاصل میشود و نقطه صه از ف و نقطه ف از خط سع چون این نقطها حاصل شده
باشد جمله بفتح آب از خط م قم شه دسه قع مخمس مرخ متساوی الاضلاع و
الزوایا بفتح آب که مثل عمود لم است حاصل شود چنانکه مصور است

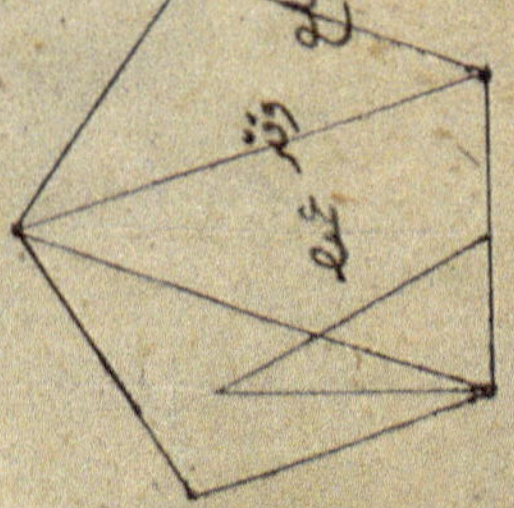

عمل آخر مخمس متساوی الاضلاع والزوایا

Fol. 186r [19 (top center) and 20 (bottom right)]

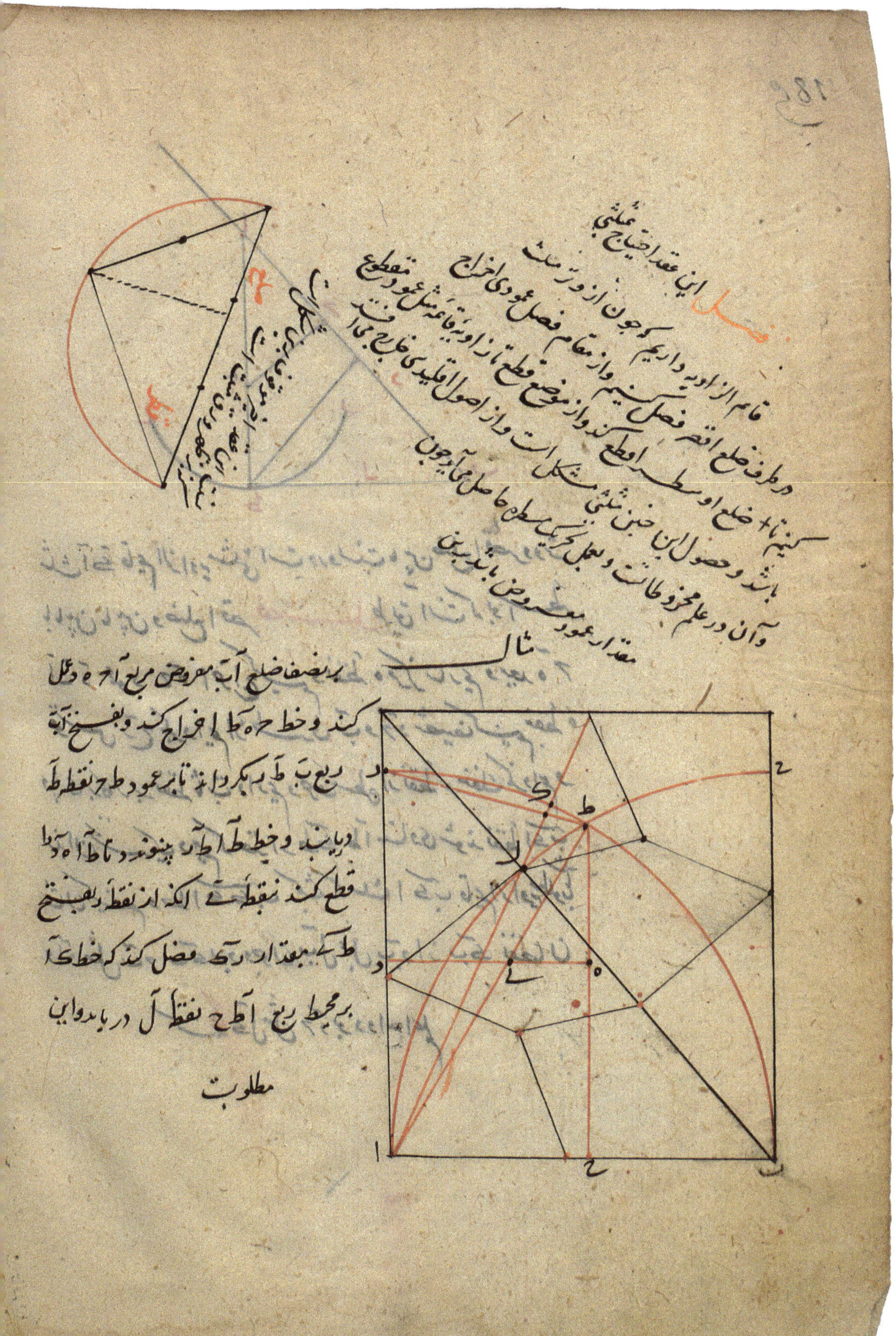

Fol. 185v [17 (top left) and 18 (bottom right)]

185

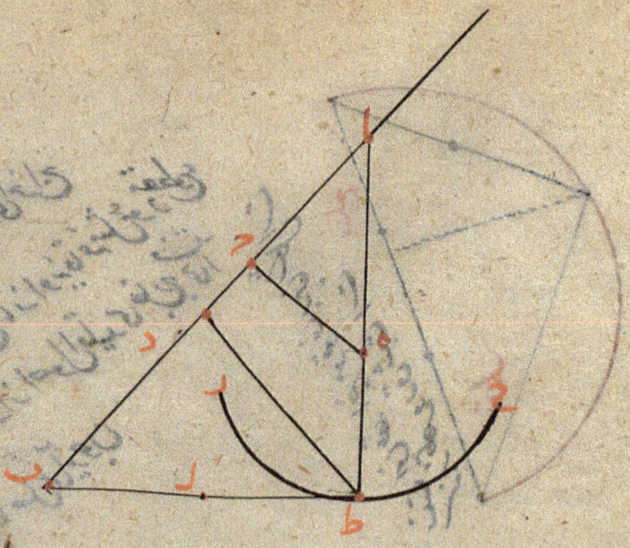

مثلث آک قایم الزاویه مثلثی است در و نسبت ما بین ضلع اقصر و وتر
با ما بین ما بین و ضلع اقصر **فصل** طریق آنست که بواسطه
آک و عمود د ح ک ا اخراج کنیم و نقطه ه مرکز سازیم و ببعد ه ح
قوس ز ح ط بگردانیم از طرفی ب و ح د تنصیف کنیم بنقطه ه
و پای پرکار بر نقطه ثابت داریم و حرف مسطر از نقطه ه منفک نگردانیم و
چندان تحریک دهیم که مقدار ط ک ط ا متساوی شوند نقطه ا ک ب
نشان کنیم و خط ا ک ک ب بکشد که مثلث ا ک ب قایم الزاویه مطلوب باشد
و آد مثل ت ح ت ک بود و چون ب ل مثل ب د از ب ک نقصان
بکنند ک ل مثل ح د بود و الله اعلم

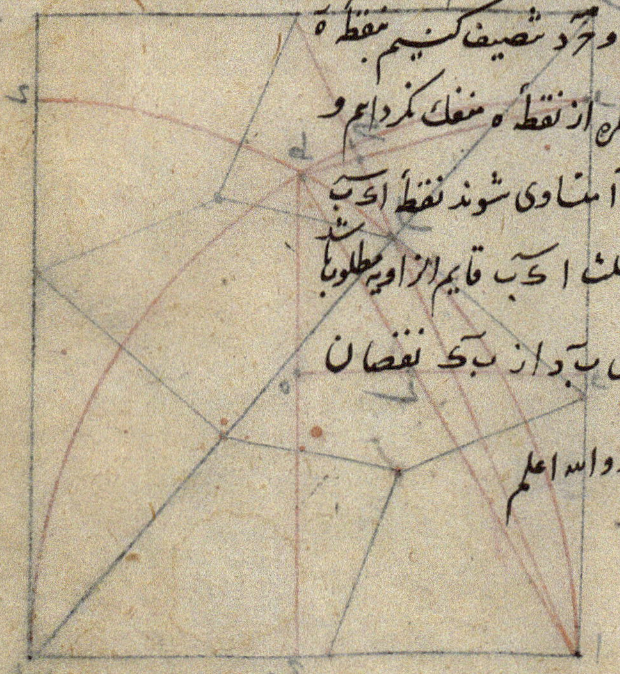

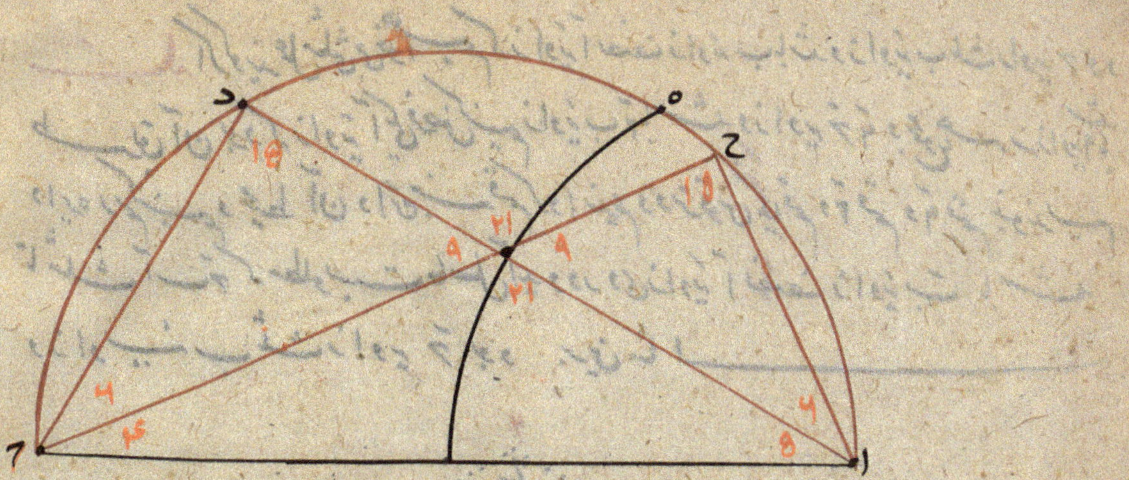

در عمل کونیا ۵ بفتح نصف قطر از کونیا ۹ و آنچنان باشد که
برخط اح آت نصف دائره ادح بگردانند بر مرکز ب آنکه نقطهٔ ا
مرکز سازد قوس وب بگرداند وباز نقطهٔ ح مرکز سازد و بر محیط قوس
نقطهٔ د معلوم کند وخط اد بنویسد تا متقاطع قوس وب شود بنقطهٔ ر
وخط رح بکشد تا منتهی شود بمحیط قوس بنقطهٔ ح وخط اح دح
بنویسد که هر یک از مثلث ادح ومثلث حرد کونیا ۵ باشد و اصل
مثلث ادح کونیا ۹ بود چنانکه
نموده شد واصل مثلث
اح ح از کونیاها مخالف
آ۵ باشد

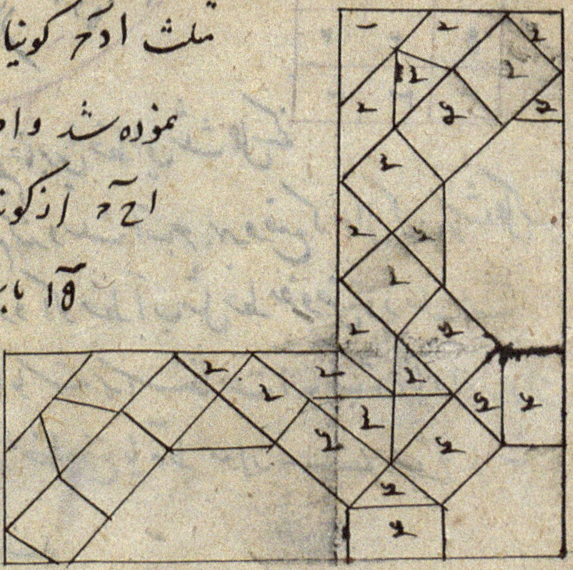

Fol. 184v [14 (top) and 15 (bottom)]

184

مسئله اگر گویند عمل مثلثی می‌خواهیم که زاویهٔ آ نصف زاویهٔ ب باشد و زاویهٔ ب ثلث زاویهٔ ج بود
طریق آن بود که زاویهٔ آ یکی فرض کنیم زاویهٔ ب ۲ باشد و زاویهٔ ج ۶ و مجموع هر سه زاویه ۹ باشد
دایره رسم کنیم و محیط آن دایره به نه قسم گردانیم و در قوس یک قسم و ۲ قسم و ۶ قسم بنویسیم
تا مثلث آ ب ج که مطلوبست حاصل آید و در وی زاویهٔ آ نصف زاویهٔ ب باشد
و زاویهٔ ب ثلث زاویهٔ ج بود برین مثال

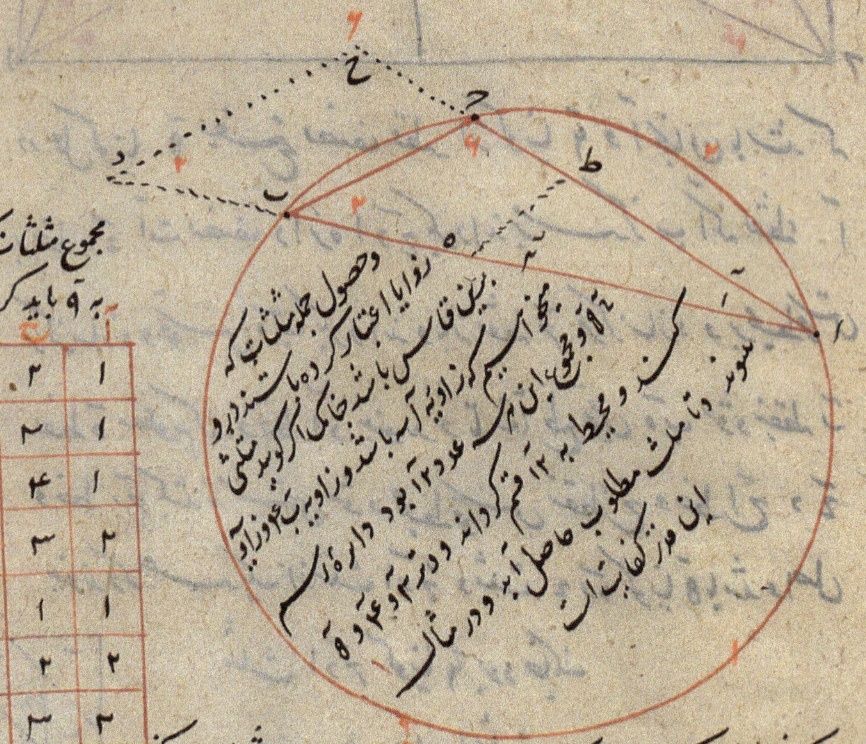

مجموع مثلثات که از بهر آنشان
به ۹ باید کرد اینست

ا	ب	ج
۱	۲	۶
۱	۳	۵
۱	۴	۴
۲	۳	۴
۱	۱	۷
۲	۲	۵
۳	۳	۳

واگر خطی فرض کنند و گویند میخواهیم تا برین خط این مثلث عمل کنیم
بهر فتح که خواهد دایره رسم کند و مثلث آ ب ج بهر وضع که سوال کرده باشد عمل کند
بدبرین مثال که گفته شد آنکه بنگرد اگر خط آ ب مثل خط مفروض بود مثلث آ ب ج
مطلوب باشد و اگر زیاده یا نقصان باشد بر استقامت آ ب ب د زیادتی بر آورد
یا ب ه ناقص بکاهد آنکه از نقطهٔ د یا ه خط د ح یا ه ط موازی ب ج باشد

والله اعلم

مقدمه هر مثلثی که در دایره واقع شود نسبت زوایای آن مثلث
با یکدیگر مثل نسبت قوس اوتار آن زوایا باشد با یکدیگر
برین مثال چون زاویهٔ آ نصف زاویهٔ ب باشد قوس ب ج
نصف قوس ج آ بود و چون زاویهٔ ب نصف زاویهٔ ج بود بضرورت

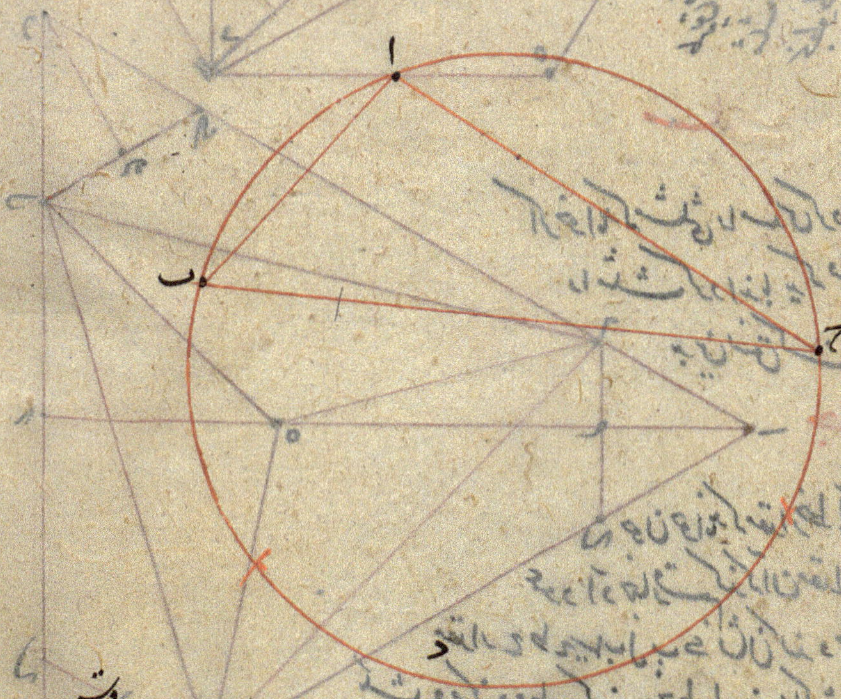

زاویهٔ ج اربعه امثال زاویهٔ آ باشد و قوس ج آ نصف قوس ا د ب بود پس قوس
ب ج ربع قوس ب د آ باشد و مجموع دایره سبعه امثال قوس ب ج باشد و
حینئذ ب ج ضلع مسبع بود و مجموع زوایای مثلث آ ب ج و محیط سبعه امثال زاویهٔ آ
باشد والله
اعلم

183

طریق قسمت مسدسی از جهت مثلثات است چون
مثلث ا آ د معلوم کند هر یک از زوایا مثلث بچهار قسم کند
بخطوط ا ط و سا د ح بکشد تا متلاقی یکدیکر شوند بنقطه ح ط ی
و مثلث ح ط ی پیدا شود این مثلث را شش قسمت کند با عمود
سا ل ح و ط م و ا ز مجموع مقطعات یک مثلث سازد
والله اعلم

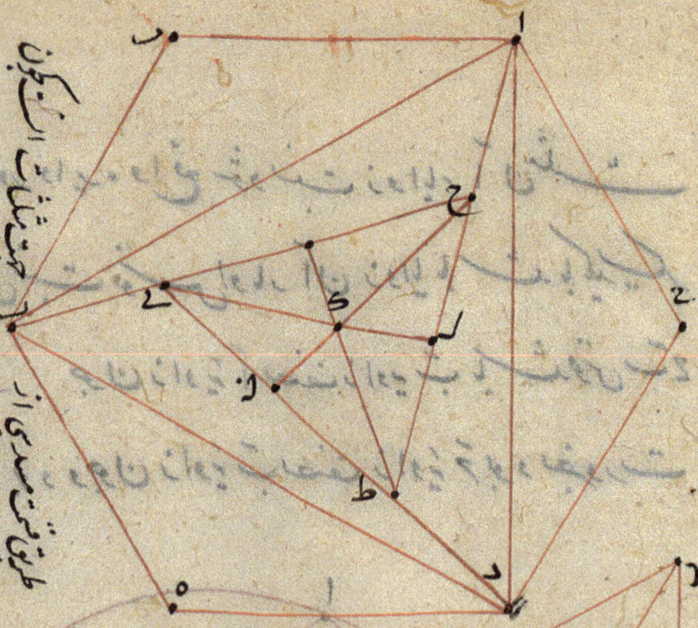

مسئله

اکر خواهد که مثلثی را مسدس کرداند و یا مسدسی
را مثلث کرداند باید که قسمت وی
بدین نسق کند تا راست آید

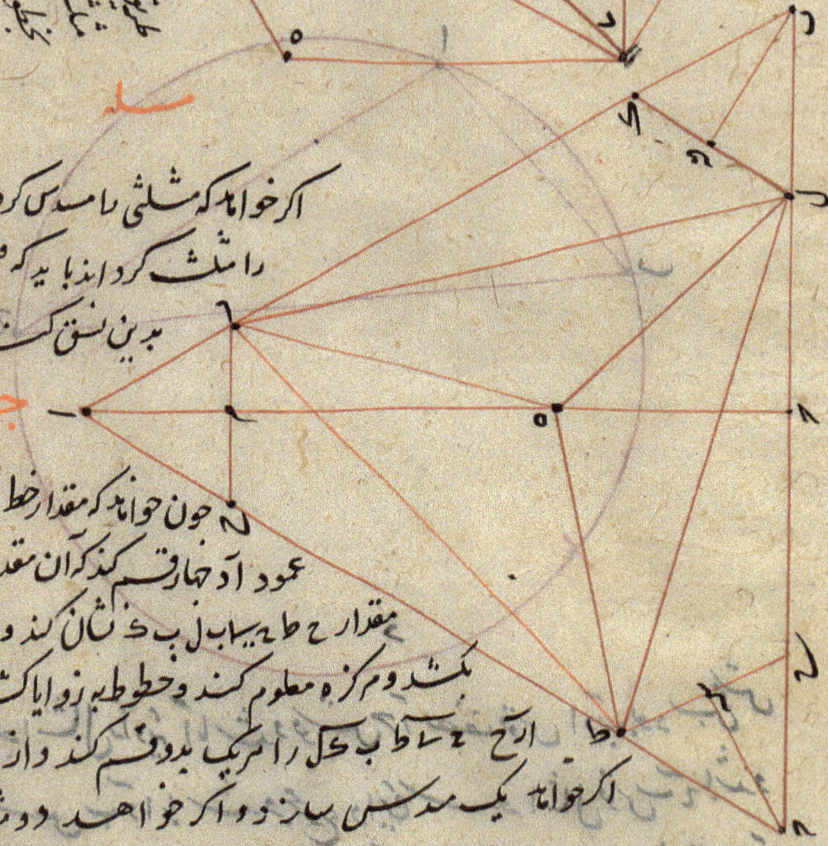

جواب

چون خواهد که مقدار خط آح معلوم کند باید که
عمود آد چهار قسم کند که آن مقدار آد آح بود بهمین
مقدار ح ط ی بیابد ل ب ذ نشان کند و خطوط مثلث را که
بکشد و مرکز ه معلوم کند و خطوط به زوایا کشد و مجموع مثلثات
ا ز ح ی ح سا ط ب کل را مرکب بود قسم کند و از مجموع مقطعات
اکر خواهد یک مسدس سازد و اکر خواهد دو مثلث سازد
والله اعلم

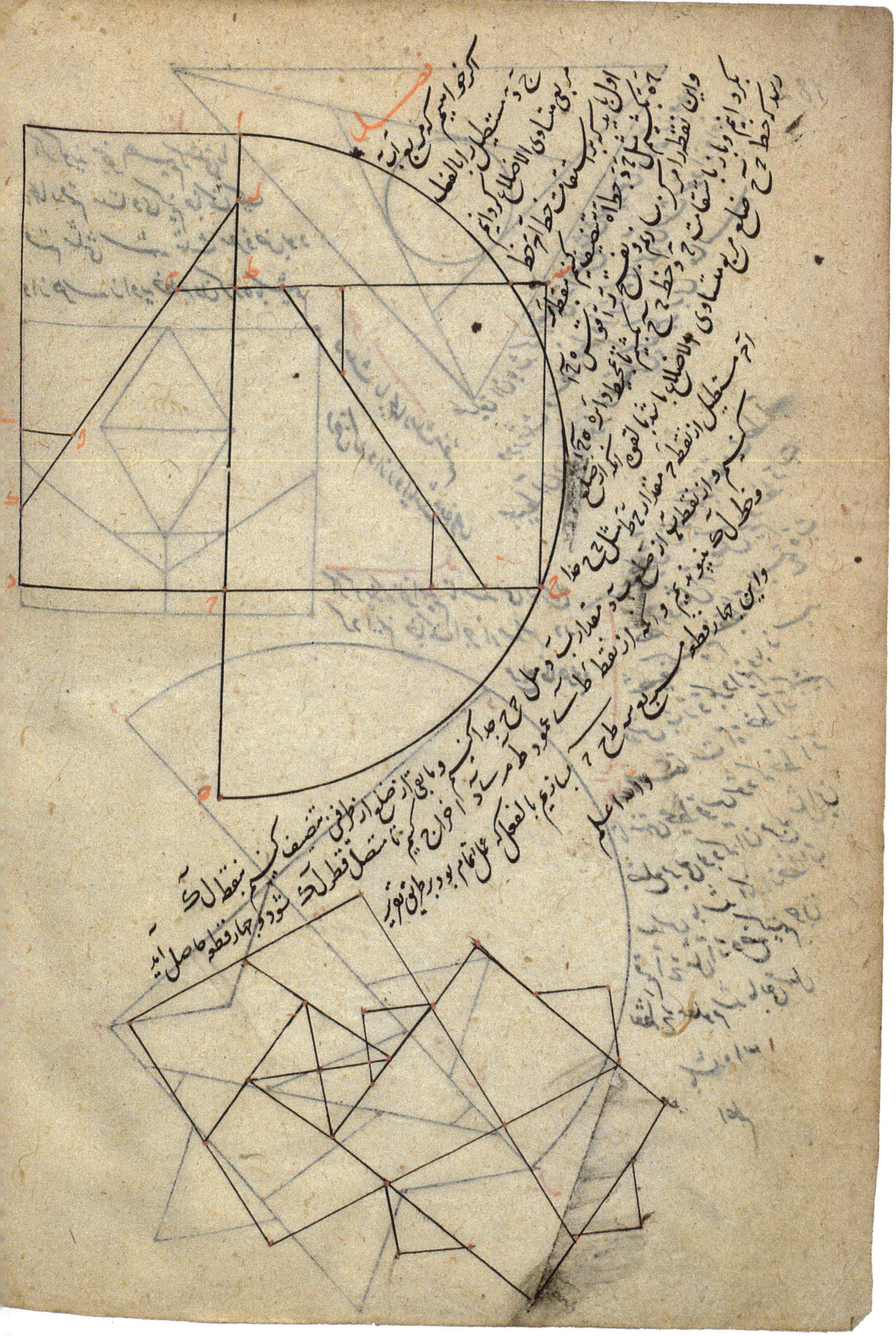

Fol. 182v [9 (top) and 10 (bottom)]

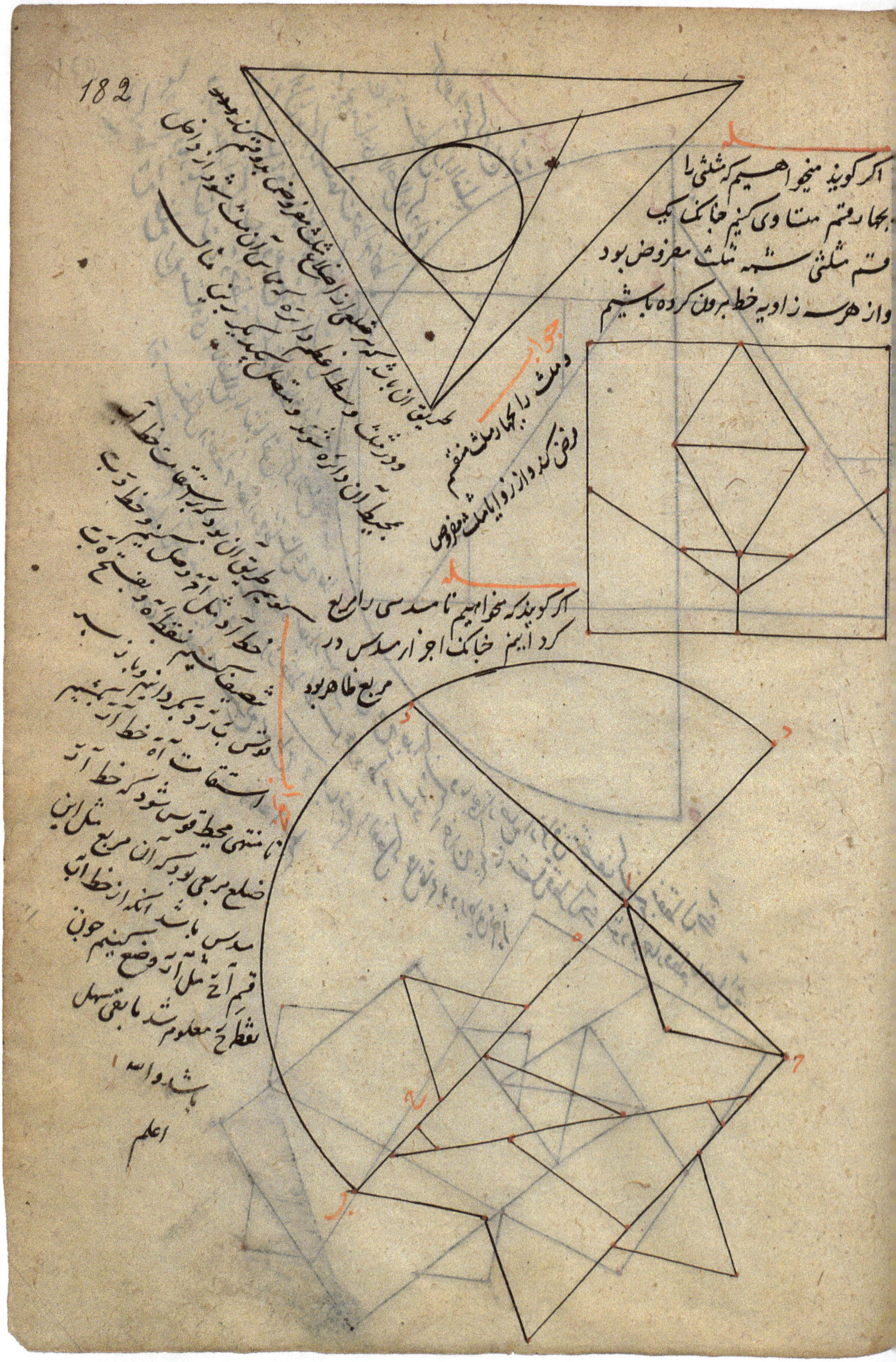

Fol. 182r [7 (top) and 8 (center right and bottom)]

فصل اگر قطعه از دایره باشد مثل قوس اجب وخواهد که آنرا دایره سازد و مرکز معلوم نبود
باید که از نقطهٔ آ ب بهر فتح که اعظم از نصف قطر باشد قوس زدح دیگر دارند تا متلاقی
شوند بنقطهٔ د و ازین نقطه بوسط قوس عمود دح اخراج کند و باز از نقطهٔ ح بفتح اول
یعنی از نقطهٔ د نشان وخط اه پیوندد و باز از نقطهٔ ه د بفتح اه قوس رح بگردانند تا از طرفین
متلاقی شود بنقطهٔ زح وخط از ب تا ح بکشد تا حاصل شود دح ا این مرکز دایره
است بفتح طح که نصف قطر دایره است والله اعلم

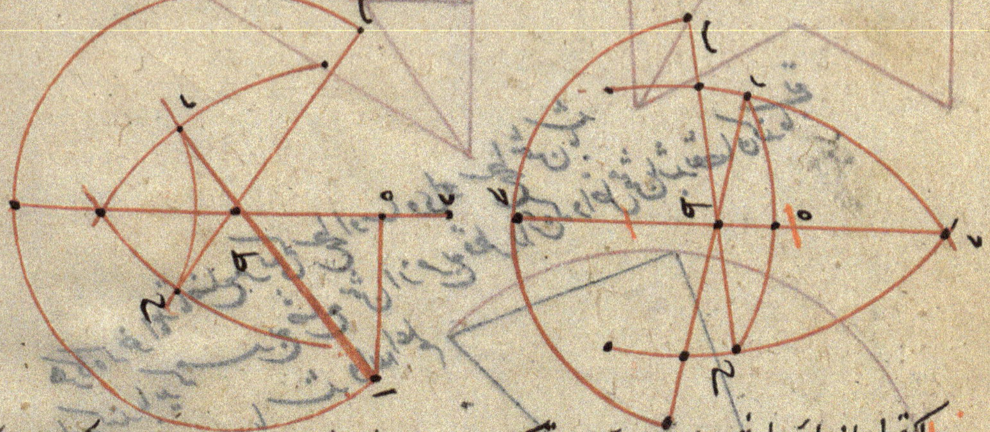

فصل اگر قطعه از دایره باشد وخواهد که آنرا تمام گردانند طریق آن بود که قوس اب را به قسم کند بنقطهٔ ج د
اقسام متساوی و از نقطهٔ اد تقاطعه زند هر فتح که خواهد و از نقطهٔ ت ج تقاطع ز پیوندد
و از تقاطع ه خط ه د بکشد و از تقاطع ز خط ز د بکشد که یک دیگر را قطع کند بنقطهٔ ح
و این نقطه مرکز دایره اصل است بفتح ح د که نصف قطرست دایره تمام کند والله اعلم

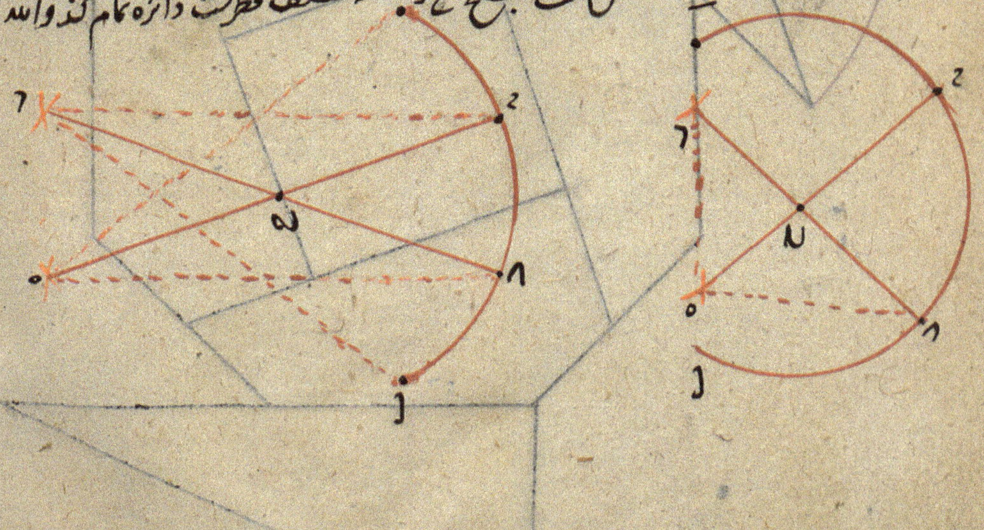

Fol. 181v [5 (top) and 6 (bottom)]

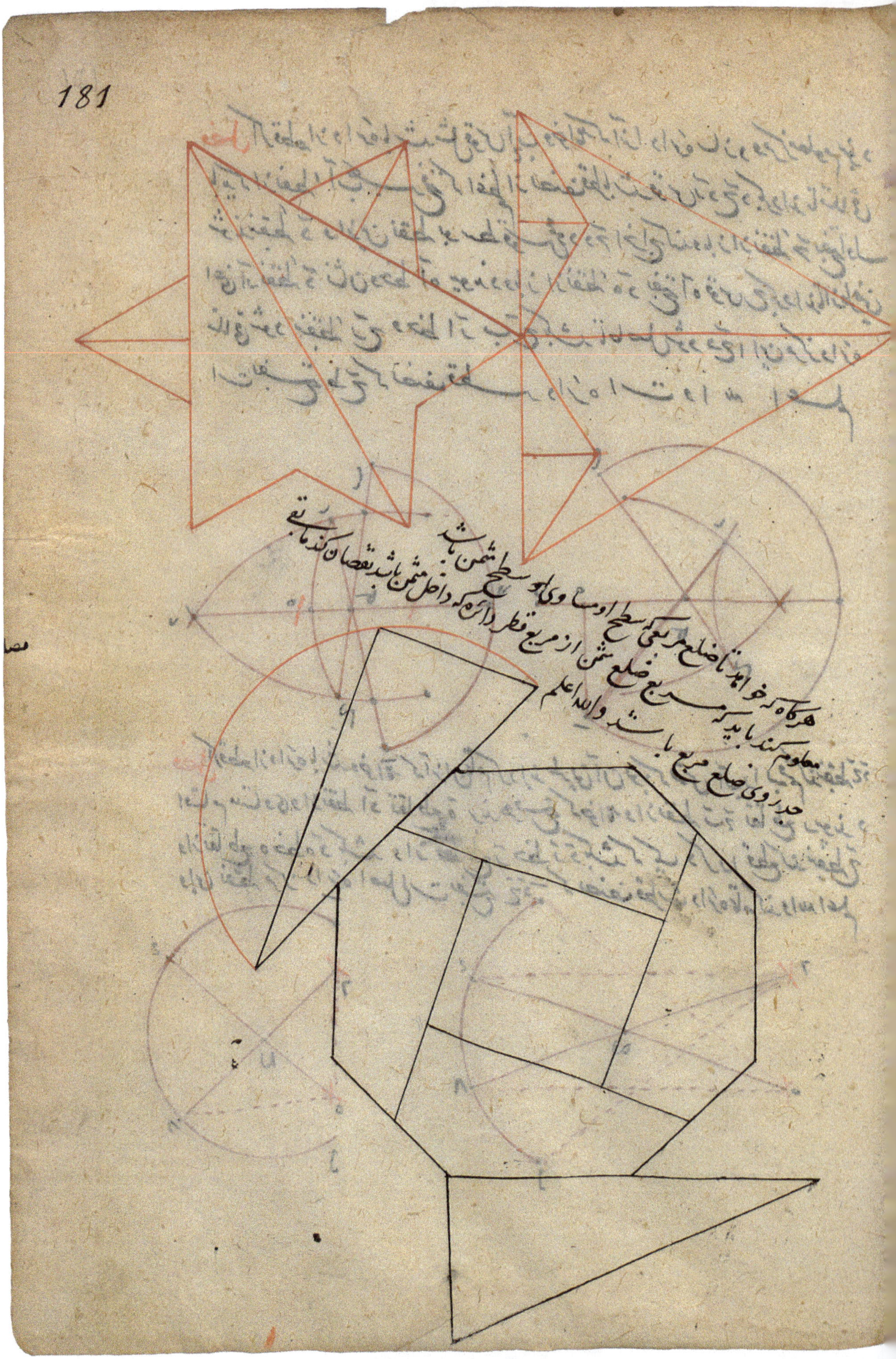

Fol. 181r [3 (top) and 4 (bottom)]

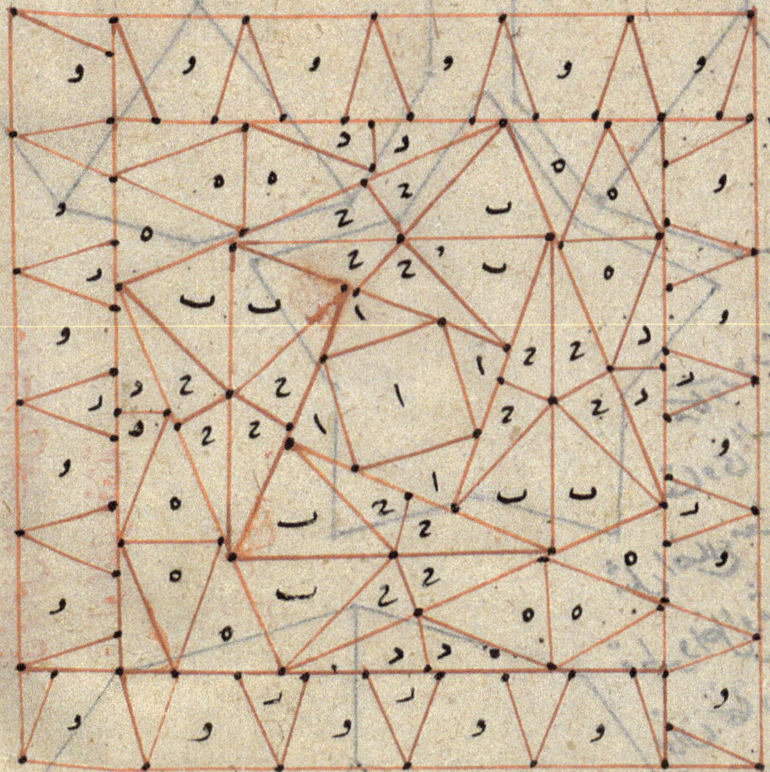

از مقطعات وآن مجموع ه یک معشر مجنح حاصل شود واگر خواهد
دو معشر مضلع سازد وباز از مقطعات ه د یک متسع
مجنح حاصل شود والا ج متسع مضلع وباز از اجزاء ح ب یک
ثمن مجنح حاصل شود والا آ ثمن مضلع ومجموع از اجزاء آ یک
مسدس مجنح یا دو مسدس مضلع حاصل شود بدین مثال

باید دانست که از اجزاء این مربع یک مربع مجنح و یک ثمن مجنح و یک متسع مجنح و یک معشر مجنح حاصل توان کرد و اگر خواهد ۴ ثمن مضلع
و ۳ ثمن مضلع و ۳ متسع مضلع و ۳ معشر مضلع ترکیب توان کرد و ۴ مربع نیز لطیف است والله اعلم

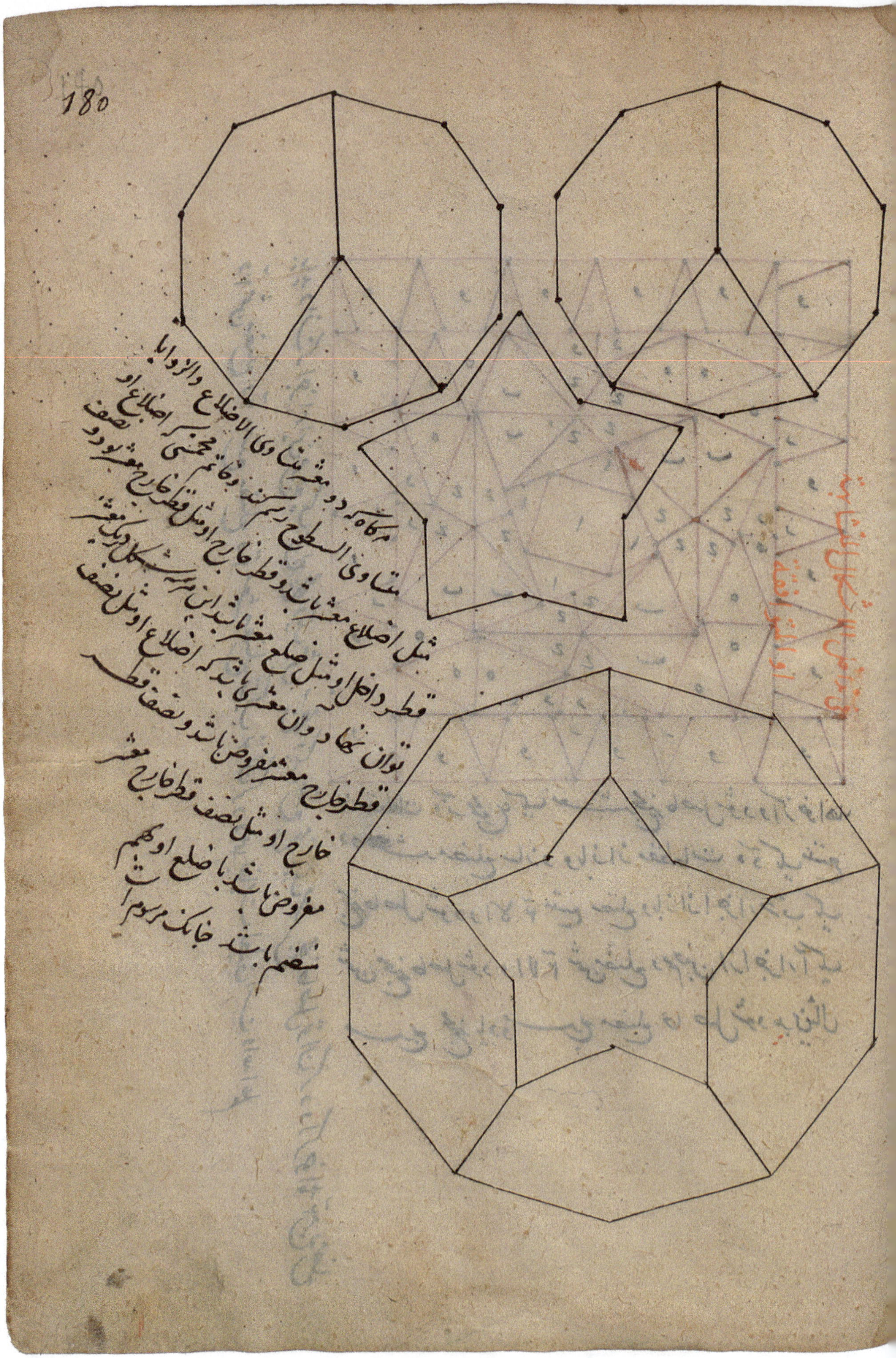

Fol. 180r [1]

FACSIMILE

فى تداخل الأشكال المتشابهة او المتوافقة

Fī tadākhul al-ashkāl al-mutashābiha aw al-mutawāfiqa,

Bibliothèque nationale de France, Ms. Persan 169, fols. 180r–199r.

Nastaʿlīq script, folio dimensions 190 × 270 mm.

(Photos: courtesy of the Bibliothèque nationale de France)

www.ingramcontent.com/pod-product-compliance
Lightning Source LLC
Chambersburg PA
CBHW061155240926
51846CB00006B/1025
* 9 7 8 9 0 0 4 6 7 9 7 1 9 *